Newton the Alchemist

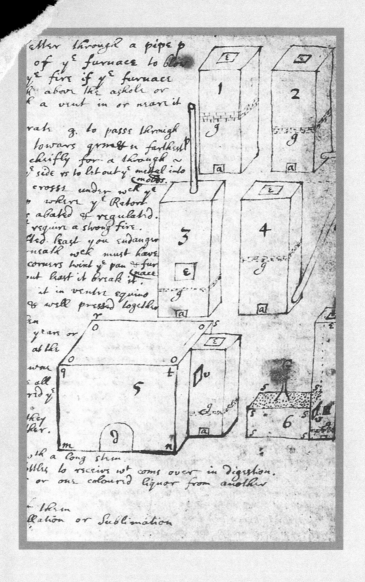

Newton the Alchemist

SCIENCE, ENIGMA, AND THE QUEST FOR
NATURE'S "SECRET FIRE"

William R. Newman

PRINCETON UNIVERSITY PRESS
Princeton & Oxford

Published by Princeton University Press

41 William Street, Princeton, New Jersey 08540
6 Oxford Street, Woodstock, Oxfordshire OX20 1TR
press.princeton.edu

LCCN 2018953066

ISBN 978-0-691-17487-7
British Library Cataloging-in-Publication Data is available

Editorial: Al Bertrand and Kristin Zodrow
Production Editorial: Brigitte Pelner
Text and Jacket/Cover Design: Pamela Schnitter
Jacket/Cover Credit: From the title page of Johann de Monte-Snyders's *Metamorphosis Planetarum*, 1964
Production: Jaqueline Poirier
Publicity: Alyssa Sanford
Copyeditor: Dawn Hall
This book has been composed in Garamond Premier Pro

Printed on acid-free paper ∞

Printed in the United States of America

10 9 8 7 6 5 4 3 2 1

FOR MARLEEN, EMILY, BEN, MARIE, AND CALLIE

Contents

Acknowledgments

Any book with a gestation period of fifteen years must necessarily owe many debts. Needless to say, human memory will fail to capture every obligation, but here I list the major contributors, beginning with the agencies that funded my research, and then passing to the individuals who have helped make *Newton the Alchemist* possible.

It is impossible to separate the research that went into *Newton the Alchemist* from the editorial work that undergirds the online *Chymistry of Isaac Newton* project (www.chymistry.org) at Indiana University, since the book could not have been written without the thorough acquaintance with Newton's chymical *Nachlass* that editing the texts has engendered. Beginning in 2003, the National Science Foundation has supported the *CIN* with four multiyear grants, NSF Awards 0324310, 0620868, 0924983, and 1556864. During this period the National Endowment for the Humanities also awarded a multiyear grant to the project. Beyond this support for the *CIN*, the immediate process of writing *Newton the Alchemist* was made possible by the following grants and fellowships, which allowed me to devote two successive academic years (2014–16) almost exclusively to research and writing: a National Humanities Center Fellowship, a Caltech/Huntington Library Searle Professorship, a Cain Fellowship at the Science History Institute in Philadelphia, an American Council of Learned Societies Fellowship, an American Philosophical Society Franklin Research Grant, an Indiana University New Frontiers Exploratory Travel Fellowship, and an Indiana University College Arts and Humanities Institute Research Travel Grant.

The present team of the *Chymistry of Isaac Newton* project, to whom I owe a heavy and ongoing debt, includes John A. Walsh, technical editor; Wallace Hooper, project manager and programmer/analyst; James R. Voelkel, senior consulting editor; Michelle Dalmau, head, digital collections services, Indiana University–Bloomington Libraries; William Cowan, head, software development (library technologies), IUB Libraries; Meagan Allen, editorial assistant and chemistry lab assistant; Meridith Beck Mink, consultant; and Cathrine Reck, consultant in chemistry. Indiana Libraries Technology Collaborators include Kara Alexander, digital media specialist; Nianli Ma, programmer/analyst, digital repository infrastructure; and Brian Wheeler, system administrator. Among the past participants in the project whose work included preliminary transcribing and encoding of the digital editions of Newton's work are Tawrin Baker, Nicolás Bamballi, Nick Best, Neil Chase, Archie Fields III, John Johnson, Joel Klein, Cesare Pastorino, Evan Ragland,

Daniel Sanford, and Whitney Sperrazza. Past participants in website design and other digital matters other than editing include Allison Benkwitt, technical editorial assistant; Timothy Bowman, web programmer and site architect; Jonathan Brinley, technical editorial assistant; Ryan Brubacher, SLIS intern; Darick Chamberlin, illustrator; Mike Durbin, infrastructure programmer/analyst; Randall Floyd, programmer/database administrator; Lawrence Glass, programmer/analyst, editorial assistant; Julie Hardesty, interface and usability specialist; Kirk Hess, programmer/analyst; David Jiao, programming consultant; Stacy Kowalczyk, associate director for projects, IU Libraries; Tamara Lopez, project programmer/analyst, Newton site architect; Dot Porter, curator of digital research services; Alan Rhoda, technical editorial assistant; Jenn Riley, metadata librarian; John Rogerson, technical editorial assistant; and Lindley Shedd, digital media specialist.

Individuals who contributed directly to the book as opposed to the *CIN* include the following three referees for Princeton University Press, who each read the entire manuscript and devoted many hours to their careful and constructive comments: Lawrence Principe, Jennifer Rampling, and Jed Buchwald. The present structure and content of *Newton the Alchemist* owes a considerable debt to these referees, each of whom made distinctive contributions. I must particularly acknowledge my frequent and ongoing consultations with Principe on the translation of alchemical processes into operable chemistry, and my debt to Rampling for her perceptive comments on restructuring my text. Buchwald, finally, offered invaluable comments on the details of Newtonian optics. Other scholars who read individual chapters and provided important advice include Niccolò Guicciardini, Mordechai Feingold, Paul Greenham, Peter Ramberg, and Alan Shapiro. Still other historians aided me in locating crucial sources, including Hiro Hirai, Michael Hunter, Didier Kahn, Scott Mandelbrote, and Steven Snobelen. Additionally, I have benefited from the advice of a multitude of other scholars, including Michael Freedman, Rob Iliffe, Domenico Bertoloni Meli, Seth Rasmussen, Anna Marie Roos, Richard Serjeantson, Alan Shapiro, George Smith, and John Young. Finally, I owe debts to David Bish, Cate Reck, Joel Klein, and Pamela Smith for facilitating my laboratory work and in some instances working with me under the hood.

Symbols and Conventions

No doubt as a result of his commanding stature at the origin of modern science, excellent biographies of Isaac Newton are easy to find.[1] All of them deal to some degree with the famous physicist's scientific discoveries, at times extensively, and they all share an expressed desire to account for his decades of alchemical research. Yet no previous study of Newton, including several devoted entirely to his alchemical quest, does full justice to the subject. The present book makes no pretense of being another biographical treatment of the famous savant; instead, it seeks to illuminate the more than thirty years that Newton spent deciphering the secrets of the sages and putting them to the test in his laboratory. Although Newton did occasionally collaborate with others at the bench, he certainly did not advertise his interest in chrysopoeia, the transmutation of metals, to the learned world. To a greater degree than is found in other areas of his scientific work, we are dependent on Newton's own manuscripts for our knowledge of his alchemical activities. The relative paucity of external events requires us to enter into our subject's private world of thought and practice to a degree that is unusual even for scholarly monographs. Fortunately, Newton left a massive corpus of around a million words documenting the evolution of his alchemical research project. But in order to cope with this daunting material, the reader must be aware of a few hurdles.

First there is the issue of alchemy's colorful language and the graphic symbols writers on the subject employed over most of its history. Throughout this book, archaic terms such as "oil of vitriol" (sulfuric acid) and "salt of

---

[1] The best known modern biography of Newton, and justifiably so, is Richard S. Westfall, *Never at Rest: A Biography of Isaac Newton* (Cambridge: Cambridge University Press, 1980). In its 908 pages of closely spaced print, Westfall covers every aspect of Newton's life and work. For readers with less time to devote to Newton, Westfall published an abridged version of the biography as well, *The Life of Isaac Newton* (Cambridge: Cambridge University Press, 1993). Of almost equal fame is Frank Manuel, *A Portrait of Isaac Newton* (Cambridge, MA: Belknap Press of Harvard University Press, 1968). Although Manuel was more interested in fleshing out Newton's character than his scientific work, his biography does contain a chapter devoted to the famous natural philosopher's alchemy. Other sometimes overlooked but still valuable modern biographies include A. Rupert Hall, *Isaac Newton: Adventurer in Thought* (Cambridge: Cambridge University Press, 1996) and Gale E. Christianson, *In the Presence of the Creator: Isaac Newton and His Times* (New York: Free Press, 1984). Another twentieth-century biography worthy of note, particularly for its open-minded treatment of Newton's alchemy, is Louis Trenchard More, *Isaac Newton: A Biography* (New York: Charles Scribner's Sons, 1934). Unfortunately, More's biography appeared before the famous Sotheby's auction of 1936, in which the stupendous volume of Newton's alchemical and religious manuscripts was revealed to the world. Popularizing biographies abound as well, the best of which is James Gleick, *Isaac Newton* (New York: Pantheon Books, 2003).

tartar" (potassium carbonate) inevitably make an appearance. I have given parenthetical explanations of such terms of art at various points in *Newton the Alchemist* in order to keep them alive in the reader's memory. But outdated terminology is only one of the linguistic difficulties presented by Newton's alchemical quest. His use of exotic *Decknamen* (cover names) such as "the net" and "Diana's doves" presents a different and more complicated problem. Arriving at the meaning of such intentionally elusive terms is in fact a central problematic of *Newton the Alchemist*, and the process of decoding them has required a combination of replication in the laboratory and sustained textual analysis, some of it aided by computational tools. Chapter two begins laying out the problems and results of this modern process of decipherment, which ironically mirrors Newton's own decades spent decrypting the works of the adepts. While issues of archaic language and willful concealment by *Decknamen* can be dealt with as they occur in the narrative, a further issue of terminology requires that we meet it head on. I refer to Newton's habitual use, and even creation, of figurative alchemical symbols.

Following a tradition popularized by the Elizabethan alchemist John Dee and developed further by the Saxon schoolmaster and writer on chymical subjects Andreas Libavius, Newton devised a series of graphic symbols that he used for his own creations in the laboratory.[2] Building on the traditional planetary symbols long used by alchemists to depict the respective metals, Newton would attach a small "o" to indicate an ore or mineral of the metal in question. Thus iron, usually represented by the symbol for Mars, ♂, became iron ore with the addition of the "o." This modification could take several different forms. The editors of the *Chymistry of Isaac Newton* project have identified three different representations Newton used for iron ore: ♂°, ⚥, and ♂⁶.

On the same principle, Newton added the traditional star symbol for sal ammoniac (ammonium chloride), ✳, in order to indicate a sublimate fabricated by means of that material. Thus he combined the symbol for copper, the planet Venus, ♀, with ✳ to become ⚢, a volatile copper compound. The clarity of this system is undercut by the fact that Newton does not restrict the ✳ symbol to ammonium chloride but employs it from 1680 onward to represent a volatile compound containing sal ammoniac and antimony, which he refers to as "sophic sal ammonic," "our sal ammoniac," or even "prepared sal ammoniac." When the traditional sal ammoniac star is combined with metals from 1680 on, it may represent either "vulgar" sal ammoniac or the sophic

---

[2] For Dee's attempt to base alchemical symbolism on his "hieroglyphic monad," a composite of the traditional planetary symbols plus a curly bracket placed horizontally at the bottom, see C. H. Josten, "A Translation of John Dee's *Monas Hieroglyphica* (Antwerp, 1564), with an Introduction and Annotations," *Ambix* 12 (1964): 84–221. An influential though dated study of Libavius may be found in Owen Hannway, *The Chemists and the Word: The Didactic Origins of Chemistry* (Baltimore: Johns Hopkins University Press, 1975). More on Libavius's use of Dee may be found in William R. Newman, "Alchemical Symbolism and Concealment: The Chemical House of Libavius," in *The Architecture of Science*, ed. Peter Galison and Emily Thompson (Cambridge, MA: MIT Press, 1999), 59–77. For a recent monograph on Libavius, see Bruce T. Moran, *Andreas Libavius and the Transformation of Alchemy: Separating Chemical Cultures with Polemical Fire* (Sagamore Beach, MA: Science History Publications, 2007). For some of Dee's alchemical sources, see Jennifer M. Rampling, "John Dee and the Alchemists: Practising and Promoting English Alchemy in the Holy Roman Empire," *Studies in History and Philosophy of Science* 43 (2012): 498–508.

variety. And as though this were not confusing enough, Newton sometimes follows other alchemical writers in employing ✳ to mean the star regulus of antimony, the crystalline form of the metalloid reduced from its ore.

Similar issues emerge with Newton's use of the traditional symbol for "antimony," ♁, or as we would say, the mineral stibnite, which is predominantly antimony sulfide in modern terminology (figure 1). The seventeenth century uniformly identified stibnite as antimony and used the term "regulus" (literally "little king") for the reduced metalloid. Newton occasionally joins the ♁ symbol with ✳ to produce ⚴, again meaning a sublimate of sal ammoniac and stibnite. More typically, he combines it with the symbol for a metal, as in ⚳, which represents a volatile compound (in the modern sense) of copper, antimony, and sal ammoniac (or sophic sal ammoniac). Further combinations can also occur, as when Newton adds the traditional symbol for salt, ⊖. Thus a volatile salt of copper containing also antimony and vulgar or sophic sal ammoniac receives the following symbol: ⚳. The same pattern is used with the other metals as well.

Below I list the alchemical symbols that occur in the present book, beginning with the more commonly used ones and then progressing to Newton's idiosyncratic versions. It is important not to be lulled into a false sense of

FIGURE 01. Stibnite from northern Romania. William R. Newman's sample. See color plate 1.

security when one encounters these glyphs. Newton was not doing modern chemistry, so one cannot expect his symbols always to refer to the same concrete, chemical referent in the way that a modern molecular formula always refers to precisely the same combination of atoms. The use of the ✳ symbol to mean both vulgar and sophic sal ammoniac is a case in point, but it is only one problem among many. Thus ☿ may refer to more than one volatile salt made from copper, antimony, and sophic or vulgar sal ammoniac. Moreover, the symbol does not reveal anything about the material's mode of production. As we will see in the later part of this book, Newton's stock laboratory reagent, "liquor of antimony," was typically employed in making his volatile salts, yet he did not incorporate a specific symbol for it, perhaps on account of his viewing it as a processing agent rather than an ingredient, or even because of its very ubiquity. In short, the symbols generally represent what Newton considered the most salient ingredients of his laboratory products, but beneath this graphic shorthand lies all the ambiguity of the experimental record. One should note also that even in the case of the simple planetary symbols, Newton often prefixes the figure with the word "our," indicating that he does not have the common, "vulgar" referent in mind. Thus "our ♀" does not mean copper but either a compound of the metal or even some other substance entirely.

Chymical Symbols Used by Newton

℔	Pound
℥	Ounce
ʒ	Drachm
℈	Scruple
Gr	Grain
✗	Crucible
♏	Retort
△	Fire
🜁	Air
▽	Water
🜃	Earth
☿	Mercury, either the supposed principle of metals and minerals, or vulgar quicksilver; also used for the sophic mercury.
🜍	Sulfur, the second principle of the metals, also vulgar brimstone.
⊖	Salt, the third metallic principle along with mercury and sulfur. Also common sea salt as well as other salts.
☉	Metallic gold, but also the putative internal sulfur of iron.
☽	Usually silver, but can also mean metallic antimony, and even sophic mercury.
♀	Copper, but it can also refer to what in modern chemical terms are copper compounds, especially, but not only, when preceded by "our." It can also refer to the "amorous," metallic component within stibnite, namely, regulus of antimony.

♂	Iron
♃	Tin
♄	Lead. Newton used this symbol mostly after the beginning of 1674. Before that time he typically used the unbarred version of it, ♄.
♄	Lead
♁	Stibnite (antimony sulfide), simply called "antimony" in the seventeenth century.
♆	Bismuth
✳	Sal ammoniac, either "vulgar" (NH_4Cl), or sophic, a compound or mixture of the former and either crude antimony or regulus of antimony. Sometimes ✳ is also used to designate the star regulus of antimony.
🜖 🜖	Vitriol, typically a sulfate in early modern chymistry, but to Newton it is used for multiple crystalline, obviously metallic salts, especially those with a styptic taste.
🝛	Amalgam
🜅 ▽F ▽F	Aqua fortis (mainly nitric acid).
🜆 ▽R ▽R	Aqua regia, in modern chemistry a mixture of nitric and hydrochloric acid, but to Newton, it is usually aqua fortis that has been "sharpened" by adding sal ammoniac.
🜈	*Spiritus vini*, that is, impure ethanol.
🜊	Vinegar
🜿	Tartar, also known as argol. Impure potassium bitrartrate deposited on the inside of wine casks.
⊖🜿ⁱ	*Sal Tartari* (salt of tartar). Mostly potassium carbonate made from tartar by calcination and leaching.
🜕	Saltpeter, mainly potassium nitrate. The same symbol is used for the Sendivogian aerial niter, a hypothetical material whose properties are modeled on those of saltpeter.
☿🝞 ☿🝞	Corrosive sublimate, that is, mercuric chloride.
♀°	Copper ore
♂° ♂° ♂°	Iron ore
♃° ♃°	Tin ore
♄° ♄°	Lead ore
♁°	Antimony ore
♆° ♆°	Bismuth ore
🜲	Regulus of antimony (reduced metallic antimony). A symbol devised by George Starkey, and used in Newton's copy of Starkey's *Clavis*, Keynes MS 18.
R	Also regulus of antimony.
✳♁ ✳♁	Sublimate of stibnite and sal ammoniac.
🜭	Salt of antimony. The crystalline material formed by crystallizing Newton's liquor of antimony; also the same salt in solution in liquor of antimony.
♀♁	Copper "antimoniate" or "antimonial." Not an antimoniate in the modern sense, but rather a so-called vitriol of copper made by imbibing the metal with Newton's liquor of antimony and then crystallizing the solution.

☿	Salt of copper antimoniate. The above vitriol of copper when filtered and allowed to crystallize separately.
☿	Sublimate of copper antimoniate.
☿ ☿	Sublimate of salt of copper antimoniate.
☿☿	Alternative symbols for antimoniate sublimates of copper.

Other Terminological, Graphic, and Chronological Issues

In addition to the problem of Newton's alchemical terms and symbols, there are several other issues of language and convention to which the reader must be introduced. First, my use of the now archaic word "chymistry" is intended to alert the reader to the fact that there was no rigid, commonly accepted distinction between "alchemy" and "chemistry" in the seventeenth century. I need not belabor the point here, for the *Oxford English Dictionary* has recently affirmed it by recognizing the capacious character of the early modern discipline comprehended under "chymistry."[3] Accordingly, throughout the present book "chymistry" and "alchemy" are synonymous, both having the sense of a field that included the attempt to transmute metals alongside the disciplines that we would today call industrial chemistry and pharmacology.

Several other terms may also confuse the reader unless they are dealt with forthrightly. The first of these, "menstruum" to mean a dissolvent, has a history in alchemy extending back at least to the early fourteenth century *Testamentum* of pseudo-Ramon Lull.[4] Hence chemists even in the nineteenth century commonly referred to menstruums when they meant the mineral acids and other corrosives or solvents. The second term, "reduction," is more problematic, as it has senses in chymistry and mineralogy that overlap and sometimes contradict its modern meaning in chemistry. The older use of reduction in chymistry simply means "to convert (a substance) into a different state or form," often with the idea that one is leading the material back to a previous, or more primitive condition.[5] This conforms to the sense of the Latin infinitive *reducere*, which means "to lead back." Hence an ore can be "reduced" to a metal by smelting, but the metal can also be "reduced" to a powdery "mineral" form by calcination. The mineralogical use of "reduction" also presents ambiguities, since metallurgical writers speak of reducing

---

[3] See the online version of the *Oxford English Dictionary*, accessed August 28, 2017, under "Chemistry." I quote the passage here: "In early use the terms 'chemistry' and 'alchemy' are often indistinguishable. Later (post-*c* 1700), *alchemy* began to be distinguished as referring to the pursuit of goals increasingly regarded as unscientific and illusory, such as the transmutation of metals into gold (see *Early Sci. & Med.* 3 32–65 [1998]). The use of the term *chemistry* to describe such practices became increasingly *arch.* and *hist.* Beginning in the late 20th cent. the otherwise obsolete spelling *chymistry* (cf. quot. 1994²) was deliberately adopted to differentiate the early, transitional science from the discipline of 'modern' chemistry as practised from the 18th cent. onward."

[4] For pseudo-Lull's use of the term *menstruum* in his own words, see Michela Pereira and Barbara Spaggiari, *Il Testamentum alchemico attribuito a Raimondo Lullo* (Florence: SISMEL, 1999), 28–29; for adjectival forms of the term consult Pereira and Spaggiari's index.

[5] *Oxford English Dictionary*, online edition, under "Reduce," III. 17. a.

both ores and metals.[6] In modern chemistry, on the other hand, the terms "oxidation" and "reduction" (paired as "redox"), refer respectively to the loss or gain of electrons. In the present book, I use reduction in the older senses unless specifically indicated.

Two additional terms of art require explanation as well. In modern English, "sublimation" refers to the passage from a solid directly to a vapor followed by its recondensation as a solid, while "distillation" designates the vaporization of a liquid followed by its return to the liquid state. In the seventeenth century, however, the two terms were often not kept rigorously distinct. Thus the 1657 *Physical Dictionary* defines "sublimation" as an operation in which "the elevated matter in distillation, being carried to the highest part of the helm, and finding no passage forth, sticks to the sides thereof."[7] In order to avoid imposing an imagined rigor on my sources, I have generally followed this period use of "sublimation." The final term that requires explanation is my use of the word "adjuvant." The English term originally meant anything that "serves to help or assist," but it has come to have a specific sense in pharmacology of "a substance added to a medicinal formulation to assist the action of the principal ingredient."[8] I use "adjuvant" to signify something similar to the latter meaning, but in the specific laboratory operation of sublimation, where Newton typically added a more volatile material to a more fixed one in order to induce the latter to sublime. The medieval alchemical author Geber referred to such aids to sublimation as "res iuvantes," which I have translated elsewhere as "adjuvants" and here employ for Newton as well.[9]

A further item requiring clarification is my way of representing Newton's scribal shorthand. Because the most important text is often found in the canceled passages of Newton's manuscripts, I have generally reproduced his chymical writings in the diplomatic form found on the *Chymistry of Isaac Newton* project (www.chymistry.org), where most of them are edited. This practice means that quotations often include struck-through text, indications of illegibility, and scribal abbreviations. It was common in the seventeenth century to use a standard set of symbols to abbreviate words. The most obvious one, perhaps, is the thorn, which looks like a "y" but represents the letter combination "th" and is normally followed by one or more superscribed letters. Thus Newton usually writes our "the" as "y^e" and our "that" as "y^t." The process of dropping the medial part of a word and presenting its terminal letter(s) in the form of a superscript appears in many other instances as well, without the thorn. Thus Newton often represents "what" as "w^t" and "which" as "w^{ch}." Another very common contraction is "sp^t" for "spirit." One could identify many other examples of this practice, but once the reader understands Newton's modus operandi, it is usually not difficult to extract his meaning. A second feature of scribal shorthand, the macron, also makes its appearance in Newton's handwriting. This consists of an overbar placed on

---

[6] *Oxford English Dictionary*, online edition, under "Reduce," III. 17. b.

[7] *A Physical Dictionary* (London: John Garfield, 1657), N$_2$.

[8] *Oxford English Dictionary*, online edition, under "Adjuvant."

[9] William R. Newman, *The Summa perfectionis of Pseudo-Geber* (Leiden: Brill, 1991), 354, 679n79.

top of a letter or letters to indicate that part of the word has been omitted. One widespread example of this practice in Newton's manuscripts appears in the contraction "Pher" for "philosopher"; slight variants of this form also occur. If the reader encounters a contracted passage that is not obvious, he or she can in most instances locate the text in the online *Chymistry of Isaac Newton* site and convert it to its normalized, expanded form by placing the cursor above the folio number and tapping the mouse. In the case of Newtonian passages edited by other scholars, as in the multivolume *Correspondence of Isaac Newton* begun by H. W. Turnbull, I have not changed the way in which the editors represent abbreviations. The conversion of Turnbull's "ye" and "yt" to their superscript forms would have required that I consult every manuscript in the original, since Newton is not consistent in his practice of superscribing the terminal letter(s) of a given contracted word.

Newton had another scribal habit that is of great significance as well, namely, his practice of reproducing his source and then placing his own interpretation of the quoted or paraphrased author within square brackets. This is often the only clue that we have to Newton's understanding of a given text, so it is obviously important to retain his brackets when quoting from his manuscripts. But this of course means that the normal use of editorial square brackets must be scrupulously avoided in order to prevent confusion between Newton's words and the editor's. Consequently, the editions of Newton's manuscripts on the *Chymistry of Isaac Newton* site employ angle brackets (< . . . >) to indicate all editorial interventions. The same practice has been adopted in this book. Moreover, in order to avoid confusion, passages from Newton's nonalchemical manuscripts that have been inserted in square brackets by other editors are here placed in angle brackets.

A final practice that requires explanation results from the confusing situation of seventeenth- and eighteenth-century British timekeeping. The British did not adopt Gregory XIII's calendrical reforms until the mid-eighteenth century, meaning that their calendar was ten days behind the one used on the European continent until 1700, on which date it fell yet another day behind. This could result in a confusion of years when a British date fell in late December. Moreover, the custom in the British Isles was to begin the new year on Lady Day, March 25, with the result that dates between our January 1 and March 24 would all fall in the previous year. In order to avoid confusing matters beyond repair, early modern British writers often gave the year in Old Style, Julian dating, followed by the New Style, Gregorian one. Thus in his laboratory notebooks, Newton refers to our January 1689 as "Ian. 1679./80." Where early modern authors employ this practice of providing both dates separated by a slash, I have reproduced it. All years that appear in the present book without a slash are Gregorian years unless noted otherwise; following the common practice, I have not modernized the dates of Julian days.

Abbreviations for Works Cited

All citations of Newton's chymical manuscripts at the Cambridge University Library (Additional MSS), Kings College (Keynes collection), the National Library of Israel (Var. and Yahuda MSS), and the Smithsonian Institution (Dibner collection) refer to the online editions published by the *Chymistry of Isaac Newton Project* (www.chymistry.org).

Babson—Huntington Library, Babson MS

BML—Boston Medical Library MS

Boyle, *Works*—Michael Hunter and Edward B. Davis, eds., *The Works of Robert Boyle* (London: Pickering and Chatto, 1999–2000), 14 vols. For citations of Boyle, I give the pagination of the modern edition followed by the date and pagination of the original printing

CIN—*Chymistry of Isaac Newton*, www.chymistry.org

CU Add.—Cambridge University Library, Additional MS (Portsmouth Collection)

Cushing—Yale University, Cushing/Whitney Medical Library MS

Dibner—Smithsonian Institution, Dibner MS

Dobbs, *FNA*—Betty Jo Teeter Dobbs, *The Foundations of Newton's Alchemy; or, "The Hunting of the Greene Lyon"* (Cambridge: Cambridge University Press, 1975)

Dobbs, *JFG*—Betty Jo Teeter Dobbs, *The Janus Faces of Genius: The Role of Alchemy in Newton's Thought* (Cambridge: Cambridge University Press, 1991)

Don.—Oxford University, Bodleian Library, Don. MS

Hall and Hall, *UPIN*—A. Rupert Hall and Marie Boas Hall, eds., *Unpublished Scientific Papers of Isaac Newton* (Cambridge: Cambridge University Press, 1962)

Harrison, *Library*—John Harrison, *The Library of Isaac Newton* (Cambridge: Cambridge University Press, 1978)

Keynes—King's College, Cambridge University, Keynes MS

Manuel, *PIN*—Frank Manuel, *A Portrait of Isaac Newton* (Cambridge, MA: Belknap Press of Harvard University Press, 1968)

Mellon—Yale University, Beinecke Library, Mellon MS

Newman, *AA*—William R. Newman, *Atoms and Alchemy* (Chicago: University of Chicago Press, 2006)

Newman, *GF*—William R. Newman, *Gehennical Fire* (Cambridge, MA: Harvard University Press, 1994)

Newman and Principe, *ATF*—William R. Newman and Lawrence M. Principe, *Alchemy Tried in the Fire* (Chicago: University of Chicago Press, 2002)

Newman and Principe, *LNC*—William R. Newman and Lawrence M. Principe, eds., *George Starkey: Alchemical Laboratory Notebooks and Correspondence* (Chicago: University of Chicago Press, 2004)

Newton, *Corr.*—H. W. Turnbull, ed., *The Correspondence of Isaac Newton* (Cambridge: Cambridge University Press, 1959–61), vols. 1–3; J. F. Scott, ed. (1967), vol. 4; vols. 5–7, A. Rupert Hall and Laura Tilling, eds. (1975–77)

Newton, *CPQ*—J. E. McGuire and Martin Tamny, eds., *Certain Philosophical Questions: Newton's Trinity Notebook* (Cambridge: Cambridge University Press, 1983)

NP—*The Newton Project*, http://www.newtonproject.ox.ac.uk/

OED—*Oxford English Dictionary*, online edition

Philalethes, *Marrow*—Eirenaeus Philalethes, *The Marrow of Alchemy* (London: Edward Brewster, 1654–55)

Philalethes, *RR*—Eirenaeus Philalethes, Ripley Reviv'd (London: William Cooper, 1678)

Philalethes, *SR*—Eirenaeus Philalethes, Secrets Reveal'd (London: William Cooper, 1669)

Principe, *AA*—Lawrence M. Principe, *The Aspiring Adept* (Princeton, NJ: Princeton University Press, 1998)

RS—Royal Society MS

Schaffner—University of Chicago, Regenstein Library, Schaffner MS

Shapiro, *FPP*—Alan Shapiro, *Fits, Passions, and Paroxysms* (Cambridge: Cambridge University Press, 1993)

Sloane—British Library, Sloane MS

Snyders, *Commentatio*—Johann de Monte-Snyders, *Commentatio de pharmaco catholico*, in Anonymous, *Chymica vannus* (Amsterdam: Joannes Janssonius à Waesberge and Elizeus Weyerstraet, 1666)

Sotheby Lot—*Catalogue of the Newton Papers Sold by the Viscount Lymington* (London: Sotheby, 1936)

Var.—National Library of Israel, Var. MS

Westfall, *NAR*—Richard Westfall, *Never at Rest: A Biography of Isaac Newton* (Cambridge: Cambridge University Press, 1980)

Yahuda—National Library of Israel, Yahuda MS

Newton the Alchemist

ONE

The Enigma of Newton's Alchemy

When Isaac Newton died in 1727, he had already become an icon of reason in an age of light. The man who discovered the laws governing gravitational attraction, who unveiled the secrets of the visible spectrum, and who laid the foundations for the branch of mathematics that today we call calculus, was enshrined at Westminster Abbey alongside the monarch who had ruled at his birth. Despite having been born the son of a yeoman farmer from the provinces, Newton was eulogized on his elaborate monument as "an ornament to the human race." Perhaps playing on the illustrious physicist's fame for his optical discoveries, the most celebrated English poet of his age, Alexander Pope, coined the famous epitaph "Nature and Nature's Laws lay hid in Night. God said, *Let Newton be!* and All was *Light*."[1] Thus God's creation of Newton became a second *fiat lux* and the man himself a literal embodiment of the Enlightenment.

Little did Pope know that in the very years when Newton was discovering the hidden structure of the spectrum, he was seeking out another sort of light as well. The "inimaginably small portion" of active material that governed growth and change in the natural world was also a spark of light, or as Newton says, nature's "secret fire," and the "material soule of all matter."[2] Written at the beginning of a generation-long quest to find the philosophers' stone, the summum bonum of alchemy, these words would guide Newton's private chymical research for decades. Even after taking charge of the Royal Mint in 1696, Newton was still actively seeking out the fiery dragon, the green lion, and the liquid that went under the name of "philosophical wine," a libation fit for transmutation rather than consumption.[3] Most compellingly of all, Newton was on the path to acquiring the scepter of Jove and the rod of Mercury, along with the twin snakes "writhen" around the staff that

---

[1] Alexander Pope, "Epitaph: Intended for Sir Isaac Newton, in Westminster Abbey," in *The Poems of Alexander Pope*, ed. John Butt (New Haven, CT: Yale University Press, 1963), 808. For Newton's eighteenth-century reputation more broadly, see Mordechai Feingold, *The Newtonian Moment* (New York: New York Public Library, 2004).

[2] Smithsonian Institution, Dibner MS 1031B, 6r, 3v.

[3] See chapter nineteen herein for Newton's late use of these terms.

would convert it into the wonder-working caduceus of the messenger god. All these exotic names referred to the material tools of the adepts, the *arcana majora* or higher secrets with whose help they hoped to transform matter from its base and fickle state into the immutable perfection of gold.

The omission of alchemy from Pope's eulogy was of course no accident. Even if the "wasp of Twickenham" had known of Newton's alchemical research, he would certainly not have used it as a means of lionizing the famous natural philosopher. By the 1720s the part of chymistry that dealt with the transmutation of metals, *chrysopoeia* (literally "gold making"), was coming under siege in many parts of Europe. But in the second half of the seventeenth century, when Newton did the bulk of his alchemical research, transmutation had formed a natural part of the chymical discipline, and indeed the term "chymistry" had long been coextensive with "alchemy." Both words had signified a comprehensive field that included the making and refining of pharmaceuticals and the production of painting pigments, fabric dyes, luminescent compounds, artificial precious stones, mineral acids, and alcoholic spirits alongside the perennial attempt to transmute one metal into another.[4] A slow process of separation was already underway by the final quarter of the century, however, and by the second and third decades of the *siècle des lumières* such chymical authorities as Georg Ernst Stahl and Herman Boerhaave, who had long upheld the traditional principles and purview of alchemy, were expressing their doubts about chrysopoeia in a highly public way.[5] Thus when the antiquarian William Stukely compiled a draft biography of Newton after his friend's death, he went so far as to suggest that Newton's work in chymistry had the potential of freeing the subject from an irrational belief in transmutation.[6] Ironically, Newton the alchemist had been transmuted into Newton the Enlightenment chemist.

Yet the celebration of the founder of classical physics as a beacon of pure reason had already begun to show signs of wear when David Brewster composed a biography in 1855 in which he was compelled to come to terms with the fact that Newton had studied alchemy. Brewster expressed his amazement that Newton "could stoop to become even the copyist of the most contemptible alchemical poetry," a fact that the Scottish scientist could only explain as the mental folly of a previous age.[7] The few lines that Brewster devoted to the topic were largely ignored until 1936, when the bulk of Newton's

[4] The archaic spelling "chymistry" has been adopted by scholars to signify this overarching field that combined medical, technical, and chrysopoetic endeavors in the early modern period. See the online *Oxford English Dictionary* under the term "chemistry," where further documentation is given (accessed June 9, 2017).

[5] For Stahl's gradual conversion to a critic of chrysopoeia, see Kevin Chang, "'The Great Philosophical Work': Georg Ernst Stahl's Early Alchemical Teaching," in *Chymia: Science and Nature in Medieval and Early Modern Europe*, ed. Miguel López Pérez, Didier Kahn, and Mar Rey Bueno (Newcastle upon Tyne: Cambridge Scholars, 2010), 386–96. For the similar process of disenchantment in the case of Boerhaave, see John Powers, *Inventing Chemistry: Herman Boerhaave and the Reform of the Chemical Arts* (Chicago: University of Chicago Press, 2012), 170–91.

[6] RS MS/142, folio 56v, from *NP* (http://www.newtonproject.sussex.ac.uk/view/texts/diplomatic /OTHE00001), accessed June 7, 2016.

[7] Sir David Brewster, *Memoirs of the Life, Writings, and Discoveries of Sir Isaac Newton* (Edinburgh: Thomas Constable, 1855), 2: 375.

surviving manuscripts on alchemy and religion were auctioned by Sotheby's in London. Suddenly a very different Newton was thrust into the light, one who had written perhaps a million words on alchemy and even more on religious subjects ranging from biblical prophecy and the dimensions of Solomon's temple to the perfidy of the orthodox doctrine of the Holy Trinity. The cognitive dissonance that these manuscripts inevitably summoned up was captured by the economist John Maynard Keynes, who collected a large number of them for King's College, Cambridge. In his famous posthumous essay "Newton, the Man," published in 1947, Keynes wrote that

> Newton was not the first of the age of reason. He was the last of the magicians, the last of the Babylonians and Sumerians, the last great mind which looked out on the visible and intellectual world with the same eyes as those who began to build our intellectual inheritance rather less than 10,000 years ago. . . . He believed that by the same powers of his introspective imagination he would read the riddle of the Godhead, the riddle of past and future events divinely fore-ordained, the riddle of the elements and their constitution from an original undifferentiated first matter, the riddle of health and of immortality.[8]

In the same article, Keynes would add that Newton's alchemical manuscripts were "wholly magical and wholly devoid of scientific value." Yet despite the pejorative tone of these comments, Keynes was not operating in a naive or unreflective way when he dismissed Newton's alchemy as magic. His 1921 *Treatise on Probability* had argued against "the excessive ridicule" that moderns tended to levy on primitive cultures, and he even went so far as to locate the origins of induction in the magician's attempt to recognize patterns in nature. Keynes would support this claim with observations drawn from the Victorian masterpiece of Sir James Frazer, *The Golden Bough*.[9] Frazer's massively influential study of mythology had used the principle of sympathy (the belief that "like acts on like") to group a wide variety of practices under the rubric of "magic."[10] A similar approach emerges in "Newton, the Man," although it is obscured by the rhetorical brilliance of the essay, with its overriding goal of toppling the traditional image of Newton the rationalist. Like Frazer, Keynes assimilated various "occult" pursuits such as alchemy and the quest for secret correspondences in nature under the same amorphous category, labeling them as magical.[11] It is highly likely that Keynes had Frazer in the back of his mind when he unselfconsciously elided the borders between magic and alchemy, two disciplines that Newton for the most part kept rigorously distinct.

[8] John Maynard Keynes, "Newton, the Man," in *Newton Tercentenary Celebrations, 15–19 July 1946* (Cambridge: University Press, 1947), 27–34, see 27.

[9] John Maynard Keynes, *A Treatise on Probability* (London: Macmillan, 1921), 245–46.

[10] Frazer's *Golden Bough* was originally published in two volumes in 1890, but eventually swelled to twelve volumes. For his treatment of the principle of sympathy, see James Frazer, *The Golden Bough* (New York: Macmillan, 1894), 9–12.

[11] For my objections to this type of lumping approach when it comes to the "occult sciences," see William R. Newman, "Brian Vickers on Alchemy and the Occult: A Response," *Perspectives on Science* 17 (2009): 482–506.

The Keynesian picture of Newton as the last of the magicians rather than as the father of the Enlightenment amounted to a radical inversion of the Augustan view: no longer a herald of light, the founder of classical physics now looked back to a dark and fabulous past. This new image of a brooding and troubled Newton buried in the decipherment of riddles "handed down by the brethren in an unbroken chain back to the original cryptic revelation in Babylonia" would go on to exercise its own attraction. One can see the influence of Keynes very clearly in the work of two eminent Newton scholars of the late twentieth century, Betty Jo Teeter Dobbs and Richard Westfall. Both Dobbs and Westfall were pioneers in the scholarly study of Newton's alchemy, and their work has provided an indispensable basis for subsequent research in the field, including my own. One cannot doubt the seriousness of their scholarship, the years that they devoted to understanding Newton, or the significance of their contributions. Yet as we shall see, their embrace of the Keynesian perspective could at times exert its own smothering grip on their critical judgment.

Dobbs, whose 1975 *The Foundations of Newton's Alchemy; or, "The Hunting of the Green Lyon"* provided the first full-length study of Newton's alchemical endeavors, came to the eventual conclusion that alchemy for Newton was above all a religious quest.[12] Although she did not endorse Keynes's blanket assertion that Newton's alchemical writings were a worthless farrago, and even criticized the famous economist for his failure to consider Newton's alchemical experiments, Dobbs built on the idea that alchemy itself incorporated a fundamentally irrational core. Her *Foundations of Newton's Alchemy* contains a largely approving exposition of the analytical psychologist Carl Jung's position that alchemical imagery embodied an "irruption" of the mind's unconscious contents and that alchemy was largely a matter of "psychic processes expressed in pseudo-chemical language," implying that something other than scientific or even material goals were the main driving force behind the aurific art.[13] Dobbs's 1991 *The Janus Faces of Genius: The Role of Alchemy in Newton's Thought* dropped this explicit adherence to Jung's analytical psychology, but nonetheless developed a favorite thesis of Jung's, namely, that the alchemical search for the philosophers' stone was primarily a quest to reunite man with the creator, a form of soteriology. Hence *The Janus Faces of Genius* gives the impression that Newton's alchemy was above all a vehicle for his heterodox religious quest, and that he thought of the philosophical mercury of the alchemists as a spirit that mediated between the physical and transcendent realms in a way analogous to the mediation of Jesus between God and man.[14]

Newton's alchemy also appears through Keynes-tinted glasses in the work of Dobbs's contemporary Westfall, though in a slightly different fashion.

[12] This is not the case in Dobbs's first book, however, where she in fact attacks Mary Churchill for overemphasizing the religious aspect of Newton's alchemy. See Dobbs, *FNA*, 15–16. As her study of Newton's alchemy extended itself over time, Dobbs came more and more to stress its putative religious goals.

[13] Dobbs, *FNA*, 25–43. Despite her affirmation of the Jungian approach to alchemy as "really promising," on page 25, Dobbs does exercise a degree of critical restraint when she correctly describes Jung's views on page 40 as "basically a-historical."

[14] Dobbs, *JFG*, 13, 243–48.

While Westfall seems to have remained impartial to the Dobbsian position that Newton's alchemy was coextensive with his private religion, he did see Newton's interest in the aurific art as a sort of romantic rebellion against the rationalist project of Cartesian physics, harking back to "the hermetic tradition" of late antiquity and the Renaissance.[15] To Westfall, alchemy and magic were characterized by a fascination with immaterial qualities, powers, sympathies, and antipathies, in short, the very antithesis of the Cartesian billiard-ball universe with its attempt to reduce nature to a succession of impact phenomena. Hence Westfall could argue that Newton's alchemy, although it lay outside the domain of rationalist natural philosophy, contributed in a major way to his mature theory of gravitation, and more broadly to his conviction that immaterial forces in general could operate at a distance. Westfall would explicitly argue that Newton's concept of force at a distance "derived initially from the world of terrestrial phenomena, especially chemical reactions." In fact, he even went so far as to claim that Newton's concept of gravitational attraction emerged only after "he applied his chemical idea of attraction to the cosmos."[16]

Westfall's claim that alchemy was behind Newton's theory of universal gravitation was adopted in turn by Dobbs in her *Foundations of Newton's Alchemy*, while her theocentric interpretation of his quest for the philosophers' stone dominated *The Janus Faces of Genius*. Largely as a result of these scholars' authoritative status, the view that Newton's theory of gravity owed a heavy debt to alchemy has become canonical in the popular literature.[17] Current scholarly treatments of the subject endorse the authoritative status of Dobbs and Westfall as well, restating the former's view that Newton aimed "to capture the essence of the Redeemer in a beaker" and asserting with both scholars that alchemy "may have helped him to conceptualize the idea of gravity."[18] It is not too much to say that the picture of Newton's alchemy as a largely theocentric pursuit that contributed to his science by allowing for a rebaptizing of magical sympathy as gravitational attraction has become the received view of the subject.

But there are compelling reasons for doubting this interpretation. The once popular notion that alchemy was inherently unscientific—already present in the work of Keynes and advanced by successive Newton scholars—has been largely debunked by historians of science over the last three decades. Indeed, the historiography of alchemy has recently undergone a sort of renaissance that

[15] In his 1971 book *Force in Newton's Physics*, Westfall explicitly linked gravitational force to alchemy and to what he called "the hermetic tradition," a term that clearly betrays the influence of Frances Yates's 1964 *Giordano Bruno and the Hermetic Tradition*. See Richard Westfall, *Force in Newton's Physics* (London: MacDonald, 1971), 369.

[16] Richard Westfall, "Newton and the Hermetic Tradition," in *Science, Medicine, and Society in the Renaissance*, ed. A. G. Debus (New York: Science History Publications, 1972), 2: 183–98, see 193–94.

[17] See for example Michael White, *Isaac Newton the Last Sorceror* (New York: Basic Books, 1997), 106, 207, and throughout. The view that Newton's concept of gravitational attraction owes an important debt to alchemy even receives support in the current Wikipedia entry on Newton. See https://en.wikipedia.org/wiki/Isaac_Newton, accessed January 22, 2016.

[18] Paul Kléber Monod, *Solomon's Secret Arts* (New Haven, CT: Yale University Press, 2013), 104.

has reversed the picture of the aurific art as an atavistic outlier.[19] It is now well known that such luminaries of the scientific revolution as Robert Boyle, G. W. Leibniz, and John Locke were all seriously involved in alchemy; Newton was no anomaly.[20] All of these figures engaged in the broad spectrum of chymical practice, seeing it as a fruitful source of pharmaceutical and technological products and yet hoping as well that it might reveal the secret of metallic transmutation. Chymistry was a natural and normal part of the progressive agenda of seventeenth-century science. Hence the need that Dobbs and others felt to locate Newton's motives for studying alchemy in extrascientific areas such as soteriology and the quest for a more primitive Christianity has lost its force. We are now free to study Newton's alchemy on its own terms and to arrive at a much clearer picture of the field's relationship to his other scientific pursuits. As I show in *Newton the Alchemist*, the claims that Westfall (and subsequently Dobbs) made for an alchemical origin to Newton's theory of gravitational attraction are actually quite weak; in reality, the connection between alchemy and Newton's better known scientific discoveries lies elsewhere, above all in the realm of optics.[21]

Nonetheless, when first confronted by the sheer volume of Newton's million or so words on alchemy, one can only sympathize with the attempts of Westfall and Dobbs to cast about for a means of interpreting this intractable material. Finding the source of Newton's belief in forces acting at a distance in alchemy or linking the subject to his Antitrinitarian Christianity are both ways of rationalizing the immense amount of time and work that he devoted to the aurific art. Nor are these the only motives that historians have claimed to lie buried within the chaotic mass of Newton's alchemical papers. Karin

[19] For a good overview of the current scholarly position of chymistry and some reflections on the earlier historiography, see the four recent essays by Lawrence M. Principe, William R. Newman, Kevin Chang, and Tara Nummedal collected and introduced by Bruce Moran for the "Focus" section of *Isis*: Bruce T. Moran, "Alchemy and the History of Science," *Isis* 102 (2011): 300–337. Additionally, one should consult Moran's *Distilling Knowledge: Alchemy, Chemistry, and the Scientific Revolution* (Cambridge, MA: Harvard University Press, 2005); Newman's *Promethean Ambitions: Alchemy and the Quest to Perfect Nature* (Chicago: University of Chicago Press, 2004); Nummedal's *Alchemy and Authority in the Holy Roman Empire* (Chicago: University of Chicago Press, 2007); and Principe's *Secrets of Alchemy* (Chicago: University of Chicago Press, 2013). Another helpful study is Jennifer M. Rampling, "From Alchemy to Chemistry," in *Brill's Encyclopedia of the Neo-Latin World*, ed. Philip Ford, Jan Bloemendal, and Charles Fantazzi (Leiden: Brill, 2014), 705–17. In the context of the recent historiography of chymistry, one cannot pass over the magisterial study of Paracelsianism in France by Didier Kahn, *Alchimie et Paracelsisme en France à la fin de la Renaissance (1567–1625)* (Geneva: Droz, 2007).

[20] Boyle's career-long involvement in the quest for chrysopoeia forms the subject of Principe, *AA*. A recent article that presents and critiques the earlier historiography of Leibniz's involvement with alchemy may be found in Anne-Lise Rey, "Leibniz on Alchemy and Chemistry," in the online *Oxford Handbook of Leibniz* (http://www.oxfordhandbooks.com/view/10.1093/oxfordhb/9780199744725.001.0001/oxfordhb -9780199744725-e-32), accessed June 9, 2017. For Locke and chrysopoeia, see Peter R. Anstey, "John Locke and Helmontian Medicine," in *The Body as Object and Instrument of Knowledge*, ed. Charles T. Wolfe and Ofer Gal (Dordrecht: Springer, 2010), 93–120. See also Guy Meynell, "Locke and Alchemy: His Notes on Basilius Valentinus and Andreas Cellarius," *Locke Studies* 2 (2002): 177–97.

[21] Dobbs herself argued for an influence from alchemy on Newton's optics, but her claims have been debunked by Alan Shapiro. See Dobbs, *FNA*, 221–25, and Shapiro, *FPP*, 116n48. The interaction between Newtonian optics and chymistry that I envision is quite distinct from the one Dobbs maintained. See the present book and also William R. Newman, "Newton's Early Optical Theory and Its Debt to Chymistry," in *Lumière et vision dans les sciences et dans les arts*, ed. Danielle Jacquart and Michel Hochmann (Geneva: Droz, 2010), 283–307.

Figala, who did exemplary work in digging up Newton's chymical collaborations and making sense of the bibliographical entries in his notes, arrived at a grand but poorly substantiated thesis that explained the bulk of Newton's alchemy in terms of specific gravity. Basing herself on Newton's view that ordinary matter consists of corpuscles that are themselves mostly made up of empty space, Figala developed mathematical schemes linking the supposed amount of void and matter in materials to the traditional alchemical principles mercury and sulfur.[22] The problem with her interesting idea is that Newton nowhere makes this linkage himself; in fact, a close reading of his alchemical laboratory notebooks shows that he rarely even mentioned specific gravity in the context of his chymical experimentation. The only way to reconstruct the supposed system Figala found is by assuming that Newton left it entirely implicit, and that the historian must reconstruct it from tacit clues by a process that altogether resembles second-guessing. But this in turn requires that we ignore more obvious approaches taken by Newton, such as his deep concern with the affinities between chemicals that guide their bonding and dissociation.

Yet another approach to Newton's alchemy may be found in *The Expanding Force in Newton's Cosmos* by David Castillejo, which provides an extreme instance of the Keynesian perspective.[23] To Castillejo, Newton's optics, dynamical physics, prophecy, and the interpretation of the dimensions in Solomon's Temple are all part and parcel of the same project as his alchemy. Here we see the Babylonian magus again regarding the cryptogram of the universe and searching for the hidden clues that God has implanted in the cosmos. Castillejo's research led him to the conclusion that Newton had discovered a "single expansive force" that contrasted with the "contractive force" of gravity and operated at all levels of being. To Castillejo's Newton, the same mathematical relations governing this expansive force are operative in the dimensions of Solomon's Temple and in the corpuscular structure of matter at the microlevel. And for Castillejo, Newton's expansive force is coterminous with the cause of fermentation, which the physicist claimed to be a fundamental force of nature in his *Opticks*. Despite several significant contributions that lie buried in *The Expanding Force in Newton's Cosmos*, much of the numerology that Castillejo claims to find in Newton's work, he has forcibly imposed on the text. It is a peculiar irony that both Castillejo and Figala seem to be unriddling Newton's alchemical papers in much the same way that Keynes claimed Newton to be unriddling the cryptogram of nature itself.

The Tower of Babel presented by the wildly divergent claims of Dobbs, Westfall, Figala, and Castillejo should alert us to the gargantuan difficulties residing in Newton's alchemical *Nachlass*. Although the material is voluminous and disordered, with few obvious indications of the times at which the different papers were composed, these are the least of the problems.

---

[22] Karin Figala, "Newton as Alchemist," *History of Science* 15 (1977): 102–37, see especially 113–28.

[23] David Castillejo, *The Expanding Force in Newton's Cosmos* (Madrid: Ediciones de arte y bibliofilia, 1981), 17–30, 105–17.

The greatest difficulty stems from the fact that Newton was writing only for himself, and as he progressed more deeply into the literature of alchemy, he assumed the voices and literary techniques of the authors he was reading. As I describe at length in the present book, he took from his sources a veritable language of cover names or *Decknamen* (to employ the German term adopted by historians of alchemy) for the materials with which he was working. Decoding these terms presents difficulties that are grueling at best, since even when we understand a particular author's original meaning, Newton's interpretation often differs strikingly from that of his source. As a result, our hard-won knowledge of other seventeenth-century chymists and their techniques can mislead us as often as it helps us in deciphering Newton's laboratory records and reading notes. A case in point may be found in Newton's pervasive use of the American chymist George Starkey, who wrote elegant Latin treatises on chrysopoeia under the pseudonym of "Eirenaeus Philalethes" (a peaceful lover of truth). Although modern scholarship has probed the depths of Starkey's alchemy and acquired a clear understanding of his processes, the celebrated physicist held an idiosyncratic interpretation of the Philalethan corpus that can only be deciphered by careful analysis of Newton's notes and experiments, and sometimes by disregarding Starkey's original sense.

The Method of the Present Work

How then can we extricate any stable meaning from the shifting and cacophonous world presented by Newton's note taking, derived as it was from the enigmatic utterances of authors whose works were written over a range of cultures and centuries? There is in fact a way, and one that previous scholars have not sufficiently used. I refer to a twofold method that incorporates rigorous textual analysis with laboratory replication of Newton's alchemical experiments. The close analysis of documents needs no justification, having a long and distinguished pedigree extending back to the philological efforts of the nineteenth century and before. "Experimental history," on the other hand, is only now coming into its own among scholars. This is the branch of historical endeavor that involves replication, or if one prefers, "reworking" or "reconstruction" of old techniques and experiments. Just as experimental archaeologists have long been reproducing the techniques that allowed premodern cultures to create the artifacts that populate current-day museums, so historians of science have in recent years come to see the need for a "hands-on" approach to the study of old experiments. The history of chemistry has proven to be a particularly rich area of study for experimental history, and it dovetails closely with the long-standing field of conservation science, a discipline that has traditionally given rigorous attention to the material composition of painters' pigments. Newton's experimental notebooks cry out for this approach, because of the wealth of technical, even artisanal detail that they contain and because of the tacit laboratory-based skill on which they rely. Without some mastery of seventeenth-century chymical techniques, the scholar simply cannot make serious headway against the flood of *termini technici* that make up Newton's notebooks. A recent issue of the journal *Ambix* devoted to experimental history indicates that reproducing experiments can result in "the uncovering of details, difficulties, and solutions left unrecorded or only hinted at by the original experimenter."[24] While endorsing this sentiment, I would go even further in the case of Newton's experimental work in alchemy. Because of his perennial use of *Decknamen* and proprietary names for materials, one cannot even identify the basic subjects of his experimentation without firsthand knowledge of the materials that were available to him. Newton's idiosyncratic terms such as "liquor of antimony" and "sophic sal ammoniac" could in principle mean many different things; only by carefully analyzing his comments and actually putting them to the test in a laboratory can we determine the precise sense of his words.

At the same time, the new digital edition of Newton's extensive alchemical laboratory records on Indiana University's *Chymistry of Isaac Newton* site (www.chymistry.org) has also allowed me to provide the first comparative, in-depth study of these essential documents. Two of them, Cambridge University Additional manuscripts 3973 and 3975, are found in the collection of Portsmouth manuscripts in the Cambridge University Library; the third is a single sheet belonging to the collections of the Boston Medical Library.[25] These remarkable notebooks chronicle Newton's laboratory experimentation for a period of at least three decades. The importance of the first two documents has long been recognized, but Newton's use of his proprietary *Decknamen* and the absence of explicit goals and conclusions in the notebooks render it extraordinarily difficult to make sense of them. Nonetheless, laboratory replications performed on a number of the experiments have led to the unraveling of many of their secrets. Understanding Newton's experiments in turn provides a link to both the Helmontian chymistry of his contemporaries such as Robert Boyle and George Starkey and to the mythological and allegorical output of chrysopoetic authors such as the obscure Johann de Monte-Snyders.

An additional key to Newton's laboratory practice is the remarkable and hitherto unstudied letter written to him by his friend and alchemical collaborator Nicolas Fatio de Duillier in August 1693.[26] In this document, Fatio quotes Newton's Latin directions for making the products that underlie the latter's famous—and famously indecipherable—*Praxis* manuscript, which is

---

[24] Hjalmar Fors, Lawrence M. Principe, and H. Otto Sibum, "From the Library to the Laboratory and Back Again: Experiment as a Tool for Historians of Science," *Ambix* 63 (2016): 85–97, see 94.

[25] One must not neglect to mention the important article by A. Rupert Hall and Marie Boas Hall, "Newton's Chemical Experiments," *Archives internationales d'histoire des sciences* 11 (1958): 113–53. The Halls analyzed CU Add. 3975 and 3973, but were unaware of the Boston Medical Library manuscript. Moreover, they were hampered by an unnecessarily negative view of alchemy and relied on purely "armchair" chemistry for their interpretations, replicating none of Newton's experiments. Their contemptuous perspective on alchemy led to a misunderstanding of Newton's goals, and their untested guesses about his laboratory work resulted in many misidentifications of his materials and products.

[26] William Andrews Clark Memorial Library, MS F253L 1693. I thank Scott Mandelbrote for originally bringing this letter to my attention.

sometimes described as his most important alchemical writing. Directions for making such desiderata as volatile Venus, sophic sal ammoniac, the scythe of Saturn, and the sword ("fauchion") of Mars all appear in Fatio's letter, but in a simplified form intended for replication by experimenters lacking Newton's years of experience with these materials. Along with Newton's laboratory notebooks, Fatio's letter makes it possible to reassemble the processes that Newton thought would lead eventually to the summum bonum of alchemy, and indeed the key to nature itself, the philosophers' stone. Using these documents as a guide, I have replicated a number of the stages in Newton's master process, and the results show why one of the most perspicacious experimenters of all time thought that his alchemical laboratory work was leading to success after decades of unremitting labor at the bench.

The physical replication of Newton's experiments is therefore a necessary tool for understanding his alchemical writings. But of course it is only one instrument among many that we must employ in a coordinated effort to extract meaning from these extraordinarily difficult texts. Another essential feature of our analysis relies on Newton's habit of providing the plain sense of a particular passage that he has extracted from his sources in square brackets or parentheses. These bracketed or parenthetical interpolations often act as a sort of Rosetta stone for arriving at Newton's understanding of a particular text. Although the Newton scholars mentioned above were all aware of this annotating practice, they did not make a systematic study of the way in which Newton's bracketed interpretations grew and developed over time. Thanks to the recent emergence of digital, searchable editions of Newton's manuscripts, however, this has become far more feasible. The *Chymistry of Isaac Newton* site has put about three-quarters of Newton's alchemical manuscripts online in edited form, and *The Newton Project* at Oxford University (http://www.newtonproject.ox.ac.uk) has performed a similar service for his religious writings. These digital editions have made it far more feasible to find bracketed expressions and detect parallel passages among widely distributed Newtonian manuscripts, thus allowing us to draw hitherto unsuspected comparisons among his writings. Advanced computational techniques available only for digital corpora such as latent semantic analysis have also facilitated this goal.[27] As a result, *Newton the Alchemist* is the first book to provide a picture of Newton's alchemy as it transformed from its earliest stages in the 1660s up to its full maturity and even after his transfer to London in 1696.

Although many problems remain, we are now well on our way to understanding why the warden and then master of the Royal Mint in his spare time jotted alchemical pseudonyms on his papers related to the Great Recoinage at the end of the seventeenth century.[28] Employing the common

[27] The *CIN* site features a Latent Semantic Analysis functionality, which allows parallel passages (even fuzzy ones) to appear automatically. See www.chymistry.org under "Online Tools." This tool was designed and implemented by Wallace Hooper.

[28] Babson 1006, 1r. It is of course possible in principle that Newton was reusing old paper on which his alchemical pseudonyms had been previously recorded. Even if that should turn out to be the case, however, we know from other sources that Newton was actively collaborating on an alchemical project with the London

early modern practice of hiding one's identity behind an anagram, Newton created two columns of alternative pseudonyms based on the Latin form of his name, "Isaacus Neuutonus." One of these, "Venus ac Jason tuus" conjures up both the classical goddess of love and the Argonaut who circled the globe in search of the golden fleece, a common symbol for the alchemical magnum opus. Although Newton famously eschewed the charms of Venus, his notes reveal that he was still dreaming of the philosophers' stone in the midst of his mission to purify the currency of England and to punish those who debased its coinage. His involvement with alchemy was still active around the time of his elevation to president of the Royal Society in 1703 and even persisted through his acquisition of a knighthood in 1705. Behind the authoritarian visage that controlled the Mint and dominated the Royal Society, the quest that ravished Newton as a young scholar intent on acquiring the caduceus of Mercury was still intact, and for all we know, his interest in the subject never died. Even in his old age, Newton told the husband of his niece, John Conduitt, that "if he was younger he would have another touch at metals."[29]

What Did Newton Want from Alchemy?
A Road Map for the Reader

The proper understanding of Newton's alchemy presents an enduring puzzle to contemporary scholarship in much the same way that the decipherment of hieroglyphics or the solution to the Greek script known as Linear B challenged Egyptologists and Hellenists in the nineteenth and twentieth centuries. Although Newton's peculiar alchemical "language" was the creation of one man building on his forebears rather than the dialect of an entire civilization, the linguistic difficulties that it presents share some similarities with these ancient scripts, particularly in Newton's creation of the idiosyncratic graphic symbols introduced in our foreword. Yet Newton's alchemy, even though it offers serious difficulties of language, cannot be deciphered by linguistic means alone; it requires a knowledge of materials, technologies, and tacit practices as well as underlying theories submerged beneath the written word. No book that does full justice to the difficulties presented by Newton's generation-long experimental research project centered on alchemy can be light reading. Newton's purpose and methods were obscure enough to mislead four dedicated scholars, as we have seen, each of them blinkered by a preconceived thesis. In order to avoid adding to the collective misunderstanding of Newton's goals and methods, I have made an effort to assess the evidence in all of its details. This is the only way to arrive at any degree of certainty as to what Newton was doing for over thirty years in his study as he devoured alchemical books and manuscripts, and then tried

---

distiller William Yworth in the first decade of the eighteenth century, well within his Mint period. See chapter nineteen of the present book as well as Karin Figala and Ulrich Petzold, "Alchemy in the Newtonian Circle," in *Renaissance and Revolution: Humanists, Scholars, Craftsmen, and Natural Philosophers in Early Modern Europe*, ed. Judith Field and Frank James (Cambridge: Cambridge University Press, 1997), 173–91.

[29] Keynes 130.05, 5v. Accessed from *NP* on January 22, 2017.

to test his understanding of them experimentally. The reader who wants to understand Newton's alchemy rather than merely assimilating one of the preexisting views on the subject must therefore be willing to engage with Newton's language, ideas, and practices over a range of genres and in considerable detail. In order to appreciate the whole we must understand its parts, even if it proves to exceed their sum.

The scope and detail of the present book call for a preliminary road map of its contents. Because of the daunting character of traditional alchemical language, which was often expressed in the form of enigmas, the next chapter begins with a consideration of literary deception in alchemy, devoting considerable space to Newton's understanding of the riddling language of the "adepts," the mysterious practitioners of alchemy who had, at least in principle, mastered the secret of chrysopoeia. This exercise requires that we understand the place occupied by the figure of the alchemical adept in the imagination of early modern Europeans and the remarkable powers that the possessors of the grand elixir were thought to possess, powers that not only included the ability to transmute base metals into noble ones but also a parallel skill in verbal deception. According to the prevailing early modern view, the very fact of their dominion over nature forced the adepts to hide behind a veil of secrecy, because of the danger that would accrue to them if the world knew of their abilities and because it was necessary to prevent the accession of the unworthy to their ranks. To the mind of Newton, the adepts were tricksters, not because they lacked the ability to carry out their marvelous transmutations, but because they veiled their knowledge under a sophisticated language of metaphor, allusion, and outright doublespeak. Not that they spoke in gibberish; to the contrary, the intelligent and properly trained student could penetrate behind their fuliginous tropes, but only if God willed it. It was Newton's belief that in his case God did so will.

But however much divine assistance might contribute to one's alchemical success, doing alchemy did not contribute to one's divinity. Newton's private belief in the infallibility and elect status of the adepts did not entail that he viewed alchemy as a path to religious salvation. In fact, references to the aurific art in the vast corpus that Newton devoted to religious topics, consisting of about four million words, are vanishingly small. And like his chymical forerunner Joan Baptista Van Helmont, Newton thought that success at chymistry must be "bought with sweat," the unavoidable, and often mundane labor of the laboratory.[30] Chapter three provides a close analysis of several related themes, considering, for example, the relationship between Newton's exegesis of biblical prophecy and his method of interpreting the textual riddles presented by writers on the philosophers' stone. At the same

---

[30] Joan Baptista Van Helmont, *Ortus medicinae* (Amsterdam: Ludovicus Elsevier, 1652), 560, #55: "Carbones emant, & vitra, discantque prius, quae nobis dedere, & vigalatae ex ordine noctes, atque nummorum dispendia, dii vendunt sudoribus, non lectoribus solis, artes." See also Newman, "Spirits in the Laboratory: Some Helmontian Collaborators of Robert Boyle," in *For the Sake of Learning: Essays in Honor of Anthony Grafton*, ed. Ann Blair and Anja-Sylvia Goeing (Leiden: Brill, 2016), 2: 621–40. For the most recent sustained look at Van Helmont's life and work, see Georgiana D. Hedesan, *An Alchemical Quest for Universal Knowledge* (London: Routledge, 2016).

time, the chapter also examines Newton's views on ancient wisdom and mythology in their relation to the aurific art, since many alchemists believed that the entertaining tales of the Greek and Roman pantheon contained veiled instructions for preparing the great arcanum. Previous scholarship has tended to assume that Newton too upheld the belief that ancient mythology was largely encoded alchemy, but as chapter three argues, this would have presented a sharp conflict with his views on ancient chronology and religious history. Further evidence shows that Newton may well have considered the mythological themes transmitted and analyzed by early modern alchemists as conventional puzzles reworked from antique sources rather than as true expressions of ancient wisdom. Nonetheless, they were conundrums to be solved if one wished to advance to the mirific tool of the adepts, the philosophers' stone.

With chapter four I also provide necessary background for the reader, but this time it concerns issues of historical context rather than language. As I argue at some length, Newton's belief that metals are not only produced within the earth but also undergo a process of decay, leading to a cycle of subterranean generation and corruption, finds its origin in the close connection between alchemy and mining that developed in central Europe during the early modern period. Alchemy itself acquired a distinct, hylozoic cast that the aurific art, at least in its more scholastic incarnation, had largely lacked in the European Middle Ages. Despite a common scholarly view that holds alchemy to have been uniformly vitalistic, the early modern emphasis on the cyclical life and death of metals was not a monolithic feature of the discipline across the whole of its history, but rather a gift of the miners and metallurgists who worked in shafts and galleries that exhibited to them the marvels of the underground world. Newton, writing for the most part in the last third of the 1600s, was the heir of a unique blend of mining lore and alchemy that had reached its efflorescence almost a century before. The fourth chapter concludes by describing additional sources used by Newton, such as his favorite chymical writer over the *longue durée*, Eirenaeus Philalethes, and also the pseudonymous early modern author masked beneath the visage of the fourteenth-century scrivener Nicolas Flamel.

In chapter five we examine the young Newton from his education at the Free Grammar School in Grantham during the 1650s up to his student years at Trinity College, Cambridge, beginning in 1661, in order to see how his interest in chymistry originated and developed. The standard view is that Newton was stimulated to his early interest in chymistry by the works of Robert Boyle. But my recent discovery of an anonymous and hitherto unexamined manuscript, *Treatise of Chymistry*, provides new evidence to show that Newton was already compiling chymical dictionaries before reading Boyle's works on the subject. Very likely his earliest chymical interests stemmed from his adolescent exposure to writers in the traditions of books of secrets and natural magic such as John Bate and John Wilkins, although he fell under Boyle's spell in due course. Chapter five then passes to what are probably Newton's earliest notes on chrysopoeia, namely, his abstracts and summaries of the works attributed to the supposed fifteenth-century

Benedictine Basilius Valentinus. Finally, the chapter tries to pin down some of the early contacts in Cambridge and London who transmitted the manuscripts and other texts to Newton that provided a major part of his alchemical knowledge. We are able to provide new information here too, although much of course remains dark.

Although Boyle's early influence on Newton already emerged briefly in the previous chapter, the next provides a sustained treatment of the self-styled English "naturalist" and his contribution to Newton's optical research. It is little appreciated that Boyle's analytical approach to chymistry had a profound impact on Newton's optics in the second half of the 1660s, the period that Newton considered "the prime of my age for invention."[31] As chapter six argues at length, Newton transferred Boyle's analysis and resynthesis or "redintegration" of materials such as niter to the realm of light. It was the decomposition of white light into its spectral colors and the subsequent recomposition of whiteness from the spectrum that provided Newton with one of his most cogent demonstrations that white light was actually a heterogeneous mixture. Chapter six establishes the influence of Boyle's chymistry on Newton's experimental methodology, using primarily terminological clues to reveal Newton's borrowings from Boyle's redintegration experiments. At the same time, the chapter also presents Boyle's and Newton's work against the backdrop of scholastic matter theory and optics in order to underscore the epoch-making character of the new color theory, which resulted in the overthrow of two millennia of research on the subject.

The seventh and eighth chapters consist of a detailed analysis of Newton's two early theoretical treatises, *Humores minerales* and *Of Natures obvious laws & processes in vegetation*, both probably written between 1670 and 1674, the very period when Newton was first making a name for himself at the Royal Society with his invention of a reflecting telescope and his controversial publication of his new optical theory. Both *Humores minerales* and *Of Natures obvious laws* employ alchemical theory to describe the process of metallic and mineral generation in the subterranean world. It is here that Newton claims in unforgettable language that the earth resembles "a great animall ^or rather inanimate vegetable" that inhales subtle ether and exhales gross vapors or "airs."[32] I argue that these works provide the theory on which he bases much of his subsequent experimental practice in the domain of chymistry. In particular, the emphasis that these two texts place on reactions in the vapor or gaseous state helps to explain the strikingly heavy emphasis that Newton gave to sublimation of various materials in his experimental practice. *Of Natures obvious laws* is also interesting for its careful attempt to disentangle natural processes that rely on mechanical interactions from those that employ "vegetation," the principle of generation, growth, and putrefaction depending on hidden *semina* or seeds buried within matter.

---

[31] CU Add. 3968.41 f.85r (= frame 1349 of http://cudl.lib.cam.ac.uk/view/MS-ADD-03968/1349, accessed May 16, 2016).
[32] Dibner 1031B, 3v.

Newton's higher goals for chymistry attempt to harness the power of these latent sources of activity for the purpose of transmutation.

With the ninth chapter we pass from theory to practice. Beginning with Newton's very early interpretations of the Polish alchemist Michael Sendivogius in the manuscripts Babson 925 and Keynes 19, the chapter shows that the brash young Cantabrigian initially thought the secret of chrysopoeia to be attainable by means of two ingredients alone, namely stibnite or crude antimony and lead. Much of his focus on antimony stems from his recent reading of the 1669 text by Philalethes, *Secrets Reveal'd*, which describes the use of that material in fairly clear terminology. The great significance that Newton idiosyncratically attaches to the metal lead in this early phase, however, has gone unnoticed by previous scholars and adds a hitherto unsuspected dimension to his aurific quest. His subsequent exposure to additional alchemical texts, especially in the extended corpus of Philalethes, soon made him understand that he had oversimplified matters. Other metals were also involved in the processes of Philalethes, especially copper. Was lead also part of the Philalethan modus operandi, or had Newton misinterpreted the American adept? In order to resolve this question, Newton turned to the same theories of metallic generation beneath the earth that had inspired *Humores minerales* and *Of Natures obvious laws*. By deepening his understanding of subterranean mineral generation, Newton believed he would be in a better position to replicate nature's processes of growth and transformation in the laboratory.

Newton's abrupt realization that his earliest understanding of the alchemical masters was erroneous also led him to adopt a form of textual interpretation that had hitherto been largely absent from his notes. In a word, he appropriated a venerable genre among medieval and early modern alchemical writers, the *florilegium* or collection and reorganization of snippets and *dicta* of the adepts for the purpose of comparing them to one another and extracting their sense. At this point, roughly corresponding to Newton's withdrawal from public scientific life between 1676 and 1684 after growing disillusioned with the public response to his radical optical theory, he had more than ample time to focus on the decryption of alchemical texts. Working through multiple treatises and winnowing out all but the information that he deemed most crucial, Newton would then group the resulting snippets with those from other texts that he thought threw light on them. This old alchemical practice has made it extremely difficult for modern scholars to determine where Newton's own beliefs begin and where those of his sources end. Patient comparison of Newtonian borrowings to the original texts and to one another, facilitated by digital searching and other computational techniques, has allowed me to obviate this problem, at least for the most part. Chapter ten provides a sustained look at an important florilegium from the period 1678–86 (Keynes 35), which shows the hitherto unsuspected influence on Newton of the German chymist Johann Grasseus.

Another author who acquires newfound significance in Newton's florilegia is Johann de Monte-Snyders, an extraordinarily obscure writer of two published texts. New information that I have unearthed on Snyders shows that he fell squarely into the mold of the self-styled wandering adept, traversing central Europe and performing demonstrations of his aurific prowess, no doubt in the hope of obtaining patronage. His life and influence serve as the subject of chapter eleven. In order to illustrate the way in which Newton tailored the writings of Snyders to fit his own conception of the alchemical magnum opus, the chapter also explores other contemporary accounts of Snyders's processes and shows that Newton's interpretation did not fit the standard view. The German adept exercised more impact on Newton the alchemist than any other author short of Philalethes. By giving a close reading to several important manuscripts, particularly Keynes 58, where Newton describes his plan for experiments that will lead to the scepter of Jove and the caduceus of Mercury, chapter twelve in turn shows how Newton combined his understanding of Snyders with motifs and practices drawn from Philalethes.

The same creative reworking of an earlier author forms the subject of chapter thirteen, which examines Newton's take on the substantial alchemical corpus ascribed to the high medieval Mallorcan philosopher Ramon Lull.[33] One can date his newfound interest in the pseudo-Lullian corpus to the publication of Edmund Dickinson's 1686 *Epistola ad Theodorum Mundanum*, which Newton read soon after its publication. This places Newton's Lullian turn to the very period when he was composing his masterwork, the 1687 *Principia*, after the astronomer Edmund Halley famously encouraged him to put his gravitational theory into written form. Influenced by the work of Dickinson, a prominent physician in Oxford and London, Newton came to believe that Lull's comprehensive description of the quintessence or spirit of wine (our ethyl alcohol) was actually an encoded discussion of the "first matter" or initial ingredient out of which the philosophers' stone, by a long and laborious process, should be made. Newton's ideas on this subject fill a complicated florilegium found in several manuscripts, which links Lull's work to that of Van Helmont, and which in turn presents detailed discussions of the alkahest or universal dissolvent. Also employing Van Helmont's foremost English expositor George Starkey, Newton attempts to determine the precise difference between the Lullian quintessence and "the immortal dissolvent," that is, the alkahest. This florilegium, simply titled *Opera* (Works) by Newton, contains hidden riches, such as a fascinating discussion of the affinities between chemical species that would undergo extensive treatment in *Query 31* of Newton's famous 1717 *Opticks*.

In chapters fourteen, fifteen, and sixteen, we arrive at Newton's experimental notebooks, containing dated chymical laboratory records from 1678 to 1696, which he kept largely distinct from his reading notes. While the two Cambridge collections, CU Add. 3973 and 3975, have been examined by previous scholars, the two sides of the single sheet composing Boston Medical Library B MS c41 c contain very early experiments that

---

[33] The extensive corpus of alchemical treatises attributed to Ramon Lull forms the subject of Michela Pereira, *The Alchemical Corpus Attributed to Raymond Lull* (London: Warburg Institute, University of London, 1989).

complement the Cambridge records in important ways.[34] All of these texts reveal Newton's extraordinary precision in experimentation and the single-minded discipline that guided his repeated variations on the same basic sets of laboratory protocols. The same exactitude in recording his experiments makes it possible to identify a number of Newton's proprietary *Decknamen* by an approach that combines textual decipherment with laboratory replication. This twofold method has allowed me to identify Newton's all important "standard reagent," the acid "menstruum" that he variously calls liquor, spirit, vinegar, and salt of antimony. With this material in hand, I have been able to produce "vitriols," that is, crystalline salts, of copper and several cupriferous minerals, in the hope of replicating Newton's "volatile Venus," a major desideratum of his alchemical research. The work of replication is ongoing, but already one can see how Newton planned his experiments and reasoned out his conclusions. His notes on the work of a contemporary chymist, David von der Becke, show that Newton was using his knowledge of chymical affinities in combination with a corpuscular theory to predict the course of reactions and to plan individual experiments. But he typically performed these operations with his chrysopoetic sources firmly in mind; in the end, most of the experiments in his laboratory notebooks consist of attempts to reverse-engineer the products allusively described in Newton's readings. Chapter sixteen concludes by examining precisely one such product, the "net of Vulcan" found in the works of Philalethes and elaborated at considerable length by Newton.

The cover names employed in Keynes 58 and the materials alluded to by Fatio also make a sustained appearance in Newton's famous *Praxis* manuscript (Huntington Library, Babson 420), which chapter eighteen analyzes

Despite the fact that Newton kept his cards close to his chest when discussing matters related to chrysopoeia, he did nonetheless engage in a variety of collaborative chymical projects. Chapter seventeen discusses one of these in considerable detail. The first of the collaborations took place in 1693, when Newton's Genevois friend Nicolas Fatio de Duillier encountered a French-speaking alchemist in London, apparently a Huguenot serving in King William's forces in the Low Countries. By examining Fatio's hitherto unstudied letter to Newton from the summer of 1693 in conjunction with Newton's manuscript "Three Mysterious Fires" (now found at Columbia University), I show that the latter text represents the fruit of an elaborate set of procedures devised by Newton in conjunction with Fatio and his Francophone friend. These processes were related to another set of operations from Newton that Fatio recapitulates in the aforementioned 1693 letter. As I argue in chapter seventeen, the procedures that Fatio quotes from Newton provide an important key for understanding both Keynes 58 and the laboratory notebooks. In a word, they are simplified procedures for making such important desiderata as the caduceus of Mercury and the scythe of Saturn, *Decknamen* that arise in the records of Newton's experimentation and reading notes.

The cover names employed in Keynes 58 and the materials alluded to by Fatio also make a sustained appearance in Newton's famous *Praxis* manuscript (Huntington Library, Babson 420), which chapter eighteen analyzes

---

[34] Boston Medical Library B MS c41 consists of three separate manuscripts, all by Newton, kept in separate envelopes. "B MS c41 c" refers to the single, folded sheet that begins "Sal per se distillari potest."

in the light of Newton's work with his young friend. Scholars have traditionally viewed *Praxis* as the culminating record of Newton's alchemical career; at the same time, some have seen its seemingly incomprehensible processes and profusion of *Decknamen* as proof that Newton was undergoing a mental crisis around the time it was written. After all, *Praxis* refers to Fatio and might even have been composed in Newton's "black year," 1693, when he angrily (if briefly) isolated himself from his friends and complained of symptoms that were subsequently interpreted as a "derangement of the intellect." Hence I devote considerable space to the analysis of this challenging text and argue that it is in reality quite comprehensible in the light of Newton's epistolary exchanges with Fatio and other collections such as Keynes 58.

Fatio was not the only chymist with whom Newton collaborated in his maturity. After his move to London in 1696, Newton was evidently approached by the obscure "Captain Hylliard," who wrote a brief alchemical manifesto that the now famous intellectual and Mint official copied. Chapter nineteen provides an extensive analysis of the episode with Hylliard and also describes Newton's extended collaboration with the Dutch distiller William Yworth, which also took place after Newton's move to London. Beyond casting new light on the processes behind Yworth's *Processus mysterii magni* and linking them to Newton's late florilegia, the chapter also uses a recently discovered manuscript in the Royal Society archives to show that the document actually contains the record of a live interview between Newton and Yworth.

The final three chapters of *Newton the Alchemist* continue the story, already begun in chapter six, of the relationship between Newton's private chrysopoetic ventures and public science in the seventeenth and early eighteenth centuries. The interaction between chymistry and optics did not end with Newton's transfer of Boyle's redintegration experiments into the realm of light and color. Chapter twenty shows that Newton developed a theory of refraction based on the chymical principle sulfur, which he described in the first edition of his famous *Opticks* (1704). The chapter also finds that the seeds of this theory extend back to Newton's 1675 *Hypothesis of Light*, where he explicitly abandons the Sendivogian theory of an aerial niter that he had affirmed in *Of Natures obvious laws*. Newton replaced the aerial niter, which had accounted for phenomena ranging from combustion and respiration to the fertilization of the earth, with a growing reliance on sulfur. Although he had reasons of his own for making this shift, Newton was also influenced by parallel developments in European chymistry, a field that was rapidly moving toward what would eventually be known as phlogiston theory. Another trend that would soon acquire great significance in Europe and England was the increasing emphasis chymists placed on affinity among different materials. Affinity also enters into Newton's sulfurous theory of combustion and into the *Opticks*' explanation of refractive power in a major way. Chapter twenty-one presents this topic by building on Newton's increasing interest in sulfur, placing his theories in the context of developments within the chymical community of the late seventeenth and early eighteenth centuries. The chapter provides a new look at Newton's developing ideas about affinity and his role in the eighteenth-century development of affinity tables, the graphic representations of selective attractions by materials that cause those with less affinity to precipitate. Finally, chapter twenty-two considers Newton's relationship with Boyle in the light of both men's attempts to arrive at a "sophic mercury" that would in principle dissolve gold into its primordial constituents and make it possible for the noble metal to "ferment," as Newton says in his short text of 1692, *De natura acidorum*. The two major English representatives of public science in the seventeenth century had very different ideas about the path to chrysopoeia, though both, in the end, were alchemists in the fullest sense of the term.

Returning then to the variations on a Keynesian theme with which I began this chapter, one can see how *Newton the Alchemist* changes our understanding of the celebrated natural philosopher. Already as a very young man, even before he had absorbed the chymical knowledge of Boyle, Newton enlisted himself in the school of the adepts. Yet alchemy was not an alternative religion for Newton, nor was it the origin of his theory of gravitation. The short-range forces operating in the chymical realm were objects of study in themselves, just as gravitational attraction was. In the later editions of the *Opticks* Newton even erects the active principle behind the phenomenon of "fermentation," by which he here means chemical reactions in general, to the status of a fundamental force like magnetism and gravitation. But these theoretical speculations, important as they were, represent very little of the immense work that Newton devoted to alchemy. To see these published ruminations as the end goal of Newton's decades of alchemical research would be a disingenuous and misleading perspective. Although he employed theories of alchemical origin as a means of understanding and enlarging natural philosophy, the countless hours he spent deciphering alchemical texts and putting his conclusions to the test in his laboratory had a more practical goal. In a word, the founder of classical physics aimed his bolt at the marvelous menstrua and volatile spirits of the sages, the instruments required for making the philosophers' stone. Difficult as it may be for moderns to accept that the most influential physicist before Einstein dreamed of becoming an alchemical adept, the gargantuan labor that Newton devoted to experimental chrysopoeia speaks for itself. The chymical tools envisaged by Newton, had he been able to acquire them, would have handed him the power to alter nature to its very heart. These were the secrets that the "true Hermetick Philosopher" must keep hidden lest they cause "immense dammage to ye world," as he said to the Secretary of the Royal Society in 1676.[35] The core of Newton's labors at deciphering the documents of the adepts lay in his own undying quest to join their number.

---

[35] Newton to Henry Oldenburg, April 26, 1676, in Newton, *Corr.*, 2: 2.

TWO

Problems of Authority and Language in Newton's Chymistry

THE CONCEPT OF THE ADEPT

Newton's engagement with chrysopoeia lasted well over thirty years and resulted in the writing of about a million words of text. His substantial chymical *Nachlass* presents interpretive difficulties that are perhaps unique within the corpus of the famous natural philosopher. In order to come to terms with this refractory material, we must first address some of the characteristics that make it unusual. Primary among them is the cluster of difficulties surrounding the concept of the "adept." Like many students of chymistry in the early modern period, Newton held an exalted view of the supposed masters of the aurific art, the adepts, or "adeptists" as they were often called in seventeenth-century English. According to a wide variety of sources, these men (for they were almost always men) were thought to hold a privileged position in the world. They made up an elect band of *filii doctrinae*, or "sons of art," who had received the philosophers' stone as a divine dispensation, a *donum dei* or "gift of god."[1] Some of this perspective seeps through, albeit in the cautious and attenuated form appropriate to public discourse, in a fragmentary passage that Newton related in old age to the husband of his niece, John Conduitt:

> They who search after the Philosopher's Stone by their own rules obliged to a strict & religious life. That Study fruitful of experiments.[2]

---

[1] An excellent synopsis of this exalted view of the adepts collected from various authors may be found in W. C., *The Philosophical Epitaph of W. C. Esquire* (London: William Cooper, 1673), 4, 6–8, 21–22, 28, 30, 32, 34, and throughout. For recent work on William Cooper and W. C., see Lauren Kassell, "Secrets Revealed: Alchemical Books in Early Modern England," *History of Science* 49 (2011): 61–87.

[2] Newton to Conduitt, as quoted in Manuel, *PIN*, 173. Manuel gives no folio number for the passage, but Scott Mandelbrote has kindly told me that it is found on folio 9r of Keynes 130.6, which has not yet appeared on the Newton Project site. It is seldom noted that Conduitt's recollections also contain some reservations, seemingly stemming from the aged Newton, regarding the quest for chrysopoeia. Keynes 130.07 twice links the "Philosopher's stone <or> Grand Elixir" to enthusiasm. It is not difficult to understand why the by-now-celebrated president of the Royal Society and master of the Royal Mint would not wish to associate himself publicly with enthusiasts, particularly since chrysopoeia was falling increasingly into disrepute across Europe by the 1720s. See Keynes 130.07, 7r, edited in *NP* at http://www.newtonproject.ox.ac.uk/view/texts

The philosophers' stone, which was the special privilege of the adepts, had astonishing powers: not only could a tiny portion of it transmute a mass of metal into gold or silver, it could also cure diseases of the most dire sort. Being the chosen sons of divine wisdom, the adepts were at heart a benevolent group, who wished to help their fellow humans. But they were continually frustrated in this wish by the venality, cruelty, and suspiciousness of humankind, which made a wholesale dispensing of their gifts impossible.[3] What would happen if the philosophers' stone were made public to the masses? The economic basis of society, gold and silver, would at once collapse, leading to chaos, war, and tyranny. As if to reinforce the baseness of human nature, it was widely believed that the mere rumor of one's being an adept could result in torture and murder from the inevitable attempt of the hoi polloi to extract the philosophers' stone by force. Being an adept was not only lonely, it was dangerous.

The privileged but precarious lives of the adepts received attention from a variety of sources. On the one hand, alchemical texts themselves, such as the popular *Secrets Reveal'd*, a translation of the Latin *Introitus apertus ad occlusum regis palatium* by the famous American adept Eirenaeus Philalethes, contained stories of persecution at the hand of the unenlightened mob.[4] And yet these accounts were not limited to narratives of special pleading by the sons of art themselves. There were numerous stories of alchemists who had really been detained by rulers in order to gain access to their technical knowledge. Perhaps the most famous of these is the veridical account of Johann Friedrich Böttger; imprisoned for at least a decade by the Elector of Saxony, August der Starke, Böttger did eventually manage to employ his chymical skills in making a highly profitable porcelain.[5] Although Böttger patently lacked the philosophers' stone, other stories of successful wandering adepts were passed on in "transmutation histories," a genre filled with seemingly verifiable names and places that could vouch for the transmutational prowess of the alchemical elixir.

From the perspective of seventeenth-century alchemical aficionados, then, the adepts occupied an isolated and problematic position in society. Forced to remain anonymous, and yet constrained by their very status as a divine elect devoted to the good of mankind, they were required to distribute their secret wisdom with the utmost care. They could of course restrict

---

/diplomatic/THEM00169, consulted June 13, 2017. Whatever Newton actually said to Conduitt, the testimony of his laboratory notebooks and correspondence shows without any possibility of doubt that he himself sought the philosophers' stone for well over three decades.

[3] See "An Essay Concerning Adepts" (1698) by the anonymous "Philadept," reprinted in Gregory Claeys, *Restoration and Augustan British Utopias* (Syracuse, NY: Syracuse University Press, 2000), 209–33, consult especially 210–11. A discussion of this treatise is found in J. C. Davis, *Utopia and the Ideal Society: A Study of English Utopian Writing, 1516–1700* (Cambridge: Cambridge University Press, 1981), 355–67.

[4] Philalethes (Starkey) referred to the *Introitus* as "my little Latin Treatise, called Introitus apertus ad occlusum Regis palatium" in his later collection, *RR*, 7. Thus although the English version of the text, *SR*, might appear at first to be the original text, it is in reality a translation and reworking of the Latin *Introitus*.

[5] Georg Lockemann, "Böttger, Johann Friedrich," *Neue Deutsche Biographie* 2 (1955), online version, at https://www.deutsche-biographie.de/gnd118512846.html#ndbcontent, accessed January 3, 2017. See also the entertaining account in Janet Gleeson, *The Arcanum* (New York: Warner Books, 1998).

the transmission of their arcane knowledge to the spoken word, but that would mean that only a handful would receive the benefit of the adepts' largesse. Thus they felt a moral duty to describe their art in writing, so that others might gain access to their secrets. But this could not be easy; as the celebrated Flemish chymist Joan Baptista Van Helmont said, the art could only be bought with sweat, the product of intense labor. There was a twofold moral imperative at play, and one that was in a state of perpetual tension. On the one hand, the adepts should make the riches of alchemy accessible in their writings, but on the other, those writings had to be so difficult to decipher that they would delude and discourage the unworthy. The adepts were forced to walk a tightrope where the abyss on one side was a misanthropic stinginess and on the other the subjection of the world to a tyranny made possible by the limitless resources of the philosophers' stone.

This was the common picture of the adepts and their mode of communication among alchemical sympathizers in early modern Europe. The very word "adept" meant one who had attained the highest understanding of nature possible; it derives from the Latin word for "having arrived" (*adeptus* from *adipiscor*). Hence to be an adept was to have arrived at an infallible comprehension of nature, even if this state of wisdom had been preceded by a long period of erroneous belief. Such an understanding required that one also be immensely intelligent, of course, which had its own ramifications in the realm of alchemical literature. Since the adepts were fantastically clever, and constrained by their vows to repulse the rabble from acquiring an entry into the secrets of the art, they developed a set of literary techniques that made it almost impossible to do so. In order to make sense of Newton's alchemical writings we will in due course acquaint ourselves with the full panoply of these techniques of concealment, since he, perhaps even more than most followers of the aurific art, believed in the tremendous powers of literary trickery that alchemy laid claim to.

But first I must address an obvious problem. Is it really the case that Newton accepted the full picture of a hidden class or stratum of adepts as I have presented it? The answer lies readily at hand, perhaps surprisingly so. Despite its daunting length, Newton's chymical corpus contains only the barest handful of criticisms directed at his sources. In one early manuscript, he mentions that the writer Bernard of Trier did not become an adept until late in life, and therefore wrote obscurely lest others attain the art at a younger age than he did. The same manuscript passes on a common criticism that Geber, the author of the high medieval *Summa perfectionis*, was so obscure that he could only be understood by fellow adepts. In an early manuscript, Newton also points out that the Italian poet Giovanni Aurelio Augurelli seemed to cast doubt on the art in the last four lines of his *Chrysopoeia*. But the soon-to-be-famous scientist adds that Augurelli's disclaimer was an intentional way of avoiding the accusation of being an adept![6] None of these

---

[6] Huntington Library, Babson MS 419, 1r–1v. Newton says the following about Augurelli: "Johannes Aurelius Augerellus ^italus poeta suavissimus Chrysopœiam scripsit in cujus 4 ultimis versiculis videtur opus falsitatis arguere, sed astute fit ne Adeptus esse suspicetur."

comments reflect a distrust of the authors' knowledge, but merely of their means of communication.

When we turn from Newton's criticisms of stylistic obscurity to those of content, the number of rebukes is so small as to be almost nonexistent. Another early manuscript, this one found in the heterogeneous collection of twelve sheaves kept in the National Library of Israel that goes by the shelf mark Var. 259, contains two negative comments. The initial one is directed at Eirenaeus Philalethes's *Marrow of Alchemy*, which Newton presents here twice in his own abridged versions. The first such synopsis bears the comment "a fals Poem" after the title, but Newton then deleted the criticism with a strike of the pen.[7] In fact, the *Marrow of Alchemy* went on to become one of his favorite and most enduring sources. The second denial of adept status is more serious. After extracting some passages from Jean Collesson's *Idea perfecta philosophiae hermeticae*, Newton struck them through and added, "I believe him not to be an adept" (*Credo hic nihil adeptus*).[8] And yet in later manuscripts, such as Newton's mature *Index chemicus*, we find him citing Collesson as an authority, suggesting that this was merely a youthful flirtation with skepticism.[9] Newton's mature manuscripts reveal only one seeming criticism of a self-styled possessor of the alchemical summum bonum. The bizarre anonymous text *Manna*, which cobbles together allegorical passages from the better known *Arca arcani* by Johann Grasseus and treats them literally, elicits only the tamest of rebukes from Newton. To *Manna*'s description of the regimens or stages required to complete the maturation of the philosophers' stone, Newton responds, "Thus this author, but something lamely."[10] Other texts equally worthy of Baron von Münchhausen extract no critical response at all. The *Epitome of the Treasure of Health*, a picaresque work in which the pseudonymous author "Edwardus Generosus" claims to have used the philosophers' stone for such noble purposes as freezing fleas in his bed and downing birds that are attracted to its chill-inducing beams, appears in Newton's *Index chemicus* and other late collections alongside such sober chymists as Jean Beguin and Nicolas Lemery, implicitly sharing their authority.[11]

What are we to make of this seemingly facile acceptance on Newton's part? It cannot be denied that in the privacy of his laboratory he admitted the reality of the philosophers' stone along with the class of enlightened individuals who possessed it. While there may have been some willing suspension of disbelief at work in Newton's note taking, it does not appear that he was troubled by exuberant claims of thaumaturgy such as those of Edwardus Generosus. Edwardus was an adept, and this meant that he should have extraordinary powers over nature. It does not follow, however, that Newton read every detail of such authors as literally true. An adept could always be

---

[7] Var. 259.7.2r.

[8] Var. 259.9.3r.

[9] Keynes 30/5, 6r, 8v, and 10r.

[10] Keynes 21, 14v.

[11] Keynes 22, 6v (freezing fleas) and 12r (downing birds); Keynes 30/1, 22r, 23r (Edwardus Generosus), 11r (Beguin), 36r, 55r (Lemery).

hiding the most important facts beneath a facade, even when the text contained no obvious allegory. Had not Geber, at the end of his *Summa perfectionis*, admitted that he had hidden the transmutative elixir "where we have spoken more openly," in other words where he employed seemingly plain speech?[12] Since the masters of the philosophers' stone could not, by virtue of their status as adepts, be wrong, it followed that apparent errors or obsolete techniques in their chymistry could only be red herrings planted in the midst of their wisdom to delude the unwary. One main purpose of Newton's remarkably exact experimental notebooks found in Cambridge University's Portsmouth collection was precisely that of arriving at a correct interpretation of the chymistry hidden beneath such delusory literary practices. This was also the primary goal of the successive drafts of the *Index chemicus* that Newton finalized around the end of the seventeenth century. Akin to a modern concordance where headwords are presented in the context of authorial snippets, the *Index chemicus* swelled to almost a hundred folios in its final version. The end of this endeavor was a tool that would allow easy comparison of different authors' views on particular *lemmata*. More often than not, Newton considered the headwords that his authors supplied him to be allusive terms hiding a secret meaning, or as historians of alchemy say, *Decknamen* (cover words).

The absolute authority of the adepts was both abetted by their practice of secrecy and diluted thereby. Apparent mistakes or outdated technologies could be written off as misleading *Decknamen*, a practice that Newton himself employed in his interpretation of Geber's *Summa perfectionis*, where he creatively transforms the medieval alchemist's mineral "marchasita" into bismuth and "magnesia" into antimony.[13] While this practice excused the adepts of any potential error or obsolescence, however, it also meant that their original meaning could easily be lost. As the present book reveals, this was very frequently the case in Newton's interpretations, sometimes amazingly elaborate in their fineness of detail, of his alchemical reading matter. Before we can proceed to the particulars of his chymistry, however, we must now look more deeply at the full armamentarium of deceptions Newton's literary sources employed.

The Tricks of the Adepts: Traditional Techniques
of Deception in Newton's Sources

One of Newton's most frequently cited authors is the American chymist George Starkey, who wrote a number of chrysopoetic treatises under the nom de guerre of Eirenaeus Philalethes (A Peaceful Lover of Truth). Born in Bermuda and educated in the 1640s at the fledgling Harvard College,

---

[12] William R. Newman, *The Summa perfectionis of Pseudo-Geber* (Leiden: Brill, 1991), 785.

[13] See Newton's copy of *Gebri Arabis Chimiae . . . a Caspare Hornio* (Leiden: Arnoldus Doude, 1668) (= Stanford University, Barchas QD 25. G367), where he has interpreted and updated Geber's minerals on the flyleaves. Neither bismuth nor antimony played a major role in medieval alchemy, but in the sixteenth and seventeenth centuries they were both subjects of great interest.

Starkey experienced an astonishing success upon his immigration to London in 1650.[14] Almost immediately, he became the client and unofficial chymical tutor of one of the best connected men in England and Ireland, the young Robert Boyle. Thanks to a succession of letters that Starkey wrote to Boyle between 1651 and 1652, we have a very clear idea of his chymical work, which ranged from attempts at chrysopoeia to the preparation of medicaments by chymical means, and even extended to the formulation of such products as perfumes and artificial ice. Among Starkey's remarkable letters is one that has achieved considerable fame in modern times precisely because Newton copied out a Latin translation of it at some point in his career. The letter, composed in April or May 1651, relates Starkey's method of providing Boyle with a "Key into Antimony" by making a "sophic mercury," that is, a special, penetrative form of quicksilver that could supposedly decompose gold into its components (sulfur, salt, and mercury), and then encourage the metal to ripen into the philosophers' stone, which Starkey believed to be gold "digested" into the final degree of its maturity. It was once thought by Newton scholars that the "Clavis" (Latin for "Key") was an original composition by Newton, and that it could therefore serve as an Ariadne's thread into his laboratory practice.[15] Although we now know that to be false, Starkey's letter to Boyle is tremendously valuable all the same for the clear way in which it decodes the works that he wrote under the sobriquet of Philalethes into replicable chymical practice.

In his 1651 letter to Boyle, Starkey describes a way of making quicksilver form an amalgam with the metalloid antimony, which is not an easy thing to do. First Starkey refines crude antimony ore, known today as stibnite, by heating it to a temperature above its melting point (620°C) with stubs of horseshoe nails and saltpeter. The iron combines with the sulfur in the stibnite to form a slag containing ferrous sulfide, and the metallic antimony sinks to the bottom of the crucible as a "regulus" (little king). If the shiny, silvery antimony is allowed to cool slowly under the slag, it can solidify as the so-called star regulus of antimony, an attractive and much-prized formation (figure 2.1). Starkey says that one part of star regulus should be fused with two parts of refined silver, which he refers to at the very end of the Latin text making up the "Clavis," as "the doves of Diana" ("Dianaes doves").[16] He then washes quicksilver with vinegar and salt to purify it and grinds the cleansed quicksilver with the silver-antimony alloy. After multiple washings and reiterate distillations, which Starkey refers to as "eagles" because they make the volatile quicksilver "fly," the sophic mercury is complete. Modern laboratory replications have shown that a small amount of gold heated with such an "acuated" or sharpened mercury will indeed form interesting dendritic formations when heated in a sealed flask, though alas, it does not become the philosophers' stone.[17]

[14] For Starkey's life, see Newman, *GF*.

[15] Dobbs, *FNA*, 133–34, 175–86, 229–30; Westfall, *Never at Rest*, 370–71. For Starkey's authorship of the *Clavis*, see William R. Newman, "Newton's Clavis as Starkey's 'Key,'" *Isis* 7 (1987): 564–74.

[16] Starkey to Boyle, April/May 1651, in Newman and Principe, *LNC*, 23.

[17] Lawrence M. Principe, *The Secrets of Alchemy* (Chicago: University of Chicago Press, 2013), 158–66.

FIGURE 2.1. The star regulus of antimony, so called because of its fern- or star-like crystalline surface. The pattern is produced when the regulus of metallic antimony is allowed to cool slowly under a thick layer of the slag left after its reduction from stibnite. Prepared by William R. Newman in the laboratory of Dr. Cathrine Reck in the Indiana University Chemistry Department.

The clarity of Starkey's 1651 letter to Boyle is matched by the obscurity in which he deliberately masked his processes in the corpus of Eirenaeus Philalethes. According to Starkey's elaborate mystification, Philalethes was a still-living adept whose abode was New England, and who had authorized Starkey to distribute his work to a small number of trusted friends. In the following, I will therefore generally refer to Philalethes instead of Starkey when speaking of the works that the Harvard graduate wrote under his chosen pseudonym. One of these works (actually a collection of disparate treatises), written under the name of Eirenaeus Philalethes, was *Ripley Reviv'd*, published in 1678—thirteen years after Starkey's death in the Great Plague of London. Philalethes gives an interesting rationalization of his concealment in the beginning of his commentary on the fifteenth-century English alchemist George Ripley's *Compound of Alchemy*. The passage is revealing for its playful yet sarcastic tone; one gets a definite sense that the adept Philalethes enjoys teasing and titillating his eager audience:

> Such passages as these we do oftentimes use when we speak of the Preparation of our *Mercury*; and this we do to deceive the simple, and it is also for no other end that we confound our operations, speaking of one, when we ought to speak of another; For if this Art were but plainly set down, our operations would be contemptible even to the foolish.[18]

Although benevolent in principle, the adepts were not easy company. As Philalethes expresses it, he has aimed his obscurity at simpletons and fools. If the would-be alchemist fails to arrive at the philosophers' stone by Philalethes's methods, the blame lies only with the practitioner's inadequate brain. By implication, more ingenious souls will be able to penetrate to the bottom of the convoluted game erected around the very processes described in Starkey's letter to Boyle.

If we examine Philalethes's work alongside several other sources used by Newton, it emerges that these authors really did write both to reveal and to conceal, as they claimed. The alchemical language of the period is often a matter of encoded meaning whose sense is conveyed by sophisticated clues rather than the meaningless and garbled farrago that it sometimes appears to be. One of the traditional techniques Philalethes made use of is the twofold expansion and compression of language that I have elsewhere given the Greek names *parathesis* and *syncope*. The first of these practices involved stuffing one's speech with unnecessary synonyms for the same materials or processes, whereas the second consists of the opposite, namely, deliberate suppression of information. An excellent example of parathesis occurs in a passage much beloved by Newton and taken from the Philalethan *Secrets Reveal'd* (1669). Like much else in the corpus of Philalethes, this paragraph describes materials that are necessary for the making of the sophic mercury, which we have encountered already:

> our Water is compounded of many things, but yet they are but one thing, made of divers created substances of one essence, that is to say, There is requisite in our Water; first of all Fire; secondly, the Liquor of the Vegetable Saturnia; thirdly, the bond of ☿: The Fire is of a Mineral Sulphur, and yet is not properly Mineral nor Metalline, but a middle betwixt a Mineral and a Metal, and neither of them partaking of both, a Chaos or Spirit;

[18] Eirenaeus Philalethes, "An Exposition upon Sir George Ripley's Epistle to King Edward IV," in *RR*, 25.

because our Fiery Dragon (who overcomes all things) is notwithstanding penetrated by the odour of the Vegetable Saturnia; whose blood concretes or grows together with the juyce of Saturnia, into one wonderful body; yet it is not a body, because it is all Volatile; nor a Spirit, because in the Fire it resembles a Molten Metal. It is therefore in very deed a Chaos, which is related to all Metals as a Mother; for out of it I know how to extract all things, even ☉ and ☽ without the transmuting Elixir: the which thing whosoever doth also see, may be able to testifie it. This Chaos is called, our Arsenick, our Air, our ☽, our Magnet, our Chalybs or Steel; but yet in divers respects, because our Matter undergoes various states before that the Kingly Diadem be brought or cast forth out of the Menstruum of our Harlot. Therefore learn to know, who the Companions of Cadmus are, and what that Serpent is which devoured them, what the hollow Oak is which Cadmus fastened the Serpent through and through unto; Learn what Diana's Doves are, which do vanquish the Lion by asswaging him: I say the Green Lion, which is in very deed the Babylonian Dragon, killing all things with his Poyson: Then at length learn to know the Caducean Rod of Mercury, with which he worketh Wonders, and what the Nymphs are, which he infects by Incantation, if thou desirest to enjoy thy wish.[19]

An acquaintance with Starkey's 1651 letter to Boyle allows us to decode this fustian passage easily. "Our water" is of course the sophic mercury itself, which is made of three things, a fire, the liquor of "Vegetable Saturnia," and "the bond of Mercury." The "fire" or "Fiery Dragon" refers to the putative sulfur contained in the iron horseshoe nails used in the refining of stibnite to arrive at the star regulus of antimony; the "Saturnia" is the stibnite itself; and the mysterious "bond of Mercury" is simply the quicksilver that must be distilled from the alloy of refined silver and antimony. The chaos, "Arsenick," air, "our ☽," magnet, and chalybs or steel all refer to the star regulus of antimony, which is a shiny, crystalline, metalloid material that volatilizes at high temperature and yet can fuse over a fire to look like a molten metal. The *Decknamen* employed here are not arbitrary: chaos refers to the idea that antimony is the Ur-mineral out of which the other metals arise, as Philalethes himself says: even Sol (gold) and Luna (silver) can be extracted out of it. Arsenic and air both connote the volatility of the antimony regulus. The Moon ("our ☽") summons up the silvery appearance of the regulus, while magnet and chalybs encode a theory that the mercurial component of the antimony attracts a sulfurous component from iron during its refinement, just as the magnet attracts steel and vice versa. The kingly diadem is also the regulus, because of its crystalline appearance, and the menstruum of the harlot is the ore of antimony, stibnite, out of which the metalloid must be smelted with the help of the iron from the horseshoe nails. In the process, the stibnite releases its slag, which Starkey implicitly compares to the harlot's catamenia. The companions of Cadmus are the horseshoe nails, and the serpent is again the stibnite that must be refined. Diana's doves are the two

---

[19] Philalethes, *SR*, 4–6.

portions of silver that must be added to the star regulus so that quicksilver will amalgamate with it, the green lion and Babylonian dragon again refer to antimony (which is poisonous), and the caducean rod of Mercury is simply the completed sophic mercury. In this passage alone, then, at least twelve different *Decknamen* are used for antimony, including both its unrefined ore and the star regulus. Since Philalethes views the regulus as existing *in potentia* in the crude antimony or stibnite, the terms for both the refined metalloid and the ore are more or less interchangeable. As the author puts it, the "Matter undergoes various states," not to mention multiple names.[20]

Despite the terminological hypertrophy of Philalethes's description, the passage from *Secrets Reveal'd* also displays the contrasting literary artifice, syncope. This is particularly evident when Philalethes claims that "our water" is made of three things—fire, Saturnia, and the bond of Mercury. Even after we have deciphered these *Decknamen* and arrived at their concrete referents, we would still be unable to make the sophic mercury. The reason for our failure would lie in the fact that Philalethes has mentioned only iron (or rather its hidden sulfur), stibnite, and quicksilver. He has intentionally left the essential ingredient silver, which must be alloyed with the star regulus in order to make the quicksilver amalgamate, out of his description.

An additional and related point of confusion emerges from Philalethes's term "mercury," which has a profusion of meanings in alchemical literature. As he says in *Ripley Reviv'd*, "Philosophers have hidden much under the *Homonymium* of Mercury."[21] The term could simply mean quicksilver, of course, but it could also refer to the mercurial principle that, along with sulfur, was traditionally thought by alchemists to compose metals. The situation became far more complex when the immensely influential Swiss chymist Paracelsus added salt to the two principles in the early sixteenth century and argued that not only metals but also all bodies were composed of mercury, sulfur, and salt, and that these three could be extracted by "anatomizing" or analyzing the materials in question.[22] In addition, "mercury" was a term used to describe a host of materials that participated in quicksilver's liquidity and volatility, such as ethyl alcohol. Nor did a material have to share those particular properties in order to qualify as a "mercury," since just as it was possible to "fix" quicksilver by rendering it solid and nonvolatile (as in "red precipitate," our mercuric oxide), so it should be possible to render other "mercuries" solid as well. The thing that is particularly interesting about Philalethes's point, however, is that he explicitly identifies "mercury" as a homonym, one of the literary devices traditionally taught in the discipline of rhetoric. Philalethes's creator Starkey was the product of a scholarly

---

[20] I count them as follows: Saturnia, chaos, arsenic, air, Luna, magnet, chalybs, harlot, diadem, serpent, green lion, Babylonian dragon.

[21] Philalethes, "An Exposition upon Sir George Ripley's Preface," in *RR*, 25.

[22] See William R. Newman, "Alchemical and Chymical Principles: Four Different Traditions," in *The Idea of Principles in Early Modern Thought: Interdisciplinary Perspectives*, ed. Peter Anstey (New York: Routledge, 2017), 77–97. The works of Paracelsus have recently become much more accessible to English speakers with the following collection of translated texts: Andrew Weeks, *Paracelsus: Essential Theoretical Writings* (Leiden: Brill, 2008).

environment that valued textual analysis to the highest degree. The son of a Scottish minister who wrote elegant Latin poetry, Starkey attended Harvard College at a time when grammar, rhetoric, and dialectic, the traditional trivium of the medieval universities, were still unchallenged in their dominance on the human intellect. Starkey's aptitude and training in these verbal arts emerges clearly from his mastery of literary artifice.

But it is possible, of course, to overstress the Daedalean gifts of Philalethes. The techniques mentioned so far, employment of *Decknamen*, parathesis, syncope, and the related verbal parsimony implied by the use of homonyms, have a long lineage in the history of alchemy. The same is true of another widely used technique explicitly employed by the Islamic writers of the Middle Ages who wrote under the collective pseudonym of Jābir ibn Ḥayyān. Originally referred to in Arabic as *tabdīd al-ʿilm* (dispersion of knowledge), this involved the splitting of a recipe or narrative into different parts, followed by its distribution over disparate sections of a book or books.[23] The practice was adopted by the Latin author of the famous *Summa perfectionis*, one of the most influential alchemy books of the European Middle Ages, who called himself Geber (after Jābir). In the *Summa perfectionis*, Geber describes the technique as follows:

> Lest we be attacked by the jealous, let us relate that we have not passed on our science in a continuity of discourse, but that we have strewn it about in diverse chapters. This is because both the tested and the untested would have been able to take it up undeservedly, if the transmission were continuous.[24]

Echoing the Latin of Geber, the practice of "dispersion of knowledge" came to be known as *dispersa intentio*. It has even been shown that Newton's correspondent Robert Boyle, that seemingly modern proponent of open speech, used *dispersa intentio* when writing about the higher secrets of chymistry such as the sophic mercury and the marvelous dissolvent or alkahest of Paracelsus and Van Helmont.[25]

We have now examined a substantial number of the techniques of concealment employed by Philalethes and other alchemists read by Newton. Understanding their use of *Decknamen*, along with parathesis, syncope, and *dispersa intentio* is not enough, however, to gain a true appreciation of the fiendish complexity in which the self-styled adepts could and did cloak their work. One of Newton's favorite sources in the late phases of his career, the well-known physician of Oxford and London Edmund Dickinson, wrote a work in 1686 consisting of an epistolary exchange between the doctor and an anonymous adept referred to as "Theodorus Mundanus" (Earthly Gift of God). Dickinson is no critic of chrysopoeia; in fact, his part of the exchange consists largely of a sustained plea that Mundanus reveal his secrets. And yet Dickinson goes on at length railing against the "jealousy" and stinginess of

---

[23] Paul Kraus, *Jābir ibn Ḥayyān: Contribution à l'histoire des idées scientifiques dans l'Islam* (Cairo: Imprimerie de l'institut français d'archéologie orientale, 1943), 1: xxxi–xxxiii.

[24] Newman, *Summa perfectionis*, 785.

[25] Principe, *AA*, 147–48.

the adepts. They invite the unwitting to their art with sweet promises, and then they obfuscate their victims with impenetrable metaphors, harsh allegories, unheard of tropes, and altogether horrid, tortuous, and barbarous locutions. With their "tropes, metaphors, allegories, enigmas, barbarous terms and neologisms," the alchemists hide their knowledge like a squid enveloped in its own ink. Using their "keen and crafty intellect" (*acutum ac subdolum ingenium*), the adepts perversely substitute words and processes for one another, creating hidden nets and snares that trap and delude the unwary. The famous thirteenth-century Mallorcan philosopher Ramon Lull (actually a school of alchemical writers using his name) is so obscure, Dickinson continues, that one needs Aristarchus to expound his work and Oedipus to hear the exposition. And yet despite the devious ingenuity of Lull and his followers, none of them has excelled at this game or imposed more cunningly and subtly on his readers than the "very celebrated philosopher Philalethes." In fact, Dickinson may well be right, for there is yet another level of concealment Philalethes used that we have not so far examined.[26]

The Higher Reaches of Literary Concealment: Graduated Iteration

The reader who has followed our discussion to this point could easily receive the impression that the literary techniques of alchemical deception were complicated and difficult, but that their fixity of meaning made them decipherable in the way that a riddle typically has but one solution. It is true that many alchemical writers had a particular process or set of operations in mind and that their texts could be decoded into a description thereof, but it does not follow that other, more misleading decipherments were impossible. To the contrary, they were encouraged. Philalethes's work again provides us with an excellent example of this point, and one that is particularly relevant to the understanding of his acolyte Newton. The following passage shows that Philalethan *Decknamen* such as "the Moon," "the doves of Diana," and "Venus" in reality had multiple chymical referents:

> In this our work, our Diana is our body when it is mixed with the water, for then all is called the Moon; for Laton is whitened, and the Woman bears rule: our Diana hath a wood, for in the first days of the Stone, our Body after it is whitened grows vegetably. In this wood are at the last found two Doves; for about the end of three weeks the Soul of the Mercury ascends with the Soul of the dissolved Gold; these are infolded in the everlasting Arms of Venus, for in this season the confections are all tincted with a pure green colour; These Doves are circulated seaven times, for in seaven is perfection, and they are left dead, for they then rise and move no more; our Body is then black like to a Crows Bill, for in this operation all is turned to Powder, blacker than the blackest.[27]

---

[26] Edmund Dickinson, *Epistola ad Theodorum Mundanum* (Oxford, 1686), 11, 34–36, 39, and 40.

[27] Philalethes, "An Exposition upon Sir George Ripley's Epistle to King Edward IV," in *RR*, 24–25.

It is the series of "regimens" that form the immediate topic of Philalethes's discussion here. In such classics as *Ripley Reviv'd* and *Secrets Reveal'd* the "American philosopher," as Philalethes was sometimes called, describes a set of stages through which the sophic mercury is supposed to pass once it has been amalgamated with gold and kept for a long while in a heated, sealed flask. Although these vary from author to author, one common early modern conception was to model the stages or "regimens" on the planets in the geocentric system. Thus *Secrets Reveal'd* indicates that there are seven regimens, each with its own characteristic color and appearance, in the order of Mercury, Saturn, Jupiter, Luna, Venus, Mars, and Sol. The two end points, Mercury and Sol, correspond to the insertion of the sophic mercury-gold amalgam into its flask, and the final production of the philosophers' stone. The regimens follow one another in a succession of color changes if the heating instructions are performed correctly. Although Philalethes speaks of many intermediate colors, Saturn is primarily black, Jupiter multicolored, Luna white, Venus green, Mars orange, and Sol red. The regimens require differing amounts of time to run their course, but on average *Secrets Reveal'd* allocates each of them about thirty to fifty days.[28]

Although these descriptions owe more to fantasy than to actual laboratory experience, they form a significant part of Philalethes's alchemy. It is therefore extremely interesting that Philalethes has here imposed an entirely new set of meanings on the Moon, the doves of Diana, and Venus, differing remarkably from those that we examined already. As we saw in his description of the chaos from *Secrets Reveal'd*, he employed the term "our Luna" there to mean the silvery regulus of antimony used to make the sophic mercury. The term "our" distinguishes the regulus from ordinary silver, which *Secrets Reveal'd* simply calls "Luna" in the way that a medieval alchemist such as Geber would have done. In the above passage from *Ripley Reviv'd*, however, the moon means neither silver nor the silver-like regulus, but something else entirely. It is now "our body when it is mixed with the water," in other words, the amalgam of the sophic mercury and gold that is sealed up and heated at the beginning of the regimens. During this stage, "Latona," an old term for "latten" or brass here used as a *Deckname* for gold because of its yellow color, is whitened in the formation of the white amalgam. So "our Diana" is here the amalgam containing gold, and most importantly, "Diana's doves" no longer refer to the two parts of silver that must be alloyed with antimony regulus so that quicksilver will amalgamate with it. Instead, the term "dove" now connotes the volatility of the heated amalgam in a sealed flask during its maturation to the philosophers' stone! Thus the doves must be circulated by reiterate distillation in their closed vessel during the course of the regimens.

In *Ripley Reviv'd*, the circulation of Diana's doves will eventually lead to the regimen of Venus with its green color, and thus the doves are "infolded in the everlasting Arms of Venus." But in *Secrets Reveal'd*, where the doves are also "folded in the everlasting Arms of ♀," Philalethes says that

---

[28] Philalethes, *SR*, 90–109.

this operation pertains to the initial making of the sophic mercury, not to the regimen of Venus.[29] At this point in *Secrets Reveal'd* it appears that "Venus" refers neither to the regimen of that planet nor to the traditional alchemical referent associated with it, namely, copper. Instead, Venus here means once again the regulus of antimony that combines with silver at high temperature in order to make a proper alloy for amalgamation in the production of the sophic mercury. The same use of Venus to mean antimonial regulus can also be found in another Philalethan treatise, *The Marrow of Alchemy*, where the combination of iron and the "reguline" component hidden within the black ore of antimony is described as a copulation of Mars with "our Venus."[30] Hence it is clear that "Venus," for Philalethes, can mean at least three things in an alchemical setting: its traditional referent, copper, the "amorous" mercurial component in stibnite that combines with the putative sulfurous ingredient in iron to yield antimony regulus by smelting, and the venereal regimen with its green coloration. Thus there is an unexpected fluidity to Philalethes's language: although his *Decknamen* are not arbitrary, they change their meaning with context.

Are there any rules or hints that govern this more advanced use of alchemical language? In fact there are, but another of Newton's alchemical sources states it more concisely than Philalethes. The learned author Alexandre-Toussaint de Limojon de Saint-Didier, a French diplomat who died by shipwreck in 1689, became one of Newton's favorites during the late part of his chrysopoetic career.[31] Limojon, or "Didier," as Newton typically calls him, describes an iterative approach where chymical processes are repeated in order to "graduate" or improve a product by further isolating it or leading it to a greater stage of maturity. Geber, for example, had spoken of three stages of transmutative perfection that were to be attained by three respective medicines or elixirs. A medicine of the first order produced a mere semblance of transmutation, as when copper is turned to gold-colored brass. A second-order perfection can induce permanent change, unlike those of the first order, but the change does not affect all of the qualities of the substance. Imagine silver, for example, that had been made to resemble gold in every quality but one—its specific gravity. Finally, a medicine of the third order can genuinely transmute a lesser metal into gold, at least according to Geber. So how does one turn a first-order medicine into a second- or third-order one? Primarily by reiterate volatilization and fixation, in other words, the same processes that were initially employed, but now repeated multiple times. The mystification enters when the same name is used for processes and products at all three levels of perfection. Thus, Didier says, in a translation from his *Lettre Aux vrays Disciples d'Hermes* made by Newton:

> The operations of ye 3 works are analogous so that Philosophers æquivocate often in speaking of one when they seem to speak of ye other. In every

[29] Philalethes, *SR*, 52.

[30] Philalethes, *Marrow*, part 2, book 1, stanza 56, p. 14.

[31] For Alexandre-Toussaint de Limojon de Saint-Didier, see Joseph-François Michaud, *Biographie universelle, ancienne et moderne* (Paris: L. G. Michaud, 1819), 24: 502.

work yͤ body must be dissolved wᵗʰ yͤ spirit & yͤ head of yͤ crow cut off, & black made white & white red.[32]

In other words, precisely the same language can be used interchangeably to describe processes and products in each of Didier's three works, which are perhaps modeled on those of Geber. As the parathesis in Philalethes's description of chaos showed, this parsimony is by no means due to the alchemists' having a limited supply of words at their disposal. It is instead a consciously employed linguistic tool. Since this technique involves repeated use of the same term at different stages in the progress toward the alchemical magnum opus, an appropriate term for it is "graduated iteration." At an early stage of the operations aiming for the philosophers' stone, namely, the preparation of the sophic mercury, the terms "Luna" or "Moon," "Diana's doves," and "Venus" have an entirely different sense from the one that they acquire after the sophic mercury has been sealed up with gold for its long digestion in a gentle heat that will lead, Philalethes says, to his summum bonum.

In *Ripley Reviv'd*, Philalethes builds on the principle of graduated iteration by employing a device from the *Compound of Alchemy* by the fifteenth-century English alchemist George Ripley.[33] The figure in question is a wheel that the alchemist must turn multiple times in order to complete his progress toward the philosophers' stone:

> Our Operation is but turning as it were of a Wheel, which runs one half of its circulation directly backwards to its first progress.... For our Wheel goes round, and when it is come thither whence it set forth, it begins again. Thus is made a third Solution, Sublimation and Calcination into a red *Elixir*, which is the Sabboth of Nature and Art; at which being arrived, there is no farther progress without a new Marriage, either by Ferment or otherwise, according to the rule of Nature and Art: so that indeed all our work is three Rotations, and every Rotation hath three Members, Solution, Sublimation, and Calcination.[34]

As Philalethes says, the wheel must be turned three times, and each rotation consists of solution, sublimation, and calcination. In the 1695 edition of Philalethes's *Opera omnia* (Complete Works), the wheel is pictured graphically as a compass-like vertical circle mounted on a tree (figure 2.2). The regimens are represented by the planetary symbols on the periphery

---

[32] Keynes 21, 1v.

[33] For Ripley, see Jennifer M. Rampling, "Transmuting Sericon: Alchemy as 'Practical Exegesis' in Early Modern England," in *Chemical Knowledge in the Early Modern World*, ed. Matthew Eddy, Seymour Mauskopf, and William Newman, *Osiris* 29 (2014): 19–34; Rampling, "Depicting the Medieval Alchemical Cosmos: George Ripley's *Wheel* of Inferior Astronomy," *Early Science and Medicine* 18 (2013): 45–86; Rampling, "Transmission and Transmutation: George Ripley and the Place of English Alchemy in Early Modern Europe," *Early Science and Medicine* 17 (2012): 477–499; Rampling, "The Catalogue of the Ripley Corpus: Alchemical Writings Attributed to George Ripley (d. ca. 1490)," *Ambix* 57 (2010): 125–201; Rampling, "Establishing the Canon: George Ripley and His Alchemical Sources," *Ambix* 55 (2008): 189–208. Rampling is currently composing a book on Ripley that will no doubt cast much new light on this influential figure.

[34] Philalethes, "An Exposition upon the First Six Gates of Sir George Ripley's Compund of Alchymie," in *RR*, 178–80.

FIGURE 2.2. Illustration of the Greek Hero Cadmus rotating George Ripley's wheel as interpreted by Eirenaeus Philalethes. Within the wheel's perimeter are the seven planets followed by a barbed triangle within a circle. The planets represent the different regimens in Philalethan alchemy and bear the following colors according to the image: Mercury, various colors; Saturn, black; Jupiter, ashen; Luna, white; Venus, green, red, blue, yellow green; Mars, dark yellow, peacock's tail (i.e., iridescent); Sol, yellow, dark purple. Reproduced from Eirenaeus Philalethes, *Anonymi Philalethae philosophi opera omnia* (Modena: Fortunianus Rosatus, 1695).

of the wheel, along with a barbed triangle within a circle, which probably represents the elixir or philosophers' stone. The implication, clearly, is that each rotation of the wheel involves a complete succession of the regimens described in *Secrets Reveal'd*, so that each is repeated three times before the process is finished. As Philalethes puts it in *Ripley Reviv'd*, the philosophers' stone results after one turning of the wheel, but it is still imperfect, corresponding only to Geber's medicine of the first order. In order to arrive at the most perfect medicine, "Imbibitions and Cibation" in the form of a second rotation must occur, followed by a third cycle consisting of "Fermentation." Only the final turning of the wheel yields the great elixir, which "tingeth *Mercury* into a Metalline Mass in the twinkling of an eye" like a basilisk dispatching its prey.[35]

Interpreting Newton's engagement with alchemy requires that we take all of the features so far described into account. The unquestioned authority of the adepts along with their almost preternatural ability to hide their true meaning form integral components of his understanding of the aurific art. Although this acknowledgment may raise a strong feeling of cognitive dissonance among those who know Newton mainly for his achievements in physics and optics, there is good reason to think that the budding savant of

[35] Philalethes, "Sir George Ripley's Recapitulation," in *RR*, 22–23.

Trinity College felt a strong sympathy with the isolated intellects making up the band of the adepts. As we shall see, his earliest alchemical writings display a remarkable confidence in his qualifications for joining this elite company. As time goes by, however, and Newton reads more deeply in the diverse corpora of alchemy, his understanding of the techniques of concealment discussed above becomes more sophisticated. In this later phase of Newton's alchemical career, from the 1680s onward, we see his full appreciation of the techniques outlined in the present chapter, particularly the practice that I have labeled graduated iteration. Despite the seemingly endless difficulties that such alchemical polysemy posed, however, Newton's suspicion that he belonged among this elevated cohort seems never to have waned. In a manuscript probably composed in or after 1689, Newton works out a series of over thirty phrases that are anagrams of his Latinized name, "Isaacus Neuutonus."[36] This is clearly an attempt to place himself in the list of adepts who, like Michael Sendivogius, employed Latin phrases to conceal their names while also providing a key to their identity for the benefit of the clever. In one text, Sendivogius was "Divi Leschi Genus Amo" (I love the race of the divine Lech), whereas another of his texts bore the authorial phrase "Angelus Mihi Doce Ius" (Teach me the law, Angel).[37] If one looks closely at Newton's list, one anagram in particular stands out—"Jeova sanctus unus" (the One Holy Jehova). This anagram begins the list and ends it, while it is accompanied by five other variations, "Javo sacus neutnus," "Venus sactnus sanctus," "Santus Iavo, Venus," "Sanctus Iavo unus e," and "Iavo sanctus unus e." Some of this makes for gibberish Latin, but what is interesting is that Newton apparently has used "Jeova sanctus unus" as a matrix on which to build further anagrams. In the final appearance of the phrase, he has even placed dots beneath its letters to mark their transposition.[38] This emphasis, along with the fact that "Jeova sanctus unus" appears on another closely related manuscript, suggests that "the One Holy Jehova" was Newton's initial choice of an anagram, and that, like Sendivogius, he intended to follow this with additional pseudonyms.

The use of such a religiously charged pseudonym by one who held the Antitrinitarian views of Newton cannot help but summon up the thought that "Jeova Sanctus Unus" was intended to encode his religious views as well as his alchemical identity as an adept. While that is certainly possible, any argument that "Jeova Sanctus Unus" was intended to signify Newton's Antitrinitarianism would have to account for the fact that most of his other anagrams are either secular or pertain to outright pagan deities. They include,

[36] Babson 1006, 1r. It is likely that this manuscript was originally part of Keynes 13, which also bears the phrase "Ieova sanctus unus" (on 4r) and like Babson 1006, consists mostly of chymical bibliography interspersed among notes pertaining to the business of the Mint. For the dating of Keynes 13, see Karin Figala, John Harrison, and Ulrich Petzold, "De Scriptoribus Chemicis: Sources for the Establishment of Isaac Newton's (Al)chemical Library," in *The Investigation of Difficult Things: Essays on Newton and the History of the Exact Sciences in Honour of D. T. Whiteside*, ed. Peter M. Harmon and Alan E. Shapiro (Cambridge: Cambridge University Press, 1992), 135–79, see 145–46.

[37] Michael Sendivogius, *De lapide philosophorum tractatus duodecim [= Novum lumen chemicum]* (s.l.: s.p., 1604); Sendivogius, *De sulphure* (Cologne: Joannes Crithius, 1616).

[38] Babson 1006, 1r.

for example, "Venus ac Iason tuus" (Venus and your Jason); "Venus Isaacus Nuto" (Venus, I, Isaac, am weak); "Novus ventus Isaac" (Isaac the new wind); "Si Venus acusat uno" (If Venus reprimands someone); and "Vniones acuat usus" (Use may sharpen unions/pearls). On balance, it seems most likely that these colorful phrases were more or less arbitrary in meaning, and that their formation was simply governed by Newton's imagination and the letters at hand in his name. Their real significance lies in the allegiance that they demonstrate between Newton and the adepts, well after his publication of the *Principia* and almost at the point of his becoming warden and master of the English Mint. It is particularly telling that in another manuscript where Newton uses "Jeova Sanctus Unus," the phrase appears on the same page as a list of the adepts with the dates at which they acquired the philosophers' stone or first committed their discoveries to writing. Thus Philalethes and Sendivogius are accompanied by "1645" and "1590," rather than the initial publication dates of their first books (1667 and 1604).[39] Was Newton perhaps wondering when his turn would come, and the adept in training would finally arrive at the success that had eluded him for over two decades?

Problems of Genre: The Alchemical Florilegium and the Conjectural Experiment

Newton's attempt to create an alchemical persona cloaking his identity leads into another problematic area of language that we have yet to examine. In the privacy of his laboratory, Newton not only adopted the view that the genuine adepts of the aurific art were infallible, he also went so far as to assume their favorite mode of exposition—the *florilegium*. Late medieval and early modern alchemy is filled with such titles as *Rosarium philosophorum* (philosophers' rose garden), *Lilium inter spinas* (lily among thorns), and *Flos florum* (flower of flowers), all names that typically connote a collection of "flowers" or a *florilegium*. Although not every florilegium openly advertised its compilatory nature in this blatant fashion, they did all share the characteristic of serving as repositories of snippets and summaries from previous authors' works. This was the root sense of the term "florilegium," which literally meant a collection of "the flowers of literature," also the original sense of the still commonly used word "anthology." The writers of these compilations had a clear idea of what they were doing, as expressed confidently in the following passage from "Toletanus," a fourteenth-century writer in the genre:

> We call this collection the *Rosarium* because we have plucked the roses out of the books of the philosophers as if freeing them from their thorns. In it we will succinctly pass on whatever we deem necessary for the attainment of this work, with clear speech and in correct order, word for word, with all its sufficient explanations.[40]

[39] Keynes 13, 4r.

[40] My translation from the *Rosarium philosophorum* of "Toletanus" as quoted in Joachim Telle, *Rosarium philosophorum: Ein alchemisches Florilegium des Spätmittelalters* (Weinheim: VCH, 1992), 2: 172.

This statement of purpose could almost pass for the method that Newton adopted for his personal chrysopoetic expositions throughout most of the 1680s and 1690s. As with Toletanus, Newton was extremely concerned to arrive at a correct order for the welter of refractory *Decknamen*, operations, and regimens that he encountered in his alchemical reading. In doing so, he was undisturbed by the polyphony ensuing from a concatenation of multiple sources, whose words he would often summarize or paraphrase. As a result of this wholesale incorporation of dicta (sayings), it is extremely easy to lose the sound of Newton's own voice among the diverse authors whose "flowers" he has plucked and sorted into a new arrangement. Fortunately, the difficulty abates somewhat when we understand that Newton often inserts his own interpretations within square brackets in the midst of the extracted dicta of the adepts. Yet even here there is room for caution. As Newton internalized ever more chymical texts over the decades of his study, his own voice merged with theirs to the point that he seems no longer to have felt the need, at least in some instances, to provide the "vulgar" or commonplace referents to *Decknamen* in his interpretive brackets. Examples of this trend can be found in Newton's late text *Praxis*, composed after 1693 and found in Babson 420. Here we find terms such as "spirit of mercury," "the extracted seed of common gold," "mercurius duplatus" (doubled mercury), "earth of Mars," and "the Caduceus and cold, saturnal fire" all enclosed within square brackets.[41] Such puzzling terms of art typically derive from Newton's chymical reading rather than being coined by him. Although these expressions may have been perfectly clear to Newton, they all refer to derived products that went through multiple stages of preparation before acquiring their names. Even if Newton may not have used them with the intention to deceive, they are every bit as unintelligible to the casual reader as the green lion and the white fume. At this mature point in his career Newton had grown so fluent in the language of alchemy that his square brackets effectively translated one *Deckname* into another *Deckname*.

In addition to Newton's bracketed comments, there is another important and less obvious feature that distinguishes his chymical florilegia from those of his forebears. Unlike the multitudes of *Rosaria* and *Lilia* that populated the chrysopoetic landscape, Newton's florilegia did not have an audience in mind other than their creator. The late medieval and early modern alchemical florilegium had become a literary genre in its own right, and one suspects that many of the compilers never saw the interior of an alchemical workspace or laboratory. Such impressive artistic productions as the anonymous sixteenth-century *Splendor solis*, itself erected on the foundation of the *Rosarium philosophorum*, provided visual and literary value independent of their ability to advise on the subject of actual experimentation.[42] This was obviously not the goal behind Newton's years of sifting and compiling texts. We must constantly bear in mind that his extracting of textual dicta

---

[41] Babson 420, 5r–7v.

[42] For *Splendor solis*, see Jörg Völlnagel, *Splendor solis oder Sonnenglanz: Studien zu einer alchemistischen Bilderhandschrift* (Altenburg: Deutscher Kunstverlag München Berlin, 2004).

went hand in hand with genuine work at the bench; in fact, the records of his chymical experimentation reveal the same unremitting commitment to exactitude that we find in other examples of Newton's scientific endeavor, such as optics.

Newton's experimental laboratory notebooks form the object of sustained study later in this book, so I will not discuss them in detail here. It is important to note, however, that impressive as his records of experimentation are, Newton did not invent the genre of the chymical laboratory notebook. In this he was preceded by others, among English-speaking authors especially by Starkey and also by Thomas Vaughan, the latter of whom wrote alchemical treatises in the 1650s under the similar-sounding pseudonym of "Eugenius Philalethes."[43] The origins of this genre would require concerted research among the chymists and physicians of the earlier seventeenth century, though it is clear that Starkey's education at Harvard College played a part in the development of his notebooks' form and style, as did his knowledge of the chymical writer Angelus Sala.[44] What is of particular interest here is Starkey's highly self-conscious method of reflecting on his chymical activities. Not only did he describe the operations that he carried out in the laboratory, he also provided systematic, dated assessments of his progress over the years. Additionally, his analyses of previous chymists' works record numbered *Observationes* (observations) accompanied by well-reasoned *Conclusiones probabiles* (probable conclusions), and even Δευτέραι Φρόντιδες or "second thoughts" emerging from repeated experimentation on the same subject.[45] But what particularly stands out for its relevance to Newton is Starkey's explicit descriptions of so-called *Processus conjecturales* (conjectural processes). Typically couched in the imperative or subjunctive mood, these are experiments that Starkey has planned, but not yet performed. Whether consisting of attempts to improve the refining of crude antimony, the sublimation of the star regulus with "stinking spirit" (an ammonia compound), or a better way to make Starkey's medicament "ens veneris" (essence of copper), these "conjectural processes" were meant to be tested; they were not themselves final products.[46]

Although Newton knew only a tiny and unrepresentative fragment of Starkey's notebooks (the *Experiments for the Preparation of the Sophick Mercury* published in 1678), he too devised conjectural processes. In fact, like the genre of the alchemical florilegium, this was a long-established practice in the discipline. The successive iterations of processes leading to "medicines" of the first, second, and third orders in the Geberian tradition represents something along the same lines as the conjectural process. If it is possible to make an ersatz silver that looks like the noble metal, and further treatment allows this product to pass certain assaying tests (for example the touchstone),

[43] For Vaughan's laboratory notebook, see Donald R. Dickson, *Thomas and Rebecca Vaughan's Aqua vitae, non vitis* (British Library MS, Sloane 1741) (Tempe: Arizona Center for Medieval and Renaissance Studies, 2001).
[44] Newman and Principe, *ATF*, 172–79.
[45] Newman and Principe, *LNC*, 331–32, 138, 142–44, 177.
[46] Newman and Principe, *LNC*, 139, 145, 166.

then the alchemist might well reason that even further laboratory procedures would lead the metal to the perfection of genuine silver. The resulting series of operations is anything but blind empiricism or copying; it represents the conscious planning and recording of processes that anticipate a very particular outcome. The difference between this practice in the published classics of chrysopoeia and in the private notebooks of Newton and Starkey is that Newton and Starkey acknowledged the incomplete status of their ongoing research projects and intended to complete and test them at a future date.

An understanding of the conjectural process is therefore a convenient—even an essential—requirement for making sense of Newton's chymical *Nachlass*. One sees this very clearly, for example, in Keynes 58, a manuscript that preserves three successive drafts of Newton's attempt to work out processes largely (though not exclusively) based on the mid-seventeenth-century German chymist Johann de Monte-Snyders. Beginning with a group of materials that have undergone previous laboratory processing, namely, salts of iron ore and copper ore, along with the green lion and its blood, Newton subjects these substances to a complicated series involving well over thirty independent operations (figure 2.3). The final results, he says, will be such desiderata as "Venus the daughter of Saturn," "Jove's eagle," "Jove's lightning bolt," "Jove's scepter," and "the rod" or caduceus of Mercury. Does this mean that Newton actually succeeded in making these exotic chymical products? A careful examination of the manuscript shows that in itself it implies nothing of the sort.

The practical part of Keynes 58 is written in the imperative language of the recipe. Newton says to dissolve and digest the salts of iron and copper and then to subject them to further operations. Nowhere does he indicate that he has already carried out this sequence of processes, nor does he describe actual products that he has made. Instead, he provides unequivocal clues to the fact that this is largely a series of conjectural processes. Thus, at an advanced stage, he says "Ioves scepter probably is Salt of his eagle extracted out of ye minera wth ye Lyons blood." What Newton is doing is deciphering a chain of operations that he believes himself to have found in his sources. This is primarily a textual procedure on his part, though aided by his actual experimental understanding, just as Starkey's conjectural processes embodied an implicit working knowledge born out of his years of experience as a practical chymist. While originating from the textual process of decipherment guided by a general, practical knowledge of chymistry, however, Newton's conjectural processes were designed to undergo specific and rigorous tests. This in fact was the primary goal of the experiments recorded in the two large collections of his laboratory records kept in the Cambridge University library, in the form of the manuscripts CU Add. 3973 and CU Add. 3975. The processes there involving such nostrums as "Vulcan's Net," "Diana," "Venus," and "the trident" all bear witness to Newton's attempts to replicate and refine the substances described by Philalethes, Sendivogius, and Snyders (as Newton called Monte-Snyders).[47]

[47] See CU Add. 3975, 43r, 54v, 71v, and 72r for examples of the net; 62r for Venus; and 138v for the trident; see CU Add. 3973, 16r–16v for Diana.

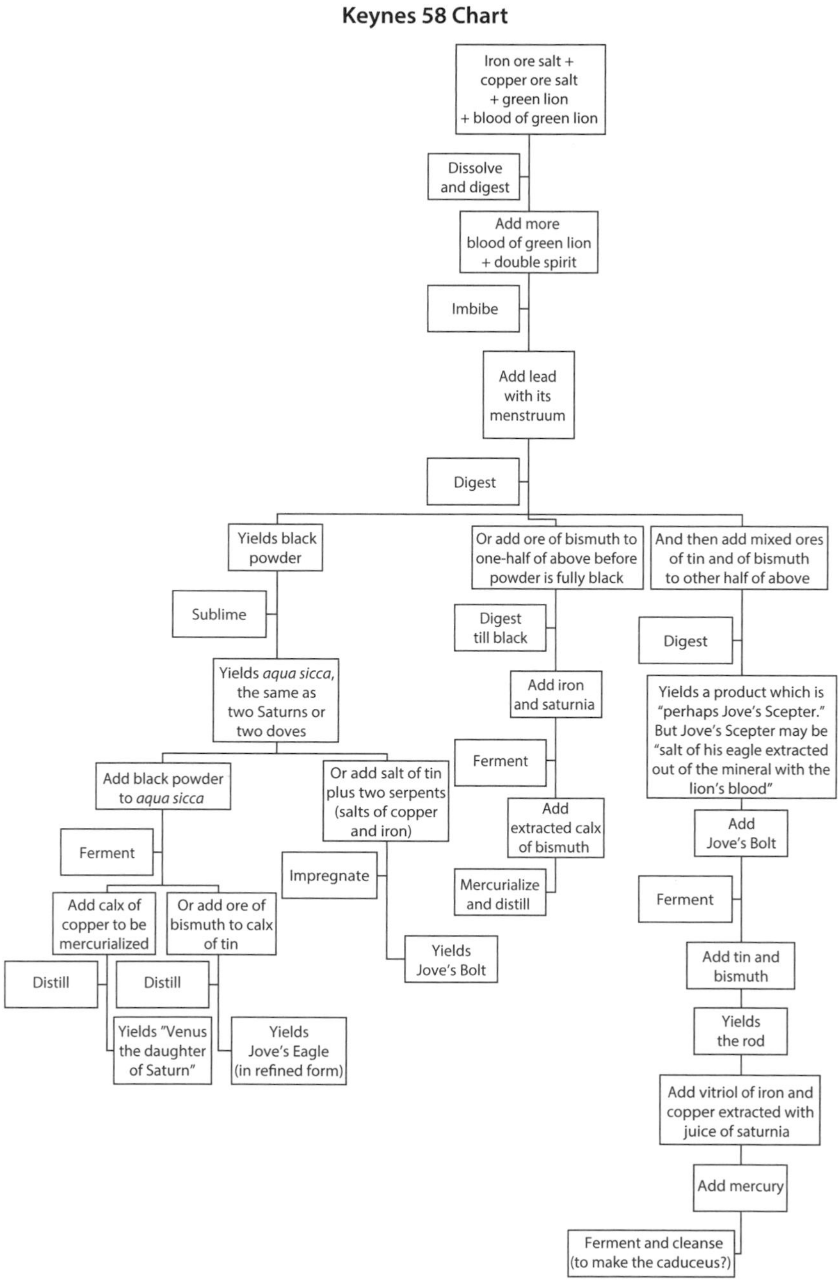

Keynes 58 Chart

Iron ore salt +
copper ore salt
+ green lion
+ blood of green lion

Dissolve
and digest

Add more
blood of green lion
+ double spirit

Imbibe

Add lead
with its
menstruum

Digest

Yields black
powder

Sublime

Yields *aqua sicca*,
the same as
two Saturns or
two doves

Add black powder
to *aqua sicca*

Ferment

Add calx of
copper to be
mercurialized

Or add ore of
bismuth to calx
of tin

Distill

Distill

Yields "Venus
the daughter
of Saturn"

Yields
Jove's Eagle
(in refined form)

Or add salt of tin
plus two serpents
(salts of copper
and iron)

Impregnate

Yields
Jove's Bolt

Or add ore of bismuth to
one-half of above before
powder is fully black

Digest
till black

Add iron
and saturnia

Ferment

Add
extracted calx
of bismuth

Mercurialize
and distill

And then add mixed ores
of tin and of bismuth
to other half of above

Digest

Yields a product which is
"perhaps Jove's Scepter."
But Jove's Scepter may be
"salt of his eagle extracted
out of the mineral with the
lion's blood"

Add
Jove's Bolt

Ferment

Add tin and
bismuth

Yields
the rod

Add vitriol of iron and
copper extracted with
juice of saturnia

Add mercury

Ferment and cleanse
(to make the caduceus?)

FIGURE 2.3. Chart showing the order of alchemical operations as conceived by Newton and described in Keynes 58.

The most highly developed extant specimen of Newton's attempt to work out the processes of the adepts is the *Praxis* text found in Babson 420, probably composed in the 1690s. Far from representing a mental or emotional breakdown on Newton's part, as Richard Westfall suggested, *Praxis* is actually an extended network of carefully constructed conjectural processes combining operational material derived from the panoply of

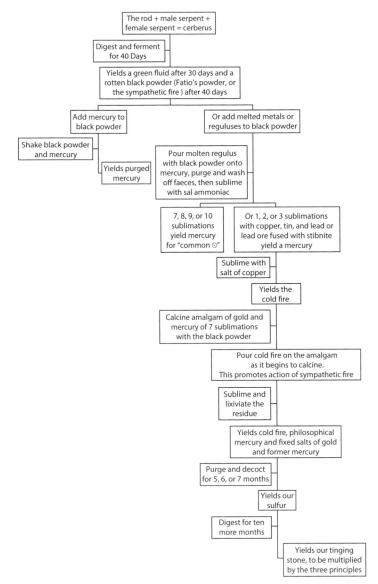

Via Sicca (The Dry Way)

FIGURE 2.4. Chart showing the order of operations in the *via sicca* as conceived by Newton in *Praxis* (Babson 420, 12r–13r).

Newton's chrysopoetic sources.[48] What is true for Keynes 58 is even more the case for *Praxis*. The charts that I have drawn up in order to provide a way into the complexity of the *via sicca* (dry way) and *via humida* (wet way) reveal more than fifty operations, and there are still others in the subsequent stage of "multiplication" that I did not represent (figures 2.4 and 2.5). Moreover, the initial ingredients of the wet and dry way include the rod of Mercury or caduceus, itself a product derived from the procedures

[48] Westfall, *NAR*, 529–30, 537–38.

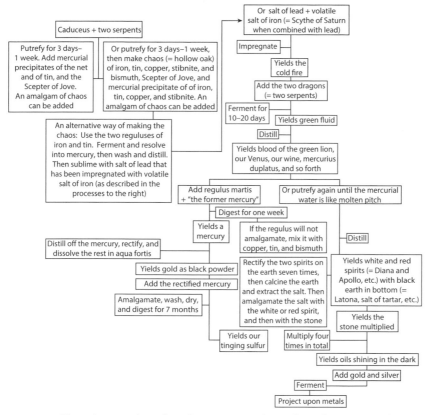

Via Humida (The Humid Way)

FIGURE 2.5. Chart showing the order of operations in the *via humida* as conceived by Newton in *Praxis* (Babson 420, 13r–14r).

outlined in Keynes 58; thus, the conjectural processes outlined in that manuscript, or something like them, implicitly precede those spelled out in *Praxis*. To say that Newton never carried out most of these processes would be a radical understatement. At the end of his dated chymical experimentation in February 1696, Newton was still trying to perfect a substance called "our Venus," which is, according to *Praxis*, at least ten steps from the final goal of the philosophers' stone.[49] Indeed, it is highly likely that "our Venus" as found in the experimental notebooks was a much more preliminary ingredient even than that, thanks to Newton's employment of the principle of graduated iteration. Just as the green lion could refer either to crude antimony or to a menstruum that included it as an ingredient, or even to the green stage among the regimens leading to the philosophers' stone, so "our Venus" seems to have had one sense in Newton's laboratory notebooks and quite another in *Praxis*. A confusing situation indeed, but then no one ever said that alchemy was easy.

[49] CU Add. 3975, 140r–140v, corresponding to CU Add. 3973, 39v. The date "Feb. 1695/6" is found in CU Add. 3973 at 30v.

In this chapter we have traced a host of related linguistic and interpretive issues that emerge from Newton's self-identification as a would-be adept. From the beginning of his serious chymical studies in the 1660s, he seems to have been confident that he belonged among the elite sons of wisdom who had been chosen to receive the philosophers' stone as a *donum dei*, a gift of the Creator himself. This does not mean that Newton ever deluded himself into believing that he had actually succeeded in attaining that summum bonum, however. The most striking thing about his chymical corpus is the remarkable contrast between the years of elaborate speculation that went into his decipherment of alchemical sources and the extraordinary rigor of his chrysopoetic experiments. Despite his private acceptance of extravagant "authorities" such as Edwardus Generosus and the author of *Manna*, Newton remained wedded to the most stringent methods of the "experimental philosophy" and refused to believe that he had succeeded at the aurific art until experiment might tell him otherwise. His growing sophistication in understanding alchemical techniques of deception such as graduated iteration, as well as his adoption of the florilegium genre and the conjectural experiment, point to Newton's remarkable ability to absorb and dominate disparate areas of activity while reserving the prerogative of critical judgment. In a word, throughout his decades-long romance with alchemy, and despite his enduring assurance that he belonged among the ranks of the adepts, there can be no doubt that Newton remained Newton.[50]

[50] In a recent article, Cornelis J. Schildt has reached a somewhat similar conclusion regarding Newton's debt to the concept of alchemical adepthood, though with different implications. Schildt argues that Newton's parsimonious method of imparting his optical discoveries was influenced by the alchemical emphasis on secrecy. As Newton described his *New Theory about Light and Colors* to Henry Oldenburg, "I designed it onely to those that know how to improve upon hints of things." See Schildt, "'To Improve upon Hints of Things': Illustrating Isaac Newton," *Nuncius* 31 (2016): 50–77.

THREE

Religion, Ancient Wisdom, and Newton's Alchemy

Introduction

Newton's long project of decoding the language of the adepts brings to mind another major undertaking on his part that also involved the "translation" of allusive, mysterious terms into their referents in the mundane sphere.[1] I refer to his extensive work on the interpretation of biblical prophecy, a topic whose consideration points to the vexed issue of the relationship between Newton's alchemy and his personal religion. Few topics in Newton scholarship have led to more misleading claims than the assertion that he viewed his alchemy as part and parcel of his heterodox, Antitrinitarian Christianity. Largely an artifact of the tendentious Jungian view that alchemy over the *longue durée* was essentially a form of soteriology, the position that Newton's alchemy was an appendage to his religion or even an alternate form of it reached its apogee in the 1990s, and has since become a received position in the literature.[2] In reality, Newton's writings on prophecy, biblical history, and the iniquity of orthodox Trinitarian doctrine contain virtually no references to chymistry.[3] Despite the fact that alchemical writings frequently contain appeals to divinity, Newton's extracts, synopses, and notes drawn from chrysopoetic writers seldom expand on religious motifs found in his sources. On the rare occasions when Newton does take up a reference to God in his alchemical notes, he makes it clear that what interests him is the

---

[1] Paul Greenham, in an interesting and sophisticated recent dissertation, has coined the expression "descriptive-translational" for Newton's approach to prophetic interpretation and has drawn an extensive comparison between this and his decipherment of alchemical imagery into laboratory practice. See Greenham, "A Concord of Alchemy with Theology: Isaac Newton's Hermeneutics of the Symbolic Texts of Chymistry and Biblical Prophecy" (PhD diss., University of Toronto, 2015), 95–229.

[2] For a sustained critique of the Jungian position regarding alchemy, see Lawrence M. Principe and William R. Newman, "Some Problems with the Historiography of Alchemy," in *Secrets of Nature: Astrology and Alchemy in Early Modern Europe*, ed. William R. Newman and Anthony Grafton (Cambridge, MA: MIT Press, 2001), 385–431. A critique of the claims of integration between alchemy and primitive Christianity made by B.J.T. Dobbs and Mary Churchill may be found in Newman, "A Preliminary Reassessment of Newton's Alchemy," *Cambridge Companion to Newton* (Cambridge: Cambridge University Press, 2016), 454–84.

[3] See the Newtonian texts on religious topics, consisting of some four million words, collected and edited by the online *NP* at http://www.newtonproject.ox.ac.uk/texts/newtons-works/religious (accessed June 13, 2017). The only clear exception to this compartmentalization is found in Huntington Library, Babson MS 420, which I discuss later in this chapter. See also Rob Iliffe, "Abstract Considerations: Disciplines and the Incoherence of Newton's Natural Philosophy," *Studies in History and Philosophy of Science* 35 (2004): 427–54.

hidden, materialist meaning of the text. Newton did not read alchemical authors as a means of acquiring spiritual truths; rather, he extracted experimental meaning from them even when they employed the idiom of divinity. A good example of this may be found in his early interpretation of the *Novum lumen chemicum* by the Polish alchemist Michael Sendivogius. If we look at the first folio of Keynes MS 19, found at King's College, Cambridge, the following excerpt taken directly from Sendivogius leaps to the eye:

> Tract 6. From one two arise, from two one. One is God, the son was born from this God: One has given two, two gave one holy spirit. b.[4]

The "b" in this extract refers to Newton's own decoding of the Polish alchemist's words. Now surely, one might think, a man of Newton's pious sensibilities would have had some reaction to this rumination on the threefold nature of God. But instead, he laconically ignores the religious sense of the passage and gives it a transparent, even prosaic chymical meaning, writing:

> b. ☿ in digesting gives ☉, ☿ & ☉ In digesting give the Elixir.[5]

Here Newton has decoded Sendivogius's Father, Son, and Holy Spirit to mean material substances, namely, mercury, gold, and the elixir or philosophers' stone. This is part of a straightforward attempt to derive a laboratory operation out of Sendivogius's obscure words, and it displays the same pattern of converting allusive texts into laboratory processes that one encounters innumerable times in Newton's notes. Very likely Newton did privately believe that the adepts had received their special gifts as a divine dispensation, but from this it does not follow either that he pursued alchemy as a means to religious salvation or as a way of demonstrating "divine activity in the world."[6] Neither consequence would have flowed as a necessary result from the Protestant ethos of Newton's upbringing, or from specific doctrines of individual election to which he may have been exposed.

But even if the excessive claims of a deep integration between Newton's alchemy and his personal religion are untenable, this does not exclude some measure of interaction between the two fields. What connections then, if any, did Newton actually advocate between biblical interpretation and his chrysopoetic quest? A deeply religious thinker, Newton expressed his views on the omnipotence and ubiquity of God in such scientific venues as the "General Scholium" to the later editions of the *Principia* and in *Query 31* of the 1717 *Opticks*. He was certainly willing to combine natural philosophy and religion in general, but does it follow that he was motivated to do so in the particular case of alchemy? The topic cannot be addressed without considering his interpretation of ancient mythology as well, since

---

[4]Keynes 19, fol. 1r: "Tract 6. Ex uno fiunt duo ex duobus unum. Vnus est Deus, ex hoc Deo filius est genitus: Vnus dedit duo duo unum dederunt spiritum sanctum. b." This is a slightly abbreviated paraphrase of Sendivogius's words from "Tractatus sextus" of the *Novum lumen chemicum* as printed in Nathan Albineus, *Bibliotheca chemica contracta* (Geneva: Jean and Samuel de Tournes, 1654), 25. At this early stage of his career, Newton was relying on Albineus's collection for the text of Sendivogius.

[5]Keynes 19, "b. ☿us digerendo dat ☉, ☿ & ☉ digerendo dant Elixar."

[6]Dobbs, *JFG*, 116.

Newton himself drew connections between sacred history and the myths of pre-Christian peoples. It is today well known that he believed in a virtuous, primitive religion shared in varying degrees by multiple ancient peoples long before the arrival of Jesus.[7] And of course early modern alchemy was replete with topoi drawn from classical myth, as in the work of the Holstein chymist Michael Maier, whose books Newton carefully read and annotated.[8] Was he therefore intent on extracting a primeval religious wisdom from alchemical texts, an age-old knowledge that had been known to those closer to the primordial revelation, but attenuated or even lost over the course of time?[9] And finally, in the event that he did not obtain specific religious doctrines from chymical writers, did he perhaps employ the same interpretive methodology tacitly when approaching biblical prophecy, ancient mythology, and alchemy?

Newton's Method of Prophetical Interpretation and Alchemy

In order to begin with the firmest evidence, we will commence with the last of the questions posed above, namely, the issue of Newton's analysis of prophetical literature, for which he actually went so far as to devise an explicit set of guidelines. Our ultimate goal will be the exploration of connections with his alchemy, but first we must examine the prophetical rules on their own terms. A well-known manuscript now found in the National Library of Israel, Yahuda MS 1, contains Newton's "Rules for interpreting and methodising the Apocalypse." A degree of controversy has emerged about these rules on account of some similarity between them and Newton's *Regulae philosophandi* (rules for philosophizing) in his *Principia*. The similarity, which may be superficial in any case, probably results from the fact that both sets of rules share a common if distal source in the scholastic and humanist techniques that Newton imbibed as part of his early education.[10] For Newton's prophetic rules, however, the influence of his Cambridge contemporary Henry More and the famous early seventeenth-century exegete of prophecy Joseph Mede were more significant sources. For our purposes, it is unnecessary to delve into these repositories, the most important of which is probably Mede's *Clavis apocalyptica* (1627) and the *Commentarius* (1632) that Mede wrote on the same subject. Under the influence of Mede, Newton drew as well on the *Oneirocriticon* or dream book of "Achmet ibn Sirin," a

---

[7] The literature on this topic has swelled to a degree that only partial justice can be done to it here. For the latest word (and additional bibliography), the reader should consult Jed Buchwald and Mordechai Feingold, *Newton and the Origin of Civilization* (Princeton, NJ: Princeton University Press, 2013), and Rob Iliffe, *Priest of Nature: The Religious Worlds of Isaac Newton* (Oxford: Oxford University Press, 2017).

[8] For Maier's influence on Newton, see above all Karin Figala, John Harrison, and Ulrich Petzold, "De Scriptoribus Chemicis: Sources for the Establishment of Isaac Newton's (Al)chemical Library," in *The Investigation of Difficult Things: Essays on Newton and the History of the Exact Sciences in Honour of D. T. Whiteside*, ed. Peter M. Harmon and Alan E. Shapiro (Cambridge: Cambridge University Press, 1992), 135–79.

[9] This is the position of Churchill and Dobbs; see Newman, "Preliminary Reassessment," 458–62.

[10] See Raquel Delgado-Moreira, "Newton's Treatise on Revelation: The Use of a Mathematical Discourse," *Historical Research* 79 (2006): 224–46.

Byzantine work that tries to arrive at simple, straightforward interpretations of prophetical images by compiling a sort of encyclopedia of them.[11] With the authority of Mede and Achmet backing him, Newton argued in Yahuda 1 that the symbolic language of scripture, and of Revelation in particular, was meant to encode specific historical events, often of a political nature.

The first of Newton's rules for interpreting prophecy reveals that the discounting of "private imagination" was one of his principal concerns. Having been brought up during the sectarian strife of the English Civil War, Newton wanted to limit the flexibility of prophetic speculation to a bare minimum. As he says, "Too much liberty in this kind savours of a luxuriant ungovernable fansy and borders on enthusiasm." How, then, should one avoid the slippery slope leading to enthusiasm and unbridled fantasy? As Newton announces at the beginning of his second rule, the answer lies in the principle of parsimony. He thus advises, "To assigne but one meaning to one place of scripture; unles it be by way of conjecture." At first this seems entirely straightforward, but the phrase "by way of conjecture" leads into a substantial qualification that he adds after the fact as an insertion on the next page. The inserted passage is a complicated one that requires our full attention:

> unless it be perhaps by way of conjecture, or where the literal sense is designed to hide the more noble mystical sense as a shell the kernel from being tasted either by unworthy persons, or untill such time as God shall think fit. In this case there may be for a blind, a true literal sense, even such as in its way may be beneficial to the church. But when we have the principal meaning: If it be mystical we can insist on a true literal sense no farther then by history or arguments drawn from circumstances it appears to be true: if literal, though there may be also a <by *redundant*> mystical sense yet we can scarce be sure there is one without some further arguments for it then a bare analogy. Much more are we to be cautious in giving a double mystical sense. There may be a double one, as where the heads of the Beast signify both mountains & Kings Apoc 17.9, 10. But without divine authority or at least some further argument then the analogy and resemblance & similitude of things, we cannot be sure that the Prophesy looks more ways then one.[12]

As one can see, Newton thinks that prophecies are written in a parabolic style in order to deceive and repel those who are unworthy of them, just as alchemical treatises employ riddles and *Decknamen* to restrict access to the "sons of wisdom" alone. The contrast that Newton erects between the "literal" sense and the "mystical" meaning of a passage does not make appeal to mysticism in the modern sense but distinguishes between the obvious or commonplace interpretation and the hidden meaning that the prophets intended. Thus Newton says, a prophetical passage may contain a literal

[11] Kristine Haugen, "Apocalypse (a User's Manual): Joseph Mede, the Interpretation of Prophecy, and the Dream Book of Achmet," *Seventeenth Century* 25 (2010): 215–39.

[12] Yahuda 1, 12r–12v. All passages from Yahuda 1 are taken from the normalized text in *NP*.

sense and yet at the same time have a deeper, "mystical" meaning. Despite the less easily demonstrable character of the "mystical" sense, Newton even goes so far as to allow the possibility of its being double. Hence the seven heads of the beast in Revelation 17:9–10 are literally heads, but "mystically" they are both mountains and kings, as the prophecy openly suggests: "The seven heads are seven mountains, on which the woman sitteth. And there are seven kings."

From rule two, then, we can gather that Newton's ideal of limiting the meaning of prophetical symbols to a bisemic relationship between the literal sense, as in the dragon's seven heads being read simply as heads, and a single "mystical" interpretation, was explicitly undercut by the text of Revelation itself. Nonetheless, wherever possible, his goal was to limit the meaning of prophecy to either the literal sense or a single "mystical" or extended one. This comes forth with particular clarity in rule four, where Newton argues that the figurative locutions of prophecy make up an actual language among the prophets "as common amongst them as any national language is amongst the people of that nation." Once the interpreter has determined "the usuall signification" of symbolic events such as the overthrow of nations connoted by hailstorms, thunder, lightning, and earthquakes, these catastrophic signifiers should be read in the same sense elsewhere unless there is some compelling reason to interpret them in another fashion. This approach allows Newton to devise a sort of dictionary of symbols, which he labels with the heading "Prophetic figures." A taste of his method can be acquired from the beginning of the section:

> The original of the figurative Language of the Prophets was the Comparison of a Kingdom to the [1] World & the parts of the one to the like parts of the other. And accordingly the [2] Sun signifies the King and Kingly power. The Moon the next in dignity that is the priestly power with the person or persons it resides in. The greater stars the rest of the Princes or inferior Kings. [3] Heaven the Throne court honours & dignities wherein these terrestrial Luminaries & stars are placed, & the [4] Earth inferior people. [5] Waters the same.[13]

Since prophetical symbolism consists of an actual "figurative Language," Newton is able to draw a one-to-one correspondence between a particular kingdom in a prophecy and different features of the cosmos. Thus "sun" means "king" and "kingly power," "moon" signifies "priestly power" and those who exercise it, the "greater stars" stand for inferior kings or princes, and "earth" and "water" represent the "inferior people." In this fashion, Newton manages to compile a lexicon of prophetical figures and their "mystical" meanings.

Is it the case then that Newton's interpretation of alchemical texts follows the same path as his rules for interpreting prophecy? It is certainly true that Newton attempted to decipher alchemical allegory into practical directions at every opportunity and that he also extracted the "mystical" sense of

---

[13] Yahuda 1, 20r. From *NP.*

prophecy by a process of "translation."[14] Nonetheless, there are significant differences in Newton's hermeneutical technique between the two realms of endeavor. The key thing to bear in mind is that Newton's decades of chymical reading gave him a sophisticated understanding of the differing uses to which his sources put their symbols. He was well aware of the fact that different authors used their *Decknamen* to mean different things. There was no single "figurative language" of the adepts as Newton believed there to be for the prophets. Instead, Newton correctly understood chymical language to be polysemic in the strict sense—a single term could be used in many different ways, by different authors and also by the same author. This point appears quite clearly if we glance at any number of entries in Newton's terminological concordance, the *Index chemicus*. Let us focus here on the term *leo viridis* or "green lion."

Keynes 30/1, the fullest surviving version of the *Index chemicus*, provides a comprehensive entry for *leo viridis* filling over a page of small script. As is typical for the *Index chemicus*, the entry does not provide a clear decipherment of the term into a single material referent. Instead, Newton's main concern is the collocation of *Decknamen* from different authors and different passages by the same author, in order to establish their meanings. He begins with the claim that the green lion is green not in color but by virtue of its crudity and vegetability. In other words, the green lion is green in the way that we might impute that quality to a novice or youth, who has not yet reached maturity but is still growing (for example, a "greenhorn"). Once color has been dispensed with as a requirement, Newton can assert that the green lion "is antimony, the crudest of all minerals" citing Philalethes's *Ripley Reviv'd* and the *Introitus apertus*. But in other passages, Newton finds Philalethes restricting the green lion to a particular form of antimony, namely, "where it is united to sulfur." Nor is this all. In other authors, the term refers to antimony that has been putrefied and turned into a menstruum or dissolvent. This salty, metallic menstruum is elicited by putrefaction and sharpened with its own sulfur. It is then the dragon, eagle, and green lion, as well as a host of other *Decknamen*, including "the doorkeeper, Maydew, secret furnace, true fire, oven, sieve, marble, poisonous dragon, and ardent wine" among other things. Well over two dozen synonyms for the green lion appear in this entry, which if nothing else is testimony to Newton's awareness of alchemical parathesis.

But Newton is still not finished with the green lion. The first entry with the headword *leo viridis* is followed by a second one where Newton gives two additional interpretations of the term:

> To be sure, Green Lion is every material brought back to its crudity (Marrow of Alk. p. 6.), as also the matter when green of color in the regimen of Jove, Ripl. p. 188.[15]

---

[14] The point that, to Newton, both alchemy and prophecy required a form of translation is ably made by Greenham in his *Concord of Alchemy with Theology*, 95–229.

[15] Keynes 30/1, 53r: "Porro Leo viridis est materia omnis incrudata (Marrow of Alk. p. 6.) ut et materia colore viridis in regimine Iovis Ripl. p. 188."

In the first instance, Newton follows up on the idea that the viridity of the green lion refers to its immaturity rather than its color, but generalizes this to every sort of matter brought back to a crude state, not just antimony. Clearly this alone opens the door to a host of chymical referents. In the second example, however, he goes even further, now taking greenness to refer to the actual color of the "lion," which here signifies the green hue that appears within the sealed flask during the regimen of Jove. Obviously, this is a radical departure from all the previous interpretations, in which "green" did not refer to a color, but to a state of immaturity. Nor is Newton done yet. He follows this with yet another entry for "Leo viridis," in which he says the reader should consult the *Index chemicus*'s entry for the term "fumus albus" (white fume). If one turns to the corresponding entry in the *Index chemicus*, yet another nest of *Decknamen* emerges, in which the green lion again appears prominently in still further contexts.[16]

None of this seems very close to the lexical approach that Newton takes for prophetical interpretation in his "Rules for interpreting and methodising the Apocalypse." Where his goal in Yahuda 1 was to arrive at a univocal or bisemic reading insofar as possible, his aim in the *Index chemicus* was something quite different. As a concordance, the *Index chemicus* was intended to gather together as many meanings for a given term as possible, not to reduce them into one. Newton knew very well that Philalethes had used the term "green lion" to mean different things in different contexts, just as he had used the terms "moon," "doves of Diana," and Venus to signify both ingredients of the sophic mercury and products that emerged later in the series of chymical operations leading to the philosophers' stone. The practice of graduated iteration alone, not to mention other forms of equivocation, made it quite literally impossible to reduce the alchemical terms of Newton's sources to single concrete referents, a fact that he obviously understood. It would therefore be misleading to suppose that Newton's rules of interpreting prophecy, at least as they are found in Yahuda 1, provide evidence for an integral relationship between his chymistry and his understanding of biblical hermeneutics.

Newton and the Mythographers

A related area where issues of authority and textual interpretation butt heads with chymistry and religion lies in Newton's interpretation of ancient mythology. Since the early Middle Ages, one current in alchemical writing had focused on the interpretation of ancient mythology as encoded alchemy.[17] Michael Maier had made a specialty of this approach, arguing that the turpitude of the Greek gods and heroes, as well as the outlandishness of their exploits, made it unlikely that the accounts of their deeds were intended as literal accounts; instead, they were allegorical descriptions of

---

[16] Keynes 30/1, 40v.
[17] Robert Halleux, *Les textes alchimiques* (Turnhout: Brepols, 1979), 144–45.

alchemy.[18] Maier's claim opened up vast landscapes for those enamored of textual decipherment, since if ancient mythology really were veiled chymistry, it would logically follow that the mythographers were employing the very techniques of deception that we have recounted in the previous chapter. At some point after 1687, Newton copied out a long passage from Maier's *Symbola aureae mensae*, a sort of bio-bibliography of chrysopoeia organized around twelve chymist-representatives of their respective nations.[19] The passage, which reappears with only minor variations multiple times in Newton's alchemical *Nachlass*, gives a good sense of Maier's approach to classical mythology:

> The ancient poets, as we elsewhere show, <when they spoke of> the descent to the deep places dedicated to Pluto and Proserpina understood nothing other than the seeking out of the metals in their hidden mines, as appears in Orpheus, Hercules, Theseus, Pirithous, and others. Thus Virgil when describing the descent of Aeneas to the underworld is imitating this, and he adds a metallic allegory to it, namely that a golden bough is hiding among dark woods, which bough has golden leaves and pliant golden twigs, that is, in the mines spread out beneath the earth in the manner of trunks, branches, and roots. A whole grove covers this because shadowy woods always surround places that are mineral-bearing unless they are chopped down. But not before it is given, etc., that is, no one can enter the depths of the earth [or the center of a metal ^by means of putrefaction] unless he has plucked this golden bough apart. Maier, *Symbola aureae mensae*, book 4, page 180.[20]

As one can see from Newton's close paraphrase, Maier interprets the descent into Hell in Book 6 of Virgil's *Aeneid* as an encoded description of the subterranean world of minerals and metals. To Maier, Virgil's golden bough is actually an allusion to massive underground formations of ores and minerals that grow in the form of branches and trees, a concept that the German chymist inherited from Paracelsus and his followers. The idea of mineralogical growth and development was encouraged, of course, by the fact that native metals are sometimes found in the form of dendrites. Newton too was enamored of this idea, but if we look at the passage more closely, it is clear that he has employed his own meta-interpretation of Virgil's text. In his usual fashion, Newton inserts his own thoughts within square brackets into

[18] Michael Maier, *Arcana arcanissima* (s.l.: 1614), A[1r]–[A4r].

[19] Keynes 48, 28v, cites the anonymous text *La lumière sortant par soi-même des ténèbres*, which was first published in 1687. Newton's pagination agrees with that of the 1687 text; his copy is found at Trinity College; see Harrison, no. 1003.

[20] Keynes 48, 21v–22r: "Antiqui Poetæ (ut alibi ostendimus) per descensum ad Infera loca Plutoni et Proserpina dicata nihil aliud intellexerunt quam metallorum in mineris suis abditis fecisse lustrationem ut patet in Orpheo, Hercule Theseo Pyrithoo et alijs. Sic Virgilius describens Æneæ descensum ad inferos id imitatur et metallicam allegoriam illi adjungit, nempe quod in arbore opaca hoc est mineris instar arborum ramorum et radicum sub terra dispersis latet aureus ramus qui et folijs et lento vimine aureolus sit. Hunc tegit omnis lucus quia semper umbrosa nemora præcingunt loca mineralium feracia nisi excisa fuerint. Sed non ante datur &c id est nemo in terræ in trina loca [seu metalli centrum ^per putrefactionem] accedere possit nisi descerpserit hunc aureum ramum. Maier Symb. aur. mens. lib. 4. p. 180."

the textual passage that he is interpreting. Thus to Newton, Aeneas's descent into Hell is not merely an allegory of the subterranean mineral world, but a veiled guide to alchemical practice. As Newton says, "no one can enter the depths of the earth [or the center of a metal ^by means of putrefaction] unless he has plucked this golden bough apart." Hence the Golden Bough is actually a *Deckname* or cover name for an alchemical substance. This secret material, moreover, is the key to decompounding metals by means of putrefaction, a conditio sine qua non in Newton's alchemy for the production of the philosophers' stone.

This raises interesting questions relating to the issues of authority and language. Did Newton really think that Virgil wrote the *Aeneid* as a way of revealing his own alchemical knowledge to the sagacious while concealing it from the vulgar masses? Or was Newton knowingly entering into a restricted, conventional genre of alchemical riddle solving that did not necessarily commit him to the belief that the ancients actually wrote their epic poems as a means of veiling their alchemical wisdom? The answer is not straightforward, and it leads to larger issues relating to the compartmentalization of Newton's thought.[21] Just as Newton employed very different approaches to the decipherment of prophecy and alchemical *Decknamen*, so he may have considered ancient myth quite differently in different contexts. To a degree that seems unusual even for the polymaths of the seventeenth century, Newton was willing to enter into different genres and adopt their mode of reasoning and presentation. Hence it does not automatically follow that a willingness on Newton's part to adopt the notion of chrysopoetic secrets buried in classical mythology extended beyond his alchemical studies to penetrate into his understanding of the ancient world more generally. As it happens, we are able to probe this issue in a rather decisive way, for alchemy was not the only area in which Newton attempted to extract the secrets of mythology.

Newton's exegetical endeavors extended well beyond alchemy to include the supposed wellsprings of his own innovations in physics. This area of Newton's thought has received considerable attention from modern historians and will therefore require that we examine their contrasting views. Since the 1960s it has been well known that Newton composed a set of mythological interpretations that he initially intended to incorporate into the second edition of the *Principia*. Newton meant for these so-called Classical Scholia to accompany propositions 4 through 9 of *Principia* Book III, and to provide evidence that the ancients, and perhaps even Aristotle, were largely in agreement with Newtonian physics.[22] Hence, Newton extracted textual material from classical mythology and the ancient doxographers to claim a widespread ancient belief in four key doctrines: (1) that matter is atomic and moves through void spaces by means of gravity; (2) that gravitational force acts universally; (3) that gravity diminishes in the ratio of the inverse square

---

[21] On this topic see Iliffe, "Abstract Considerations," 427–54.

[22] Niccolò Guicciardini has kindly alerted me to a passage in CU Add. 3970 where Newton attributes an understanding of inertia to Aristotle. The passage is reproduced in Hall and Hall, *UPIN*, 310–11.

of the distance between bodies; and (4) that the true cause of gravity lies in the direct action of God. The "Classical Scholia" received their first extensive modern scrutiny in "Newton and the Pipes of Pan," an influential and brilliantly written article published in 1966 by J. E. McGuire and P. M. Rattansi.

Here we must recapitulate Newton's discussion of the harmony of the world that gave the two authors their title. Relying partly on Natale Conti's sixteenth-century *Mythologiae*, Newton discusses the seven pitches of the ancient pipes supposedly invented by Pan and notes that each pitch was assigned to a planet. But then he turns the myth to his own purposes by linking the ancient tradition of *musica mundana* (celestial harmony) to the principle that gravitational attraction between bodies diminishes in proportion to the square of their distance from one another. Reading Book II of Macrobius's commentary on the *Somnium Scipionis*, Newton encountered the arresting but erroneous story that Pythagoras discovered the mathematical basis of the octave, fourth, and fifth by passing a blacksmith's shop where a group of smiths were beating the same piece of metal with hammers whose weights were in the ratios of the musical intervals—one, two, three, and four. The regular succession of the pitches supposedly led Pythagoras to the discovery of an inverse proportionality between pitch and weight such that two hammers, one weighing twice as much as the other, would produce the interval of an octave when struck on the same metal. In reality Newton knew perfectly well that no such simple proportionality would exist in the case of successively striking hammers of different weights. But weight did enter into the production of harmonic intervals in a different way. Following contemporary work in acoustics, Newton realized that in the case of strings stretched by hanging weights, the pitch was proportional to the square root of the weight.[23] From his perspective, the garbled account of Pythagoras's discovery of the musical intervals was an excellent example of the ancient *sapientes* (wisemen) hiding their wisdom from the vulgar. Pythagoras and his followers deliberately introduced error into their experimental report in order to delude the unworthy. Similarly, when Macrobius and other doxographers reported that the harmonic intervals could be found by relating a central earth to the moon, sun, and other planets, they were hiding their genuine heliocentric knowledge beneath the delusory veil of geocentric astronomy. To Newton, Pythagoras and his followers were writing all of this to drop hints of the inverse square law. As the English natural philosopher puts it:

Therefore, by means of such experiments he \<Pythagoras\> ascertained that the weights by which all tones on equal strings \<were made audible (*audirentur*),\> were reciprocally as the squares of the lengths of the string by which the musical instrument emits the same tones. But the proportion discovered by these experiments, on the evidence of Macrobius, he

[23] For Newton's knowledge of contemporary harmonics, see the careful and lucid article by Niccolò Guicciardini, "The Role of Musical Analogies in Newton's Optical and Cosmological Work," *Journal of the History of Ideas* 74 (2013): 45–67, see 62–65.

applied to the heavens and consequently by comparing those weights with the weights of the Planets and the lengths of the strings with the distances of the Planets, he understood by means of the harmony of the heavens that the weights of the Planets towards the Sun were reciprocally as the squares of their distances from the Sun.[24]

While focusing mainly on the "Classical Scholia" and Newton's claim that his discoveries were practically as old as the human race itself, McGuire and Rattansi also presented the argument that the English natural philosopher saw ancient wisdom as a unified whole in the tradition of the Italian Neoplatonists and their heirs at Cambridge, particularly Henry More and Ralph Cudworth. Hence for McGuire and Rattansi, Newton's exegetical efforts were aimed at extracting and reassembling a holistic and primal wisdom, essentially the *prisca sapientia* of the Neoplatonic tradition. Alchemy, natural philosophy, and biblical hermeneutics were all paths to the recovery of this ancient wisdom. Like John Maynard Keynes in his essay "Newton the Man," McGuire and Rattansi saw Newton as "the last of the magicians," not as an early modern natural philosopher trying to find a distinguished ancient pedigree for his work.

The position of McGuire and Rattansi has been challenged more recently by Paolo Casini, who argues that the two scholars failed to recognize a *specific* tradition of mythological interpretation to which the "Classical Scholia" belong. Instead of seeing the "Classical Scholia" as the work of "a theosophist and a neo-Platonist," to use his terminology, Casini situates them in the tradition that led Copernicus to see Pythagoras and Philolaus as his heliocentric forebears in the famous *De revolutionibus orbium caelestium*. This astronomical tradition was still alive and well in the seventeenth century, a fact made evident by such synthetic depictions as Giovanni Battista Riccioli's presentation of the heliocentric cosmos as the *Systema Philolai, Aristarchi, et Copernici* (World System of Philolaus, Aristarchus, and Copernicus) in his *Almagestum novum* of 1651.[25] In Casini's view, then, Newton's "Classical Scholia" belong to "a particular tradition" that is not that of the traditional *prisca sapientia* in the broad sense, but rather a "Copernican" variant already being used by astronomers to "vindicate the validity" of their alternatives to the geocentric universe of Ptolemaic astronomy.[26] Thus Casini argues forcefully that the function of mythology in the "Classical Scholia" was primarily one of legitimation by means of invoking ancient authority. He is eager to clear Newton of any deep-seated interest in the mythology that might seem to appear there.[27]

---

[24] J. E. McGuire and P. M. Rattansi, "Newton and the 'Pipes of Pan,'" *Notes and Records of the Royal Society of London* 21 (1966): 108–43, see 116–17. The translation is a slightly modified version of the one given by McGuire and Rattansi. There is a long treatment of this theme in Yahuda 17.3, complete with a discussion of the inverse square law, which was unknown to McGuire and Rattansi.

[25] Giovanni Battista Riccioli, *Almagestum novum* (Bologna: Benatius, 1651), 102.

[26] Paolo Casini, "Newton: The Classical Scholia," *History of Science* 22 (1984): 1–58, see 10.

[27] See Casini, "Newton: The Classical Scholia," 15, where the Italian scholar explicitly sets out his goal of clearing Newton of the imputation of being a "charlatan."

Casini thus explicitly sets himself at odds with the Keynesian view of the "Classical Scholia" expressed by McGuire and Rattansi. The role of Newton's mythological interpretation was that of legitimizing the *Principia* by placing it in an authorial context already established by Copernicus and carried forward by subsequent astronomers. Obviously Casini has a point if he is merely arguing that Newton's theories of gravitational attraction did not derive from ancient myth, but he seems to go far beyond this trivial claim in restricting Newton's use of mythology to one of cloaking his scientific discoveries in the mantle of authority. In making his claim, Casini seems completely to ignore other documents, in particular Newton's *Theologiae gentilis origines philosophicae* (Philosophical Origins of Pagan Theology), a long but unfinished text that outlines his theory of the descent of the ancient and rational Noachian religion into idolatry over successive generations. Several of the points found in the "Classical Scholia" are adumbrated in other manuscripts now kept at the National Library of Israel, namely, Yahuda manuscripts 16.2 and 17.2, which contain various versions of the *Theologiae gentilis origines philosophicae* and associated notes. We learn there, for example, that Pythagoras was actually a heliocentricist and that he devised the music of the spheres as a means of deluding the unwary, all while teaching his true disciples about the sun-centered cosmos.[28] In the *Theologiae gentilis origines philosophicae* and its accompanying notes, this claim forms part of Newton's argument that the wisdom of Noah and his descendants and of the Egyptians was originally based on a schema that equated the twelve fundamental gods with features of the natural world. As Newton says in Yahuda 17.2, the Egyptian natural philosophy based itself on these twelve natural objects:

> The Egyptians named the Planets and the elements in this order: Saturn, Jupiter, Mars, Venus, Mercury, Sol, Luna, Fire, Air, Water, Earth; Tellus, which is represented by the four Elements, completes the tally of twelve. The whole of Philosophy is comprehended in these twelve, while the Stars represent Astronomy, and the four Elements the rest of Physiology.[29]

In a word, then, Casini seems to do scant justice to the deep interest in ancient mythology that Newton exhibited in the *Theologiae gentilis origines philosophicae* and elsewhere. Newton's attempt to find his own science veiled in the enigmata of the ancients cannot be reduced to a mere attempt to cloak his discoveries in the authority of the ancients.

Looking at the "Classical Scholia" has therefore revealed a stark fault line between two competing interpretations of Newton's approach to mythology—the holistic, Keynesian view of McGuire and Rattansi, and the compartmentalizing perspective of Casini. Neither camp presents a completely acceptable position—McGuire and Rattansi have overlooked the astronomical tradition of mythological exegesis to which Casini alludes, while Casini himself has failed to see that the "Classical Scholia" only scratch the surface of Newton's ongoing attempt to decipher the riddles of antiquity.

[28] Yahuda 17.2, fol. 18v. From *NP*, accessed April 26, 2016.
[29] Yahuda 17.2, fol. 20r, in the translation given by *NP* (accessed April 26, 2016).

Moreover, neither side really helps with Newton's treatment of myth and alchemy: McGuire and Rattansi make a few hand-waving motions toward Michael Maier and his alchemical interpretation of myth as a supposed complement to the approach of the "Classical Scholia," while Casini simply dismisses alchemy as "rubbish."[30] In reality, Newton's use of mythology in alchemy was very different from the legitimizing approach that Casini finds in the "Classical Scholia." This does not mean, however, that Newton's alchemical use of myth conformed to the holistic model proposed by McGuire and Rattansi. In order to drive this point home, we must also consult the work of Betty Jo Dobbs, who built extensively on the viewpoint of McGuire and Rattansi. This reliance is particularly evident in *The Janus Faces of Genius*, where Dobbs argues that Newton's alchemy was primarily the expression of his heterodox religious quest, and that he equated the wisdom of the alchemists with the "true primitive religion" professed by the earliest men in the *Theologiae gentilis origines philosophicae* and elsewhere.[31] As we have already seen, the devotees of Newton's original religion upheld a heliocentric worldview, and fragments of this *prisca sapientia* lay buried in the pre-Aristotelian wisdom of the Greeks, a claim that we have already encountered in his analysis of the Pythagorean tradition. Newton expands on this idea in his theological and chronological manuscripts, where he argues that the men of the first religion acknowledged their heliocentric belief by worshiping around a prytaneum, a structure with a fiery altar at the center. Additionally, they honored their heroes by deifying them: following the ancient *Sacred History* of Euhemerus of Messene, Newton believed this honorific celebration to be the origin of the pagan pantheon.[32] The religion of the ancients grew increasingly corrupt when they began worshipping the fire itself, the cosmos that the prytaneum was meant to represent, and the multiple gods that had originally been exemplary men.

According to Dobbs, Newton's alchemical reading of ancient mythology was part of the same impulse to retrieve the *prisca sapientia* as it existed before its corruption at the hands of idolaters. Dobbs particularly emphasizes the connection between a cosmic, vegetative spirit that Newton sometimes equated with the alchemists' philosophical mercury and the spirit of the world. According to Dobbs, Newton even thought of the philosophical mercury as a spirit that mediated between the physical and transcendent realms in a way analogous to the mediation of Jesus between God and man.[33] One of the key pieces that Dobbs adduces for the claim that Newton's alchemy was closely linked to his interest in the original religion of mankind lies in her analysis of a table that Newton composed in the 1690s, now found

---

[30] McGuire and Rattansi, "Newton and the 'Pipes of Pan,'" 136–37; Casini, "Newton: The Classical Scholia," 15.

[31] Dobbs, *JFG*, 150–68.

[32] Buchwald and Feingold, *Newton and the Origin of Civilization*, 146, see 141–63 for Newton's theories of the primitive religion. For Euhemerus, see Marek Winiarczyk, *The Sacred History of Euhemerus of Messene* (Berlin: Walter de Gruyter, 2013).

[33] Dobbs, *JFG*, 13, see also 243–48, where she again stresses the role of "the alchemical vegetable spirit" as a mediator between God and man and associates this with "the Arian Christ."

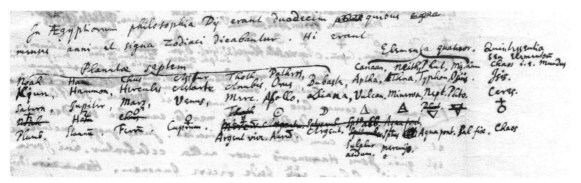

FIGURE 3.1. Detail from Huntington Library, MS Babson 420, 1v. Newton's chart labeled "Seven Planets, Four Elements, [and] Quintessence" above, followed by five successive horizontal rows of correspondences respectively showing Old Testament figures; Egyptian gods; Greco-Roman gods; the seven Ptolemaic planets, four elements, and Earth; and the seven metals, along with "acid sulfur, spirit of mercury, pontic water, fixed salt [and] chaos."

in the Huntington Library (figure 3.1).[34] This tabular representation connects the twelve gods of the Egyptians and Greeks with the twelve signs of the zodiac. The complex web of correspondences also includes Noah, Ham and his four sons, Canaan's sister-wife Astarte, Mizraim's consort Isis, her three children, and the goddess Neith.[35] These in turn are linked to the seven planets, four elements, and the planet Earth. Finally, at the bottom one sees the seven metals known to the ancients along with five specifically alchemical materials—"sulphur acidum, Spiritus mercurii, Aqua pontica, Sal fixus, Chaos" (acid sulfur, spirit of mercury, pontic water, fixed salt, chaos). In Dobbs's interpretation of this image, the column at the far right is uniquely privileged, as the Quintessence is another name for the philosophers' mercury, the Christ-like spirit that unites the cosmos in her analysis.[36] The key fact for Dobbs is that in the elemental world, the symbol for this material is the *salvator mundi* symbol of Christ, the redeemer of the fallen world. Of course the circle surmounted by a cross is also a traditional symbol for antimony, but to Dobbs this merely cements the strong association that she sees between Newton's alchemy and his religion. For Dobbs, then, the Quintessence was for Newton both "the fire at the heart of the world" and "the creative fire at the heart of matter" acting in accordance with "the Arian Logos still active in the creation of the world."

All of this might seem compelling were it not for an additional factor that Dobbs overlooks. As Jed Buchwald and Mordechai Feingold correctly point out in their recent *Newton and the Origin of Civilization*, the euhemerist reading of ancient mythology that permeates the *Theologiae gentilis origines philosophicae* is at odds with the alchemical reading of myth conveyed by Newton's sources, in particular Michael Maier.[37] Maier rejected the claim that Osiris had really been an Egyptian king or deity, whereas Newton

---

[34] Babson 420, 1v.

[35] For these biblical figures, see Buchwald and Feingold, *Newton and the Origin of Civilization*, 147.

[36] Dobbs, *JFG*, 162.

[37] Buchwald and Feingold, *Newton and the Origin of Civilization*, 148.

in the *Theologiae gentilis origines philosophicae* and elsewhere accepted his historicity. It is true that Newton's *Index chemicus* paraphrases Maier on the subject of the Egyptian king by saying "Osiris, Isis, and Typhon are a fixed salt, white spirit, and red spirit," but here Newton is trying to get to the bottom of alchemical processes, not reconstruct ancient history.[38] In short, Newton's alchemical reading of mythology in the *Index chemicus* and elsewhere in his alchemical corpus was a different project from his reconstruction of ancient history and even distinct from his decipherment of Pythagorean enigmata as prefigurations of early modern physics and astronomy. Thus there is a general problem inherent in Dobbs's approach, which employs the Keynes-tinted spectacles donned by McGuire and Rattansi: although the same "Sumerian" magus may be peering out and unraveling the secret of the universe, he is coming to radically different conclusions when he employs alchemical interpreters of myth as opposed to chronologizing ones.

Nonetheless, by focusing on Babson 420, Dobbs provides a challenge. In concentrating on this manuscript, she presents us with one of very few instances where Newton's historico-mythological studies actually do intersect with his alchemy. And yet if we examine Babson 420 more closely, this instance also fails to support Dobbs's claim that Newton's alchemy formed an integral part of his interpretation of ancient religion.

Let us begin with the first words on Babson 420, at the very top of folio 1r:

The Elements of Metals are Red Spirit	White Spirit	Pontic Water	Fixed Salt
The Elements of Minerals are Sulfur	Arsenic	Tutia	Red Earth
Vitriol	Marchasite	Zinc	
	Bismuth[39]		

Given the small size of Newton's hand here and the cramped character of the text, it is quite possible that he added these words after writing the heading below, "In Aegyptiorum Philosophia, Dii erant Duodecim nempe . . ." (In the philosophy of the Egyptians there were twelve gods, namely . . .). In fact, it may well be that Newton began this page as a summary of the material on the twelve great gods of the ancients that occupies much of the *Theologiae gentilis origines philosophicae*, and then decided later that he needed to look more deeply into the nature of the four elements and the quintessence that correspond to the Egyptian gods Aptha, Neith, Typhon, Osiris, and Isis. Immediately after this heading announcing that there were twelve Egyptian gods, Newton gives two versions of the table not reproduced by Dobbs, one of which he has crossed out.[40] These are almost identical to the version published by Dobbs, but followed by six concluding lines, which consist of

---

[38] Keynes 30/1, fol. 87r: "Sunt igitur Osyris Isis et Typhon sal fixum & spiritus albus et rubrus."

[39] Babson 420, 1r top.

Elementa metallorum sunt spiritus ruber	spiritus albus	aqua pontica	sal fixus
Elementorum minerae sulph.	arsen.	tutia	terra rubra
vitriol	marcasit.	zinetum	
	bismuth.		

[40] Babson 420, 1r middle.

further observations about chymistry.[41] One should note that these six lines are smaller and lighter than the text and tables that precede them, and might well have been written around the same time as the chymical comments at the very top of the page. Newton in fact explains in these final lines why he has allocated different minerals to each of the four elements and to the quintessence. As he puts it there,

> Sulfur and vitriol abound in the same fiery spirit, which spirit is a chymical fire; arsenic is ^highly volatile; this and marcasite are the minerals of bismuth, which is referred to Jove, god of the air. Tutia is the mineral of zinc, which is referred to Venus or the philosophical water. For it is easily resolved into a water, and that water is quite fluid and penetrating. Adam is a ^subtle and fixed earth but is not every earth. Magnesia is not fire, air, water, or earth, but is all of these. It is fiery, airy, watery, and earthy; ^it is hot, dry, wet, and cold. It is a watery fire and a fiery water. It is a bodily spirit and a spiritual body. It is the condensed spirit of the world and the noblest quintessence of all things and therefore it is customarily signified with the character of the world.[42]

Newton's main intention here is to group different minerals under the four elements and quintessence. Hence sulfur and vitriol are igneous, because they are both sources of a fiery spirit; what Newton probably has in mind is sulfuric acid. Arsenic and marcasite are airy because they contain volatile components; Tutia, the mineral of zinc, is watery, because zinc is a highly reactive substance that can be dissolved easily in various menstrua or acids. The earth called "Adam," a traditional name for red clay, is fixed and hence referred to the element earth. We should note in particular what Newton has to say about magnesia, namely, antimony. It can be grouped under none of the individual elements because it has properties of them all: hence it is properly a fifth element unto itself, a quintessence. Like other alchemists of the time, Newton sees antimony as a primordial tellurian material from which other substances derive, but there is nothing in his comments about the primitive religion, the prytaneum, the Arian *logos*, or the redeemer. In short, where Newton had the opportunity to bring these topics into the discussion, he pointedly neglected to do so.

In a word, Newton's comments are undoubtedly alchemical and they do place alchemical ideas and material in the context of his discussion of the ancient religion. But what was his purpose in doing this? Was he trying to arrive at a unified picture of a theocentric cosmos where alchemy served as a key to understanding the relationship between god and man, as Dobbs

[41] Babson 420, 1r bottom.

[42] Babson 420, 1r–1v: "Sulphur et Vitriolum eodem spiritu igneo abundant qui spiritus est ignis Chmicus <*sic*>. ~~Marcasita et~~ Arsenicum ^est maxime volatile. Hoc et Marcasita sunt mineræ Bismuti quod ad Iovem ^Deum aeris refertur. Tutia est minera Zineti quod ad Venerem seu aquam philosophicam refertur. Nam et in aquam ~~penetrantem~~ facile resolvitur, et aqua illa est maximè fluida et penetrans. Adam terra ^subtilis et fixa est sed non omnis terra. Magnesia nec ignis est nec aer nec aqua nec terra sed omnia. Est igneus aereus, aqueus terreus. ^est calidus, et siccus ~~humi frigidus et~~ humidus et frigidus. Est ignis aquosus et aqua ignea ~~quare coropora uruntur et lavantur.~~ Est spiritus corporalis et corpus spirituale. Est condensatus spiritus mundi, ^et rerü oium quintessentia nobilissima ideoq, charactere mundi ~~insignitur~~ et insigniri solet."

argues? Why did Newton go to the trouble of composing these correspondence charts in the context of an alchemical manuscript, if not to argue that alchemy could be used to arrive at the primitive, uncorrupted Christianity of the ancients?

In reality, Newton probably had a much more modest goal for his alchemical jottings on the first folio of Babson 420 than the above questions might suggest. Let us return briefly to the *Theologiae gentilis origines philosophicae* and consider the way Newton's thoughts about Pythagoras evolved over time. In the notes to the document found in Yahuda 17.2, Newton says that Pythagoras created the music of the spheres merely in order to delude the vulgar and to spread heliocentric astronomy secretly to his acolytes. There is nothing here about the inverse square law that features so prominently in Newton's interpretation of the Pythagorean *musica mundana* in the "Classical Scholia" or in some of Newton's other notes (for example, Yahuda 17.3). The idea that Macrobius's recounting of the relationship between weight and pitch was really about the inverse square law is clearly a later lucubration on Newton's part inserted into his interpretation of Pythagoras after he had composed the *Principia*. As his own scientific discoveries progressed, so did his interpretation of ancient wisdom. We see a similar phenomenon occurring in Newton's materialist interpretation of the twelve great gods of antiquity. The very beginning of the *Theologiae gentilis origines philosophicae* found in Yahuda 16.2 announces that "Dij duodecim majorum Gentium sunt Planetæ septem cum quatuor elementis et quintessentia Terra" (the twelve greater gods of the pagans are the seven planets with the four elements and the quintessence earth). There is nothing here about the detailed chymical topics found in Babson 420, only the seven planets, four elements, and quintessential earth. The same thing is true throughout the document, though on 3v Newton uses the Latin term "Tellus" for the earth to indicate that he means the planet rather than the element. The case is the same for the notes found in Yahuda 17.2; again there are twelve physical bodies including the seven planets, four elements, and "tellus," no red and white spirits, pontic water, or fixed salt. I propose, then, that the first folio of Babson 420 represents a late stage in the evolution of Newton's thought, where he believed that he could squeeze out more information from the four elements and quintessence than he had been able to do in the *Theologiae gentilis origines philosophicae*. Whether this new alchemical interpretation was merely due to his reading of sources and ruminating on their meaning or owed a debt to the ongoing chymical research that Newton did in his laboratory is a question for future research. It seems clear, however, that his understanding of the ancient enigmata was deepening, at least in his own mind, in the same way that he was gaining an ever deeper understanding of the achievements of Pythagoras. To reiterate, Newton was using his alchemical studies in the service of his research on primitive religion in the same way that he used his physics and astronomy to flesh out the meaning of ancient mythology. This was a natural and obvious move for him to make, and it clearly does not support the view that Newton equated the antimonial quintessence of the final column of the genealogy of the gods with an Arian Christ.

Newton's interpretation of myth in the context of alchemy was not an integral part of his quest to arrive at the uncorrupted wisdom and religion of the ancients, at least not in the fashion Dobbs proposed. Admittedly, the first folio of Babson 420 finds him using alchemy as one of many tools to probe the religion of the ancients. But this is a very different matter from Newton's interpretation of myth as a succession of *Decknamen* in the *Index chemicus* and throughout his alchemical corpus more broadly. In his chryso-poetic interpretation of myth, Newton very rarely turns from the early modern chymists to their ancient sources. It is true, of course, that Newton may seem at times to be laboring to wrest the secrets of the ancients directly from their tightly clenched fists even if this means joining Aeneas in his hellish descent. As he put it in another manuscript:

> In nothing do they strive so bitterly as in hiding their golden bough, which the whole grove covers; nor does it yield to just any powers but it easily and willingly will follow him who knows the maternal doves.[43]

And yet a closer inspection shows that this is not an original observation of Newton's; rather, it is a verbatim extract from Jean d'Espagnet's 1623 *Arcanum hermeticae philosophiae*. Like most of the passages where Newton is interpreting ancient mythology alchemically, he is actually deciphering sixteenth- or seventeenth-century alchemists who had already done the mythological spadework. This is the same impulse that we examined earlier, where Newton's reading of Michael Maier led him to the conclusion that Aeneas's golden bough was a substance that would induce putrefaction in metals and cause them radically to dissolve. The reference in d'Espagnet's passage is to the two doves of Venus who revealed the golden bough to Aeneas by landing on it. Like the bough itself, the doves were thought by many early modern alchemists to stand for materials that were necessary to have in order to make the philosophers' stone. They become the two doves of Diana in the work of Philalethes, to which Newton dedicated untold hours of interpretation. Newton's golden bough is testimony to his ability to submerge himself in the thought-world of the alchemists and to become one of their number. But it is one thing to decipher self-styled adepts who were using mythology as a means of writing alchemical riddles, and quite another to believe that the bulk of classical mythology was itself encoded alchemy. Once we step outside Newton's chymical corpus, the evidence does not testify to a broader commitment on his part to the decryption of mythology as a quest for the elixir. Unlike Maier and various other contemporaries, Newton does not employ the alchemical reading of myth as a tool for understanding ancient religion, science, or chronology more widely.

To conclude this chapter, then, we saw first that Newton's decipherment of alchemical *Decknamen* was far more open-ended than his interpretations

---

[43] Keynes 59, 1r: "In nullo tam acriter contendunt quam in celando ramo ipsorum aureo, quem tetigit omnis lucus nec ullis cedit viribus, sed facilis volensque sequetur eum qui maternas agnoscit aves et geminæ cui forte columbæ, ipsa sub ora viri venere volantes. Arc. Herm. c 15." See Jean d'Espagnet, *Arcanum hermeticae philosophiae*, in [d'Espagnet], *Enchiridion physicae restitutae* (Paris: Nicolaus Buon, 1623), 17–18.

of biblical prophecy. The cryptic terms of the alchemists did not decode into unique physical referents, whereas Newton hoped to link specific prophetical topoi to particular historical events. Nor did Newton expect to extract the secrets of ancient culture and chronology from the alchemical interpretation of classical myth. On the occasions when he did draw on mythology for a detailed ordering of ancient events, his interpretations of specific personalities were at odds with the decoding of the same figures when he discussed them in the context of alchemy. His euhemerist reading of pagan wisdom, where ancient divinities represented long-dead heroes, did not permit him to extract chymical meaning from ancient sources while at the same time interpreting them chronologically. Osiris could either have meant an actual king to the ancients or a fixed salt: one could not simultaneously follow Maier's interpretation, which denied the historicity of the Egyptian pantheon in favor of a materialist reading, while also accepting the history that Newton believed himself to have derived from studying the ancient records. In practically every case in which Newton read ancient myth alchemically, he was simply interpreting early modern sources such as Maier rather than returning to the antique sources themselves. Hence Newton's references to classical mythology in an alchemical context almost always derive from other early modern alchemists, not from original ancient authors or even from compilations such as Natale Conti's *Mythologiae*. Although he knew such sources and used them in his "Classical Scholia," Newton did not go to similar lengths in order to make sense of alchemical authors, whom he understood to be using ancient mythology as a conventional vehicle for encoding their alchemical practice. The evidence therefore supports the view that for Newton, prophecy, mythology, and alchemy were separate areas of endeavor with their own distinct hermeneutical methods and goals. Although the three domains might interact at times, as in the case of Babson 420, such interpenetrations do not provide evidence of a fundamental relationship between Newton's alchemy and either of the two other fields.

FOUR

Early Modern Alchemical Theory

THE CAST OF CHARACTERS

Sendivogius, Grasseus, and the Hidden Life of Metals: Newton's Sources and Alchemical Theories of the Subterranean World

At some point in the early 1670s, around the time of publicly announcing his momentous discovery that white light is actually a mixture of unaltered spectral colors, the young Newton made an equally stupendous, if lesser known, finding. In a private notebook devoted to chymical and physical topics, he entered the revelation that the globe of the earth resembles a "great animall" or rather an "inanimate vegetable" that breathes in ether for its "refreshment" and the maintenance of its life. When it exhales its subtle, ethereal breath, the material is transformed and condensed, whereupon it must rise up again to be replenished in the higher regions.[1] Here Newton paints an unforgettable picture of our planet as a biological organism, inhaling a material ether for its breath and exhaling it continually over the entire course of its lifespan. Although deprived of an animal soul, Newton's earth is a living being in a far more literal sense than the self-regulating system of modern Gaia hypothesists. Not only does the world as a whole experience its origin, respiration, and eventual death, its internal parts are also constantly undergoing generation and corruption along with the growth and diminution characteristic of living things.

In another text composed around the same time, Newton argued that even materials as durable as metals experience their own life cycle, being generated and destroyed beneath the surface of the earth.[2] In Newton's theory a continual process of tellurian circulation occurs: metals and minerals are generated out of subterranean sulfurous and mercurial fumes; the fully formed metallic materials in turn eventually decay into their primordial constituents under the influence of heat and powerful menstrua, or solvents, more powerful than the mineral acids known to man. Once they are broken down into their primitive ingredients, which are volatile, these mercurial and sulfurous fumes rise up within the earth and recombine to regenerate the metals. Thus a continual cycle of metallic birth, death, and rebirth is always taking place within the organismic structure of our living planet. And behind this

[1] Dibner 1031B, fol. 3v.
[2] Dibner 1031B, 6r–6v.

perpetual circulation lie the two traditional principles of alchemy—mercury and sulfur, which Newton seems to view as grosser forms of the very ether that preserves and refreshes the earth as a whole. As he puts it, the two "spirits," sulfur and mercury in a volatile form, "wander over the earth" and provide life to "animals and vegetables, and they make stones, salts, and so forth."

What is the origin of Newton's strange and visually striking theory? His use of the terms "mercury" and "sulfur" for the constituents of metals suggests that his sources lie in the literature of alchemy, and this of course comes as no surprise. We now know that Newton engaged in chymical research for over thirty years and that he transcribed and composed about a million words on the subject. The present chapter identifies his major sources and provides the dramatis personae for Newton's alchemical ideas more generally. But this consideration also allows us to make some general remarks on the development of alchemy from the Middle Ages up to Newton's time. The organismic theory Newton expressed was by no means characteristic of alchemy over its entire history. It was instead a product of the Renaissance. Those who have studied the subject of alchemy in the High Middle Ages will be more familiar with the simple sublimation-based theory of metallic generation that modeled metallogenesis on the reaction between sulfur and mercury that yields vermilion. Consider the following passage from the *De aluminibus et salibus*, a popular alchemical *practica* attributed to Rhazes that circulated widely in the thirteenth century and later:

> You should know that the mineral bodies are vapors which are thickened and coagulated according to the working of nature over a long time. What is first coagulated is mercury and sulfur. And these two are the elements of the mineral. And they are "the water" and "the oil," upon which a temperate concoction works with heat and humidity until they are congealed. And from them the <mineral> bodies are generated, and they are permuted until they become silver and gold in thousands of years.[3]

There is nothing here of the earth inhaling and exhaling, nor of a tellurian life cycle, nor even the idea that metals live, much less die, beneath the terrestrial surface. Instead, sulfur and mercury react with each other and thicken to produce mineral bodies, and eventually metals. One could adduce many other examples of this mechanistic approach to metallic generation in medieval alchemy, especially prominent in the Rhazean tradition and also in the works ascribed to Geber and Albertus Magnus. But instead, let us return to Newton in order to determine the sources of his view that the earth is a living—and ultimately dying—being. Here we will examine evidence

---

[3] Robert Steele, "Practical Chemistry in the Twelfth Century," *Isis* 12 (1929): 27: "Scias quod corpora mineralia sunt vapores qui inspissantur et coagulantur secundum mensuram servitutis nature in spatio longe. Et primum quidem quod coagulatur est mercurius et sulphur. Et sunt duo elementa minere. Et <non *delendum est*> sunt aqua et oleum, set unum generatur ab aqua et aliud ab oleo super quibus assiduat decoctio equaliter cum caliditate et humiditate donec congelata sunt. Et ex eis generantur corpora, et permutantur gradatim donec fiant argentum et aurum in millibus annorum." See also Julius Ruska, *Das Buch der Alaune und Salze* (Berlin: Verlag Chemie, 1935), 62, 95. There is no fully adequate edition of *De aluminibus et salibus* at present.

that Newton's sources for an earth that is constantly undergoing a cycle of birth and death do not stem from some timeless idea essential to alchemy but rather from the evolving beliefs of people associated with the central European mining explosion of the early modern period.

The protoindustrial revolution of mining and metallurgy during the fifteenth and sixteenth centuries in the Erzgebirge mountains of central Europe and elsewhere generated a literature of influential printed how-to books stretching from Ulrich Rülein von Kalbe's *Bergbüchlein* (Mining Booklet) of 1505 up to Georg Agricola's 1556 *De re metallica* (On Metallic Material) and beyond.[4] Only recently have scholars come to stress the fact that there was a fruitful interchange going on between alchemists and miners from the very beginning of the *Berg-* and *Probirbüchlein* (Mining and Assaying Booklet) genres. Rülein von Kalbe's *Bergbüchlein* already employs the sulfur-mercury theory, and this appears alongside other borrowings from alchemy in later booklets such as the *Rechter Gebrauch d'Alchimei* (the Correct Use of Alchemy) of 1531 and the *Alchimi und Bergwerck* (Alchemy and Mining) of 1534. But this interchange was far from being a one-way street. Not only did writers on mining and metallurgy borrow from alchemists, the chymists themselves also incorporated material from the rapidly expanding knowledge of subterranean processes that accompanied the European mining boom. It was the porous boundary between alchemy and the world of mining that led, I believe, to the new emphasis on a subterranean realm that experienced birth, death, decay, and rebirth just like the earthly surface early modern Europeans inhabited.

Among Newton's early modern sources there are many that describe the subterranean origin of the metals in terms that resonate with his own hylozoism. Newton was heavily influenced by the work of Michael Sendivogius, a Polish courtier and mining official in the entourage of the Habsburg Emperor Rudolf II, whose small but widely read literary corpus also imputes great significance to generative vapors circulating within the earth.[5] Sendivogius's earliest work, the 1604 *De lapide philosophorum tractatus duodecim* (Twelve Tracts on the Philosophers' Stone) was republished many times with

[4] For the early modern central European mining boom, see Adolf Laube, *Studien über den erzgebirgischen Silberbergbau von 1470 bis 1546* (Berlin: Akademie-Verlag, 1974). A still useful study of the early genre of German mining, assaying, and technical manuals may be found in Ernst Darmstaedter, "Berg-, Probir- und Kunstbüchlein," *Münchener Beiträge zur Geschichte und Literatur der Naturwissenschaften und Medizin* 2/3 (1926). More recent studies include Urs Leo Gantenbein, "Die Beziehungen zwischen Alchemie und Hüttenwesen im frühen 16. Jahrhundert, insbesondere bei Paracelsus und Georgius Agricola," *Mitteilungen, Gesellschaft Deutscher Chemiker / Fachgruppe Geschichte der Chemie* 15 (2000): 11–31; Christoph Bartels, "The Production of Silver, Copper, and Lead in the Harz Mountains from Late Medieval Times to the Onset of Industrialization," in *Materials and Expertise in Early Modern Europe*, ed. Ursula Klein and E. C. Spary (Chicago: University of Chicago Press, 2010), 71–100. For more on the connections between alchemy and practical metallurgy, see also Tara Nummedal, "Practical Alchemy and Commercial Exchange in the Holy Roman Empire," in *Merchants and Marvels: Commerce, Science, and Art in Early Modern Europe*, ed. Pamela H. Smith and Paula Findlen (New York: Routledge, 2002), 201–22.

[5] Rafał T. Prinke, "New Light on the Alchemical Writings of Michael Sendivogius (1566–1636)," *Ambix* 63 (2016): 217–43; see also Prinke, "The Twelfth Adept," in *The Rosicrucian Enlightenment Revisited*, ed. Ralph White (Hudson, NY: Lindisfarne, 1999), 141–92. This should be supplemented by Julian Paulus's entry on Alexander Seton, with whom Sendivogius is often confused, in Priesner and Figala, *Alchemie*, 335–36.

his humorous 1607 *Dialogus Mercurii, alchymistae et Naturae* (Dialogue of Mercury, an Alchemist and Nature), in combined form as the *Novum lumen chemicum* (New Light of Chymistry); he also wrote a well-received *Tractatus de sulphure* (Tract on Sulfur) in 1616, which is often collected with the foregoing titles. During his long and colorful life, Sendivogius managed to work his way up from an obscure, possibly peasant birth to become a respected counselor of two Holy Roman emperors, Rudolf II and Ferdinand II, as well as the Polish King Sigismund III. Not only did he perform public transmutations of metals, he was also employed as a metallurgical expert by the Polish magnate Mikołaj Wolski in an ambitious venture involving ironworks, and he may have been brought back to the imperial seat at the behest of Ferdinand II to oversee lead mines.[6]

Sendivogius developed an influential theory in the *Novum lumen chemicum*, in which saltpeter (*sal nitrum*) is used as a sort of model substance for explaining mineral growth and generation more generally.[7] The material that we now refer to as potassium nitrate (saltpeter or niter) does in fact effloresce on some soils and on cellar walls, so it was not an unreasonable exemplar for discussing mineral growth. Moreover, Sendivogius argues that saltpeter or niter within the earth attracts a celestial analogue, an "aerial niter" from the heavens in the same fashion that hygroscopic calcined tartar (anhydrous potassium carbonate) attracts humidity from moist air to form "oil of tartar." Sendivogius employed magnetic metaphors to make this attractive power of the *sal nitrum* still more compelling; thus he speaks elsewhere of the attracting sulfurous fatness as a *chalybs* (Latin for "steel"), which draws the mercurial moisture out of the air just as an ordinary piece of steel attracts and is attracted by a magnet (*magnes* in Latin). One could also argue, as Newton later did, that "spirit of niter," or nitric acid distilled out of saltpeter with the help of sulfates, gets its ability to dissolve metals from its attractive power.

The Polish chymist thought that *sal nitrum* contained a principle of life because of its absorption of a vital material from the heavens. This claim too could be justified by considering the properties of ordinary saltpeter. On the one hand, the substance can indeed be made to release the material that we now refer to as oxygen by means of moderate heating. On the other hand, the vital power imbedded in niter could also be used to explain the effectiveness of saltpeter in preserving meats. The idea that what keeps the body from decay after death must exercise the same agency during life has a long history in European alchemical literature, going back at least as far as the distillation of ethanol in the High Middle Ages. Finally, it was known in the seventeenth century that niter could be used as a fertilizer, a fact that we now impute to its high nitrogen content. But to Sendivogius, the ability

---

[6] Rafał T. Prinke, "Beyond Patronage: Michael Sendivogius and the Meanings of Success in Alchemy," in *Chymia: Science and Nature in Medieval and Early Modern Europe*, ed. Miguel López Pérez, Didier Kahn, and Mar Rey Bueno (Newcastle upon Tyne: Cambridge Scholars Publishing, 2010), 175–231, see 205–8.

[7] For an excellent treatment of Sendivogius's theories and their sources, see Didier Kahn, "Le Tractatus de sulphure de Michaël Sendivogius (1616), une alchimie entre philosophie naturelle et mystique," in *L'Écriture du texte scientifique au Moyen Âge*, ed. Claude Thomasset (Paris: Presses de l'Université de Paris-Sorbonne, 2006), 193–221.

of saltpeter to stimulate plant growth was one more indication of its vital power: obtained from the heavens and transmitted by rain to the earth, the fertilizing agency was acquired and absorbed by terrestrial saltpeter to be passed on in turn to the vegetable realm.

A final advantage of Sendivogius's theory lay in its ability to explain the striking combustibility of gunpowder, a fact that early modern warfare had made its mainstay. From the Sendivogian perspective, gunpowder can deflagrate without the help of ambient air because of its high content of *sal nitrum*. Since *sal nitrum* was believed to exist in the atmosphere, and since it was an essential ingredient of gunpowder, it was an easy extension of the theory to suppose that thunder and lightning were also caused by the explosion of the aerial niter in the atmosphere. Newton himself would claim in his *Opticks* that atmospheric sulfur combined with the airborne "nitrous acids" to "cause Lightning and Thunder, and fiery Meteors."

While emphasizing the role of the aerial niter, Sendivogius's theory employs the traditional alchemical principles of mercury, sulfur, and salt to explain metallic generation, but he typically interprets these as *Decknamen* (cover names), referring to various stages in the maturity of his "philosophical *sal nitrum*." Thus Sendivogius thinks of the alchemical principle sulfur as being a more active, mature form of his philosophical mercury, which is itself identical with the sophic niter. According to the *Novum lumen chemicum*, every body has a center, a "point of seed or sperm," which is always that body's "1/8200 part." The elements project their *sperma* (literally "sperm"), the bearer of their virtues, into the earth's center, which is a hollow place rather like a womb. This *sperma* is the "mercury of the philosophers," given that name because of its heaviness, fluidity, and ability to conjoin with all things, just as common quicksilver amalgamates with other metals. Following the alchemical custom of employing many names for the "first matter," Sendivogius also calls this sperm the "central salt" or *sal nitrum*. The womblike hollow at the earth's center then digests the seed of the elements, ejecting their superfluity in the form of stones. This expulsion is due to the fact that at the center of the earth there exist a *sol centralis* and a *luna centralis*, another sun and moon, which have a force driving matter outward toward the earth's surface, just as the celestial sun and moon project their own rays down to the earth.[8] Thus the elemental sperm after digestion is driven upward through the pores of the earth in the form of a vapor; there it combines with a philosophical sulfur resident in the soil. Depending on the impurities and the degree of heat encountered there, different metals and minerals are formed: the less impurities, the nobler the metal. But where the pores of the earth are open, and there is an absence of "fat" or sulfur in the earth to combine with the philosophical mercury, the vapor passes out to the surface

[8] Michael Sendivogius, *Novum lumen chemicum*, in Nathan Albineus, *Bibliotheca chemica contracta* (Geneva: Jean Antoine and Samuel des Tournes, 1654), 25, 39. Although Sendivogius here mentions two subterranean luminaries, the central moon plays practically no part in his further discussion. The central sun, on the other hand, reappears at pp. 39, 40, 42, and elsewhere. Throughout the present book, I rely mainly on Albineus's edition of Sendivogius rather than the *editio princeps*, since Newton himself employed Albineus extensively.

and serves to nourish plants. Having passed through the pores of the earth, the vaporous sperm of the elements congeals into "a water, from which all things are born."

On the earth's surface, Sendivogius says, the elemental sperm imbued with the virtues of the central sun receives the powers of its celestial counterpart, and the two in combination are responsible for life and generation in general. The philosophical mercury, or "water," is driven into the atmosphere, where it receives a vital power from the air:

> On the surface of the earth, rays are joined to rays, and they produce flowers and all things. When rain comes to pass, it receives the power of life from the air, and combines that with the *sal nitrum* of the earth (because the *sal nitri* of the earth is like calcined tartar, attracting air to itself by its dryness, which air is resolved in it into water: this *sal nitri* of the earth, which was itself an air, and is conjoined to the fatness of the earth, has such an attractive power) and the more abundantly the solar rays strike it, the greater the quantity of *sal nitrum* is produced, and consequently a greater crop grows, and this occurs continually.[9]

The *sal nitrum* joins with the "power of life" imparted to the atmosphere by the celestial rays, returns to earth in this activated form, and in turn combines with "the fatness of the earth" to yield ordinary niter. Thus the aerial form of the niter bonds with the sulfurous fatness in ordinary humus to form solid niter. The growth of metals in their mines is due to the same process as that of plants on the surface of the earth. Both depend on the descent of a vital power brought down by rain, which joins with the volatilized *sal nitrum*: the combination of this vital power and the *sal nitrum* acts like a sort of universal fertilizer. In this fashion, Sendivogius devised a cosmic system in which chymistry played the central role. The circulation of the aerial niter and its regeneration of the earth surely lie behind Newton's view of the tellurian globe as a living creature.

In addition to Sendivogius, Newton's language betrays the influence of Johann Grasseus, a German lawyer and advisor to the powerful patron of alchemists, the bishop-prince Ernst von Bayern. Newton heavily annotated Grasseus's *Arca arcani* (Arc of the Secret) in various manuscripts including his *Index chemicus*, the comprehensive concordance that he compiled over a number of years.[10] A vivid picture of Grasseus is painted by his contemporary, the chymist Michael Maier, who had firsthand experience with the author of the *Arca arcani*. Maier complains that his countryman fashioned himself as a visible model of success with his sartorial splendor while

[9] Sendivogius, *Novum lumen chemicum*, in Albineus, *Bibliotheca chemica contracta*, 51–52.

[10] Keynes 30/1, passim, and Keynes 35, folios 2r ff. (the manuscript lacks reliable foliation). For Grasseus, see Thomas Lederer, "Leben, Werk und Wirkung des Stralsunder Fachschriftstellers Johann Grasse (nach 1560–1618)," in *Pommern in der Frühen Neuzeit*, ed. Wilhelm Kühlmann and Horst Langer (Tübingen: Max Niemeyer, 1994), 227–37; and Lederer, "Der Kölner Kurfürst Herzog Ernst von Bayern (1554–1612) und Sein Rat Johann Grasse (um 1560–1618) als Alchemiker der Frühen Neuzeit: Ein Beitrag zur Geschichte des Paracelsismus" (Inaugural diss., Ruprecht-Karls-Universität Heidelberg, 1992). I thank Hiro Hirai for alerting me to Lederer's dissertation. See also Claus Priesner and Karin Figala, *Alchemie: Lexikon einer hermetischen Wissenschaft* (Munich: C. H. Beck, 1998), 165–66.

cheating many aspirants to the aurific art by offering pedestrian products as great secrets.[11] Yet Maier's bitter comments bear witness as much to the rivalries among alchemists in quest of patrons as they do to Grasseus's character.

Grasseus, like Sendivogius, had connections with the central European mining industry. At one point in the *Arca* he reproduces the mineral stamps of high-grade lead ore from various mines in ascending order of their silver content: these include Joachimsthal in the Bohemian Erzgebirge, Olkusz in Poland, Freiberg in Saxony, the area near Bratislava in "Hungary" (modern Slovakia), Villach in Carinthia, and Annaberg in the German Erzgebirge.[12] Also like Sendivogius, Grasseus employs the terms "sulfur" and "mercury" for the primordial constituents of metals, even though the Polish alchemist differed from him by introducing the theory of the aerial niter. Grasseus and Sendivogius were effectively appropriating and updating the medieval theory of metallic generation according to which the metals were formed within the earth by the combination of ascending fumes of sulfur and mercury, much in the way that cinnabar can be made by subliming those two materials in a flask. The earliest form of the sulfur-mercury theory had appeared hundreds of years before in the *Book of the Secret of Creation*, a work written in Arabic, possibly in the eighth century, and ascribed to one Balīnās.[13] This fundamental doctrine, probably based on the observation that most of the then-known metals would amalgamate with mercury and that the common sulfide ores of metals tend to deposit sublimed sulfur in the flues of refining furnaces, was accepted in altered form until the end of the eighteenth century.

Unlike Sendivogius, however, Grasseus did not assimilate the traditional principles of mercury and sulfur to a geochemical theory modeled on the properties of saltpeter. Instead, his *Arca arcani*, the work that Newton copiously annotated, adds another step to the sublimation-process forming the empirical basis of the sulfur-mercury theory. The *Arca arcani* in effect fuses the classical exhalation theory in which sulfur and mercury vapors combine directly to form the metals with a solution theory whose ultimate source was a text that exercised considerable influence among the medieval alchemists, namely, the *Summa perfectionis* of Geber, written around the end of the thirteenth century by an occidental author. The *Summa* accepts the basic concept of the sulfur-mercury theory but adds that the sulfurous and mercurial vapors must first cool and be dissolved in a circulating subterranean humidity that transports the dissolved principles away from their respective points of origin by flowing through subterranean passages, and is then sublimed, cooled, and gradually converted into various metallic ores, depending on a variety of factors.[14] This theory had the advantage of explaining the otherwise embarrassing fact that metal ores are not usually found in conjunction with large deposits of mercury and sulfur, a condition that one would otherwise expect to follow from the sulfur-mercury theory in its usual form.

[11] Lederer, "Der Kölner Kurfürst Herzog Ernst von Bayern," 52–56.
[12] Lederer, "Der Kölner Kurfürst Herzog Ernst von Bayern," 70–71.
[13] Ursula Weisser, *Das "Buch über das Geheimnis der Schöpfung" von Pseudo-Apollonius von Tyana* (Berlin: Walter de Gruyter, 1980), 9, 106–9.
[14] William R. Newman, *The Summa perfectionis of Pseudo-Geber* (Leiden: Brill, 1991), 664–65.

In his *Arca arcani* Grasseus argues like Geber that the metallic veins within the earth drip down (*stillant*) sharp, salty, vitriolic waters, which can be observed in mines. These waters, which also contain a hidden mercury, sink downward within the earth, where they encounter the sulfurous vapors that are always rising up from the earth's core. This can lead directly to the formation of metals, as Grasseus puts it:

> That sharp and salty waters are always dripping down in metal-mines is open to view. Thus while these waters drip down from above (for all heavy things are borne downwards), at the same time sulfurous vapors ascending from the center of the earth encounter them. But if the salty waters are pure and clear, and the sulfurous vapors pure, and they embrace one another upon meeting, a pure metal is thereupon generated.[15]

Things are not so simple when the initial ingredients are less pure, however. In such a case, the mercurial substance within the sharp, salty water and the sulfurous exhalations gradually coalesce within subterranean interstices and emit a vapor. This vapor eventually thickens to become an immature "muci-laginous and unctuous" material called "Gur" (probably from the German "Gärung"—a ferment), a term that Grasseus borrowed from the well-known Lutheran pastor of Joachimsthal, Johann Mathesius, to whom we will return shortly.[16] According to Grasseus, Gur looks at first like soft, white butter, but eventually matures into ores. Grasseus argues that the ores themselves gradually ripen into the noblest metal, gold, but that in their immature form, they all begin as lead ore, which is therefore the closest of the ores to the primordial Gur. Hence one can see that Grasseus's system, unlike the rather mechanical one typically presented by the medieval sulfur-mercury theory, added a pervasive hylozoic content to the theory of metallic generation. Along with the theory of cosmic regeneration Sendivogius proposed, this emphasis on the life and growth of metals would have a pronounced effect on Newton.

Both Sendivogius and Grasseus thus conceived of the earth as a living whole filled with active spirits that continually led to the generation and growth of ores and metals. This view received support from the fact that many minerals do actually seem to grow within the earth. Saltpeter is known to replenish its supply after having been collected by miners. Alum too is often found to be replenished in nature, thanks to the action of sulfurous fumes in the volcanic areas called *solfataras*; it can also rapidly crystallize out of solutions in caves and mines. Growth and replenishment also was known to occur with vitriols: iron and copper vitriols, which we now call sulfates, were found adhering to the walls within mines as green or blue crystals that grew and changed with time (figure 4.1). In fact, it is not just unrefined

---

[15] Johannes Grasseus, *Arca arcani artificiosissimi de summis naturae mysteriis*, in *Theatrum chemicum* (Strasbourg: Haeredes Eberhardi Zetzneri, 1661), 6: 294–381, see 305: "In metalli fodinis enim semper aquas acres & salsas destillare visu deprehenditur. Dum itaque illae aquae desuper destillant (omnia gravia enim deorsum feruntur) tunc vapores sulphurei ex centro terrae ascendentes ipsis in occursum veniunt. Quod si igitur aquae salsae purae & clarae, & vapors sulphurei puri fuerint, & se in occursu amplectantur, metallum inde generatur purum."

[16] See Grasseus, *Arca arcani*, 306, where he cites Mathesius on the subject of Gur.

FIGURE 4.1. A mine in Cornwall where blue vitriol (copper sulfate) has permeated the shaft. This highly soluble material can accumulate and form stalactites when it drips down from the upper walls; dissolved in runoff, it forms the vitriol pools whose transmutative powers Newton wanted his friend Francis Aston to investigate in Europe. Photo courtesy of Simon Bone Photography. See color plate 2.

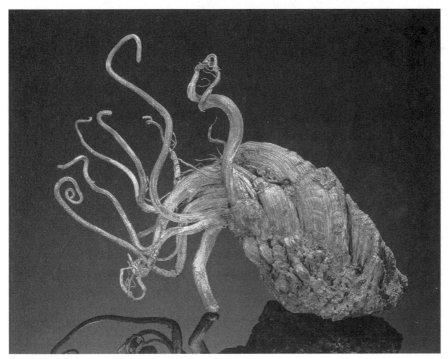

FIGURE 4.2. Native Wire Silver from Himmelsfurst Mine, Freiberg. Courtesy of Kevin Ward. See color plate 3.

minerals but pure metals themselves that appear to grow or vegetate in nature. Native silver, for example, is often found in the form of twisted stalks and branches beneath the earth (figure 4.2). Copper too can form branching formations in its native state. All of this evidence and more was available to early modern alchemists, and the influential sixteenth-century chymical writer Paracelsus used it along with the existence of mineral veins to claim that metallic ores grow from massive underground trees that can ramify and re-ramify for twenty, forty, or even sixty miles. These mineral trees fill up the empty pores within the earth, growing, maturing, and dying, just like their surface counterparts, and the fruits that they bear are the metals.[17] Hence it is no great surprise that early modern alchemists arrived at a notion of the earth as teeming with life.

Senescence and Death within the Earth: Mathesius, Solea, and Basilius Valentinus

But neither Sendivogius nor Grasseus, nor for that matter the bulk of the myriad alchemists whom Newton read, put much if any emphasis on the parallel senescence, death, and decay of metals and minerals. Like most

---

[17] See Paracelsus's *De mineralibus*, for example, which Karl Sudhoff dated to the period 1526–27, in Karl Sudhoff, *Theophrast von Hohenheim, Sämtliche Werke* (Munich: R. Oldenbourg, 1930), series 1, vol. 3, pp. 37, 40.

alchemical authors, Sendivogius and Grasseus were content to describe the formation of metals and their transformations within the earth. By focusing on the divergent quality and relative quantities of their primordial sulfur and mercury, and by considering the heat or cold of the subterranean interstices where these vapors congealed, such authors were able to provide plausible explanations for the generation of one metal as opposed to another. Where then did Newton encounter the idea that metals were not only born beneath the earth, but also that they met their death there as well?

An important clue to this puzzle lies in the work of another favorite of Newton's, namely, Michael Maier, the cosmopolitan physician and alchemist whom we have already encountered as a critic of Grasseus. Maier's 1618 *Atalanta fugiens* (Atalanta Fleeing) provides famous testimony to the emblematic side of early modern alchemy with its elegant engravings of alchemical topoi juxtaposed against scores for musical fugues. Like Sendivogius, Maier was a man of humble birth who managed to attend multiple universities and acquire the education that would allow him to write alchemical works in elegant Latin and enter the rarefied space of the Imperial Court; he would in fact become a physician to Rudolf II.[18] Although less involved with the world of mining and metallurgy than Sendivogius, Maier took the trouble to learn assaying and refining techniques while practicing as a physician in Königsberg.[19] As I mentioned in the previous chapter, Maier also developed a theory in his *Arcana arcanissima* (Most Secret Secrets) of 1614 that the mythology of the ancients could not possibly be meant literally, because it imputed scandalous and perfidious deeds to the divinities and because it placed their activities in a chronological period that would have extended back beyond the agreed-on Christian origin of time.[20] In reality, ancient mythology was not to be taken literally; rather, it was encoded alchemy, an idea that Maier did not invent, but which he developed in graphic form in *Atalanta fugiens* and elsewhere.

But it is not *Atalanta fugiens* that concerns us here—rather it is Maier's 1618 *Viatorium*, or *De montibus planetarum septem* (Concerning the Mountains of the Seven Planets), a work that Newton heavily annotated in his Keynes MS 32. One of the passages that Newton transcribed was the following:

---

[18] A great deal of information has emerged about Michael Maier's life over the last generation, largely as a result of the detective work of Karin Figala and Ulrich Neumann. See their "Ein früher Brief Michael Maiers an Heinrich Rantzau," *Archives internationales d'histoire des sciences* 35 (1985): 303–29; Figala and Neumann, "Michael Maier (1569–1622): New Bio-Bibliographical Material," in *Alchemy Revisited: Proceedings of the International Conference on the History of Alchemy at the University of Groningen*, April 17–19, 1989, ed. Zweder R.W.M. von Martels (Leiden: Brill, 1990), 34–50; Figala and Neumann, "'Author Cui Nomen Hermes Malavici': New Light on the Bio-Bibliography of Michael Maier (1569–1622)," in *Alchemy and Chemistry in the 16th and 17th Centuries*, ed. Piyo Rattansi and Antonio Clericuzio (Dordrecht: Kluwer, 1994), 121 147. See also the nine very interesting letters of Maier's from the last years of his life, published in Nils Lenke, Nicolas Roudet, and Hereward Tilton, "Michael Maier—Nine Newly Discovered Letters," *Ambix* 61 (2014): 1–47.

[19] Hereward Tilton, *The Quest for the Phoenix: Spiritual Alchemy and Rosicrucianism in the Work of Count Michael Maier (1569–1622)* (Berlin: Walter de Gruyter, 2003), 60–61, where passages from Maier's rare, autobiographical treatise *De medicina regia et vere heroica, Coelidonia* are reproduced.

[20] Tilton, *Quest for the Phoenix*, 80–86.

If miners <*metallarii*> hit upon a mineral that is burned into a black matter, they gather from practically indubitable signs that the mineral, once it attained its perfection, was consumed by the subterranean heat and it expired; and they justly say that they arrived too late.[21]

We should pay attention to the fact that Maier explicitly ascribes the belief that minerals die to miners—*metallarii*—rather than to alchemists. In the *Viatorium*, he even gives his precise sources: "Mathesius in his *Sarepta* and Solea in his *Septuriae*." The first of these figures is very easy to identify, the second less so. In short, Maier's first reference belongs to Johann Mathesius, the aforementioned Lutheran pastor of Joachimsthal, whose *Bergpostill oder Sarepta* first appeared in 1562; this weighty tome consists of sermons that Mathesius delivered to the miners and other *Bergleute* in the mining boomtown of Joachimsthal. The sermons are filled with detailed information about mining and minerals; it has recently been shown that Mathesius supplements this rich material with additional doctrines taken from alchemy.[22]

As for the *Septuriae* of Solea, this refers to the sevenfold division of the *Büchlein von dem Bergwergk* (Mining Booklet) of one Nicolaus Solea, which was printed by Elias Montanus in 1600.[23] The work has an interesting history in terms of its origin and its fate. First, an early manuscript copy of the *Büchlein von dem Bergwergk* once belonging to the chymical Maecenas Count Wolfgang II von Hohenlohe exists in the library of the University of Hamburg. The title page of the manuscript reveals that it was composed "by N. Solea Bohemian" (durch N. Soleam Boemium), and adds that it was completed "in the month of March, 1569, in Königsberg in Prussia."[24] Hence Solea was a Bohemian, seemingly active in Königsberg during the 1560s. If this Nicolaus Solea was identical to a certain Lutheran preacher named Nikolaus Solia of Altenstein, he has the distinction of having taught the rudiments of the aurific art to the notorious alchemical *Betrüger* Phillip Sömmering—who would himself go on to be drawn and quartered by Duke Julius of Braunschweig in 1575.[25] The colorful details of Sömmering's career and involvement with the likewise executed Anna Maria Zieglerin, a female

---

[21] Keynes 32, 30v: "Metallarij si mineram offendant combustam in materiam nigram <*illeg.*> ex signis haud dubijs colligunt, a perfectione occupata calore subterraneo eam consumptam expirasse & dicunt se justo tardius advenisse." For this passage in Michael Maier, see his *Viatorium, hoc est, de montibus planetarum septem seu metallorum* (Oppenheim: Johann Theodor de Bry, 1618), 96.

[22] John Norris, "*Auß Quecksilber und Schwefel Rein*: Johann Mathesius (1504–65) and *Sulfur-Mercurius* in the Silver Mine of Joachimsthal," in *Chemical Knowledge in the Early Modern World*, ed. Matthew Daniel Eddy, Seymour H. Mauskopf, and William R. Newman, *Osiris* 29 (2014): 35–48. See also Norris, "Early Theories of Aqueous Mineral Genesis in the Sixteenth Century," *Ambix* 54 (2007): 69–86, and Norris, "The Mineral Exhalation Theory of Metallogenesis in Pre-Modern Mineral Science," *Ambix* 53 (2006): 43–65.

[23] For further information on Elias Montanus, see Wilhelm Kühlmann and Joachim Telle, *Der Frühparacelsismus* (Berlin: Walter de Gruyter, 2013), part 3, pp. 927–33.

[24] University of Hamburg *Codex Alchimicus 192*, folio 323r. The date and place of completion could refer to the copying of the manuscript rather than the finishing of the text by Solea, of course.

[25] Jost Weyer, *Graf Wolfgang II. Von Hohenlohe und die Alchemie* (Jan Thorbecke: Sigmarinen, 1992), 283–85. See also Kühlmann and Telle, *Der Frühparacelsismus*, part 3, pp. 937–38.

alchemist who professed to be receiving secrets from a descendant of Paracelsus, have recently been the subject of considerable study.[26]

Nor is his perhaps regrettable impact on Sömmering the end of Solea's influence. It has been known since the early eighteenth century that the early part of the *Leztes Testament* (Last Testament) published in 1626 and later as a work of Basilius Valentinus was actually an abridged version of Solea's *Büchlein von dem Bergwergk*. Not surprisingly, since the legendary Basilius was supposed to be a fifteenth-century Benedictine monk, the occasional references to Paracelsus and Georg Agricola Solea made had to be removed from the book by its editors in order to make it seem an authentic Basilian work. Nonetheless, major portions of Solea's *Büchlein* and the Basilian *Testament* are verbatim identical. Hence when Isaac Newton acquired the 1657 English edition of *Basilius Valentinus Friar of the Order of St. Benedict His Last Will and Testament*, he was indirectly exposing himself to doctrines that were current in the German mining communities of the mid- to late sixteenth century. Let us now return to the beliefs Mathesius and Solea expressed in order to see how they square with the claim of Michael Maier that German miners believed in the death of metals as well as their birth.

I begin with Mathesius, since his text is earlier than Solea's and it is far from impossible that the latter was influenced by the Joachimsthal pastor. The third sermon in Mathesius's *Sarepta* is titled "Of the Origin, Growth, and Decline of Metals, Minerals, and Ores" (*Von ursprung / zu und abnemen der Metallen / und Minerischen Bergarten und Ertzen*). Most of the sermon deals with the formation and growth of metals, and Mathesius introduces his influential theory here that they stem from a pasty, fermenting protometallic material called *Guhr*, which is the same as the Gur mentioned by Grasseus in his *Arca arcani*.[27] Mathesius actually thought of this material as being either a sulfur-mercury compound or a type of altered mercury rather than a substitute for the two alchemical principles. Yet despite its status as a second-order product of the alchemical sulfur and mercury principles, Mathesius's *Guhr* was a concrete mineral product, in our terms "a mildly acidic mud, containing dissolved metallic salts, with fragments of ore minerals and metal," produced mainly by the weathering of sulfide ores to soluble sulfates such as the blue and green vitriols often found in mines.[28] Given its apparent observational origin in mineral works, Mathesius's *Guhr* concept probably stemmed from the mining industry rather than from the literature of alchemy.

But what of the death of metals? Mathesius first states that there is no consensus among the learned as to whether metals are destroyed within the earth after having attained their maturity. Nonetheless, he points out that no

[26] On Sömmering and Zieglerin, see Tara Nummedal, *Alchemy and Authority in the Holy Roman Empire* (Chicago: University of Chicago Press, 2007). See also Nummedal, "Alchemical Reproduction and the Career of Anna Maria Zieglerin," *Ambix* 48 (2001): 56–68.

[27] Johann Mathesius, *Bergpostilla oder Sarepta* (Nuremberg: s.e., 1578), 37v.

[28] Norris, "Auß Quecksilber und Schwefel Rein," 43. See also Anna Marie Roos, *The Salt of the Earth* (Leiden: Brill, 2007), 41, 46, 68; and Ana Maria Alfonso-Goldfarb and Marcia H. M. Ferraz, "Gur, Ghur, Guhr, or Bur? The Quest for a Metalliferous Prime Matter in Early Modern Times," *British Journal for the History of Science* 46 (2013): 23–37.

created thing is eternal, and that metals, being ultimately composed of the four elements, must decompose into them. Yet the Lutheran pastor is not content to stand on such general principles alone. In a fascinating passage that is probably the source of Michael Maier's comment in the *Viatorium*, Mathesius says the following:

> I hear some clever miners who can do more than produce *Guldengroschen* and dig a shaft, when they hit a burnt out type <of mineral> or encounter a large passage and fall upon a powerful mine-damp <*Witterung*>, and find only dust or powder in <the cavity> which holds no silver, or one sees well that the silver there has passed off in the subterranean fire, they are accustomed to say "we have come too late." Likewise, if they touch an ore that has done fermenting, which is depleted as if bees had been through it, and as though it never had a body, and is as light as burnt up kitchen ashes in an oven, they conclude that good ore may well have been there but that the natural heat in the mountain burnt it up, and additionally dried out the mountain so that great hollows, caverns, and passages were left there.[29]

What is particularly striking here is that Mathesius carefully ascribes the empirical observation that metals die and decompose to miners, not to alchemists. The same thing occurs in other passages where he refers to this belief, despite the fact that he explicitly attributes other views, such as the claim that each planet produces a particular metal, to the *Alchimisten*. Mathesius concludes his biological treatment of metallic death by saying that the decomposition of subterranean metals occurs when they lose their *humidum radicale* or radical moisture as a result of their *Nahrung* or *Speise*—their food—being driven off by too excessive an underground heat. In a word, the metals return to a useless dust or powder as a result of their slow starvation.[30]

Similar ideas about the life and death of metals are expressed in Solea's *Büchlein von dem Bergwergk*, though here they are presented in far greater detail. Solea begins his treatise with the claim that metals, like other creations of God, have their own life. In the case of metals, Solea consciously employs the archaic term *Ferch* for this principle of life. This Old High German word originally meant blood, soul, or life according to the *Wörterbuch*

[29]Mathesius, *Sarepta*, 36r: "Ich höre etliche vernünfftige bergleut / die mehr können als güldengroschen zelen / und ein Schacht fassen / wenn sie inn ein verbrennt art oder grosse drusen erschlagen und treffen ein mechtige witterung / und finden noch staub oder gemülb drinne / das noch silber helt / oder da man fein sihet / daß dem silber im erdbrand abgangen ist / pflegen sie auch zu sagen: Wir sind zu spat kommen. Dergleichen wenn sie ein ergesen ertz berüren / das außgesogen ist / als weren die bienen drüber gewest / und das nimmer am leib hat / und ist so leicht als ain verbrandter aschekuchen im stuben ofen / so schliessen sie es sey wol gut ertz da gewesen / aber die natürliche hitz im berge hab es verbrandt / unnd darneben den berg außgederzt / das grosse hölen / klüfft unnd drusen da worden sein."

[30]For the medieval theory of the radical humidity, see the classic study in Michael R. McVaugh, "The 'Humidum Radicale' in Thirteenth-Century Medicine," *Traditio* 30 (1974): 259–83. A more recent treatment may be found in Arnald of Villanova, *Tractatus de humido radicali*, in *Arnaldi de Villanova opera medica omnia*, ed. Michael R. McVaugh, Chiara Crisciani, and Giovanna Ferrari (Barcelona: Universitat de Barcelona, 2010), see the "Introduzione," particularly 323–571. For the alchemical corpus ascribed to Arnald of Villanova, see Antoine Calvet, *Les oeuvres alchimiques attribuées à Arnaud de Villeneuve: Grand oeuvre, médecine et prophétie au Moyen-Âge*, Textes et Travaux de Chrysopoeia 11 (Paris: S.É.H.A., 2011).

of the brothers Grimm, so Solea's usage is far from arbitrary.[31] He then points out that metals beneath the earth are highly mobile: they experience a constant *Wegen und Regen* (moving and stirring), words for which Solea also substitutes the Latin *lubricum* (slippery) and *volatile* (flying). These terms apply to two types of motion, both of which give evidence of the life of metals, their inner *Ferch*. Solea seems to be trying to account for the fact that subterranean metals can move about in their deposits either in the dissolved, liquid form of their *lubricum* or as sublimed vapors, in the form of their *volatile*. In such conditions where the *Ferch* is in a state of *Wegen und Regen*, it is fully awake and the metal needs to feed. If the awakened metal does not receive its proper sustenance, Solea says, it will begin to consume itself and enter into a declining state of health.

Solea then incorporates these ideas, which seem to correspond to the beliefs ascribed by Mathesius to miners, with themes that are clearly drawn from the literature of alchemy. Solea argues that the traditional alchemical principle mercury is actually the feminine seed of the metals. They have a male seed as well, which is more properly their food, and this is of course the alchemical principle sulfur. Solea then launches into a complicated theory involving multiple mercuries corresponding to the different metals, which we need not pursue here. What is important for our purposes is merely the fact that fully formed metals die and decay if their *Ferch* has been aroused and the metallic substance encounters no food that it can ingest. As in Mathesius, Solea says that such metals starve to death beneath the earth. Let us consult him here in the 1657 English version of the pseudo-Basilius's *Last Testament* through which Newton encountered Solea's views. Since the translation is often inexact, I have compared it to Solea's German and made some tacit changes, though some problems remain, thanks in large part to Solea's very specialized vocabulary:

Metals have their set time as all other creatures, they decay and dye <i.e., die> when their appointed time comes. For when Nature hath brought the metalline body unto *Sol* <i.e., gold>, then by reason it wanteth nourishment, and is starving, then it descends, experiences a stronger exhaling <*Vonwitterung*>, and the inhaling <*Zuwitterung*> becomes an exhaling <*Vonwitterung*>, and an air-exhalation <*Lufftwitterung*> becomes a fire-exhalation <*Fewerwitterung*>. If the exhalation groweth stronger in a metal than its inhalation is, then it descendeth by degrees, and decayeth, and then is it called a dead ore or metal; for one external body <i.e. metal> dies after the other, at last in one place or another it maketh a total egression with its *Ferch* and seed. This breathing is known by the particular Rod of each.[32]

[31] See "Ferch" in Jacob Grimm und Wilhelm Grimm, *Das Deutsche Wörterbuch*, in the digital version published online at http://dwb.uni-trier.de/de/, consulted July 1, 2016. For Solea's use of the term, see Nicolaus Solea, *Büchlein von dem Bergwergk* (Zerbst: Elias Montanus, 1600), especially 2.

[32] Basilius Valentinus, *Basilius Valentinus Friar of the Order of St. Benedict His Last Will and Testament* (London: s.l., 1657), 21. For the German, see Solea, *Büchlein von dem Bergwergk*, 30.

The leading idea here is that when a metal begins to starve, its breathing within the mine grows weak, and it begins to exhale more than it inhales. It then sinks down within the mine, and the metal—in this case gold—experiences a reverse transmutation. Hence, as Solea reveals a few lines later, gold first loses its color, becoming initially electrum, and then declining through the series of ever baser metals until eventually no metal remains at all. What Solea is leading up to, of course, is a comprehensive discussion of mineral exhalations, the *Witterungen* that fascinated early modern German miners and mineralogists.

Such *Witterungen* or vapors were thought by many to glow and emit light of various colors that depended on the particular metal that was growing beneath the ground; they therefore provided one of the tools that prospectors could use to find ore deposits. Mine exhalations had formed a special subject of the pseudo-Paracelsian *De natura rerum* (On the Nature of Things), a wide-ranging and influential text that dealt extensively with the intersection of mining and alchemy. In the English translation of *De natura rerum* that Newton read, the pseudo-Paracelsian text refers to mine exhalations as "coruscations" in reference to their supposed flashing. White coruscations were supposed to reveal the *primum ens* or immature matter of tin, lead, or silver; red flashing detected the presence of copper or iron; and yellow provided evidence of gold.[33] Thus in an early manuscript devoted primarily to lead ores, Newton mentions that "Corruscation like Gunpouder running along is a signe of metals unripe & in primo ente."[34] Unlikely as such phenomena may appear to modern readers, similar reports of strangely illuminated mineral works can be found even in the literature of American mining in the nineteenth century.[35] Although Solea is less interested than pseudo-Paracelsus in the colors of different exhalations, he too has techniques for exploiting their ability to reveal different types of ores. Much of Solea's text is dominated by his treatment of specialized divining rods that are supposed to respond to different types of *Witterungen* and lead mining prospectors to the locations of different ore and metal deposits. We need not follow him further in this discussion as it has been dealt with by scholars concerned with the history of the divining rod in early modern Germany.[36]

This brief excursion into the rich and difficult texts of Solea and Mathesius supports the likelihood that these authors derived their belief in the decline and death of metals not from the traditional literature of alchemy that had been circulating in Latin and vernacular European languages for several centuries, but from direct interactions with miners and metallurgists. We must not erect an artificial barrier between early modern alchemy

---

[33] Pseudo-Paracelsus, *Of the Nature of Things Nine Books: Written by Philipp Theophrastus of Hohenheim, Called Paracelsus* (London: Thomas Williams, 1650), 129–30.

[34] Mellon 79, 1v.

[35] Dan de Quille (William Wright), *History of the Big Bonanza* (Hartford: American Publishing, 1877), 172–74.

[36] See, for example, Warren Alexander Dym, *Divining Science: Treasure Hunting and Earth Science in Early Modern Germany* (Leiden: Brill, 2010).

and mining, of course, since very often the same individuals were pursuing both the extraction and the transmutation of metals. We saw this dual range of activity already in the case of Sendivogius, who was involved in multiple mineralogical and metallurgical activities for his powerful patrons, but it is perhaps even more obvious in the case of the Basilius Valentinus corpus. Not only was the *Büchlein von dem Bergwergk* of Solea incorporated wholesale into the *Last Will and Testament* of the legendary alchemist Basilius Valentinus, but also the corpus of Basilius as a whole displays an integration of artisanal metallurgical and mineralogical concerns with traditional chymical pursuits such as chrysopoeia. Since the pseudonymous Basilius Valentinus was an important source for Newton, it is important here to say a few words about the writings attributed to him.

The *Last Will and Testament*, first published in German in 1626, represents a rather late phase in the development of the corpus ascribed to Basilius Valentinus. The original member of the Basilius corpus was the *Kurtz Summarischer Tractat, Fratris Basilii Valentini Benedicter Ordens Von dem grossen Stein der Uralten* (Brief, Summary Tract of Basilius Valentinus of the Benedictine Order concerning the Great Stone of the Ancients), published by Johann Thölde in 1599.[37] Thölde was descended from a family of *Pfannenherren*—masters of saltworks—in Allendorf an der Werra, near the principality of Hessen-Kassel, which under Landgraf Moritz I would become a major magnet for alchemists in the early seventeenth century.[38] Marrying into a prominent family in Frankenhausen am Kyffhäuser in Thuringia, Thölde moved there and acquired a succession of positions including *Berghauptmann* (mining official) alongside his roles as *Pfannenherr* and *Ratskämmerer* (member of the Chamber of Councilors).[39] Having attended the University of Erfurt in the 1580s, Thölde was educated both in the world of books and in the commercial extraction and refining of minerals. His Erfurt connection is significant, since in a manuscript *Proces Buch* (Book of Processes) that he wrote in 1594 and dedicated to Moritz of Hessen-Kassel, Thölde describes a recipe for an antimonial tincture that he found in the Benedictine monastery in that city (zu Erffurtt im Closter uff dem Petersberge). This discovery took on a life of its own: long after Thölde's death, the *Last Will and Testament* reported that Basilius Valentinus's works were hidden by the putative monk under a marble table in the "high altar" in Erfurt to be discovered later by posterity.[40]

---

[37] Basilius Valentinus, *Ein Kurtz Summarischer Tractat, Fratris Basilii Valentini Benedicter Ordens Von dem grossen Stein der Uralten* (Eißleben: Bartholomaeus Hornigk, 1599).

[38] For Moritz of Hessen-Kassel and alchemy, see Bruce T. Moran, *The Alchemical World of the German Court: Occult Philosophy and Chemical Medicine in the Circle of Moritz of Hessen (1572–1632)* (Stuttgart: Sudhoffs Archiv Beiheft, 1991).

[39] Hans Gerhard Lenz, ed., *Triumphwagen des Antimons* (Elberfeld: Oliver Humberg, 2004), 291. Claus Priesner, "Johann Thoelde und die Schriften des Basilius Valentinus," in *Die Alchemie in der europäischen Kultur- und Wissenschaftsgeschichte*, ed. Christoph Meinel (Wiesbaden: Otto Harrasowitz, 1986), 107–18, see 110–11.

[40] Lenz, *Triumphwagen des Antimons*, 211–12, 335. Thölde appears to have been deceased by 1614; see the appendix by Oliver Humberg, "Neues Licht auf die Lebensgeschichte des Johann Thölde," in Lenz, "*Triumphwagen des Antimons*, 373.

Under the supposed authorship of Basilius Valentinus, a macaronic pseudonym that literally means "Mighty King" but which may combine the given names of Thölde's grandfather and brother Valtin or Valentin and his father Bastin, the Frankenhausen *Pfannenherr* published a number of works in the first decade of the seventeenth century.[41] The best known of these Basilian works are the 1604 *Triumphwagen des Antimonii* (Triumphal Chariot of Antimony) and a 1602 reimpression of *Von dem grossen Stein*.[42] The former concerned itself primarily with chymical medicine, while the latter contained "Twelve Keys" that consisted of riddling, metaphorical descriptions of operations for producing the philosophers' stone. These "Keys" are also found in the 1599 edition, but the 1602 printing adds woodcuts to each of the stages, thus inaugurating an important iconographical tradition that would culminate in the elegant copper engravings found in a Latin translation of the text made by Michael Maier and published in 1618. The first of the keys, as though to advertise Thölde's competency in the realm of metallurgy and assaying, shows a king and a queen standing behind a wolf, which is jumping over a fiery furnace (figure 4.3). To the right of the wolf stands a one-legged figure holding a scythe, a traditional representation of Saturn. All of this encodes Basilius's instructions in the first key to cleanse the body of the king with a ravenous gray wolf that is "subject to valorous *Mars*" and the "Son of old *Saturn*."[43] The king refers to the noblest of metals, gold, a fact that no educated reader would have missed. The wolf is the common ore of antimony, stibnite (antimony trisulfide), which was used by early modern assayers to refine gold by melting the metal with it. The wolf "devours" the base metals and other impurities mixed with gold: it is subject to Mars, the traditional *Deckname* for iron, because metallic antimony can be reduced out of the trisulfide by fusing it with bits of iron. Finally, the stibnite or crude antimony is the son of Saturn because it was commonly thought to be related to the traditional referent of Saturn himself, namely, lead.

With Thölde in his Basilian costume we come to our last representative of the remarkable fusion of mineralogical and alchemical knowledge that central Europe produced at the juncture between the sixteenth and seventeenth centuries. Michael Sendivogius, Johann Grasseus, Michael Maier, and Johann Thölde, as well as their predecessors Johann Mathesius and Nicolaus Solea, were all in varying degrees members of two worlds, the learned realm of the early modern university and the hardscrabble domain of the mine and the refining yard. All of these figures provide evidence that European alchemy underwent a major transformation between the Middle Ages and the early modern period, culminating in the hylozoist picture of a restless

---

[41] Lenz, *Triumphwagen des Antimons*, 338. The fact that Thölde loved double entendre, made evident by the riddling nature of his *Von dem grossen Stein*, suggests that the name "Basilius Valentinus" may have been coined both in reference to Thölde's relatives and to the combination of the Latinized form of the Greek term "βασιλεύς" (king) and Latin "Valentinus" (a personal name formed from "valens"—mighty or powerful). The two derivations are by no means mutually exclusive.

[42] Basilius Valentinus, *Ein kurtzer summarischer Tractat, Fratris Basilii Valentini Benedicter Ordens Von dem grossen Stein der uhralten* (Leipzig: Jacob Apel, 1602).

[43] For a recent and illuminating treatment of these images, see Lawrence M. Principe, *The Secrets of Alchemy* (Chicago: University of Chicago Press, 2012), 137–72.

FIGURE 4.3. Key One from the *Twelve Keys* of Basilius Valentinus. Reproduced from Basilius Valentinus, *Practica cum duodecim clavibus et appendice* in *Musaeum hermeticum reformatum et amplificatum* (Frankfurt: Hermann à Sande, 1678).

chthonic kingdom growing, maturing, evolving, and dying beneath our feet. As knowledge of the subterranean world and its processes increased, largely because of the expanding domain of mineral extraction and refining, alchemical literature absorbed the hard-won experience of miners along with their beliefs in a dynamic realm below the tellurian surface. Isaac Newton's ruminations on the birth and death of metals would have been entirely believable to the *Bergleute* who made up Johann Mathesius's Joachimsthal audience, as the Cantabrigian savant's ideas descended from the very beliefs popularized by those denizens of the underworld.

Eirenaeus Philalethes, Nicolas Flamel, and the Passage from Pseudonym to Myth

So far we have considered Newton's sources among the chymical, metallurgical, and mineralogical writers of the sixteenth and early seventeenth centuries. These authors were the major fonts of Newton's early beliefs about the life and death of metals, though by no means did their works exhaust his omnivorous alchemical reading. We must now consider two further chymists who were less obviously connected to the world of mining than the foregoing, but who were also deeply important to Newton. Both writers shared another characteristic feature of early modern alchemists, namely, their frequent use of pseudonyms when writing on the delicate subject of chrysopoeia or other *arcana majora* (the "greater secrets" of chymistry). Sendivogius published his important aurific works under anagrams of his real name, such as *Divi Leschi Genus Amo* (I Love the Race of the Divine Lech), whereas Thölde of course hid behind the colorful name of Basilius Valentinus, a figure who in time developed into a veritable fictive personage. When we come to the realm of mythic adepts, however, few writers could compete with Eirenaeus Philalethes, the "peaceful lover of truth" whom we have encountered already, and whose mysterious writings fanned the fires of chrysopoetic furnaces throughout the second half of the seventeenth century. Eirenaeus Philalethes was a well-known name among the major figures of early modern science: Boyle, Locke, and Leibniz are all known to have read his works, among countless other chymists. Philalethes was undoubtedly Newton's favorite chymical author over the *longue durée*, and at the end of his alchemical career, the "American philosopher's" only rival for this honor was Johann de Monte-Snyders, a dark star in the already dimly lit heavens of early modern alchemy to whom we devote a separate chapter later in this book.[44]

The most famous of the Philalethan treatises, his posthumously published *Introitus apertus ad occlusum regis palatium* (Open Entrance to the Closed Palace of the King), claims that the adept acquired the secret of the philosophers' stone in 1645, when he was only twenty-three.[45] We now know that the child prodigy Philalethes was actually a brainchild of George Starkey, born in Bermuda in 1628 and educated at Harvard College, who immigrated to London in 1650. It is astonishing that Starkey, a product of the barely hewn wilderness of the Massachusetts Bay Colony, would become the chymical tutor of Robert Boyle almost upon his arrival in London, a fact certified at once by their extant correspondence of 1651–52 and by a document in the Bodleian Library containing both their hands. The manuscript conveys Latin recipes written by Starkey, along with a translation by Boyle of one of the recipes as well as his additional notes.[46]

Nor was the chymical education of Boyle the end of Starkey's successes. In the second half of the 1650s, he became the leading representative of the medical reformer Joan Baptista Van Helmont in the Anglo-Saxon world, penning such widely read books of pharmaceutical chymistry as *Natures*

[44] For Boyle and Leibniz, see Newman, *GF*, 2, and Principe, *AA*, passim. John Locke's copy of Philalethes's *Introitus apertus ad occlusum regis palatium* is extant at Oxford as Bodleian Library, Locke MS 7.404. For Locke's involvement in Helmontian chymistry, see Peter Anstey, "John Locke and Helmontian Medicine," in *The Body as Object and Instrument of Knowledge: Embodied Empiricism in Early Modern Science*, ed. Charles T. Wolfe and Ofer Gal (Dordrecht: Springer, 2010), 93–117. As for Newton, Westfall counted an astonishing 302 references to Philalethes and his works in the forty-six longest entries of Keynes 30/1, the largest version of Newton's *Index chemicus*. By contrast, he found only 140 references to the next runner-up, Michael Maier, in the same entries. See Richard Westfall, "Isaac Newton's *Index Chemicus*," *Ambix* 22 (1975): 174–85, see especially 182–85.

[45] Eirenaeus Philalethes, *Introitus apertus ad occlusum regis palatium* (Amsterdam: Joannes Janssonius à Waesberge and Vidua ac Haeredes Elizei Weyerstraet, 1667), 1.

[46] The manuscript is Oxford, Bodleian Library Locke MS C29. This text is reproduced, along with Starkey's extant letters to Boyle, in Newman and Principe, *LNC*, 3–31, and 49–83.

Explication and Helmont's Vindication (1657) and *Pyrotechny Asserted and Illustrated* (1658), as well as a number of medical pamphlets. Although the works that Starkey published under his own name dealt more with *chymiatria* or chymical medicine than with chrysopoeia, the border between the two areas is less distinct than one might suppose. The quest for the Helmontian alkahest or universal dissolvent, an important desideratum for preparing medicines, was explicitly linked by the Flemish chymist to the traditional philosophical mercury of the medieval alchemists, which was widely thought to be a precursor to the philosophers' stone. Both the alkahest and the sophic mercury were Helmontian *arcana majora*—the higher secrets of the hermetic art restricted to adepts and inaccessible to the tyros or street-corner hawkers of proprietary trochisks and strong waters.

How did this backwoods colonial, educated at a provincial outpost on the edge of the known world, manage to become the teacher of Boyle, an aristocratic heir to one of the largest fortunes in England who would later come to be known as the "father of modern chemistry?" The answer seems to lie in the very real knowledge that Starkey managed to acquire while still a resident of New England. Not only was Starkey able to obtain sophisticated chymical theory and practice in the environs of Harvard College and the Boston area, he also befriended various members of the fledgling ironworks founded by the younger John Winthrop on the Saugus River. Like Johann Thölde and the other alchemist-mining experts earlier in the century, Starkey managed to combine Latin learning with hands-on metallurgical expertise, a fact that his laboratory notebooks abundantly demonstrate.[47]

Starkey's mastery of chymical and metallurgical knowledge also underwrote the striking success of the works attributed to Philalethes, in the form of his process for making a sophic mercury by "cleansing" quicksilver with star regulus of antimony, as we have already seen. In addition, the Philalethes treatises received a major boost from the developing myth of the youthful adept who supposedly wrote them. Immediately on landing in London in late 1650, Starkey began spreading rumors of an anonymous New England adept to the members of the protoscientific, technical, and utopian circle surrounding the German "intelligencer" Samuel Hartlib. The "American philosopher," as the adept came to be known, had performed marvels in New England, restoring a withered peach tree to its fruit-bearing prime and regenerating the teeth and hair of an elderly woman. Starkey claimed that the adept gave him a quantity of sophic mercury and alchemical manuscripts that he could circulate among worthy friends. Although this information was spread by word of mouth, the myth of Philalethes entered print in 1654–55 with Starkey's publication of *The Marrow of Alchemy*, a text supposedly written by "Eirenaeus Philoponus Philalethes," who presents himself as a student of the anonymous New England adept. The *Marrow* provides detailed directions for preparing the philosophers' stone while also cleverly portraying Philalethes as an adept in training who has not yet achieved the final success of the "red tincture" that could supposedly produce gold. The full success of

[47] For Starkey's life in New England, see Newman, *GF*, 14–53, and Newman and Principe, *ATF*, 156–61.

Philalethes as an adept had to await the publication of the *Introitus apertus* in 1667, two years after Starkey's death in the Great Plague of London.[48] In the meantime, Starkey had managed to establish himself in London as an essential middleman with special access to the New England adept but without having made the overt and perhaps dangerous claim that he could prepare the great transmutational elixir himself. A similar motivation may have been the original inspiration behind the equally fictive Basilius Valentinus, and one cannot help but wonder how many of the pseudonyms that populate the history of alchemy stem from the recognition that it was easier and safer to occupy such an intermediary position rather than laying overt claim to the status of the adept.

The creation of fictive adepts in the examples of Eirenaeus Philalethes and Basilius Valentinus is matched by the mythic embellishment of an actual historical figure in the case of the corpus attributed to another of Newton's favorites, namely, Nicolas Flamel and his putative source of knowledge, "Abraham the Jew." Flamel was a genuine historical figure, a Parisian scribe who died in Paris in 1418. Marrying a rich widow and investing in real estate allowed Flamel to acquire a comfortable fortune, some of which he used to have an elaborate monument erected for himself and his wife at the Cemetery of the Holy Innocents in Paris. The monument or "charnel house" (*charnier*), covered with painted bas-reliefs of mostly religious themes, became the object of speculation among alchemists of the sixteenth and early seventeenth centuries. Among the images were several fantastic animals that could be interpreted to be dragons, a favorite beast among alchemists since the origin of the aurific art in late antiquity. As a result, after several preliminary attempts by various authors to interpret these images, a little-known gentleman of Poitou, one Pierre Arnauld de la Chevallerie, published a *Livre des figures hiéroglyphiques de Nicolas Flamel* in 1612.[49] This would be translated in 1624 as *Nicholas Flammel, his exposition of the hieroglyphicall figures*; the work would go on to generate considerable interest in Britain and would even find an eager and devoted audience in New England.[50]

The practical meaning underlying Flamel's *Exposition of the Hieroglyphicall Figures*—assuming that there is one—is so well buried as to make Philalethes and Basilius look like models of openness and clarity. The interest of Arnauld, if in fact he was the pseudepigrapher behind the *Exposition* fostered on Flamel, seems to have resided more in the creation of the Flamel legend than in actual work in the laboratory. The pseudonymous author was not satisfied with merely decoding the charnel house of the scribe but created an entire legend explaining how Flamel acquired the alchemical knowledge that led to his wealth. According to the story presented in the *Exposition*, Flamel managed to acquire a wonderful book after the death of his parents,

---

[48] Newman, *GF*, 2, 58–62.

[49] Robert Halleux, "Le mythe de Nicolas Flamel ou les mécanismes de la pseudépigraphie alchimique," *Archives internationales d'histoire des sciences* 33 (1983): 234–55.

[50] Eirenaeus Orandus, trans., *Nicholas Flammel, his exposition of the hieroglyphicall figures which he caused to bee painted vpon an arch in St. Innocents Church-yard, in Paris. Together with the secret booke of Artephius, and the epistle of Iohn Pontanus* (London: Thomas Walkley, 1624).

which bore the inscription of "Abraham the Jew," a self-styled prince, priest, levite, astrologer, and philosopher to the "nation of the Jews." The remarkable book of Abraham receives a plenary description in the *Exposition*, and since we will encounter it again in Newton's chymical notes, Flamel deserves to be quoted here:

> It was not of Paper, nor Parchment, as other Bookes bee, but was onely made of delicate Rindes (as it seemed unto me) of tender yong trees: The cover of it was of brasse, well bound, all engraven with letters, or strange figures; and for my part, I thinke they might well be *Greeke Characters*, or some such like ancient language: Sure I am, I could not read them, and I know well they were not notes nor letters of the *Latine* nor of the *Gaule*, for of them wee understand a little. As for that which was within it, the leaves of barke or rinde, were ingraven, and with admirable diligence written, with a point of *Iron*, in faire and neate Latine letters coloured. It contained thrice seven leaves, for so were they counted in the top of the leaves, and always every seventh leafe was without any writing, but in stead thereof, upon the first seventh leafe, there was painted a *Virgin*, and *Serpents* swallowing her up; In the second seventh, a *Crosse* where a *Serpent* was crucified; and in the last seventh there were painted *Desarts*, or *Wildernesses*, in the middest whereof ran many faire fountaines, from whence there issued out a number of *Serpents*, which ran up and downe here and there.[51]

Unable to decipher this marvelous document without the help of a master, the scribe made a pilgrimage to Santiago de Compostela. On the return home, he met a Jewish physician named Master Canches, to whom Flamel showed an extract taken from Abraham's book. Master Canches was ecstatic and at once began to decipher the passage. Since Flamel had not brought the book itself, but only a copied fragment, Canches undertook to accompany him to Paris, but died en route of an illness. Yet the entry that he had provided Flamel, along with fervent and frequent prayer, led to the latter's complete decipherment of the text, so that he was able finally to transmute other metals into pure gold, "better assuredly than common Golde, more soft, and more plyable."[52]

Fascinated by Flamel's description of the "Book of Abraham" and its mysterious "hieroglyphs," Newton would attempt to supply a practical, laboratory practice to the elusive meaning behind the exotic images described in the text. As it happens, the *Exposition* was also a favorite of Philalethes, whose *De metallorum metamorphosi* includes the Parisian scribe among "the most candid authors."[53] This approval of the *Exposition* made it even more compelling from Newton's perspective, and the imprimatur of Philalethes meant that it was possible for Newton to assimilate the work of Flamel to that of the

[51] Flamel, *Exposition*, 6–8.

[52] Flamel, *Exposition*, 29.

[53] Eirenaeus Philalethes, *De metallorum metamorphosi*, in Philalethes, *Tres tractatus de metallorum transmutatione* (Amsterdam: Johannes Janssonius à Waisberge and the Widow of Elizeus Weyerstraedt, 1668), 19. Newton recapitulates Philalethes's approving words about Flamel in Jerusalem Var. 259.8.2v.

"American philosopher." Already in the collection of manuscripts gathered in the Jerusalem manuscript Var. 259, some of them quite early, Newton extracted extensive passages from the printed *Exposition*, and he was still actively interpreting Abraham's images in his important *Praxis* manuscript composed during or after 1693.[54] We will consider his interpretation of Flamel at length when we arrive at our analysis of *Praxis* later in the present book.

The colorful cast of characters that we have assembled in the present chapter gives a powerful sense of the allure that chymistry held for early modern figures across a wide range of disciplines that are today distinct. The *Novum lumen chemicum* of Sendivogius focused on chrysopoeia but presented the subject as an elegant, Latin riddle worthy of a sophisticated man of letters. In his *Tractatus de sulphure*, Sendivogius even went so far as to write a satire on the quest for the philosophers' stone, a form and topic favored by early modern literati; at the same time his multifarious activities included the life of a courtier and advisor in mining ventures.[55] Grasseus had a similar range of interests, while his countryman Maier, a trained physician, leaned more toward chymical medicine than either his compatriot or the noble Pole. In the works of Thölde and Starkey, medicine, metallurgy, and practical chymistry worked in unison with chrysopoetic themes to produce the full blend of interests characteristic of early modern chymistry as a whole. Finally, the man behind the Flamel text, Pierre Arnauld de la Chevallerie, is a dark star about whom little can be said except that he produced an alchemical romance worthy of the *Hypnerotomachia* of Poliphilo and other esoteric adventures of Renaissance literature. It would be no exaggeration to say that Newton too was involved in each of these pursuits, either as a full and active participant recording iatrochemical recipes, repeating chymical experiments, and creating pseudonyms for himself in the style of Sendivogius, or as an eager consumer and decoder of alchemical riddles. How then did the young Cantabrigian with his commitment to the early modern alchemical vision of a living earth, host to an internal forest of metallic fruits growing, maturing, and dying beneath its surface, acquire his knowledge of the hermetic art? In the next chapter we will examine Newton's growing involvement with the aurific art and some of the conduits by which he received his chymical knowledge while still a student and fellow at Trinity College.

[54] Var. 259.3.1r–4r.

[55] For the involvement of Petrarch, Erasmus, and other humanist authors in alchemical satire, see Tara Nummedal, *Alchemy and Authority in the Holy Roman Empire* (Chicago: University of Chicago Press, 2007), 40–72.

FIVE

The Young Thaumaturge

In June 1727, less than three months after Newton's death, his friend William Stukely completed a draft memoir describing the life of "the greatest genius of <the> human race." Among other things Stukely reports the marvelous feats of the young Newton, whose mechanical inventions aroused the wonder of his neighbors when he was enrolled at the Free Grammar School in Grantham in the 1650s. Stukely speaks of model windmills and waterclocks built by the teenager, and even describes one mill that was apparently powered by a mouse, "which he calld the miller," adding "he would joke too upon the miller eating the corn that was put in." Newton also made kites and illuminated them with candles, to the dismay of the Lincolnshire country people, who may have seen prodigies in these skyborne lights, "thinking they were comets."[1] In passing, Stukely points out that Newton lodged "at Mr Clarks house, an apothecary," but makes little of the fact that this would no doubt have exposed young Isaac to at least some of the paraphernalia and operations of seventeenth-century chymistry. Perhaps this omission stems from Stukely's own rather low regard for contemporary chymists, and his desire to present the budding scientist as a child prodigy.[2] In a later draft of his memoir, Stukely would fall victim to an unwitting irony when he tried to disabuse Newton of any interest in chrysopoeia, saying that chymistry "had need enough of his masterly skill, to rescue it from superstition, from vanity, & imposture; and from the fond inquiry of alchymy, & transmutation."[3]

We now know, of course, that Stukely's attempt to rescue Newton's reputation from the stain of transmutational alchemy led him to a direct inversion of the truth. But when did Newton first develop a serious interest in chymistry? Was it part of the adolescent thaumaturgy vividly remembered

---

[1] Keynes MS 136.03, 3–4, as reproduced by *NP*, accessed June 7, 2016.

[2] Newton's friend John Conduitt, the husband of his niece Catherine Barton, rectified this omission on Stukely's part with the following observation in his own memoirs of Newton's life: "His natural curiosity & inquisitive temper put him upon observing the composition of the medicines & the whole business of the shop where he lived, wch gave his mind the first turn to Chymistry & an early inclination to that mistress [wch jilts so many but proved a convenient handmaid to him in his other great designs]. See Keynes MS 130.02, 20–21, as reproduced by *NP*, accessed June 7, 2016.

[3] William Stukely, RS MS/142, folio 56v, from *NP*, accessed June 7, 2016.

and recounted by Newton's childhood acquaintances? Or did it emerge only under the influence of the mechanical philosophy when Newton was a student at Cambridge, during his intensive reading of Robert Boyle that the existing scholarship tells us led to his compiling of chymical glossaries? In the present chapter we consider evidence mainly from the decade between 1659 and 1669, which encompasses the final period of Newton's education in Grantham as well as his life at Trinity College, Cambridge, from his matriculation in 1661 through the awarding of his master's degree in 1668. This is the period when Newton made some of his most important discoveries in mathematics and optics, and in the next chapter we will look in detail at the relationship between the latter and Newton's chymistry. At present, however, we must restrict ourselves to his earliest jottings on chymistry, which are unremittingly practical in nature. From these testimonies we then turn to what is probably the first record of Newton's interest in chrysopoeia, namely, the reading notes that he left from his perusal of the early seventeenth-century alchemical pseudepigrapher, Basilius Valentinus. Finally, the chapter terminates with a consideration of the personal contacts that Newton had with other chymists in the 1660s and 1670s in order to throw light on the conduits through which he received his alchemical manuscripts and books. But first let us return to Newton's adolescence and examine the sources of his wonder-working youth.

Fortunately, Newton was assiduous in saving his handwritten documents from all periods. As a result, we even have the notebook from his last two years at Grantham, directly before his matriculation at Cambridge. The notebook, now found in the Pierpont Morgan Library, extends from 1659 up to the first few years of Newton's life as a student at Trinity College. Two of its sources, however, closely map on to the recollections of Newton's neighbors. The first, John Bate's *Mysteries of Nature and Art*, was initially published in 1634 and reissued in 1654, just a few years before Newton copied passages out of it for the Pierpont Morgan notebook. Bate's work falls squarely within the traditions of natural magic and books of secrets as they were conceived in the late sixteenth and early seventeenth centuries, although he prefers to speak of mysteries rather than magic.[4] Just as the later editions of Giambattista della Porta's much more famous *Magia naturalis* contained directions on everything from the making of the camera obscura to how one can "generate pretty little dogs to play with," so Bate's work was a potpourri of practical operations intended to produce astounding results.[5] The *Mysteries of Nature and Art* consists of four sections dealing with "water works," "fire works," "drawing, coloring, painting, and engraving," and "diverse experiments."

Newton's Pierpont Morgan notebook begins with four folios on pigment mixing, drawing, and painting that derive from the third section of the

[4] For early modern books of secrets, see William Eamon, *Science and the Secrets of Nature* (Princeton, NJ: Princeton University Press, 1994).
[5] Giambattista della Porta, *Natural magick by John Baptista Porta, a Neapolitane; in twenty books* (London: Thomas Young and Samuel Speed, 1658), 37, 363–64.

Mysteries. Bate's text includes some practical, "vulgar chymistry," as the technical side of the discipline was called in Newton's day, but the book does not proceed as far as the actual manufacture of the artists' colors described there. Apparently it was assumed that the reader would purchase his or her vermilion, verdigris, minium, and other pigments already made. Newton's notes from Bate include simple directions for making "alum water," "gum water," "lime water," and water of soap ashes, which precede extensive directions for mixing artists' colors. This material is followed by directions for inebriating birds in order to trap them, in turn succeeded by directions for the old alchemical project of making ersatz pearls from cheaper materials, in this case chalk. In addition to further material on pigments and inks, Newton copies out directions for simple medicines. Much of the Pierpont Morgan notebook stems from Bate, and scholars have long suspected that the adolescent Newton's mechanical and "pyrotechnic" inventions also find their ultimate source in the extensive sections that the *Mysteries* devotes to mills and fireworks.[6] It is noteworthy that Newton also supplemented his Bate with notes on "Certaine tricks" taken from an unidentified source. These tricks include turning water into various colors of wine and even a method of curing the ague by carrying on one's person a sort of amulet consisting of certain words uttered by Jesus.[7] With these instructions Newton passed beyond the mere powers of nature to a more transgressive area of endeavor, a fact that he seems to have acknowledged by concealing the directions in Sheltonian shorthand.[8]

A second source emerges from a later part of the Pierpont Morgan notebook, namely, the *Mathematicall Magick* written in 1648 by John Wilkins, the imaginative warden of Wadham College, Oxford. Wilkins's book recounts "the wonders that may be performed by mechanicall geometry," as taken from the Alexandrian engineers of antiquity such as Ctesibius and Hero of Alexandria as well as from more modern sources. Hence Wilkins describes his topic as *thaumatopoiētikē*, a variation on the more usual term *thaumaturgy* (wonder working).[9] The title, *Mathematicall Magick*, hearkens back to the second book of Heinrich Cornelius Agrippa von Nettesheim's celebrated *De occulta philosophia libri tres* (Three Books on Occult Philosophy), which also dealt with the magic of numbers. Like Agrippa, Wilkins considers such topics as humanoid automata, flying machines, perpetual lamps, and other engineering marvels, though the English scholar focuses more on actual mechanisms than does his German predecessor. In reality,

[6]E. N. Da C. Andrade, "Two Historical Notes," *Nature* 135 (1935): 359–60; G. L. Huxley, "Newton's Boyhood Interests," *Harvard Library Bulletin* 13 (1959): 348–54; Westfall, *NAR*, 60–62.

[7]Pierpont Morgan MS, 12r–13r, as transcribed by *NP*, accessed June 7, 2016.

[8]Richard Westfall deciphered this passage, which he views as a sort of humorous excuse for Newton to practice his Sheltonian shorthand. Westfall believes the passage was composed in 1662, but an examination of his argument shows that that date is actually only a terminus ante quem. In reality, Newton's reason may well have to do with his religious scrupulosity, which would no doubt have given him qualms about the possibly supernatural claim made for the amulet. See Westfall, "Short-Writing and the State of Newton's Conscience, 1662," *Notes and Records of the Royal Society* 18 (1963): 10–16, especially 132.

[9]John Wilkins, *Mathematicall Magick; or, The wonders that may be performed by mechanicall geometry in two books* (London: Sa. Gellibrand, 1648), [A5r].

however, it is not these mechanical prodigies that aroused Newton's interest. Instead, the Pierpont Morgan manuscript recapitulates material from Wilkins that is entirely chymical in nature.

The first of Newton's borrowings consists of a recipe for perpetual motion that works by mixing and distilling quicksilver, tin, and corrosive sublimate (mercuric chloride). The result, in Wilkins's words, will be "divers small atomes" that retain a perpetual motion.[10] Newton immediately follows this information with a claim deriving from the early seventeenth-century chymist Thomas Tymme that perpetual motion can also be achieved by means of "a fiery spirit out of ye Minerall matter joyning ye same wth his proper aire." This information also comes from Wilkins, though the Oxford scholar is relying on Tymme's account of the famous cosmological automaton of Cornelius Drebbel, which the Dutch inventor made for James I.[11] Finally, Newton passes from this material on perpetual motion induced by chymical means to perpetual lamps. A substantial portion of *Mathematicall Magick* is devoted to these marvels of art, which Wilkins assures us have continued burning "many hundred yeares" in the sepulchers of the ancients. He explicitly says that such wonderful luminaries are the fruit of "Chymicall experiments" rather than mechanical ingenuity, and his only excuse for including the subject in a work ostensibly devoted to mathematics is that "the subtilty and curiosity of it, may abundantly requite the impertinency." From this section Newton extracts two pages of information on permanent wicks and on oils and liquors drawn from minerals and other substances by chymical techniques.[12]

Newton was clearly not put off by the impertinency of these chymical inclusions in Wilkins's *Mathematicall Magick*. To the contrary, they were the very sections of the text to catch his eye. Had the young thaumaturge moved beyond the mechanical marvels of his Grantham years to acquire a more profound knowledge of chymical secrets than Bate could offer? It is likely that these notes do in fact represent a deepening interest on Newton's part in chymistry during the early 1660s, when this part of the Pierpont Morgan manuscript was apparently composed.[13] There is also further evidence for Newton's growing involvement with chymistry that has not received a full treatment by other scholars, but this presents a puzzle requiring a separate section of the present chapter. Let us therefore leave the adolescent Newton to his "tricks" and pass to a slightly more mature phase of his development.

[10] Pierpont Morgan MS, 18r (from *NP*). Wilkins, *Mathematicall Magick*, 228.

[11] Wilkins, *Mathematicall Magick*, 230. See Thomas Tymme, *A Dialogue Philosophicall Wherein Natures Secret Closet Is Opened* (London: Clement Knight, 1612), 60–62. For more on Tymme, see Bruce Janacek, *Alchemical Belief: Occultism in the Religious Culture of Early Modern England* (University Park: Pennsylvania State University Press, 2011). For one recent example of the extensive literature on Drebbel, see Vera Keller, "Drebbels' Living Instruments, Hartmann's Microcosm, and Libavius' Thelesmos: Epistemic Machines before Descartes," *History of Science* 48 (2010): 39–74.

[12] Wilkins, *Mathematicall Magick*, 232–56. Pierpont Morgan MS, 18v–19r, from *NP*, accessed August 31, 2017.

[13] Richard Westfall dates the central part of the Pierpont Morgan notebook to the period between 1662 and 1664. See Westfall, *NAR*, 61n54.

Newton's Chymical Dictionaries

We enter now into an area that requires a thorough reassessment of the received view. The two authorities who have dealt most with Newton's developing interest in chymistry, Dobbs and Westfall, both suppose that he had little interest in the subject until the period around 1666 to 1668, and that when he did begin to study the field seriously, his interest stemmed from his exposure to Boyle's mechanical philosophy. As Westfall put it, "he started with sober chemistry and gave it up rather quickly for what he took to be the greater profundity of alchemy."[14] The problem with this position is twofold. First, it assumes a clear distinction between alchemy and chemistry; this is in fact anachronistic for the seventeenth century, which still for the most part grouped these endeavors under the overarching label of "chymistry," in which the fields of technical chemistry, iatrochemistry or the laboratory-based production of pharmaceuticals, and transmutatory alchemy or chrysopoeia, were all included. It was widely acknowledged to be impossible to arrive at the philosophers' stone without a competence in practical chymistry. One did not give up "sober chemistry" in order to practice "alchemy." It was perfectly natural to undergo training in chymical course books or to take lessons in the basic operations before attempting the *arcana majora* such as the philosophers' stone or the alkahest. The fact that Newton tried to attain a basic practical knowledge of chymistry as a first step tells us nothing of his initial motives.

This leads to the second problem for the trajectory from "chemistry" to "alchemy" proposed by Dobbs and Westfall: their assumption that Newton's first serious interest in chymistry stemmed from his reading of Boyle not only ignores his earlier exposure to authors such as Bate and Wilkins, it also assumes that Newton was not reading still other works that were more directly focused on chymistry. As it turns out, however, both historians overlooked Newton's use of another text that was crucial to his early development. In order to present this important new information, which is presently unknown to the scholarly world, we must first review the evidence. Both Westfall and Dobbs base their position on the fact that Newton compiled an early chymical dictionary that drew partly on the work of Robert Boyle. The manuscript in question, Don. b. 15, now found in the Bodleian Library at Oxford University, relies in part on Boyle's *Origin of Forms and Qualities*, published in 1666. Because Newton spent much of the period from 1665 to 1667 on his ancestral estate in Woolsthorpe to escape the plague then raging in Cambridge, Dobbs assumes that he only obtained Boyle's book in 1667 and used it to fill out his chymical glossary soon thereafter. In reality, he may have obtained it as early as 1666, thanks to a return of several months in that year to Cambridge.[15] Westfall goes so far as to say that "Boyle

---

[14] Westfall, *NAR*, 285. The same view is expressed by Dobbs when she refers to "the threshold between Newton's straightforward chemistry and his alchemy." See Dobbs, *FNA*, 124.

[15] Dobbs, *FNA*, 121. As Westfall argues, Newton actually returned to Cambridge on March 20, 1666, and remained there until June. For Newton and the plague years, see Westfall, *NAR*, 141–44.

FIGURE 5.1. Detail from Newton's chymical dictionary found in University of Chicago, MS Schaffner Box 3 Folder 9, showing the initial divisions into bracketed dichotomies.

supplied his introduction to the subject" of chymistry, and argues that Don. b. 15 is "based largely on Boyle." At the same time, both scholars insist that the notebook displays considerable sophistication in the practice of "vulgar chymistry," and Dobbs asserts that the level of detail, which "could probably not be found in any contemporary literature," implied that Newton "had himself handled the apparatus and worked through the processes."[16] Hence we receive the impression that Don. b. 15 was exclusively based on the combination of Newton's own laboratory experience and his reading of Boyle.

An additional piece of the puzzle is found today in the Schaffner collection at the University of Chicago. The manuscript in question, Box 3, Folder 9, consists of a single sheet measuring about twelve by sixteen inches, on which Newton has written out a chymical dictionary in minute lettering.[17] The text is accompanied by carefully drawn images of furnaces, and most importantly, it is arranged in the form of a dichotomy chart. Each heading is divided into two subheadings by braces, and very often the individual subheadings are themselves broken down into further dichotomies (figure 5.1). The Chicago manuscript was unknown to Dobbs, at least at the time of writing her study of Don. b. 15. Westfall was aware of it but seems to have considered it to be merely another copy of the Oxford glossary.[18] Yet a careful examination of the Schaffner manuscript shows that while it conveys many passages that are verbatim identical to those in Don. b. 15, it contains no explicit references to Boyle whatsoever, and the Boylean passages in the Oxford manuscript do not appear there even in unattributed form. In a word, the Schaffner manuscript appears to be an earlier, pre-Boylean version of the same basic text that Newton wrote out in the form of Don. b. 15.

What are we to make of this interesting fact? If we follow the hypothesis of Dobbs and Westfall to its conclusion, the absence of Boyle in Schaffner Box 3, Folder 9 could be taken to mean that the manuscript represents Newton's own work, based on laboratory experience that he had already acquired before his introduction to *The Origin of Forms and Qualities*. But in reality this avenue is closed by the discovery of the hitherto unknown text that I alluded to a few lines ago. As it turns out, the glossary found on Newton's

---

[16] Westfall, *NAR*, 282. Dobbs, *FNA*, 121–22.

[17] Schaffner Box 3, Folder 9 also contains a second sheet with four lines of text and an image of a crudely drawn furnace. This sheet does not appear to be related to the chymical dictionary on the first sheet.

[18] Westfall, *NAR*, 282–84.

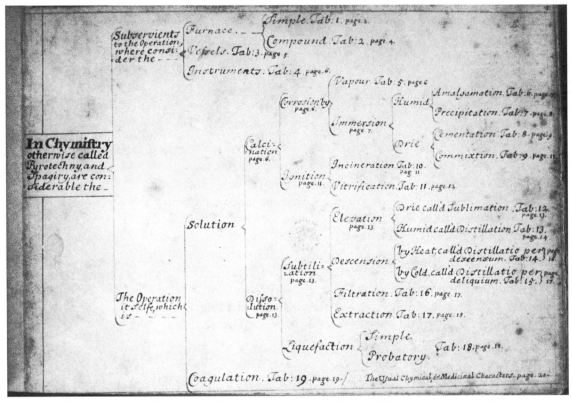

FIGURE 5.2. British Library, MS Sloane 2206, folio 2r, showing the division of the entire text into dichotomies. The actual entries appear on subsequent folios.

single sheet in the Schaffner collection is actually an incomplete copy of a work that he must have acquired in the period before he had read Boyle's *Origin of Forms and Qualities*. I refer to an anonymous text in an unknown hand that is now found in manuscript form in the British Library, where it bears the shelf mark Sloane 2206 and the rather unenlightening title, *A Treatise of Chymistry*. Like the Schaffner manuscript, this *Treatise of Chymistry* is arranged in the form of bracketed dichotomies (along with the occasional trichotomy and tetrachotomy). But the anonymous author, unlike the young Newton, wisely chose to forego the possibility of cramming all these brackets and their accompanying information onto one sheet. Hence the division into brackets occurs solely at the beginning of the manuscript as a sort of table of contents, where each entry is called a "Table" (Tab:). Paging through the successive folios reveals each of these "Tables" in the order that they appear in the initial dichotomy chart; what Newton tried to do on the front and back of one sheet, the anonymous author accomplishes in twenty-one folios (figure 5.2).

How do we know that the *Treatise of Chymistry* found in Sloane 2206 is not an original composition by Newton that somehow entered into general circulation and was copied by another hand? First, the composition of texts in the form of bracketed dichotomies, while popular among other seventeenth-century scholars, was highly uncharacteristic for Newton. Even more compellingly, several errors appearing in the Schaffner glossary militate against his having been the author. The most obvious of these lies in the trouble that Newton created for himself by attempting to reproduce the bracketed format of the *Treatise of Chymistry* on a single sheet of paper. He quickly found that this was impractical, but not before he made an abortive attempt to copy out the initial dichotomy of the *Treatise*, which begins with the words: "In Chymistry otherwise called Pyrotechny and Spagiry, are considerable the—." The anonymous author then divided this heading into the following bifurcation: "Subservients to the Operation, where consider the—" and "The Operation it selfe, which is—." Newton attempted to reproduce this most general division of chymistry into apparatus (the "subservients") and laboratory operations as found in the Schaffner manuscript but immediately discovered that there would be insufficient room on the sheet; hence, he struck the first dichotomy through in its entirety and moved to the next one. He also neglected to copy the final five folios of the *Treatise*, although much of the omitted material reappears in Don. b. 15. For more evidence that the Schaffner manuscript is an apograph of a work by another author, the reader may consult appendix one herein. At the moment it is enough to say that Newton made rather a hash of his copy, and on the basis of textual criticism, the *Treatise of Chymistry* cannot possibly be its descendant.

This interesting discovery has several immediate implications. First, the Schaffner glossary contains no explicit debt to Boyle and was very likely composed before Newton was exposed to the chymical work of the famous "naturalist," as Boyle styled himself. The fact that Newton subsequently incorporated Boylean material into his other early chymical dictionary, Don. b. 15, suggests strongly that once he was exposed to the English natural philosopher's chymical work, he made every effort to put it to use. But at the same time, the fact that the Schaffner text is a close adaptation of the anonymous *Treatise of Chymistry* also means that neither it nor Don. b. 15 tell us anything about Newton's skill in practical chymistry at this early phase of his career. We can neither assume that Newton's first interest in chymistry stemmed from his exposure to the mechanical philosophy, nor can we argue that he was an accomplished practitioner in the period before 1669. In short, his knowledge at this period appears to have been mostly bookish, and to have stemmed initially from sources other than Boyle.

Can we provide a more precise date for Newton's copying of the anonymous *Treatise of Chymistry*? I believe that we can, though it will require that we consider another early composition by Newton. Between 1664 and 1665, when he was an undergraduate at Trinity College, the young undergraduate compiled a comprehensive commonplace book devoted mostly to natural philosophy, titled *Certain Philosophical Questions* (Quaedam quaestiones philosophicae).[19] Among Newton's readings as recorded in that manuscript there are a number of texts by Boyle. It is likely, in fact, that *Certain Philosophical Questions* contains Newton's very first written reference to the English "naturalist." Early in the manuscript one finds the following, rather

[19] Newton, *CPQ*; see pp. 8–9 for the dating of the manuscript.

formally worded, citation to the phenomenon of air pressure: "as appears by the experiments of Robert Boyle, Esquire."[20] The fact that Newton felt compelled to identify Boyle by rank in this passage strongly suggests that he was newly acquainted with the man and his work. As one progresses through the notebook, an increasing number of Boyle's works are cited, including his *Experiments Touching Colours* of 1664. This text deals extensively with chymistry and the colors of bodies, a topic to which we will return. It is enough for now to make the point that by 1665 at the latest, Newton had a working knowledge of Boyle's publications in chymistry.

From these facts a conservative argument can be made that Newton's Schaffner manuscript was copied from an exemplar of the *Treatise of Chymistry* at some point during or before 1665, the date when *Certain Philosophical Questions* was finished. The considerable amount of chymistry in *Experiments Touching Colours* tipped him off to the fact that Boyle was a chymical writer of the first rank, and when Newton managed to obtain a copy of *The Origin of Forms and Qualities*, soon after its publication, he used that text to supplement the material that he had already gleaned from the anonymous *Treatise of Chymistry*. The Boylean additions to Don. b. 15, made in or around 1666, represent a second stage in Newton's early chymistry, not his initial exposure to the subject. It is therefore likely that Newton's copying of the *Treatise of Chymistry* stemmed from an earlier interest in the subject aroused by his reading of authors like Bate, Wilkins, and probably others in the same tradition. If this is so, then the view of Westfall and Dobbs that Newton began with "sober" or "straightforward chemistry" and only later plunged into the "greater profundity" of alchemy seems misleading at best. It may well have been the mysteries of Bate and magic lamps of Wilkins that steered the young undergraduate to immerse himself in the marvels of chymistry, rather than the charms of the mechanical philosophy. Nonetheless, the mechanical philosophy did have an undeniable allure, as Newton's student notebook *Certain Philosophical Questions* forcefully reveals. In the next chapter we will pass to that subject by examining the relationship in Newton's mind among the topics of corpuscular matter, light, color, and chymistry.

Newton's Early Readings in Chrysopoeia: Basilius Valentinus

With a few possible exceptions, Newton's chymical work throughout most of the 1660s consisted more of turning pages than tending furnaces.[21] This is confirmed not only by the fact that he relies exclusively on Boyle for his discussion of color indicators in his early studies of colors, but also by notes that appear to describe his earliest chymical experimentation. As Dobbs has pointed

---

[20] Newton, *CPQ*, 349.

[21] One possible exception, at least in the realm of alloying metals, may lie in the recipe for making an alloy for mirrors found in Newton, *CPQ*, 402–3. But it is far from impossible that Newton may merely be recording information here that he found elsewhere.

out, Newton's reading of Boyle's *Certain Physiological Essays* in its second edition led him to carry out experiments for the extracting of "mercuries" from various metals, that is, the isolation of their putative mercurial constituent. In fact, these experiments provide the earliest definite record of Newton's experimentation in chymistry: they appear in CU Add. 3975, the same laboratory notebook in which his first full depiction of his new theory of colors is found. Since only the second edition of Boyle's *Essays* (1669) describes the mercurial experiments, we have a well-established *terminus post quem* for the beginning of Newton's actual experimental practice in chymistry—1669.[22] In April of the same year, Newton also recorded his first definite purchases of chymical apparatus and materials. A well-known notebook found in the Fitzwilliam Museum in Cambridge records the prices Newton paid for "glasses" purchased at Cambridge and London, a furnace, a "tin ffurnace," and "Aqua ffortis, sublimate, oyle perle, fine silver, Antimony, vinegar Spirit of Wine, White lead, Allome Niter, Tartar, Salt of Tartar ☿." Along with this extensive list of reagents, enough to equip a basic laboratory, Newton purchased the six-volume *Theatrum chemicum*, a comprehensive collection of alchemical treatises containing a wealth of medieval and early modern authors in Latin.[23] These purchases in 1669 represent Newton's leap into the realm of chymical experimentation, which would soon become a wholesale immersion. May 1669 also sees him advising Francis Aston, a Fellow of Trinity College who was planning a trip to the Continent, to look into vitriol springs in central Europe, and to determine the possibility of transmuting iron into copper thereby. Newton's inspiration for this was Michael Maier's 1617 *Symbola aureae mensae duodecim nationum* (Symbols of the Golden Table of the Twelve Nations), an extended bio-bibliography of chymistry. The same letter asks Aston to investigate the doings of Giuseppe Francesco Borri, a charismatic alchemist who had strong connections with Queen Christina of Sweden.[24] Despite this sudden burst of enthusiasm for the aurific art, we have now established that Newton had been a bookish student of chymistry throughout much of the decade. Well before his descent into the realm of practice, he had eagerly read works such as the anonymous *Treatise of Chymistry*, and his supplementing of that text with borrowings from Boyle's *Origin of Forms and Qualities* led him to other authors. It is in the context of these early attempts to learn about chymistry through the medium of dictionaries that we encounter Newton's first introduction to the literature of transmutation.

The Oxford manuscript that contains Newton's Boylean additions to the *Treatise of Chymistry*, Don. b. 15, also presents several entries for materials that belong among the *arcana majora*, the higher secrets of chymistry.

[22] Dobbs, *FNA*, 139–41.

[23] Fitzwilliam Notebook, 8r–8v. Accessed from *NP*, June 19, 2016. The editors of the *Newton Project* read "oyle {y}erbe" where I read "oyle perle." Neither reading inspires vast confidence, but oil of pearl or *oleum perlarum* was a product discussed in works of early modern chymistry. See Samuel Norton, *Metamorphosis lapidum ignobilium in gemmas* (Frankfurt: Caspar Rötelius, 1630), 4. Newton acquired this book at some point in his career; see Harrison no. 1184.

[24] Newton to Aston, May 18, 1669, in Newton, *Corr.*, 9–13. On Borri and Queen Christina, see Susanna Åkerman, "Queen Christina's Esoteric Interests as a Background to Her Platonic Academies," *Scripta Instituti Donneriani Aboensis* 20 (2008): 17–36.

Along with a heading for the "alkahest," the marvelous universal dissolvent of Paracelsus and J. B. Van Helmont, for example, one finds Newton's description for the *menstruum peracutum* described in Boyle's *Origin of Forms and Qualities*.[25] This was a particularly powerful "menstruum" that Boyle believed to be capable of transmuting gold into silver, and with which he managed to volatilize the former metal. Boyle draws a connection between this unusual menstruum and the *aqua pugilum* (water of the duellers) described by Basilius Valentinus, the German author—or rather authors—whose macaronic pseudonym means "Mighty King" in English.[26] This hint may well have stimulated Newton to look into the works ascribed to the putative Benedictine monk: at any rate, several of his very early surviving manuscripts devote themselves to deciphering the chymistry of Basilius.[27] Because Newton often followed the seventeenth-century pedagogical habit of making "digests" or synopses of individual books or authors, the absence of references to other writers in these notes cannot be taken as evidence that he was unaware of other chrysopoetic writers at the time of his Basilian note-taking.[28] Nonetheless, the obviously juvenile handwriting supports the claim that they are among the first records of Newton's exposure to the aurific art, and indeed, we can probably date some of these notes to a period before 1669 because he relies on them for another composition that he almost certainly composed in that year, namely, an early commentary on Sendivogius.[29]

The fruits of Newton's first serious exposure to Basilius Valentinus's chrysopoetic practice may possibly be found in a large collection of undated

[25] Don. b. 15, 1r and 4r.

[26] Boyle, *Origin of Forms and Qualities*, in *Works*, 418–21; 1666, 351–58; for Basilius Valentinus, see p. 370. Boyle's *menstruum peracutum* is discussed in Lawrence M. Principe, "The Gold Process: Directions in the Study of Robert Boyle's Alchemy," in *Alchemy Revisited*, ed. Z.R.W.M. von Martels (Leiden: Brill, 1990), 200–205.

[27] I refer to the following manuscripts: Var. 259, Keynes 64, and British Library Additional 44888. Of these, BL Add. 44888 is definitely the latest, as it employs the barred Saturn symbol (on 6r) and refers to Eirenaeus Philalethes's *Secrets Reveal'd* (on 8v), which was only published in 1669. As for Var. 259 and Keynes 64, both employ the unbarred Saturn symbol exclusively and refer to no works outside the Basilian corpus, making it harder to judge which is earlier. At any rate, Keynes 64 consists of a digest of the *Currus triumphalis antimonii fratris Basilii Valentini* (Toulouse: Petrus Bosc., 1646), a translation of Basilius's *Triumphwagen des Antimonii* made by Pierre Jean Fabre, along with some works by other authors. Fabre's *Currus triumphalis* is not found in Harrison's *Library of Isaac Newton*. Newton may well have borrowed the book from an acquaintance; at any rate his digest of it found in Keynes 64 probably reflects Newton's earliest reading of the *Triumphwagen*.

[28] For this reason I refrain from judging whether Keynes 64 is older than Var. 259 or vice versa, even though it is true that Var. 259 refers to the Fabre edition of the *Currus triumphalis* (at 11.7r, 11.7v, and 11.8r), whereas Keynes 64, which is a digest of that book, makes no reference to the major source of Var. 259, Basilius's *Last Will and Testament* (London: s.l., 1656–57) = Harrison 128. One might normally take this to mean that Keynes 64 was the older manuscript, but Newton's practice on various occasions of "digesting" or condensing the contents of individual texts makes this assumption unreliable.

[29] Keynes 64 contains the following passage from Fabre's edition of the Basilian *Currus triumphalis* (p. 117) on 4v: "Aqua in ventre Arietis est ☿ in Antimonio Nam ~~Magi in~~ (ut magis patet) <*illeg.*> ~~☉ incipit exaltari in primo cæli~~ ♈ est primum cæli signum in quo ☉ incipit exaltari, & Antim̄ in quo aurum." This passage is reproduced without statement of source in Keynes 19 on 3r: "Antimonium enim apud veteres dicebatur Aries Quioniam <*sic*> Aries est primū Signum Zodiaci in quo Sol incipit exaltari & Aurum maxime exaltatur in Antimonio." Since Keynes 19 reproduces passages from Philalethes's 1669 *SR* and displays no awareness of the *Theatrum chemicum* that Newton purchased in the same year, it was probably composed in 1669. See Dobbs, *FNA*, 152, for the dating of Keynes 19.

notes now held in Jerusalem at the National Library of Israel with the shelf mark Var. 259.[30] The collection consists of twelve documents from various dates that Newton grouped together after their composition; we are concerned here only with the eleventh—Var. 259.11. This manuscript contains thirteen folios in which Newton has carefully written multiple versions of notes on Basilius's "Twelve Keys," reflecting at least three successive readings of that text in the 1657 English edition of *Basilius Valentinus Friar of the Order of St. Benedict His Last Will and Testament*: Each set of notes in Var. 259.11 bears its own title, which I provide here:

1v "B. Valentines process"
2r "F. B. Valentines 12 Keys"
6r "B. Valentines 12 Keys"

In addition, Newton's notes include two readings of commentary based mainly on the "Elucidation" or "Declaration of the XII. Keys" also found in the 1657 printing, which he entitles "B. Valentines process" (1v) and "B. Valentines process described in his 12 Keys & other writings" (10r). Finally, the manuscript presents Newton's numbered analysis of the *Last Will and Testament* in the form of an *index rerum* titled "References to B. Valentines works" (1r), and a section called "Things remarkable in B. Valentines works" (6r).

Three striking features of Var. 259.11 deserve immediate comment. First, the highly formal character of the notes strongly suggests the influence of Newton's experience as a student at Cambridge. The three successive series of notes on Basilius's "Twelve Keys," the two on the "Elucidation," and especially the numbered list of "Things remarkable" or observations are all redolent of the structured way in which English undergraduates of the seventeenth century were taught to organize their readings. A generation earlier, Richard Holdsworth, master of Emmanuel College at Cambridge, had composed a popular set of "Directions for a Student in the Universitie." Holdsworth's "Directions" present detailed instructions for taking notes in "paper books," where the scholar should "abbreviate and contract the sence" of the author being studied.[31] Newton's early synopses fulfill that mandate practically to the letter, but his mature alchemical notes do not display this overtly scholastic character. Instead, as he delved ever more deeply into the world of chrysopoeia, Newton came to adopt the very form and structure of the alchemical treatises themselves. In lieu of numbered lists of observations, we find the mature Newton writing full-fledged *florilegia*, the collections of alchemical *dicta* or sayings favored by late medieval and early modern alchemists. Newton's later adoption of alchemical genres as his own favored style of note-taking surely reflects his growing comfort with the riddling language of the adepts and an eagerness to be included in their ranks, as I have argued in a previous chapter.

---

[30] Although Keynes 64 may possibly be older than Var. 259, the former manuscript focuses on the Basilian *Currus triumphalis*, which is mainly concerned with iatrochemistry rather than chrysopoeia.

[31] Harris Franklin Fletcher, *The Intellectual Development of John Milton* (Urbana: University of Illinois Press, 1961), 650–52. See also Westfall, *Never at Rest*, 81–83.

The second obvious characteristic of the Basilian notes in Var. 259 follows on the heels of the first; as one might expect of notes taken in the style taught in the bookish atmosphere of seventeenth-century Cambridge, they are entirely focused on the literary decipherment of text and display no evidence of experimental practice on the part of the young Newton. Admittedly, the absence of experimental practice presents us with a potentially serious interpretive difficulty. Even in the period after 1669 for which we have ample laboratory records of Newton's work in chymistry, the bulk of his reading notes do not describe his own work at the bench. His method of deciphering the enigmatic texts at his disposal consisted first of decoding their meaning based on a given author's words alone, which typically meant treating them as purely verbal riddles. Even in his mature notes Newton usually did not overtly employ his general knowledge of chymistry and its laboratory techniques in order to make a first run at understanding an author's meaning, never mind having recourse to experimental testing. Only after he had made initial sense of the text on its own terms did he turn to the laboratory in order to determine the legitimacy of his interpretation. This disciplined approach to textual analysis, with its rigorous focus on the exact wording of a given author, characterizes Newton's interpretive notes throughout the thirty years or more of his involvement in alchemy. This raises the obvious question: how then can we know that Newton's early notes on Basilius, like his chymical dictionaries, are purely records of reading unaccompanied by experimental practice? The answer can only come from an analysis of other dated material, such as Newton's records of his experimentation, particularly from the laboratory notebooks CU Add. 3973 and 3975, which contain both explicitly dated experiments and others that can be dated indirectly by references to Boyle and other authors. As we saw earlier, Newton's experimental work in chymistry only began in 1669, a fact that is confirmed by the laboratory reports in CU Add. 3975 and by the records of his first chymical purchases found in the Fitzwilliam notebook. Since we know from other evidence that Newton's earliest notes on Basilius precede that date, they almost certainly represent the fruit of reading alone.

The third feature of Var. 259.11 that commands attention is the unremittingly practical character of Newton's notes. As we saw in chapter four, the corpus ascribed to Basilius Valentinus had grown by the 1620s to include the Lutheran pastor Nicolaus Solea's *Büchlein von dem Bergwerck*, which had been carefully tailored by the editor to make it look like a genuine work of the putative Benedictine monk. Solea's *Büchlein* occupies the first two parts of the *Last Will and Testament* ascribed to Basilius in both its German and English printings. It is a highly speculative work describing the life and death of minerals within the hidden recesses of the earth, and in its first Basilian printing it extends to almost three hundred pages. And yet Newton chooses to ignore it almost entirely in his early notes to the *Last Will and Testament*, focusing instead on the chrysopoetic puzzles presented by Basilius's "Twelve Keys," which are found in part four of the book.[32] This suggests rather strongly that Newton's early interest in Basilius Valentinus followed the same path as his reading of Bate, Wilkins, and the anonymous *Treatise of Chymistry*. Although the young Newton's notes from Boyle and his use of chymistry in optics show that he was already open to chymical theory and its probative structures, his initial interest appears to have stemmed from more practical concerns. To Newton, in the 1660s chymistry—including its chrysopoetic branch—was primarily a tool for exploiting the hidden products of nature and art. His view of these mysteries should be seen as an outgrowth of his early interest in the "Books of Secrets" tradition, which included the *arcana majora* of transmutation along with more mundane pursuits such as making mineral acids and refining metals.

Newton's notes to the *Last Will and Testament* quickly expose the rudimentary character of his practical knowledge at about the time when he received his master's degree from Cambridge. The first section of Var. 259.11, titled "References to B. Valentines works," consists of an *index rerum* to the book with headings for such basic materials as *aqua fortis* or impure nitric acid, and *aqua regis* or *aqua regia*, the mixture of hydrochloric and nitric acid that can dissolve gold. The inclusion of these commonplace products suggests the same impulse that drove Newton to compile his early chymical dictionaries. At the same time, Newton's Basilian notes reveal the great interpretive difficulties offered by early modern chymical literature. Alongside the staples of the seventeenth-century laboratory already mentioned, one also finds more obscure substances, such as "spirit of mercury." Determining the precise identity and uses of this material would provide Newton with a puzzle that would last for decades, thanks to the deliberate obfuscations of the Basilian corpus. "Spirit of mercury" plays a major part in the *Last Will and Testament* and has nothing to do with mercury in the modern sense of that word. As Newton says in his notes, It is made from "white sp^t of ⊕ & digests y^e ♃ of ☉ & ☽ & other metalls to \<sic\> to potability as also to particular medicines for metals."[33] In other words, the spirit of mercury is here derived from vitriol (copper or iron sulfate) by a process that involves distillation, and it can lead the sulfurous component of gold, silver, and other metallic materials to a state where they can heal both the ills of humans and the imperfections of base metals.

The Basilian term "spirit of mercury" obviously relies on the Paracelsian theory of three principles, according to which everything is made from mercury, sulfur, and salt. Hence vitriol too should be divisible into three principles: the "Elucidation" says these are a white, mercurial spirit, a red, oily sulfur, and a clarified salt.[34] If we jump to the section of Newton's notes titled "Things remarkable in B. Valentines works," we find the neophyte chymist

---

[32] Only two lines of the thirteen folios that Var. 259.11 devotes to the *Last Will and Testament* seem to come from the part of the book pirated from Solea. They are at folio 8r, and run as follows: "That to putrefy metalls you must raise the ferch," and "That a mixture of mineralls with metals makes them brittle."

[33] Var. 259.11.1r.

[34] Basilius Valentinus, *Last Will and Testament*, 133–35.

relaying the theory of three principles as though it were a novel discovery. The first of the numbered entries states, "1. That common Gold may by heterogeneous corrosives duely prepared be separated into 🜍 ⊖ & ☿." Here and elsewhere in his notes on Basilius, Newton relies heavily on the "Elucidation," a text that purports to be by the same author as the "Twelve Keys," but which is actually a commentary by another author that the German editor of the *Last Will and Testament* saw fit to present as a genuine work by Basilius, much as Solea's work came to be absorbed into the Basilian corpus.[35] Following this source, Newton decides that the chrysopoetic practice obscurely described by Basilius in his "Twelve Keys" involves a division of gold into its three principles, followed by their purification, exaltation, and recombination. In all of this practice the spirit of mercury plays a considerable role, as the following comments reveal:

> 2. That this solar 🜍 & ⊖ may be digested wth ye mercuriall sp^t of ♂ to an Elixir. Which intimates that that sp^t is substantially ☿ because substituted in lieu of y^e ☿ of ☉.
> 3. That ye white body of ☉ wch conteins y^e ⊖ & ☿ after y^e 🜍 is abstracted may be digested wth red philosophick 🜍 or red oyle (extracted wth ye sp^t of ☿) so as to becom ☉ again. Which shows that oyle to be of ye same substance wth the anima or 🜍 of ☉ becaus substituted in its stead. And consequently the fixed substance remaining after the sp^t & oyle are abstracted is of ye substance of y^e ⊖ of ☉. p 155.[36]

Confusingly, the "spirit of mercury" described in this pair of numbered headings does not seem to be the same as the one derived from vitriol. In entry number two, Newton speaks of substituting a "mercurial spirit" of antimony for the mercury principle in gold once the noble metal has been divided into its three ingredients. The remaining sulfur and salt will be digested and ennobled by the mercurial spirit to the point of becoming an elixir. The third entry suggests a similar process to be carried out on the divided salt and mercury of the gold by a "red philosophic oil" that has been produced along with or perhaps by means of the spirit of mercury. This ambiguity in the sense of the so-called spirit of mercury does not stem from a misunderstanding on Newton's part, but rather from the "Elucidation." The "Elucidation" describes how the philosophers' stone or elixir, which the author equates with "the best purified gold," may be made from sulfur and salt "with the help of the spirit of Mercury, which must be drawn from a crude unmelted Minera."[37] Newton has divined, reasonably enough, that this unrefined mineral (the above "Minera") is crude antimony or stibnite, the ore made up primarily of what we now call antimony trisulfide. He confirms this at another point in Var. 259.11:

---

[35] Basilius Valentinus, *Fratris Basilii Valentini Bendedicter Ordens Geheime Bücher oder letztes Testament* (Straßburg: Caspar Dietzel, 1645). The 1626 edition of the *Letztes Testament* edited by Georgius Claromontanus and published in Jena announces in its table of contents that it contains an "Erklerung der 12. Schlüssel," but the several copies that I have seen actually end with the pirated *Büchlein* of Solea on page 272.

[36] Var. 259.11.6v.

[37] Basilius Valentinus, *Last Will and Testament*, 118.

the Magnet is a deep glittering ^unmelted minerall, Saturns ofspring, subjet by name to ♂, conteining yᵉ matter of wᶜʰ all metalls are made wᶜʰ is a mercuriall spᵗ made up of three principles, is the golden seed &c p 117, 118, 119, 127 & Key 1 & this is Antimony; compare pag 12 wth Key 1.[38]

The interpretive problem that Newton faced stems ultimately from the fact that the "Elucidation" presents itself as a commentary by Basilius Valentinus himself on the "Twelve Keys," though it is actually by another author. The "Twelve Keys," despite its obscurity on many points, clearly does begin with a process that involves metallic gold. The noble metal is refined and purified with crude antimony according to the best metallurgical practice of the day, and then it is dissolved in the "water of the duelers," or *aqua pugilum* mentioned by Boyle, which is a form of aqua regia. By means of reiterate cohobation of the gold in this menstruum, Basilius manages to bring it "over the helm," that is, distill it. Since these processes are described in a way that is clearly decipherable in the "Twelve Keys," the author of the "Elucidation" realized that he could not avoid explaining them. The real purpose of his commentary, however, was to steer the reader away from the "Twelve Keys'" work with metallic gold and to substitute a parallel set of processes employing vitriol instead. The "spirit of mercury" made from antimony is possibly a relic of the gold-based processes making up the original "Twelve Keys." The vitriol-based "spirit of mercury" represents the commentator's new interpretation of Basilius.

One can see quite clearly how the commentator tries to distance himself from metallic gold in multiple passages of the "Elucidation." There are cheaper and easier ways to extract the essence of gold than the one that involves destroying the noble metal. Thus he argues that the "Astrum," or sulfur of gold, in which its color is found, exists not only in the noble metal but also in copper and steel, "two immature Metals," both of which "as male and female have red tinging qualities, as well Gold it self."[39] Moreover, this "soul" or sulfur "of the best Gold" is found in the vitriols of the two base metals; as the "Elucidation" says:

> Besides, this Mineral in our Mothers tongue is a Mineral, called *Copper* water, and of broken, or digged Verdigreece, or Copper there can be made a Vitriol, in all which is found gloriously a Soul of the best Gold, and come well to passe very profitably many wayes, no Countrey clown can believe it.

Hence the sulfur or "soul" of gold, in which its tincture resides, can be extracted from copper vitriol or "Copper water," which in turn is produced from mineral verdigris or from copper. And although this passage makes no mention of iron, its vitriol also contains the golden *anima* that can be extracted and transplanted to other metals in order to ennoble them.

The Basilian commentator hiding behind the "Elucidation" is quite explicit in touting his vitriolic process over the auric operations of the "Twelve

[38] Var. 259.11.7r.
[39] Basilius Valentinus, *Last Will and Testament*, 128.

Keys." As he puts it, "let none be so over witty, as to make our stone onely of dry and fully digested Gold." Yet he is clearly aware of the incongruity of his newfound chariness toward the noble metal. In the same breath he openly defends himself against the apparent inconsistency, saying, "be not offended at my former writings if they seem to run contrary against this."[40] The commentator's twofold desire to appropriate the authority of Basilius, which requires him to accept the processes underlying the "Twelve Keys," and to supplant them with his own processes based on vitriol, leads to the confusing situation that both the extraction of the golden sulfur from vitriol and the extraction of the sulfur from gold, now relegated to a secondary position, underlie the text. For the modern reader, the experience of reading the "Elucidation" conjures up a situation rather like watching a cinematic production where the same scene is replayed with the same characters, but with subtle differences in dialogue that lead to an entirely new meaning.

It is this double "plot line" of the "Elucidation" that leads to the commentator's ambiguous use of the expression "spirit of mercury" to mean either a product of antimony or one of vitriol. As we saw above, the "Elucidation" at one point says that this spirit must be drawn from a "crude, unmelted minera," which could well be stibnite, as Newton suspected. Later in the text a process is given for dissolving "the purple Cloak of the King," namely, "the sulphur of Sol" after it has been extracted from gold. This involves subliming a feathery product from crude antimony ground with tile meal or bole; the sublimate is supposed to resolve over time to form a menstruum. Newton again equates this with spirit of mercury in Var. 259.11:

> That if ♂ after its preparation be set in a strong sublimation mixed wth thre ♄^ts of bole or tile meal there riseth a sublimate like feathers or Alumen plumosum wch in due time resolveth into a strong effectuall water (ye sp^t of ☿) to putrefy thy seed in. p 127.[41]

In other words, the "strong effectual water" or menstruum arrived at by subliming crude antimony with ground bole or tiles provides the means of dissolving and putrefying the sulfur of gold.

Although there is no evidence that Newton tried this process for making a menstruum in the 1660s, it is worth examining briefly, for it reappears in his more mature alchemical manuscripts such as the famous *Praxis* probably stemming from the 1690s. Operations that involved subliming a product from crude antimony would receive considerable discussion from the well-known French academician and chymist Nicolas Lemery in his early eighteenth-century *Traité de l'antimoine*. As Lemery points out there, it is possible to collect both a sublimate and a distillate from the unrefined antimony ore when heating it in the presence of air. The former product consists of white, red, or yellow "flowers," presumably a mixture of antimony trisulfide, trioxide, and free sulfur. As for the liquid distillate, Lemery says that

---

[40] Basilius Valentinus, *Last Will and Testament*, 124.
[41] Var. 259.11.8r.

when he heated crude antimony mixed with sand to prevent it from melting into a mass, he managed to produce a slightly acid liquid smelling of sulfur. He then dephlegmed this liquid and arrived at a more acidic solution that appeared to him to be nothing more than "spirit of sulfur," in other words our sulfuric acid. Although it is not entirely clear that the "Elucidation's" product would be identical to either Lemery's sublimate or distillate, since the commentator does not describe the initial preparation that is supposed to precede the main operation, Lemery's experience at least confirms that a dissolving menstruum can be obtained by distilling crude antimony.[42]

As for the spirit of mercury extracted from vitriol, Newton's list of "Things remarkable in B. Valentines works" provide a means of preparing this as well:

> 11. That the sp^t of ☿ is prepared out of sp^t of ⊕ digested wth white calcined Tartar & then distilled. This sp^t riseth 1st in a white form & leaves a red ponderous water behind it wch is ye philosophicall ♃ & ⊖ mixed. p 141, 128, 134.[43]

This recipe also comes directly out of the "Elucidation." First, spirit of vitriol (sulfuric acid) is produced by destructive distillation of copper vitriol or iron vitriol. This is then added to "calcined tartar" (potassium carbonate), whereupon an effervescent action ensues. On distillation of the product, a white spirit rises up, leaving behind a more fixed, red liquid; the "Elucidation" says that this too will distill off at a higher temperature.[44] At first face this process seems nonsensical. Why would anyone neutralize sulfuric acid by combining it with potassium carbonate and then distill off the volatile product, which would seem to consist only of water? The answer may lie in the interesting ability of sulfuric acid to produce not only potassium sulfate (K_2SO_4) but also potassium bisulfate ($KHSO_4$) when an excess of the acid is added. The potassium bisulfate can in turn be heated at a high temperature to release gaseous compounds of sulfur and oxygen as well as water vapor (H_2O).[45] In a sealed receiver, these materials would combine to yield sulfuric acid again. To the author of the "Elucidation," this process may well have seemed a convenient path to the purification of his white spirit and red oil obtained by destructive distillation of vitriol.

The Basilian "Elucidation's" inclusion of two distinct "spirits of mercury" derived from the very different starting materials of crude antimony and vitriol must have been as perplexing to its contemporary readers as it is today. At least eight of Newton's numbered headings in his "Things remarkable in B. Valentines works" concern spirit of mercury, which he identifies in some cases with the antimonal product and sometimes with the vitriolic one. The material was necessary in order to make the universal elixir or philosophers' stone and also for the production of lesser, "particular tinctures" that could

---

[42] Nicolas Lemery, *Traité de l'antimoine* (Paris: Jean Baudot, 1707), 32–37 and 69–73.

[43] Var. 259.11.7v.

[44] Basilius Valentinus, *Last Will and Testament*, 141.

[45] Harvey W. Wiley, *Principles and Practice of Agricultural Analysis* (Easton, PA: Chemical Publishing, 1895), 218.

be employed on specific metals.[46] Although Newton makes no comment as to the confusing state of affairs provided by the "Elucidation," the antimonial and vitriolic varieties of this substance appear in his later treatments of Basilius as well.[47] Indeed, throughout the three decades or more of his chrysopoetic research, Newton would return repeatedly to the use of crude antimony and its products in conjunction with copper and iron vitriol. Because the "Elucidation" never explicitly says that the antimonial spirit of mercury is a vestige of the gold-based processes found in the "Twelve Keys," it appeared to Newton that both antimony and the vitriols of copper and iron were all necessary in order to arrive at the philosophers' stone. The precise way in which these three materials should be used with one another remained a puzzle to him throughout his three decades of laboratory work in chymistry, especially after his intensive study of the Basilius-inspired works of Johann de Monte-Snyders beginning in the 1670s. Yet as we have just seen, the seeds of this riddle were already planted by the young Newton's early exposure to the supposed Benedictine monk and the equally pseudonymous commentator of the "Elucidation."

Newton's Early Chymical Contacts: The Testimony of the Manuscripts

We have now given a brief picture of the origin and early evolution of Newton's chymical interests from his initial exposure to texts in the books of secrets genre, through his acquisition and revising of chymical dictionaries, and up to his first attempts at deciphering the alchemical corpus of Basilius Valentinus. Although Newton had a strong predilection for burying himself in the solitary attempt to decode enigmatic authors like Basilius, his use of manuscripts indicates that he received texts from often unnamed chymical acquaintances. Indeed, the sweeping command that Newton rapidly acquired over the entire literature of chymistry, as well as his growing expertise in chymical experimentation from 1669 onward, both suggest that he must have been in contact with other experts in the field from the earliest phases of his research. But it is remarkable how little he left in terms of concrete evidence identifying his chymical associates in the million or so words that he wrote on the subject. Whether this is because the alchemical Newton dwelt in an ethereal space occupied only by him and the disembodied riddles of the adepts, or whether it is the product of a deliberate discretion on his part, cannot be known. Identifying the circle of Newton's alchemical acquaintances at Cambridge and London therefore presents a serious problem to the historian. But it is not an entirely insoluble one. In order to do proper justice to the evidence, I here present the clues left by Newton's manuscripts and books that strongly implicate particular individuals. Even here, as we will see, there is considerable room for error. We are now in the

---

[46] Var. 259.11.6v–8r. See the headings numbered two, three, nine, ten, eleven, twelve, thirteen, and twenty-six.

[47] As in British Library Add. 44888, at 6r, 7r, and 7v.

ghostly realm of the possible and the contingent, even if this domain possesses varying degrees of palpability.

One of the very few clues of provenance that Newton left among his chymical manuscripts follows a transcript in another hand of the peculiar treatise *Manna*, as found in Keynes 33.[48] Written in an effusively pious style, this little text claims to teach not only the preparation of the philosophers' stone but also its uses in "natural magick." Yet despite the author's protestations, the magic described seems closer to the supernatural realm than the natural one. *Manna* describes a succession of wonders that can be performed with the philosophers' stone, beginning with its ability to represent the six days of creation merely by letting a few drops of the liquefied stone fall into water that has previously been purified by deposition of sediment. First a dark mist will rise up, followed by a separation of light from the darkness; dripping in more of the "blessed Stone" will produce a vision of each day of the creation. After this come directions for producing a simulacrum of the heavens within the alchemical laboratory by melting the seven metals in the order of their corresponding planets and then adding seven drops of the philosophers' stone. As a result of these operations, the room will be bathed in the light of the sun and moon, and the seven planets will appear in the starry firmament, moving in their accustomed courses. As though these marvels were not extravagant enough, the author then advises that some of the philosophers' stone smeared on the alchemist's temples will allow him to attract other adepts by a sort of oneiric telepathy: first the anointee must go to sleep and have a vision, the memory of which will remain on waking, along with the name and address of the sought-for "Good Company." A final wonder leads directly to the realm of the supernatural. Although the author does not reveal the precise method, he says that the stone will allow one "to converse with spirits." Additionally, this "Angelical wisdom" will allow the practitioner to learn "Astronomy, astrology, & al the arts of the mathematitians" without labor or expense. As the author assures us, "nether is schollarship required, it is the Gift of God."[49]

Manna's unceasing appeals to piety and the author's continual invocation of God's love should not blind us to the fact that this text is actually a cynical forgery, most of which the author has cribbed wholesale from Johann Grasseus's *Arca aperta*. In Grasseus's text, however, the sequence of apparitions and claims to long-distance communication are related by a riddling, grizzled, dwarf-like figure who is an obvious narrator of allegories and *aenigmata*.[50] It is not at all clear whether we are meant to take these visionary accounts literally in the *Arca aperta*, whereas *Manna* leaves us with the impression that the pious author has actually experienced them. None of this seems to have aroused any suspicion on Newton's part as to the authenticity of *Manna*: his *Index chemicus* and other manuscripts cite the text as

[48] *Manna* is found in print form in John Frederick Houpreght, *Aurifontina chymica* (London: William Cooper, 1680), 109–43. There it bears the title "Tractatus de Lapide, Manna benedicto, &c."

[49] For more on the putative ability of the philosophers' stone to attract spirits and allow communication between them and the adept, see Principe, *AA*, 194–201.

[50] Grasseus, *Arca arcani*, 336–38.

an authoritative source. In Keynes 33, Newton's only comments have to do with the collation of the text against another manuscript, whose provenance provides him with the occasion to mention his source for the second document. Thus he says the following:

> Here follow several notes & different readings collected out of a M. S. communicated to Mr F. by W. S. 1670, & by Mr F. to me 1675.

Who was this "Mr F." who was responsible for transmitting the variant form of *Manna* to Newton in 1675? The answer is shrouded in multiple ambiguities. Beyond the obvious difficulty of determining surnames from only their initial letter, the abbreviated title "M^r" could in the seventeenth century refer either to "Mister" or to "Master," with quite different meanings. An inspired guess led Dobbs in 1975 to suggest that Mr F. was Ezekiel (or Ezechiel) Foxcroft, a graduate of Eton who subsequently received a master's degree from King's College, Cambridge, and resided there as a fellow from 1652 until his death in 1674 or 1675. Foxcroft in fact translated *The hermetick romance; or, The chymical wedding* of Christian Rosencreutz from German; the translation was published in 1690, long after his death. As Dobbs points out, a late list of alchemical books in Newton's hand ("De scriptoribus chemicis") identifies "Mr F." as Foxcroft.[51] Obviously Newton could have encountered more than one "Mr F." over the fifteen years between the two entries, however. In addition, Karin Figala raised the objection that the Eton College *Registers* clearly state that Foxcroft was dead by 1674, a year before the date of 1675 in which Newton says he received the variant text of *Manna*. Yet the case against Foxcroft as "Mr F." is by no means closed. As Figala herself pointed out, Eton College employed the Old Style, Julian system used then in England, where the New Year began on Lady Day, March 25. Only in the mid-eighteenth century did England switch to the Gregorian calendar, thereby losing eleven days and officially beginning the year on January 1. Hence, if Foxcroft died between January 1 and March 24 in the Gregorian year of 1675, his death would have been recorded as taking place in 1674.[52]

There is further evidence unmentioned by Dobbs that supports a connection between Foxcroft and Newton, but first let us say a few words about the life of the King's fellow. As Dobbs points out, Foxcroft was the son of Elizabeth Foxcroft, who in turn was the sister of Benjamin Whichcote, a member of the group of Cambridge Platonists whose most famous representative was Henry More. Elizabeth Foxcroft was a learned woman who served as the amanuensis or secretary of Lady Anne Conway, the famous focal point of the eponymous "Conway Circle" at Ragley. Lady Conway interacted as a philosopher with intellectuals including More and other Cambridge

---

[51] Dobbs, *FNA*, 112.

[52] There is some confusion in Figala's account, for she writes "Old Style" twice where she means "New Style." See Karin Figala, "Newton as Alchemist," *History of Science* 15 (1977): 102–37, especially 103–4. She corrects her error in Karin Figala, John Harrison, and Ulrich Petzold, "*De Scriptoribus Chemicis*: Sources for the Establishment of Isaac Newton's (Al)chemical Library," in *The Investigation of Difficult Things: Essays on Newton and the History of the Exact Sciences in Honour of D. T. Whiteside*, ed. Alan E. Shapiro and P. M. Harman (Cambridge: Cambridge University Press, 1992), 135–79, see 146.

Platonists, including Ralph Cudworth and John Worthington.[53] Another member of the Conway Circle was Francis Mercurius Van Helmont, the son of the celebrated Flemish chymist Joan Baptista Van Helmont. Ezekiel Foxcroft and a certain "Mr Doyly" were introduced to Francis Mercurius by More on account of their "genius to Chymistry."[54] Perhaps the best known feature of Foxcroft's life, however, is his support for Valentine Greatrakes, the Irish "Stroker" who performed marvelous cures by the application of his hands to the patient.[55] Foxcroft wrote a fourteen-page recollection of Greatrakes's cures for Henry Stubbe's 1666 *Miraculous Conformist*, a work dedicated to vindicating the authenticity of Greatrakes as a healer.[56] Certainly a man with Foxcroft's interests would have found a kindred text in *Manna*.

It is because of his connection with the younger Van Helmont, however, that we definitely know Foxcroft to have shared chymical knowledge with Newton. Through a communication by John Woodward, the famous mineral and fossil collector whose cabinet went on to form the nucleus of the Sedgwick Museum of Cambridge, we receive the information that Foxcroft gave an exotic mineral with renowned chymical properties to Newton. Describing the mineral, Woodward says:

> 'Twas found at —— *in Germany*: brought over by *Fr. M. Van Helmont*, and given as his Father's *Ludus*; to Mr. *Foxcraft*, fellow of King's-College, in *Cambridge*. The latter gave it to Sir *Isaac Newton*, and he to me.[57]

The mineral that Foxcroft gave Newton was a piece of the elder Van Helmont's famous *ludus*, of which two Helmontian samples are still found at the Sedgwick Museum.[58] Van Helmont had subjected this cubic pyritic mineral to various chymical operations that were supposed to render it capable of dissolving bladder stone, a prevalent scourge of the seventeenth century.[59] We know that Newton was keenly interested in Helmontian chymistry, so it is entirely understandable that Foxcroft would have shared this arcanum with the young scholar at some point before the former's untimely passing in 1674/5. All of this certainly increases the plausibility of Dobbs's identification of Foxcroft with the "M[r] F." of 1675, though of course the matter is still not entirely resolved.

In the same breath as his reference to Foxcroft and Van Helmont, More mentions a "Mr Doyly," whom he also introduced to the Flemish savant.[60]

---

[53] Sarah Hutton, *Anne Conway: A Woman Philosopher* (Cambridge: Cambridge University Press, 2004).

[54] A. Rupert Hall, *Henry More and the Scientific Revolution* (Cambridge: Cambridge University Press, 1990), 100.

[55] Peter Elmer, *The Miraculous Conformist: Valentine Greatrakes* (Oxford: Oxford University Press, 2013).

[56] Henry Stubbe, *The Miraculous Conformist* (Oxford: Richard Davis, 1666), 31–44.

[57] John Woodward, "A Catalogue of the Foreign Fossils," in *An Attempt Towards a Natural History of the Fossils of England* (London: F. Fayram, 1729), 8 (separately paginated).

[58] Ana Maria Alfonso-Goldfarb, Márcia Helena Mendes Ferraz, and Piyo M. Rattansi, "Seventeenth-Century 'Treasure' Found in Royal Society Archives: The *Ludus helmontii* and the Stone Disease," *Notes and Records of the Royal Society* 68 (2014): 227–43.

[59] Joan Baptista Van Helmont, "De lithiasi," in *Opuscula medica inaudita* (Amsterdam: Ludovicus Elzevir, 1648), chapter 7.

[60] Henry More to Lady Anne Conway, October 13, 1670, in Sarah Hutton, ed., *The Conway Letters: The Correspondence of Anne, Viscountess Conway, Henry More, and Their Friends, 1642–1684* (Oxford: Clarendon Press, 1992), 323.

This was surely Oliver Doyley or Doiley, a fellow of King's College who received his master's degree there in 1642, and who has previously received no notice as a source for Newton's alchemical reading. He became a senior proctor at King's and belatedly received his doctor of laws degree in 1690, only three years before his death.[61] Doyley is known today for the part he played in the English reception of Spinoza, particularly in sending More's works to the Dutch Remonstrant and anti-Spinozist Philipp van Limborch.[62] But like Foxcroft, he gave a gift to Newton, for on the latter's copy of *The fame and confession of the fraternity of R: C: Commonly, of the Rosie Cross* by Thomas Vaughan is written "Is. Newton. Donum Mᵣᵢ Doyley" (Isaac Newton. Gift of Master Doyley).[63] Thomas Vaughan, twin brother of the renowned Metaphysical poet Henry Vaughan, was a well-known Welsh chymist who published a series of influential pamphlets in the 1650s.[64] Although Newton was not particularly interested in his work, it is certainly significant that Doyley took it on himself to bestow Vaughan's animated description of Rosicrucian chymistry on his Cambridge colleague. The fact that Doyley is sometimes described as a minor member of the "neo-Platonic circle of Cambridge" suggests that both he and Foxcroft represented an undercurrent of interest in chymistry and perhaps Rosicrucianism among the Cambridge Platonists.[65] Yet More himself had engaged in a bitter controversy with Vaughan over the enthusiasm that he detected in the Welshman's chymical writings, and despite a serious attempt to portray More as a possible source of alchemical manuscripts for Newton, Dobbs was unable to provide convincing evidence in support of her view.[66] The examples of Foxcroft and Doyley make it more likely that second-tier members of More's circle rather than the redoubtable Platonist himself were circulating such texts.

A third early source of chymical texts for Newton has also, like Doyley, been overlooked by the previous scholarship. This is surprising, since his initials appear in an early manuscript studied by both Dobbs and Westfall, namely, Keynes 52, which consists mainly of "Sᵣ George Ripley his Epistle to K. Edward unfolded." The transcript of the text in Keynes 52 does not correspond closely to either of the printed versions of this work by Eirenaeus Philalethes, and the manuscript is in Newton's early hand. Although I am less certain than Dobbs and Westfall that Keynes 52 is as early as the 1660s,

[61] John Venn and J. A. Venn, *Alumni Cantabrigienses* (Cambridge: Cambridge University Press, 1922), part 1, vol. 2, p. 63.

[62] Lisa Simonutti, "Reason and Toleration: Henry More and Philip van Limborch," in *Henry More (1614–1687): Tercentennial Studies*, ed. Sarah Hutton (Dordrecht: Kluwer, 1990), 201–17.

[63] Harrison, no. 605. See Harrison, p. 142, where he reproduces Newton's inscription.

[64] For Vaughan, see William R. Newman, "Thomas Vaughan as an Interpreter of Agrippa von Nettesheim," *Ambix* 29 (1982): 125–40; and Alan Rudrum, ed., *The Works of Thomas Vaughan* (Oxford: Oxford University Press, 1984).

[65] Lisa Simonutti, "Spinoza and the English Thinkers: Criticism on Prophecies and Miracles; Blount, Gildon, Earbery," in *Disguised and Overt Spinozism around 1700*, ed. Wiep van Bunge and Wim Klever (Leiden: Brill, 1990), 191–212, see 192.

[66] Dobbs, *FNA*, 112–21. For the debate between More and Vaughan, see Arlene Miller Guinsburg, "Henry More, Thomas Vaughan, and the Late Renaissance Magical Tradition," *Ambix* 27 (1980): 36–58, and the more recent article by Robert Crocker, "Mysticism and Enthusiasm in Henry More," in *Henry More (1614–1687) Tercentenary Studies*, ed. Sarah Hutton (Dordrecht: Kluwer, 1990), 137–56.

Newton's evolving use of chymical symbols indicates that it was probably written by 1673 or early 1674 at the latest.[67] It is clear that Newton had copied a manuscript there that was circulating before the final form of Philalethes's *Exposition upon Sir George Ripley's Epistle to King Edward IV* was printed in 1678.[68] At any rate, near the end of the manuscript one finds three closely related Latin paragraphs that reveal their source. Newton has written "from the papers of Mʳ Sl." (Ex chartis Mʳ Sl.).[69] Alas, this reference is even more ambiguous than "Mʳ F.," for I have found no passage in Newton's chymical corpus that expands the abbreviation. Yet two possibilities immediately come to mind—Hans Sloane and Frederic Slare, the first a future president of the Royal Society and the second an active member of that organization from 1680 until his death in 1727. The first can be excluded easily, even though the overwhelming number of alchemical manuscripts found in the Sloane Collection at the British Library today indicate that Sloane must have had an interest in the subject. Having been born in 1660, however, he was far too young to be the unidentified "Mʳ Sl." The possibility that Slare, on the other hand, might be "Mʳ Sl." requires more thought.

Frederic Slare (1646/47–1727), the son of a German immigrant to England who received his doctor of medicine degree in 1679 and became a Fellow of the Royal Society in 1680, was certainly known to Newton by the first decade of the eighteenth century and probably before. The 1706 *Optice*, Newton's Latin reworking of his *Opticks*, draws heavily on experiments Slare published in the *Philosophical Transactions* of 1694. There Slare describes the explosive reactions produced when oil of caraway, turpentine, and other essential oils are mixed with a "compound" spirit of niter made by adding saltpeter to oil of vitrol (sulfuric acid) and distilling the product. This experiment is accompanied by others that also involve the production of flame and detonations from the combination of two cold ingredients, a phenomenon that Slare attributed to latent fire hiding in the ingredients to be mixed. Newton's 1706 *Query 23* borrows the matters of fact elicited from Slare's dangerous and spectacular experiments without attribution, even employing the same quantities specified by the younger researcher.[70] But this late knowledge of Slare's chymistry does not mean that Newton

[67] Keynes 52 contains eight instances of the unbarred Saturn symbol and no examples of the barred version. Newton had abandoned the unbarred form of the symbol by 1674, as can be determined by examining CU Add. 3975. Disregarding some later additions at the very beginning of the manuscript, the barred Saturn comes into use in CU Add. 3975 only on 43v; that folio also reproduces extracts from David von der Becke's letter to Joel Langelot, a text that appeared in the form of a digest in *Philosophical Transactions* 8 (1673/4). Although von der Becke's Latin *Epistola ad praecellentissimum virum Joelem Langelottum* had been published in Hamburg in 1672, it is likely that Newton's access was through the *Philosophical Transactions*.

[68] Dobbs, *FNA*, 88n153, and 113; Westfall, *NAR*, 286–88. For more on the variations among manuscripts of Philalethes's *Exposition upon Sir George Ripley's Epistle to King Edward IV*, see Ronald Sterne Wilkinson, "Some Bibliographical Puzzles concerning George Starkey," *Ambix* 20 (1973): 235–44. See also Newman *GF*, 263–70, no. 16 and 27.

[69] Keynes 52, 7v.

[70] Frederic Slare, "An Account of Some Experiments Relating to the Production of Fire and Flame, Together with an Explosion; Made by the Mixture of Two Liquors Actually Cold," *Philosophical Transactions* 18 (1694): 201–18. For Newton's borrowings, compare pages 202, 209, 211, and 212 of Slare's article with Newton *Optice* (1706), 324–25.

knew anything of the man or his work by the early 1670s, the time by which he must have recorded his debt to "M^r Sl." In order to judge the likelihood of an earlier encounter between the two men, we must briefly review the early career of Slare.

It is well known that Slare was a chymical assistant to Robert Boyle by the early 1670s, though the precise date at which he undertook this employment is undetermined.[71] Among other things, he would work on Boyle's successful project to prepare phosphorous, which led to public demonstrations of the wonderful material beginning in the late 1670s and no doubt contributed to the choice of Slare as curator of experiments for the Royal Society in 1682/3.[72] But did Slare share his employer's interest in chrysopoeia, and was he the sort of person to transmit Philalethan material to Newton? Slare's publications reveal no obvious linkage to the quest for transmutation, but there is evidence to indicate that he was an active participant in the trading of chymical arcana during the early 1670s. He engaged in an epistolary exchange with G. W. Leibniz during 1673 in which he reports on the activities in London of one Schroeder, possibly the German alchemist and cameralist Wilhelm von Schröder, who spent extensive time in England during this period.[73] In April of the same year, Slare offered chymical secrets to Leibniz, particularly the way of making a "menstruum Stanni," either a corrosive for tin or one in which the dissolved metal was supposed to play a part in acting on other materials. Slare made sure to indicate that he was under no obligation to Boyle for this arcanum, and that he therefore had the "freedome and readinesse" to communicate it.[74] Another letter from Slare reports on the success of Thomas Willis in producing a chymical medicament from amber and sal ammoniac and offers Leibniz access to additional products.[75]

Slare's involvement in the culture of secrets and his proximity to Boyle, the patron of George Starkey and possessor of rare documents pertaining to Philalethes, make the young chymist a reasonable candidate for the "M^r Sl.," who passed on at least one alchemical manuscript to Newton. Moreover, Slare's connections with Boyle and Willis suggest that he was in London during the period when he might have transmitted the antigraph of the Philalethan manuscript Keynes 52 to Newton. As Westfall has pointed out, Newton visited the metropolis at least five times between 1668 and 1677, sometimes resulting in an absence of weeks at a time from Cambridge.[76]

---

[71] Michael Hunter, "Boyle and the Early Royal Society," in *Boyle Studies: Aspects of the Life and Thought of Robert Boyle (1627–91)* (Milton Park, Abingdon: Routledge, 2016), 62n33. See also the entry for Slare by Lawrence Principe in the *Oxford Dictionary of National Biography* (online version, consulted July 23, 2016). I thank Principe for additional suggestions regarding Slare made in private communication.

[72] Marie Boas Hall, "Frederick Slare, F.R.S. (1648–1727)," *Notes and Records of the Royal Society of London* 46 (1992): 23–41, see 25–28.

[73] For Schröder see Rudolf Werner Soukup, *Chemie in Österreich: Von den Anfängen bis zum Ende des 18. Jahrhunderts* (Vienna: Böhlau Verlag, 2007), 451–55. For Schröder's presence in England during this period, see Pamela Smith, *The Business of Alchemy* (Princeton, NJ: Princeton University Press, 1994), 76n67, and 253n20. Evidence of Schröder's relationship with Boyle is documented in Principe, *AA*, 200 and 298.

[74] Frederic Slare to G. W. Leibniz, April 10, 1673, in Gottfried Wilhelm Leibniz, *Sämtliche Schriften und Briefe* (Berlin: Akademie-Verlag, 1988), series 3, vol. 1, p. 80.

[75] Slare to Leibniz, July 17, 1673, in Leibniz, *Sämtliche Schriften und Briefe*, series 3, vol. 1, p. 100.

[76] Westfall, *NAR*, 196n60, and 290.

Hence an encounter with Slare or other London chymists is far from unlikely even before Newton's transfer to London in 1696 as warden of the Mint. Until further knowledge of Slare's activities in the early 1670s emerges, however, the matter must remain conjectural.

Conclusion

The present chapter began by following Newton's chymical interests from his youthful enthusiasm for producing marvels in the tradition of books of secrets and texts on natural magic up to the early years at Trinity College, where he deepened his knowledge of the subject by copying and compiling chymical dictionaries such as the anonymous *Treatise of Chymistry*. Contrary to the prevailing scholarly view, it was only after this initial, self-directed exposure to the literature of chymistry that Newton began reading Robert Boyle's natural philosophy, where contemporary chymical doctrines were integrated with the mechanical philosophy. He was already copying out chymical literature even before his exposure to Boyle's work in the field. The practical chymistry of the neophyte alchemist in the mid-1660s should be viewed as a more mature version of the same interest in the wonders of art and nature that drove his initial reading of authors like Bate and Wilkins, not as a radical departure brought on by his exposure to the mechanical philosophy. This discovery has important implications, for it means that we can no longer say with confidence that the young Newton made any passage at all from "sober chemistry" to the *arcana majora* of alchemy. It is a small step indeed from the perpetual lamps and self-moving compounds of Wilkins to the sophic mercury and the philosophers' stone, and Newton's interest in the *Treatise of Chymistry* may well have stemmed from the desire to enter the ranks of the adepts. Certainly this was his goal by the time of his first attempts to decode the corpus of Basilius Valentinus at some point before 1669. Yet Newton's early interest in chymistry was not limited to the production of effects, important as that was. Throughout his career as an experimental scientist, he would pursue both the practical fruits of chymistry and the implications of the discipline for natural philosophy more broadly; these were two distinct though interrelated projects. It is true that most of the gargantuan experimental effort Newton devoted to the subject from 1669 up to his departure for London to become warden of the Mint in 1696 focused on the practical attempt to arrive at the techniques and reagents required for chrysopoeia; yet even after his retirement from active experimental practice, he would continue using chymistry in the service of natural philosophy, above all in *Query 31* of the 1717 *Opticks*. Nowhere is the integration of Newton's alchemy with his better known scientific discoveries more evident than in his work on optics. In the next chapter we will see how Newton repurposed a classic chymical technique of analysis and resynthesis in order to demonstrate the composite nature of white light, thereby overturning some two thousand years of optical doctrine.

SIX

Optics and Matter: Newton, Boyle, and Scholastic Mixture Theory

Although Boyle's work by no means served as Newton's introduction to chymistry, his youthful notebook *Certain Philosophical Questions* reveals the striking influence that the author of *The Sceptical Chymist* had on the impressionable undergraduate. Newton devoured Boyle's *Spring of the Air* (1660), *Experiments Touching Colours* (1664), *New Experiments Touching Cold* (1665), and probably his *Defence of the Doctrine Touching the Spring and Weight of the Air* (1662), recording his reaction to these texts in the student notebook. Of these texts it is only *Experiments Touching Colours* that is of concern to us, however, in part because that work deals at length with chymistry, especially in the context of producing and destroying colors. Indeed, the book contains some of Boyle's most sustained discussions of color indicators and uses them to distinguish matter into the different classes of "alcalizate salts" (mostly carbonates), "urinous salts" (primarily ammonia compounds), and "acid salts" (our acids, both strong and weak).[1] Newton summarizes much of this material on "tinctures" or colors and the way of changing their appearance by means of these classes of salts in the last folios of *Certain Philosophical Questions*. This section takes up part of Newton's famous early experimentation with prisms, showing a thematic relationship between his chymistry and optics.[2] But the link between the two fields goes much further than the mere fact that both prisms (or lenses) and chymical operations can make things take on different colors. In reality, a consideration of Newton's early optics together with his chymistry opens the door to a rich and understudied area with broad implications for the seventeenth-century's break with scholastic natural philosophy in general. As we will now show, the domains of chymistry and optics in mutual

---

[1] Boyle, *Experiments Touching Colours*, in *Works*, 109, 125, 129–30, 154–57; 1664, 205–6, 246–48, 257–59, 313–21.

[2] Newton, *CPQ*, 453–62. This section is a continuation of Newton's famous entry "Of Colours," which occurs earlier in the notebook. See p. 453, where he indicates the point earlier in the text that this part picks up ("vide pag 69").

combination formed one of the principal tools by which Newton was able to effect that rupture.

In the years immediately preceding Newton's use of prisms to demonstrate that sunlight is actually a composition of heterogeneous spectral rays rather than being perfectly homogeneous, a sophisticated methodology based on the analysis and resynthesis of gross matter had entered the province of natural philosophy and decisively shown that what we today call "chemical compounds" were also made up of heterogeneous components. Analysis and synthesis had long been known to experimentally minded alchemists even though their real significance was largely lost on academic natural philosophers until the seventeenth century. An extensive alchemical tradition extending from the High Middle Ages up to Boyle's immediate predecessors had long been using the analytic retrievability of the constituents of compounds to argue for the permanence of the ingredients that went into them. Boyle was the direct heir of this lengthy alchemical tradition, especially in his use of the atomistic writings of the Wittenberg medical professor and chymist Daniel Sennert.[3] It is well known that the young Newton was heavily influenced by Boyle, but here I argue for a deeper significance to his intellectual debt: there was a direct transfer from Boyle's work on chymical analysis and synthesis to the optical analyses and syntheses that formed the bases of Newton's early work with light and colors.[4]

The peripatetic theory of "perfect mixture," according to which the process of mixing produces a homogeneous material product in which the ingredients no longer remain as such, had held sway among Aristotle commentators for almost two millennia when Newton was born in 1642/3.[5] Although a widespread school of alchemical argumentation had long opposed the strong form of the theory, the fruits of this tradition only entered the mainstream of English natural philosophy in the early works of Boyle. During Newton's years as an undergraduate in the early 1660s, Boyle employed existing alchemical arguments to wage a successful war against the Aristotelian theory of mixture, culminating in a series of publications that appeared almost exactly at the time when Newton first argued that white light too was heterogeneous. Less than a decade after Boyle's first publications on the corpuscular nature of matter, Newton arrived at his own theory, also based on implicitly corpuscular presuppositions, that white light is composed of immutable rays of differing refrangibility.

---

[3] For Sennert and his important role in the development of Boyle's corpuscular philosophy, see Newman, *AA*, especially 85–189.

[4] This is not to say that no other scholars have noticed the parallelism between Newton's analyses and syntheses of white light and chemical analysis and synthesis. See for example Noretta Koertge, "Analysis as a Method of Discovery during the Scientific Revolution," in *Scientific Discovery, Logic, and Rationality*, ed. Thomas Nickles (Dordrecht: Reidel, 1980), 139–57, see 151–52.

[5] It is important to distinguish between the views of Aristotle himself on mixture and the tradition inaugurated by Thomas Aquinas in the thirteenth century. Although Aristotle believed that "perfect mixture" implied homogeneity, he did not deny that the ingredients of such a mixture could be regained. Thomas and his followers in this matter, such as John Duns Scotus, held that the forms of the initial ingredients were destroyed by the process of mixture itself; hence, the ingredients as such (i.e., numerically identical ingredients) could not be recaptured. See Newman, *AA*, 23–44.

The congruence of these discoveries is a striking fact, but in and of itself, their timing could of course be coincidence. In order to demonstrate that more than mere simultaneity is involved, we must therefore explore the similarities between Newton's demonstrations that white light is a mixture of unchanged colorfacient rays and Boyle's demonstrations that seemingly homogeneous mixtures are really composed of unchanged corpuscles. Restricting ourselves here to cases where Newton was explicitly borrowing from Boyle's written work, we can provide linguistic evidence that the young savant was in fact applying Boylean terminology about chymical compounds to the mixture of light. Although Newton employed this terminology in a cautious and heuristic fashion, it provides evidence, nonetheless, of his debt to the newly triumphant chymical corpuscularism of the seventeenth century.

Alchemy versus Perfect Mixture

As we discussed, chymistry in the seventeenth century comprehended a wide and diverse variety of activities ranging from such technological pursuits as the making of alcoholic beverages, pigments, and salts, to the manufacture of drugs and the performing of iatrochemical cures, and finally, to the attempted transmutation of metals. One thing that characterized the theory espoused by almost all alchemists from the Middle Ages onward, however, was a belief that the metals were composed of two principles, mercury and sulfur, to which Paracelsus in the early sixteenth century added the third principle, salt. By and large, alchemists had long believed that analytical processes such as calcination, sublimation, and dissolution in solvents could resolve minerals and metals into their preexistent components, namely, their sulfur and mercury, or after Paracelsus, their mercury, sulfur, and salt.

This traditional alchemical emphasis on the analytic retrievability of the principles put alchemists at odds with a range of scholastic positions arguing for the impossibility of separating the ingredients from a genuine mixture. In a word, the most widespread interpretations of Aristotelian matter theory in this period stated that it was not possible to reisolate the initial constituents of a homogeneous substance once those constituents had combined to form a mixture, and such homogeneous "mixts" were widely thought to include materials as commonplace as metals, flesh, wood, milk, and wine. During the Late Middle Ages and the early modern period, this theory came increasingly into conflict with a host of empirical examples supplied above all by chymistry, a field where corpuscular theories of matter had been circulating in the Latin West since the thirteenth century. Indeed, alchemical writers were the first to provide matter theories of any sort, including optical theories, based on experimental demonstrations of paired analysis and resynthesis.[6] It was no accident that Boyle, the famous seventeenth-century popularizer of the mechanical philosophy and debunker of Aristotelian

[6] Newman, *AA*, 23–44.

mixture, was himself a chymist. He was, in fact, giving further articulation and modifications to the views of alchemists as expressed over a period of several hundred years. To make matters short, it was the field of chymistry that supplied Boyle's primary ammunition against early modern scholastic matter theory as taught in the universities. Chymistry provided a way out of the impasse resulting from a strict interpretation of substance and mixture first promulgated by Thomas Aquinas and later adopted by other scholastic schools that had forbidden the persistence and retrievability of ingredients within a mixture.[7]

The degree to which early modern scholasticism was committed to the position that ingredients could not be retrieved from a genuine mixture has been largely overlooked in the modern literature on the Scientific Revolution. By a "genuine mixture," I refer to the Aristotelian concept of *mixis*—an absolutely homogeneous combination of ingredients, often called a "perfect mixture" by the scholastics. In order to understand the meaning of Aristotelian *mixis*, the contemporary reader must make a conscious effort to forget the terminology of modern chemistry, which refers to mechanical juxtapositions of particles as "mixtures" and distinguishes such uncombined ingredients from those that have entered into a "chemical compound" joined by "chemical bonds." The language chemists employ today reverses the terminology of Aristotle, for whom "mixture" meant a homogeneous combining of ingredients and "compound" or "composition" meant a mere juxtaposition of uncombined parts. Aristotle had claimed in Book I, Chapter 10 (328a10–12) of his *De generatione et corruptione* that genuine *mixis* occurred only when the ingredients of a mixture acted on one another to produce a state of absolute homogeneity. Otherwise, he asserted, a sufficiently keen-sighted person, such as the classical hero Lynceus, would be able to see the heterogeneous particles that made up what had seemed to be a genuinely uniform substance. Aristotle's predecessor Empedocles had espoused precisely the sort of theory that Aristotle was here debunking. Empedocles had maintained a century before Aristotle that the four elements were composed at the microlevel of immutable particles, which lay side by side to form compounds (what chemists today would call "mixtures"). Aristotle argued that such corpuscles could only form an apparent mixture, like wheat and barley in a jar; he dubbed such illusory mixture synthesis—literally "setting-together." Aristotle himself did not believe that the ingredients of a genuine mixture were incapable of retrieval. At *De generatione et corruptione* I 10 327b27–29 he argues the contrary, and his ancient followers, especially John Philoponus, spoke of separating mixtures by means of oiled sponges, river lettuce, and the like.[8] Boyle was not responding to

[7] Newman, *AA*, 38–43, 85–125.

[8] Aristotle, *De generatione et corruptione* at I 10 327b27–29, that "it is clear that the ingredients of a mixture first come together after having been separate and can be separated again" (in the translation of E. S. Forster). For Philoponos, see Frans A. J. De Haas, "Mixture in Philoponus: An Encounter with a Third Kind of Potentiality," in *The Commentary Tradition on Aristotle's* De generatione et corruptione, ed. J.M.M.H. Thijssen and H.A.G. Braakhuis (Turnhout: Brepols, 1999), 21–46, especially 26n22.

ancient commentators, however, but rather to the medieval scholastics and their early modern heirs, who had their own views on the matter.

The Jesuits, to name one early modern current, had adopted Thomas Aquinas as their master in theology, at the urging of Roberto Bellarmino in the 1590s.[9] Hence it is no surprise to find that the great Jesuit *De generatione et corruptione* commentaries, such as those of Franciscus Toletus and the Coimbrans, assume an explicitly Thomistic position on the subject of mixture. Even before the Jesuits appeared on the scene, the Thomistic view had become, as the historian of scholastic natural philosophy Anneliese Maier argued, the dominant view among scholastics.[10] Like all scholastic Aristotelians, Thomas viewed matter as consisting of the four elements, fire, air, water, and earth. These in turn contained four "primary qualities"—hot and dry in fire, wet and hot in air, cold and wet in water, and dry and cold in earth. Although the pairs of these qualities along with an undifferentiated "prime matter" (*materia prima*) constituted the fundamental stage of material analysis, the primary qualities were not immutable, for the hot could pass away and be replaced by cold, just as the wet could pass away and be replaced by dry. This opened the door to the possibility of elemental transmutation; if, for example, the hot and dry in a sample of fire were replaced by cold and wet, that portion of fire would be transmuted into water.[11]

But the situation was still more complicated than this, for Thomas's hylomorphism insisted that Aristotelian *mixis*, the one type of mixture that led to a genuinely homogeneous product, could only occur if a new substantial form, called the "form of the mixture" (*forma mixti*), was imposed on the four elements.[12] This process occurred in a well-defined series of steps. First, the four primary qualities of the elements produced, as a result of their mutual action and passion, a single medial quality preserving something of the extremes; this medial quality then provided the disposition necessary for the induction of the new substantial form, the form of the mixture. Yet in such a case, Thomas insisted, the imposition of the new form of the mixture meant that the four antecedent elements would be destroyed—the generation of the one entailed the corruption of the other. All that remained of the fire, air, water, and earth would be the primary qualities, the hot, cold, wet, and dry that had been paired within the elements before their destruction, and which were somehow responsible for the dispositive medial quality that

[9] Sylvain Matton, "Les théologiens de la Compagnie de Jésus et l'alchimie," in *Aspects de la tradition alchimique au XVII* siècle, ed. Frank Greiner (Paris: S.É.H.A., 1998), 383–501, see 383.

[10] Anneliese Maier, *An Der Grenze von Scholastik und Naturwissenschaft*, 2nd ed. (Roma: Edizioni di Storia e Letteratura, 1952), 89.

[11] Aristotle points out that this process has a cyclical character: if the dry in fire passes away and is replaced by wet, the fire will become air; if the hot in air is replaced by cold, the air will become water; if the wet in water is replaced by dry, the water will become earth; and if the cold in earth is replaced by hot, the earth will become fire. See Aristotle, *De generatione et corruptione* II 3–4 330a30–332a2, especially II 4 331b2–4.

[12] Maier, *An Der Grenze*, 31–35. A much inferior study to Maier's, though still useful on certain points, is Xaver Pfeifer, *Die Controverse über das Beharren der Elemente in den Verbindungen von Aristoteles bis zur Gegenwart, Programm zum Schlusse des Studienjahrs 1878/79* (Dillingen: Adalbert Kolb, 1879). Thomas's discussion of mixture may be found in Thomas Aquinas, *De mixtione elementorum* in *Sancti Thomae de Aquino opera omnia* (Rome: Editori di San Tommaso, 1976), 43: 127–30. As Maier points out, the corresponding section of Thomas's *De generatione et corruptione* commentary is interpolated. See Maier, *An Der Grenze*, 31–32.

prepared the way for the form of the mixture. Even here it is not clear that the four qualities that remained were the original ones underlying the elements or rather similar ones that had been newly generated, for in general Thomas insisted that the primary qualities were accidents of the substantial form. If the substantial form itself had been newly introduced to the ingredients, then how could its accidents be the same ones that had been present before in the preexistent elements (which had now been destroyed)? As for the elements themselves, they were now present within the mixture only *in virtute* or *virtualiter*—"virtually"—as a result of the said primary qualities.[13]

To employ a distinction made in many later scholastic treatments of mixture (though not in that of Thomas), one could not get the original ingredients back out again in number (*in numero*), since they had been destroyed by the very act of mixing. If one could perhaps retrieve fire, air, water, and earth that were the same as the original elements in species (*in specie*), there was no guarantee that they would return in the same relative quantities in which they had entered the mixture.[14] After all, the original fire, air, water, and earth had been destroyed by the process of mixture, and there was no reason to think that the primary qualities would reassemble into exactly the same pairings in proportions identical to those that they originally possessed. Hence the empirical correlation between input and output had been severed—mixture was effectively a black box linking substances with no shared material identity.

Newton, Boyle, and the Chymical Tradition of the Reduction to the Pristine State

With this overview of scholastic mixture theory at our disposal, we now turn to the period from about 1664 up to the publication and responses to Newton's famous *New Theory about Light and Colors* published by Henry Oldenburg in the *Philosophical Transactions of the Royal Society* in 1672. As we will see, chymistry provided the young Newton with an important heuristic in his unfolding theory that white light is a heterogeneous mixture composed of immutable spectral colors. I do not mean to say that Newton found anything approximating this optical theory in his chymical sources,

---

[13] Maier, *An der Grenze*, 33–35. Thomas's position on mixture fit very nicely with his view that every substance could have only one substantial form (the so-called unity of forms theory). Nonetheless, the "unity of forms" theory did not follow necessarily from Thomas's theory of mixture, since many scholastic authors believed that one substantial form could be subordinated to another, even in a single substance. Those authors who maintained a plurality of substantial forms in a given substance often invoked the human body and soul as a case of such subordination. Although the soul was the substantial form of man per se, the body had its own subordinate form, which accounted for its ability to resist decomposition into the elements for some time after death. See Roberto Zavalloni, O.F.M., *Richard de Mediavilla et la controverse sur la pluralité des formes: Textes inédits et étude critique*, Philosophes medievaux 2 (Louvain: Éditions de l'institut supérieur de philosophie, 1951), 303–81.

[14] A good account of the *in numero/in specie* distinction is found in the *De generatione et corruptione* commentary of Franciscus Toletus. Toletus makes it clear that the scholastic distinction hinged on the absence or presence of substantial corruption. See Toletus, *Commentaria, una cum quaestionibus, in duos libros Aristotelis, de generatione et coruptione* (Venice: Juntas, 1603), fol. 93v.

or even that the earliest phases of his discovery owed a significant debt to chymistry. To the contrary, Newton's early and serendipitous discovery that different colors are produced by rays of different refrangibility owes no obvious debt to chymical theory or practice. What is incontestable, however, is that the earliest descriptions of Newton's theory occur imbedded among the extensive notes on chymistry taken by him from Boyle's *Experiments Touching Colours*, in *Certain Philosophical Questions*, and in his more developed treatise found in Cambridge University Additional MS 3975, probably from around 1666 or 1667. Newton labeled both of these short treatises "Of Colours." For the sake of simplicity, I will call the version in *Certain Philosophical Questions* "Of Colours I" and the version in CU Add. 3975 "Of Colours II."[15] In *Experiments Touching Colours*, Boyle tentatively proposes a theory that white light is modified by reflection and refraction to produce colors, performs experiments in color mixing by projecting one prism's spectrum on that of another, and advises future researchers to carry out more extensive experiments with prisms.[16] While Boyle does not arrive at anything resembling Newton's bold claim that white light is actually a mixture of unaltered heterogeneous colors, the bulk of *Experiments Touching Colours* is in fact taken up with chymical processes that lead to color change as a result of minute corpuscles aggregating with one another and separating from one another. Boyle's other treatises of the period, such as *Certain Physiological Essays* (1661) and *The Origin of Forms and Qualities* (1666), employ extensive use of analysis and resynthesis to demonstrate the corpuscular nature of matter, a feature that is less prominent in *Experiments Touching Colours*.

It is further significant that Newton's early optical theory underwent major changes between "Of Colours I" and "Of Colours II." In the first treatise, Newton relied solely on observations of the colors produced when one looks at bodies through a prism. He interprets the differing refrangibility of the red and blue rays as being due to a difference in the speed of the light corpuscles. Furthermore, in "Of Colours I" he thinks that this speed can change, so that color mutation remains a possibility. All of this has changed by the time of "Of Colours II." In this treatise, Newton has begun experimenting with sunlight projected through prisms. He has observed the oblong shape of a beam projected by a prism on a wall about twenty-one feet distant, he has devised several experiments for resynthesizing the white light divided by the prism, and he has observed that a body of a given color will appear brighter when illuminated by a ray of the same color, whereas a body of a different color will appear fainter. Most importantly, in "Of Colours II" there is no more discussion of light corpuscles that change their speed, and

[15] Both "Of Colours I" and "Of Colours II" have been edited with valuable commentary in McGuire and Martin Tamny, *Certain Philosophical Questions*, 431–42 and 466–89. The reader who wishes to see the chymical text in which "Of Colours II" was imbedded by Newton, however, will have to consult the edition of CU Add. 3975 in *CIN*.

[16] For Boyle's general influence on Newton's optics, see Alan E. Shapiro, *The Optical Papers of Isaac Newton*, vol. 1, *The Optical Lectures, 1670–1672* (Cambridge: Cambridge University Press, 1984), 4–7, and Shapiro, *Fits, Passions, and Paroxysms* (Cambridge: Cambridge University Press, 1993), 99–102, 120.

indeed the evidence is that Newton had by this time come to the view that colors are immutable, though without stating this as a formal principle.[17]

There is another very significant feature of CU Add. 3975, the manuscript in which "Of Colours II" is found. This manuscript, unlike *Certain Philosophical Questions*, contains important notes explicitly taken from Boyle's *Certain Physiological Essays* and *The Origin of Forms and Qualities*, works in which Boyle described chymical analysis and synthesis at great length. Is it not then possible that Newton's research on light, which he considered from the time of his earliest recorded optical experiments to consist of material globules, transported some of Boyle's matter theory into the realm of optics? Can one perhaps even argue that Boyle's treatment of chymical analysis and synthesis encouraged Newton to move from a semi-Cartesian view of light corpuscles that can change their speed and hence the color that they produce to his mature position that colors are immutable, like the corpuscles arrived at by chymical analysis?

These questions are particularly significant in the light of research over the last two decades, which has revealed that Boyle was not so much the father of modern chemistry, as he is often depicted, as he was a committed Helmontian chymist with a powerful and lifelong interest in chrysopoeia, the transmutation of base metals into precious ones.[18] Additionally, new work has revealed alchemical sources behind Boyle's famous corpuscular theory of matter. According to Boyle's corpuscular theory, particles of the smallest sort called *prima naturalia* combine to form larger aggregate corpuscles called *prima mixta or prima mista*, "primary clusters," which can in turn recombine to form still larger clusters called "decompounded" or twice compounded particles—resembling what we would today call molecules. The odd term "decompounded"—having the sense of "further compounded" rather than "uncompounded"—is borrowed via Latin from the Greek grammatical term *parasynthetos*, which means "formed or derived from a compound word." Hence "to decompound" meant "to compound further," as in the case where the preposition *super* is added to the Latin infinitive *exaltare* (which already contains the preposition *ex*).[19] Boyle's hierarchical matter theory was heavily dependent on traditional alchemical theories with roots that lie in the medieval alchemical author Geber, who conceived of elementary corpuscles combining to form larger particles of sulfur and mercury, which in turn recombined to make up the minute corpuscles of metals per se.

These theories were transmitted to Boyle by a variety of sources, but chief among them seems to have been the German academic Daniel Sennert, who was the direct source for Boyle's term *prima mixta*.[20] Sennert embedded his corpuscularism within a sustained attack on the Aristotelian theory of perfect mixture, as it had been transmitted by the medieval and early

---

[17] A detailed discussion of Newton's evolving optical theory between "Of Colours I" and "Of Colours II" may be found in Newton, *CPQ*, 241–74.

[18] Newman and Principe, *ATF*. See also Principe, *AA*.

[19] See the online *OED*, consulted June 10, 2016. "Superexaltare" means "to exalt further."

[20] William R. Newman, "The Alchemical Sources of Robert Boyle's Corpuscular Philosophy," *Annals of Science* 53 (1996): 567–85, see 583.

modern scholastics. At this point, chymistry entered the picture in a highly significant way. As Sennert and Boyle argued, some of Aristotle's so-called perfect mixtures—such as blood and wine—could be subjected to distillation to yield their components. Even more importantly, the chymist could himself make seemingly perfect mixtures by dissolving metals in acid—after the violent dissolution of the metal, the perfectly clear solution could even be poured through filter paper without leaving any residue. Surely such a mixture of metal and acid was at least as homogeneous as Aristotle's examples of wine and blood. And yet, after dissolving his metal in acid, the chymist could then precipitate the metal out unchanged merely by adding an alkali, such as salt of tartar (potassium carbonate). These "reductions to the pristine state" provided direct evidence against the Thomistic claim that the ingredients of a mixture could not be recaptured intact—the obvious conclusion to draw was that the bits of metal had simply been hidden within the solution all along in the form of indissoluble corpuscles or atoms. Hence the homogeneity of a host of seemingly uniform material substances was called into question by means of chymical experimentation. Indeed, it is no exaggeration to say that defeating the Aristotelian theory of perfect mixture in favor of corpuscularism with its emphasis on heterogeneity was an idée fixe with Boyle, which occupied an important place in his mechanical philosophy from his earliest works on natural science until his death in 1691. It is highly significant that Boyle's most important works debunking scholastic mixture theory, the *Certain Physiological Essays* of 1661, *The Sceptical Chymist* of the same year, and *The Origin of Forms and Qualities* of 1666, were all in print in the years when Newton was formulating his theory that white light is a compound of immutable spectral colors. In fact, the first and last of these three works definitely served as sources for Newton in CU Add. 3975, the manuscript that contains the important second draft of his early treatise "Of Colours."

Indeed, CU Add. 3975 contains an extract that recounts one of Boyle's most important reductions to the pristine state, where Boyle explicitly uses it to criticize the Thomistic theory of mixture.[21] The passage describes the dissolution of camphor in nitric or sulfuric acid. If sulfuric acid is used, the camphor forms a deep reddish solution and loses its odor. Hence the camphor becomes unrecognizable as camphor and seems to be perfectly mixed in the solution. But the mere addition of water will cause the camphor to return to its former state, including the reacquisition of its powerful scent. Boyle points out that this experiment throws considerable doubt on the scholastic theory that mixture entailed the loss of the initial ingredients. As he puts it:

> This Experiment may serve to countenance what we elsewhere argue against the Schools, touching the Controversie about Mistion. For whereas though some of them dissent, yet most of them maintain, that the Elements alwaies loose their Forms in the mix'd Bodies they constitute; and though if they had dexterously propos'd their Opinion, and limited

---

[21] Newton, CU Add. 3975, 32v–33r, from the Chymistry of Isaac Newton, http://webapp1.dlib.indiana.edu/newton/mss/dipl/ALCH00110/.

their Assertions to some cases, perhaps the Doctrine might be tolerated: yet since they are wont to propose it crudely and universally, I cannot but take notice, how little tis favour'd by this Experiment; wherein even a mix'd Body (for such is Camphire) doth, in a further mistion, retain its Form and Nature, and may be immediately so divorced from the Body, to which it was united, as to turn, in a trice, to the manifest Exercise of its former Qualities.[22]

Boyle views the camphor as having remained intact within the sulfuric acid, which merely caused it to alter its texture. The addition of water weakened the sulfuric acid, making it release the camphor, on which the latter regained its usual qualities of whiteness and penetrating smell. Let us step back for a moment and consider the general form of Boyle's demonstration. First, one substance is mixed with another so that it loses its perceptible qualities—that is, the camphor loses its whiteness and its smell when mixed with the sulfuric acid. Then the camphor is reduced to its pristine state by adding water, whereon it regains its original qualities. To Boyle, this demonstrates that the camphor was present all along in the mixture, in the form of intact corpuscles. The mixture, in Aristotelian terms, was not a true mixture at all, but a compounding or juxtaposition of corpuscles.

There are many interesting features to Boyle's argument, and several that are pertinent to Newton. But for the moment I want to focus on Boyle's assumption that just because the camphor can be retrieved intact, it follows that the camphor was present in unaltered form all along in the sulfuric acid. Nowhere does Boyle explain why this must be the case. Why could the camphor not be regenerated from its ingredients rather than lurking in the mixture all along, in unchanged form? One needs no reminder of the fact that similar problems dogged Newton in his oft-repeated claim that white light consisted of unaltered and immutable colorfacient rays, which were merely separated by the prism on account of their unequal refrangibility. This problem was already raised by Robert Hooke in a letter to Oldenburg written only a week after Newton first presented his *New Theory about Light and Colours* in February 1672. Hooke argues that there is no more reason to suppose that white light consists of immutable colorfacient rays than there is to suppose that the sounds made by an organ already exist in the air of its bellows.[23] Even though Newton had described the recombination of spectral colors to regain the white light from which they had been divided, Hooke felt no compulsion to accept that the rays responsible for the colors retained their integrity within the seemingly homogeneous white light before its refraction by a prism. Instead, he argued that the colors could have been manufactured by the initial act of refraction, as was the case in his own theory.

This makes one wonder why Hooke did not raise similar objections about Boyle's reductions of metals and camphor to the pristine state. Boyle's

[22] Boyle, *Origin of Forms and Qualities*, in *Works*, 5: 396.
[23] Hooke to Oldenburg, responding to Newton's New Theory, February 15, 1671/72, in Newton, *Corr.*, vol. 1, letter 44, p. 111.

arguments for the permanence of metals and of camphor in acid solutions were structurally identical to those of Newton for the persistence of colorfacient rays in white light. In each case analysis provided evidence for the persistence of the ingredients within a mixture. Shouldn't Hooke have evinced the same skepticism toward Boyle's demonstrations that he did toward Newton's? Whatever Hooke's position should have been, the reality is that he did not doubt Boyle's claims about the persistence of ingredients dissolved in powerful acids. Hooke's 1665 *Micrographia* is replete with comments about the particles of metals that he believes to remain intact in acid solutions, even though they are disguised within the liquid until they are precipitated. Hooke in fact goes so far as to argue that because the compounds crystallized out of metallic solutions are transparent, the individual metallic particles themselves must therefore be transparent.[24] So why, then, did Hooke and others give analysis and synthesis such credence for determining the nature of the ingredients of a mixture in the case of material bodies and yet deny its validity in the case of light?[25]

One could perhaps argue that the phenomena themselves were much better known in the case of metals than in that of light. Every metallurgist knew that one can recapture the dissolved metals from acids unchanged, but phenomena such as the elongated dispersion of a projected spectrum or the resynthesis of white light from spectral colors were, to put it mildly, not widely known before Newton (if known at all). Nonetheless, the commonplace nature of acid solutions does not in itself address the issue. Even if one knew that the metal could always be regained intact from the solution, it did not automatically follow that the metal was in the solution all along rather than being regenerated from more primitive ingredients. Although one could detect the bitter taste of dissolved silver or the blue color of the solution that it typically made, the only way one had of knowing that these properties were ordinarily associated with silver was by comparing the solution either with the initial silver before it was dissolved or with the silver precipitated out of the solution. The properties of silver dissolved in acid are no more properties of ordinary undissolved silver than the spectral colors are perceptible properties of unrefracted white light.

In their argument that metals and other substances retained their nature in compounds and solutions, however, Boyle and his predecessor Sennert had one great advantage over Newton. They were arguing against scholastic authors who had accepted as a matter of faith that the ingredients had been destroyed in the process of mixing them. Hence it was possible to turn the scholastics' own arguments against them. How could one reasonably argue

---

[24] Robert Hooke, *Micrographia* (London: Royal Society, 1665), c[2v] (where Hooke discusses the taste of metals dissolved in acids), and 72–73 (where Hooke argues that the particles of metals are transparent since their solutions and crystals are transparent).

[25] At first face one might suppose that Hooke could have argued for a significant difference between the cases of metals and light in that an acid solution holds a fixed quantity of metal, whereas light is passing through a prism in a continuous stream. But the objection would easily have been countered by Newton's experimental demonstrations even as early as those in "Of Colours II," where the very same white light that is analyzed into spectral colors by an array of three prisms is recombined merely by shining the rays of respective prisms on the same section of a wall. See CU Add. 3975, 7v.

that the dissolved silver had been regenerated de novo by the mere addition of potassium carbonate if one was committed to an Aristotelian theory that all metals had to be generated out of fumes beneath the surface of the earth? The scholastic authors would have had to abandon one important peripatetic theory in order to accommodate the other. And furthermore, if potassium carbonate could generate silver out of a silver solution, why could it not generate silver out of a solution of dissolved copper or iron? For that matter, since the initial ingredients had been destroyed, why should a powdered metal emerge from the solution instead of aardvarks or artichokes?

Thomistic authors had no satisfactory answers to these or a number of other, more technical objections raised by the reduction to the pristine state.[26] This was entirely unlike the situation with white light, where Newton's experiments with analysis and synthesis had essentially no precedent. Although Aristotle had held a mutation theory of color, of course, there was no preexisting body of scholastic literature arguing against the persistence of the colors in white light because it had not occurred to scholastic authors that white light was a mixture, homogeneous or otherwise. In addition, Newton's own strong claims made it possible for his main opponents, such as Hooke, to shift the burden of proof onto him. Hooke was particularly adept at this, for he was content to call his own theory of color a hypothesis so long as Newton would do the same for his. Here Newton balked, however, for he believed that he had proven beyond any doubt that white light is composed of colorfacient rays that remain unaltered in the mixture. Although Newton acknowledged that he could not prove the corpuscular nature of light, which was hypothetical, he asserted that he could prove with mathematical certainty that white light contained the spectral rays *in actu*. How did Newton go about doing this? Once again he turned to chymistry, but to a slightly different type of experiment from that of the reduction to the pristine state. The reduction to the pristine state usually proceeded by first synthesizing a seemingly perfect mixture (such as that of silver and nitric acid) and then isolating one of its components by means of analysis (as in the reduction of silver by means of salt of tartar). Newton, however, would follow another type of chymical demonstration that inverted this order by starting with analysis and then passing to resynthesis. Let us begin with Newton's analysis of white light.

Newton's Resynthesis of White Light and Chymical Redintegration

Newton's 1672 *New Theory about Light and Colours* is famous for its inclusion of the *experimentum crucis*, the experiment using two prisms with two pierced boards between them to demonstrate that the rays producing individual spectral colors are always refracted at the same angle (figure 6.1).

[26] For these objections, see Newman, *AA*, 106–23. The precipitate would actually be silver carbonate rather than powdered metallic silver, but since silver carbonate reduces to silver on simple heating, Sennert reasonably supposed that the precipitate was merely finely divided silver.

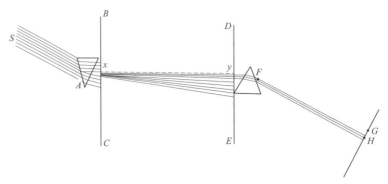

FIGURE 6.1. Newton's *experimentum crucis* from his second letter to Pardies (from Newton, *Corr.*, vol. 1).

The unequal yet fixed refrangibility of the spectral rays led Newton to the claim, as he puts it, that "the species of colour, and degree of Refrangibility proper to any particular sort of Rays, is not mutable by Refraction, nor by Reflection from natural bodies, nor by any other cause, that I could yet observe."[27] A great deal has been written about the *experimentum crucis*, but what I want to focus on here is another experiment that appears at the end of the *New Theory*. There Newton advises that sunlight be passed through a single prism so that the oblong spectrum is projected on the opposite wall. After one has observed the spectrum, a lens is interposed between the prism and the wall, so that the refraction induced by the prism is reversed. The result is that the spectral colors recombine to form white light again.[28] Although this experiment has not received the same degree of scrutiny as the *experimentum crucis*, it would serve an important role in Newton's subsequent arguments with Hooke and Christiaan Huygens.

Various passages in Newton's responses to his critics, as well as in the *Lectiones opticae* and the *Optica*, the extensive optical treatises that Newton composed after his appointment to Lucasian professor in 1669, but before the *New Theory about Light and Colours* submitted to Oldenburg in 1672, reveal the function that Newton intended resynthesis to serve in his argument. The *experimentum crucis*, as Alan Shapiro has pointed out, was intended primarily to demonstrate the unequal refrangibility of the colorfacient rays, not to demonstrate color immutability.[29] Already in the early *Optica*, however, Newton had devised an experiment for proving the proposition that the spectral colors were immutable, by interposing a lens immediately after the first prism, which allowed one to focus the spectrum onto the second prism and thereby obtain a clearer separation of the spectral colors than the *experimentum crucis* allowed. The purer spectral colors that emerged from the second prism were incapable of analysis into more basic colors, did not act

[27] Newton to Oldenburg, February 6, 1671/72, Newton, *Corr.*, vol. 1, letter 40, p. 97.

[28] Newton to Oldenburg, February 6, 1671/72, Newton, *Corr.*, vol. 1, letter 40, p. 101.

[29] Alan E. Shapiro, "The Evolving Structure of Newton's Theory of White Light and Color," *Isis* 71 (1980): 213–14.

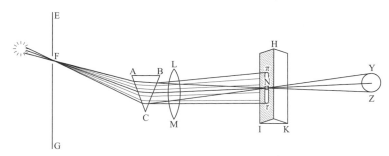

FIGURE 6.2. Newton's method of isolating the individual spectral colors by means of a lens placed before a second prism (from Alan Shapiro, *The Optical Papers of Isaac Newton*).

on one another, and could not be changed by reflection from colored bodies, so Newton viewed them as absolutely immutable (figure 6.2).

Once immutability was demonstrated to Newton's satisfaction, he then passed to his next proposition, that white light is a compounding of immutable spectral colors. The situation is actually more complicated than Newton envisioned it, if we take into account more modern wave theories of light that rely on Fourier analysis and other techniques unavailable to either Newton or his opponents.[30] But it is important to understand how the compounding of spectral rays was linked in Newton's mind with the issue of immutability. As he conceived it, if one grants that the colorfacient rays are unconditionally immutable, in accordance with his experimental evidence, they must therefore continue to be immutable once they are reassembled to form white light. There is no halfway house between strict unchangeability and change. Since the spectral rays cannot be altered by any means, they must remain in act within the compound that we perceive as uniform white light. This point is worth reiterating in a slightly different way. Suppose that a critic argued the opposite of Newton's position, asserting that the prism's refraction does not merely separate the preexisting colorfacient rays, but actually generates them out of white light that is itself homogeneous and uniform. Then let the critic concede that the newly generated colorfacient rays, once produced, are absolutely immutable, as Newton's experiments seemed to show. Here Newton's opponent would have made a potentially fatal concession. If the opponent further admitted that the combined spectral rays could now generate white light, in accordance with the phenomena displayed by Newton's experiments, he would be conceding the fact that he first denied, namely, that the spectral rays exist unchanged within white light. Once unconditional immutability is granted, even if it is induced by an initial refraction, the resynthesis of white light can only lead to the conclusion that the colorfacient rays exist in act within the white light that is produced.[31] One can therefore see the critical role that the resynthesis of

[30] See A. I. Sabra, *Theories of Light from Descartes to Newton* (Cambridge: Cambridge University Press, 1981), 261, 280–81.

[31] These points have already been made, albeit in more concise form, by Alan E. Shapiro in his magisterial article "The Gradual Acceptance of Newton's Theory of Light and Color, 1672–1727," *Perspectives on Science* 4 (1996): 59–140, see 106–7.

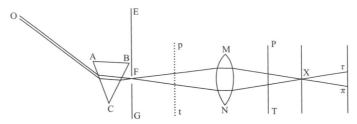

FIGURE 6.3. Newton's resynthesis of white light from the spectral colors by means of a lens (from Alan Shapiro, *The Optical Papers of Isaac Newton*).

white light, easily effected by means of a lens placed at a point where it could capture the analyzed spectral colors, played in Newton's thought (figure 6.3).

This, then, was the general argumentative role that Newton allocated to resynthesis. If one allowed that the spectral rays separated by a prism were indeed immutable, then the production of white light from those unchangeable rays would show that that white light is a mere compounding of them. Unfortunately for Newton, however, the argument required that his opponents first admit the immutability of the spectral colors, a condition that some refused to acknowledge. As Shapiro has shown at length, Newton's obscure comments about the production of pure spectral colors in the "New Theory" led to considerable confusion that undercut his expectations of immediate success. Because Newton did not describe a clear method of separating the spectral colors there in pure form, his opponents were able to devise methods that seemed to reveal that further colors could be derived from them.[32] Even before such demonstrations had been formulated, however, Hooke had already shown his unwillingness to take the bait. Already in his initial response to Newton's "New Theory," which appeared only a week after Newton presented his paper to the Royal Society, Hooke compared the generation of spectral colors from white light to the production of musical tones from strings and from the air within the bellows of a pipe organ. Hooke did not deny Newton's claim that the prism divides light into its spectral colors, but he saw no necessity to grant the existence of heterogeneous colorfacient rays already existent in white light before it encounters a prism.[33] In his second response to Newton's theory, written a few months later, Hooke elaborated further on the string and pipe-organ comparisons:

> I have only this to say that he doth not bring any argument to prove that all colours were actually in every ray of light before it has sufferd a refraction, nor does his experimentum Crucis as he calls it prove that those proprietys of colourd rayes, which we find they have after their first Refraction, were Not generated by the said Refraction. for I may as well conclude that all the sounds that were produced by the motion of the <? strings> of a Lute were

---

[32] Shapiro, "Gradual Acceptance," 73–80, 107–19.
[33] Hooke to Oldenburg, February 15, 1671/72, responding to Newton's New Theory, in Newton, *Corr.*, vol. 1, letter 44, p. 111.

in the motion of the musitians fingers before he struck them, as that all co-lours wch are sensible after refraction were actually in the ray of light before Refraction. All that he doth prove by his Experimentum Crucis is that the coloured Radiations doe incline to ye Ray of light wth Divers angles, and that they doe persevere to be afterwards by succeeding mediums diversly refracted one from an other in the same proportion as at first, all wch may be, and yet noe coloured ray in the light before refraction; noe more then there is sound in the air of the bellows before it passt through the pipes of ye organ—for A ray of light may receive such an impression from the Refracting medium as may distinctly characterize it in after Refractions, in the same manner as the air of the bellows does receive a distinct tone from each pipe, each of which has afterwards a power of moving an harmonious body, and not of moving bodies of Differing tones.[34]

It is noteworthy that neither in his first response nor in this short elabora-tion did Hooke address the issue of resynthesis. He simply refused ab initio to accept that Newton had provided evidence for the immutability of the colorfacient rays before their initial exposure to a prism, while also ignor-ing the fact that the rays could be reassembled to form white light. In this fashion Hooke managed to evade the conclusion that would follow from ac-knowledging that white light had been resynthesized from immutable spec-tral rays. One can begin to understand Newton's frustration with Hooke and his other opponents when one considers their unwillingness to consider both the analytic and synthetic halves of his demonstration that white light is composed of heterogeneous rays.

But there were other ways for Newton's opponents to respond, even if they did take resynthesis into account. One response, quite simply, could have been that the light that one produces by resynthesis is not the same light that comes from the sun. Why was it necessary to assume that the re-synthesized white light was identical to the original sunlight that entered the prism? Could it not simply have been regenerated from spectral rays that coalesced and lost their individual identity once they came into contact with one another? In such a case, both the resynthesis of the white light and the repeated analysis of the spectral rays from it would yield products that were at best identical *in specie*, like the transmutable elements of Aristotelian nat-ural philosophy. The resynthesized white light would in fact be regenerated from ingredients that were themselves generated de novo on each succes-sive analysis by the prism. This was precisely the position that Thomistic and Scotist authors had taken when they discussed the production of mixtures from the four elements. According to this reasoning, the spectral colors, like the scholastic elements or elemental qualities, would lose their individual identity to be replaced by whiteness, just as the four elements were replaced by the form of the mixture. A similar concern about the force of arguments based on resynthesis clearly occurred to Newton, for in his 1672 response to Hooke he described a method of excluding the possibility of a transmutation

[34] Hooke, apparently to Brouncker, ca. June 1672, in Newton, *Corr.*, vol. 1, letter 71, pp. 202–3.

"wrought in the colours by their mutuall acting on one another, untill, like contrary Peripatetic Qualities, they become assimilated."[35] Newton's evidence consisted of a rotating wheel that allowed only one spectral color to be perceived at a time. By turning the wheel rapidly and letting the spectral colors fall in swift succession on the eye of the viewer, the illusion of whiteness inevitably ensued. Since the spectral colors were never simultaneously perceptible to the viewer, Newton was able to decouple the production of whiteness from the necessity of mixture in a conclusive fashion.

Despite the cleverness and demonstrative force of Newton's color-wheel experiment, he had still not proven that the white light resynthesized from refracted sunlight was identical to white light tout court. The very fact that our perception of whiteness on the strength of Newton's own theory was somehow illusory could weaken the claim that it must always be caused in the same way and by the same factors. As if to acknowledge this fact, Huygens suggested in 1673 that a light perceived as white might well arise from the combination of blue and yellow alone. Huygens's suggestion would lead Newton eventually to modify his theory, and to admit that he had not synthesized white light *simpliciter*, but merely sunlight.[36] Even before this, Newton had been aware of the fact that the green produced from the refraction of sunlight was not the same as the green made by mixing blue and yellow, since the former green was indecomposable.[37] The inability of human vision to distinguish such composite and simple colors clearly made an approach based on "maker's knowledge," where the production of an effect acted as a warrant for the correct knowledge of its principles, suspect at best.[38]

But this argument, when extended to the resynthesis of sunlight, would fly in the face of the empiricist principles that Newton's major early source, the mechanical philosopher Boyle, held most dear. In a very important passage of the *Optica*, Newton responds to this type of objection at some length. After pointing out that sunlight is constantly refracted by the atmosphere and reflected by clouds, not to mention the refraction that it must suffer on entering our eyes, Newton says the following:

> Yet, since the sun's direct light is perceived to be white, and that color is not one of the primitives but may be shown to be generated by a mixture; and since there is no sensible difference between original light and that which is compounded from diversely colored rays, it must not be doubted that both are of the same nature.[39]

In short, the perceptible identity of the whiteness of sunlight and of the resynthesized white light acts as a warrant of their real identity. The facts

---

[35] Newton to Oldenburg, June 11, 1672, in Newton, *Corr.*, vol. 1, letter 67, p. 182.

[36] Oldenburg to Newton, January 18, 1672/73, in Newton, *Corr.*, vol. 1, letter 99, pp. 255–56. See Shapiro, "Evolving Structure," 211–35, 215–16, 222.

[37] Newton to Oldenburg, June 11, 1672, in Newton, *Corr.*, vol. 1, p. 181. See Shapiro, "Evolving Structure," 222.

[38] For "maker's knowledge," see Antonio Pérez-Ramos, *Francis Bacon's Idea of Science and the Maker's Knowledge Tradition* (Oxford: Clarendon Press, 1988).

[39] Shapiro, *Optical Papers*, 1: 505.

that both the direct white light of the sun and the artificially recompounded white light color bodies with the same colors, refract into the same spectrum, and cannot be sensibly distinguished from one another provide sufficient evidence that they are indeed identical.[40] To dispute this position would be to argue explicitly against principles that lay at the basis of the mechanical philosophy, at least in the form that Boyle enunciated it. Consider, for example, Boyle's comments, without doubt of alchemical origin, about the possibile identity of natural and artificial gold:

> And therefore not onely the Generality of Chymists, but diverse Philosophers, and, what is more, some Schoolmen themselves, maintain it to be possible to Transmute the ignobler Mettals into Gold; which argues, that if a Man could bring any Parcel of Matter to be Yellow, and Malleable, and Ponderous, and Fixt in the Fire, and upon the Test, and indissoluble in Aqua Fortis, and in some to have a concurrence of all those Accidents, by which Men try True Gold from False, they would take it for True Gold without scruple. And in this case the generality of Mankind would leave the School-Doctors to dispute, whether being a Factitious Body, (as made by the Chymists art,) it have the Substantial Form of Gold. . . . And indeed, since to every Determinate Species of Bodies, there doth belong more then One Quality, and for the most part a concurrence of Many is so Essential to That sort of Bodies, that the want of any of them is sufficient to exclude it from belonging to that Species: there needs no more to discriminate sufficiently any One kind of Bodies from all the Bodies in the World, that are not of that kind.[41]

Newton's early argument that natural sunlight and resynthesized sunlight make the same colors appear in bodies, refract the same spectral colors, and cannot be otherwise distinguished from one another finds its analogue in the various metallurgical tests that Boyle suggests should be used to determine the identity of natural and artificial gold. Just as Newton was content to argue that the white light produced by resynthesis was identical to natural sunlight before its analysis, so Boyle was happy to claim that a synthetic gold that passed all the assaying tests for natural gold would be identical to that natural gold. To argue otherwise would have been to invite back the imperceptible substantial forms of the scholastics, unknowable entities that were responsible for the different species into which natural things fell. Substantial forms underwrote the distinction between artificial and natural entities in a way that no mechanical philosoper could tolerate. To Boyle, on the contrary, it made no difference whether a substance had been broken down into its primitive constituents and then built back up again artificially, so long as the substance retained those properties that were deemed to be essential to it. This principle permeates Boyle's works, particularly *The Origin of Forms*

[40] Shapiro, *Optical Papers*, 1: 143: "Any one falling upon the same body, whatever it be, colors it with the same colors; any one, if it is transmitted through a prism, shows the same colors and performs the same way in every respect."

[41] Boyle, *Origin of Forms and Qualities*, in *Works*, 5: 322–23; 1666, 61–63.

and Qualities, the very work that Newton was extracting in 1666 while devising his own experiments to demonstrate color immutability and the mixing of the colorfacient rays to make white light.

Let us return briefly to another feature of Boyle's experimentation that may well have served as Newton's inspiration for his important experiments with the resynthesis of white light. This method of decomposition followed by recomposition is precisely the method that Boyle called "redintegration" of a body by chymical means, only here Newton has transferred this chymical method to the analysis and synthesis of sunlight. The classic Boylean description of redintegration had appeared already in his *Certain Physiological Essays* of 1661, where Boyle describes the dissolution of saltpeter into its ingredients and the subsequent recombination of those ingredients to arrive once more at saltpeter.[42] In simplest terms, Boyle's experiment worked by injecting burning charcoal into molten saltpeter, and thus igniting it. This resulted in the release of nitrogen and carbon in combination with oxygen, leaving a nonvolatile residue of "fixed niter" that resembled salt of tartar (potassium carbonate—in reality it *was* potassium carbonate). Knowing that spirit of niter (nitric acid) could be produced by the thermal decomposition of niter, Boyle then added spirit of niter to the tartar-like residue and acquired a product that resembled the original saltpeter in all its significant properties. Employing the principle of substantial identity based on identity of sensible properties that we encountered in the case of gold, Boyle argued that the product was genuine niter. He was then able to conclude that niter itself is merely a compound of two very different materials, namely, spirit of niter and fixed niter, which we would today call an acid and a base.[43] In *The Origin of Forms and Qualities*, Boyle would elaborate on this redintegration further and also describe experiments aimed at redintegrating turpentine and stibnite, the ore of antimony.

Now let us return to Newton. The fact that Newton was thinking about the composition of white light in Boylean terms is not just borne out by the structural similarity of his prism experiments and Boyle's redintegration of saltpeter, but also by the terminology that Newton employs when describing this series of experiments in his optical lectures. Both in the *Lectiones opticae* and the *Optica*, Newton speaks of the sunlight reconstituted from spectral colors as being an *albedo redintegrata*—quite literally a redintegrated whiteness.[44] In the *Optica*, as I have pointed out, he explicitly argues that it is the redintegration of the white light that proves beyond any reasonable doubt that it is actually composed of a mixture of colorfacient rays.[45] Although one might argue that this agreement of Newton's terminology with that of

[42] See Newman and Principe, *ATF*, chapter five, for Worsley. See also John T. Young, *Faith, Medical Alchemy, and Natural Philosophy: Johann Moriaen, Reformed Intelligencer and the Hartlib Circle* (Brookfield, VT: Ashgate, 1998), 183–216, esp. 198–200.

[43] The experiment is clearly described by Boyle, *Certain Physiological Essays*, in *Works*, 2: 93–96; 1661, 108–13.

[44] Shapiro, *Optical Papers*, vol. 1, p. 162, line 9; and p. 516, line 16.

[45] Shapiro, *Optical Papers*, vol. 1, p. 504: "Et eadem ratione constat reflexam albedinem similiter compositam esse, siquidem (ut dixi) redintegrata est."

Boyle is mere coincidence, there are other clues that Newton had already read about Boyle's experiments with redintegration before composing either the *Lectiones opticae* or the *Optica*. At the same time as Newton's famous *annus mirabilis*, 1666, the year in which he claimed to have discovered the heterogeneity of white light, Boyle had published his *Origin of Forms and Qualities*. Indeed, the very manuscript in which Newton recorded his first experiments with the resynthesis of white light from the spectral colors, CU Add. 3975, also contains extensive notes drawn from Boyle's *Origin of Forms* on the redintegration of stibnite and turpentine.[46] It is clear, then, that chymical redintegration was a phenomenon that interested Newton, and one that he could easily have adapted to his optics from his reading in Boyle's chymistry.

If we now briefly consider Newton's April 1673 reply to his critic Huygens, we will find other important clues, also of a terminological nature, that reveal a Boylean influence. Shapiro has argued in a persuasive article that Newton's conception of white light as a mixture of immutable color-producing rays owes an important debt to comments that Boyle made in his *Experiments Touching Colors* about the so-called painters' primaries—blue, red, and yellow.[47] The theory that all other colors originate from these three was not old in Newton's day, and he seems to have derived it partly from a direct reading of Boyle's work. The mixing of pigments acquired particular significance for Newton in the response to Huygens.

What is interesting in this for us is Newton's use of Robert Boyle's peculiar corpuscular terminology. In arguing against Huygens's view that only yellow and blue may be responsible for the production of white light, Newton says that even if experiment revealed this result it would not be significant. The yellow and blue would themselves have to be compound colors, or as Newton says:

> But what Mr. Hugens can deduce from hence I see not. For the two colours <i.e. yellow and blue> were compounded of all others, & so the resulting white to speake properly was compounded of them all & onely decompounded of those two.[48]

As we can see, Newton has borrowed Boyle's characteristic terminology whereby preliminary mixtures are "compounded" from simple ingredients, and these compounds are in turn recombined or "decompounded" to make more complex mixtures. Huygens's white can be produced from blue and yellow only if the blue and yellow are already compounds rather than simple

[46] Newton, CU Add. 3975, fol. 32v: "The purenesse of this ^redintigrated Antimony seemed to proceede from yᵉ recesse of so much Sulphur wᶜʰ is not at all necessary to yᵉ constitution of Antimony though perhaps too yᵉ vitrum a top might proceede from yᵉ avolation of two much Antimony from yᵉ superficiall parts. pag 265
But redintegration of Bodys succeded best f<illeg.> in Turpentine for a very cleare liquor being distilld from it <illeg.> was againe put to yᵉ caput Mortuum (wᶜʰ was very dry brittle Transparent sleeke & red but purely yellow when poudered) it was immediatly dissolved part of it into a deepe red Balsome. And by further disgestion in a large well stopt Glasse became perfect Turpentine againe both as all men judgd by yᵉ smell & Taste. pag 268 of for<ms>."

[47] Alan E. Shapiro, "Artists' Colors and Newton's Colors," *Isis* 85 (1994): 600–630, see 614–15.

[48] Newton to Oldenburg, responding to Huygens, April 3, 1673, in Newton, *Corr.*, vol. 1, letter 103, p. 265.

colors, so that the white is actually a decompounded color containing all the spectral primaries. Newton then goes on to demonstrate the force of his argument by analogy between the composition of white light from the spectral colors and the making of a gray powder by mixing variously colored powders. Here too he employs Boyle's compositional stages of mixture, saying that a decompounded gray can be made from an orange and blue that are themselves compounded colors composed of simpler ones. As in the case of Newton's use of the Boylean term "redintegration," Newton has here adopted an unusual terminology from his older compatriot along with the underlying idea that it encapsulates.

Conclusion

To summarize, Newton's principal object of attack in much of the *New Theory about Light and Colours* and the optical lectures was the idea that white light is "transmuted" into the spectral colors by refraction in the same way that the Aristotelian elements could be transmuted to yield an entirely new product. Instead of this being the case, he wanted to show that the colorfacient rays are themselves immutable and retain their "form" or "disposition" to produce the sensation of distinct colors within the eye.[49] At the same time, he wished to show that white light is a mixture of these immutable spectral rays, which do not affect one another when they are compounded, but only act on the sense of sight to produce the sensation of whiteness. Newton's principal way of demonstrating this was by means of repeated analyses and syntheses of light—exactly the method that Boyle used in the chymical realm for showing that saltpeter, stibnite, turpentine, and other substances were produced out of unchanging corpuscles that could be disassembled and reassembled like the parts of a watch. Boyle's redintegration experiments in turn derive from the tradition of the reduction to the pristine state that stemmed ultimately from medieval alchemy and its need to demonstrate that metals and minerals are composed of heterogeneous particles retaining their substantial identity while undergoing the separation and recombination that results in phenomenal change. Newton's experimental decomposition and redintegration of white light owed a significant debt to a practical and theoretical tradition of chymical analysis and synthesis whose origins recede well into the Middle Ages. Although Newton was deeply influenced by the optics of Descartes and Hooke, we must not ignore his transformation of paired chymical analysis and synthesis, long used to reveal the heterogeneity of material substances, into a tool for demonstrating the same fact in the realm of light and color.

The profound relationship between Newton's early optical theories and his chymistry reveals that even in his youth he saw the art of Hermes from multiple vantage points. The long alchemical tradition of corpuscular reasoning or "chymical atomism" transmitted through Boyle provided Newton

---

[49] For the term "transmutation," see Shapiro, *Optical Papers*, 1: 472. For "forms" and "dispositions," see 505.

with his innovative use of resynthesis after analysis, hence yielding a critical part of his demonstration that white light is a mixture of unaltered spectral colors.[50] At the same time, alchemy provided Newton with important theories of the subterranean generation and decay of minerals, as we have already seen. Yet Newton also envisioned deeply practical goals for chymistry, ranging from the technical production of mineral acids and salts and the assaying and purification of metals on the one hand to the production of the philosophers' stone and other *arcana majora* on the other. He also devoted significant time to *chymiatria*, or chymical medicine.[51] In a word, the young Newton was involved in virtually every aspect of early modern chymistry, seeing it both as a source of desirable products and as a potent means of enriching natural philosophy. When we view Newton's involvement in alchemy from the perspective of the seventeenth century, the long-held modern astonishment at his devotion to the aurific art melts away to be replaced by a fine-grained image of a scientist alive to the promise of chymistry across the full range of the discipline.

[50] For a justification of the term "chymical atomism," see William R. Newman, "The Significance of 'Chymical Atomism,'" in *Evidence and Interpretation: Studies on Early Science and Medicine in Honor of John E. Murdoch* (a special issue of *Early Science and Medicine* edited by Newman and Edith Dudley Sylla), *Early Science and Medicine* 14 (2009): 248–64.

[51] For Newton's interest in chymical medicine, see William R. Newman, "Newton's Reputation as an Alchemist and the Tradition of Chymiatria," in Elizabethanne Boran and Mordechai Feingold, eds., *Reading Newton in Early Modern Europe* (Leiden: Brill, 2027), 311–27.

SEVEN

Newton's Early Alchemical Theoricae

PRELIMINARY CONSIDERATIONS

Introduction: Alchemy and the Imitation of Nature

We have now followed Newton's alchemical career from his earliest interests in the mysterious phenomena of chymistry taken from books of secrets and natural magic through the even more impressive wonders proffered by the corpus of Basilius Valentinus. While Newton's interest and expertise in practical alchemy was deepening, he was also repurposing the demonstrative arguments chymists employed in support of their matter theory so that he could provide a proof of the composite nature of white light. The image of Newton simultaneously developing his geometrical optics in "Of Colours II" while deciphering the extravagant riddles of Basil Valentine may present an element of cognitive dissonance to the modern reader, but I have now shown how the domains of optics and chymistry were connected in his mind. The linkage between these two fields was no anomaly; it had already formed the subject of Robert Boyle's influential *Experiments Touching Colours*, though Boyle was unaware of the composite nature of white light, of course. Yet there are other areas as well where alchemical theory played a part in Newton's overall reform of natural science. One of these points of intersection lies in Newton's developing thoughts about the generation of metals and minerals within the earth, and the ways in which different materials are related to one another. As we already saw in chapter four, mineralogical theories had long formed an integral part of alchemy. But why was this the case, and what would drive Newton to make the subterranean activities of our planet an important part of his alchemical quest?

The answers to these questions lie quick to hand if we consult Newton's sources among the adepts. At some point in the late 1660s Newton acquired a copy of Michael Sendivogius's *Novum lumen chemicum*, one of the most widely read alchemical books of the seventeenth century. The first part of Sendivogius's work consists of twelve "tractates," which convey both the theory and practice of alchemy. Sendivogius is keen to explain the ways in which the earth produces minerals, which he views in organismic terms as a process beginning with invisibly small "seeds" (*semina*) and leading to the perfect maturity of the precious metals when nature is unimpeded by adventitious circumstances. In order to succeed at chrysopoeia, the sons of art must mimic the subterranean activities of nature and remove the accidental obstacles that have prevented base metals from maturing fully into their noble counterparts. To employ techniques alien to nature's own methods would be to effect a mere counterfeit, an ersatz substitute rather than a genuinely natural product. This idea had long permeated alchemical texts and finds its origin in the Aristotelian and Galenic concept that art must imitate or aid nature wherever possible.[1] But Sendivogius gives the idea a particular emphasis, as in an early passage where he advises alchemists to understand nature so that they may mimic its processes:

> Then let them diligently consider, whether their purpose be agreeable to Nature; whether it be possible, let them learn by clear examples, *viz*. Out of what things any thing may be made, how, and in what Vessel Nature works. For if thou wilt do any thing plainly, as Nature her self doth do it, follow Nature; but if thou wilt attempt to do a thing better than Nature hath done it, consider well in what, and by what it is bettered, and let it always be done in its own like.[2]

In order to produce a genuine transmutation of base metals into gold, the alchemist must therefore learn the hidden operations of nature, and even devise methods for correcting its occasional failures. Hence the *Novum lumen chemicum* follows this advice with page after page on the four elements, the three Paracelsian principles, the structure of caves and other subterranean formations, and yet more information pertinent to mineralogy. The chymist must acquire this information if he is going to follow nature in a successful fashion.

Similar advice is given by another of Newton's early favorites, Eirenaeus Philalethes, in his *Brevis manuductio ad caelestem rubinum* (Brief Guide to the Celestial Ruby). The "American philosopher" explicitly states that the art of alchemy must imitate the subterranean actions of nature, even if clever chymists have come up with ways to abbreviate nature's workmanship:

> But all this Work, very well Answers to the *subterraneal Operations* of *Nature*, from whence the *Work* is deservedly called *Natural*. For Nature doth produce *Metals*, according to their *species* out of *Mercury* alone, *Cold* and *moist*, by a daily *Digestion & Coction* in the Veins of the *Earth*. But *Art* to shorten the *Work*, hath found out a far more subtile Operation, yet like to this: For it *Conjoyns* with *Crude Mercury*, *cold* and *moist*, ripe *Gold*, and both of these by *Commixtion*, and *secret Conjunction*, makes one *Mercury*, which they Name *Aqua Vitae*, which *Mercury* at last they Decoct into *Gold* not *Vulgar*, but far more *Noble*: which falls upon all *Imperfect Metals*, and *tings* them into *tryed Gold*, exposed to all *Tryals*.[3]

---

[1] On this point, see William R. Newman, *Promethean Ambitions* (Chicago: University of Chicago Press, 2004), chapter two.

[2] Michael Sendivogius, *A New Light of Alchymy* (London: Tho. Williams, 1674), 5. At this point in his career, Newton was probably reading Sendivogius in the edition found in the *Bibliotheca chemica contracta* edited by Nathan Albineus and printed multiple times in the seventeenth century. See Karin Figala, John Harrison, and Ulrich Petzoldt, "*De scriptoribus chemicis*: Sources for the Establishment of Isaac Newton's (Al) chemical Library," in *The Investigation of Difficult Things*, ed. P. M. Harmon and Alan Shapiro (Cambridge: Cambridge University Press, 1992), 135–79, especially 159–60n87.

[3] Eirenaeus Philalethes, *A Short Manuduction to the Caelestial Ruby*, in *Three Tracts of the Great Medicine of Philosophers* (London: T. Sowle, 1694), 108–9.

Just as nature makes metals out of a primordial mercury digested and cooked within the earth, so art produces a sophic mercury that can digest metallic gold into the philosophers' stone. Once this marvelous agent of change has been acquired, the slow ripening processes of nature can be compressed into the time required to melt a base metal and add a fragment of the transmutatory stone. Philalethes sees all of this in terms of removing nature's impediments and liberating it from extraneous obstruction.

But the necessity of imitating nature's activity was not the only reason for early modern alchemy's stress on geochemical themes. One must never forget that texts in the genre of the *Novum lumen chemicum* and *Brevis manuductio* were intended both to reveal and to conceal.[4] If we consider only the *Novum lumen chemicum*, not only is the work provided with a self-standing "Philosophical Ænigma or ridle" after the twelve treatises, but the bulk of the text is also written in a riddling style. Indeed, Sendivogius explains that he has sometimes hidden his true meaning behind intentionally misleading words, and that the only way for the adept-in-training to acquire his genuine sense is to understand "the possibility of Nature," the ways in which nature operates in the subterranean generation of metals. As he says:

> I would have the Courteous Reader be here admonished, that he understand my Writings not so much from the outside of my words as from the possibility of Nature; lest afterward he bewail his time, pains and costs, all spent in vain.[5]

From this we see that Newton's sources provided him with a twofold goal for studying the generation of metals from an alchemical perspective: on the one hand he had every reason to think that such knowledge was a requirement if one was going to succeed where nature itself had failed to arrive at complete perfection, and on the other, the very meaning of chrysopoetic chymists such as Sendivogius was inaccessible without a deep understanding of nature's methods. When we couple these practical imperatives with Newton's desire to build a comprehensive new natural philosophy, the diverse aims of his alchemical enterprise emerge in their full complexity. The multifarious character of Newton's chymical quest already appears quite clearly by the beginning of the next decade.

Newton's Two Early Theoricae

In the period between 1670 and 1674, several years after his first exposure to the chrysopoetic writings of Basilius Valentinus, Newton wrote two remarkable treatises that would provide a theoretical basis for much of his

---

[4] This point is made very clearly in Didier Kahn, "Le Tractatus de sulphure de Michaël Sendivogius (1616), une alchimie entre philosophie naturelle et mystique," in *L'Écriture du texte scientifique au Moyen Âge*, ed. Claude Thomasset (Paris: Presses de l'Université de Paris-Sorbonne, 2006), 193–221.

[5] Sendivogius, *New Light of Alchymy*, f. [A4v].

later practical work in chymistry.[6] The importance of these short documents cannot be overstated, as they furnish an unparalleled glimpse into certain features of Newton's thought that implicitly underlie his chaotic corpus of reading notes, synopses, anthologies, indexes, and experimental records. They are also unique among the million or so words making up Newton's alchemical corpus in that they consist of original cosmological and geological speculations inspired by alchemy, while most of his manuscripts are aimed at deciphering the practical meaning intentionally buried by chymical authors. Moreover, the two treatises in question display a format long favored among alchemists. Just as medieval alchemical texts were often neatly divided into a theoretical part followed by a practical one that aimed to capitalize on the foregoing theories, so the two texts to which I am referring together act as a sort of alchemical *theorica* to the *practica* making up the bulk of his alchemical *Nachlass*.[7]

The two opuscula also serve to link Newton's chymistry explicitly to the better known scientific work that he was doing in the second half of the 1660s and early 1670s in a way that most of his other alchemical manuscripts do not. They were composed at a time when Newton was actively framing a cosmological system that would pull together elements from his optics, mechanics, chymistry, and other scientific pursuits. In short, they belong to the most fruitful period of his life, around the time when Newton said that he was in "the prime of my age for invention."[8] Both of these documents are found in the same manuscript, Smithsonian Institution Dibner 1031B, and the first, usually referred to by its first line *Of Natures obvious laws & processes in vegetation*, is well known. The second, which was only recently published, can also be referred to by its first words, which consist of the canceled Latin phrase "Humores minerales continuò decidunt" (Mineral humors continuously descend). Although *Humores minerales* is written upside down at the end of the manuscript, it is quite possible that Newton composed it first and

---

[6] A textual locus in *Of Natures obvious laws & processes in vegetation* reveals that Newton could almost certainly not have composed that text before early 1670. I refer to Newton's apparent reference on 3v to an experiment made on ice in December and January 1669 (Old Style), which is described in CU Add. 3975 on 20v–21v. As for *Humores minerales*, Newton's evident use of texts found in the *Theatrum chemicum*, which he acquired in 1669, provides a terminus post quem for that document. Newton's use of Bernhard Varenius's *Geographia generalis,* which he published in 1672, might also provide a chronological marker except for the fact that we do not know how long before publishing the text Newton worked through it. According to a later editor of Varenius, Newton edited the text for his Lucasian lectures, which he began to deliver in 1670 (See Newton, *Corr.*, vol. 2, p. 264, n. 1. At any rate, it is likely that *Of Natures obvious laws & processes in vegetation* was composed before Newton had read Robert Boyle's 1673 *New Experiments to Make Fire and Flame Stable and Ponderable*. Given the considerable use that Newton's slightly later treatise *De aere et aethere* makes of Boyle's calcination experiments in *New experiments*, which seemed to argue for a subtle material that could penetrate glass and calcine metals, it is likely that he would also have employed this text when drawing up *Of Natures obvious laws* if it had been available. A definite terminus ante quem for both *Humores minerales* and *Of Natures obvious laws* is provided by the 1675 *Hypothesis of Light*, which presents more developed versions of the ideas found in the previous two texts. For *De aere et aethere*, see A. Rupert Hall and Marie Boas Hall, *Unpublished Scientific Papers of Isaac Newton* (Cambridge: Cambridge University Press, 1962), 187–88.

[7] For the *theorica-practica* division, see Richard L. Kremer, "Incunable Almanacs and *Practica* as Practical Knowledge Produced in Trading Zones," in *The Structures of Practical Knowledge*, ed. Matteo Valleriani (Cham: Springer, 2017), 333–69.

[8] Cambridge University Library, MS Add MS 3968.41 f.85r (= frame 1349 of http://cudl.lib.cam.ac.uk /view/MS-ADD-03968/1349, accessed May 16, 2016).

then reversed the manuscript in order to write *Of Natures obvious laws*. We need not insist on this point, but it will be useful to examine *Humores minerales* before the other text, since it deals more directly with materials that we have already examined in the context of contemporary chymistry, namely, the generation and decay of metals and minerals within the earth.

In addition to their close connection to Newton's chymistry, what makes these opuscula particularly interesting is the fact that they were framed in response to specific questions. *Humores minerales* deals with the problem of mineral erosion and decomposition within the earth, as well as the generation of metals. The ever quantitative mind of Newton poses the following query. If metals and minerals are constantly being corroded by subterranean acids and converted into soluble products carried downward in solutions, how is it that the supply of ores and metals has not been exhausted over time? Newton supplies an answer to this problem in the form of a fascinating theory whereby the metals are constantly regenerated. Drawing on contemporary chymical beliefs that we have already encountered in chapter four of the present book, Newton explains this regeneration in terms that employ the language of organic vegetation or growth and putrefaction.

Of Natures obvious laws, on the other hand, uses these issues, particularly in the realm of salts, to develop a means of distinguishing between mere mechanism and "vegetability" (the ability of some things to assimilate nutriment and grow) and to determine the intersecting borders between the two types of activity. The distinction was a key one for Newton, since even in his undergraduate days he was already searching out the flaws in Cartesian physics, a system that of course left no space for vegetation as a nonmechanical process. *Certain Philosophical Questions*, Newton's student notebook from the 1660s, is filled with criticisms and corrections to the system of natural philosophy erected by the French savant.[9] *Of Natures obvious laws* should be viewed as a bridge between the more orthodox mechanical philosophy of *Certain Philosophical Questions* and Newton's radical 1675 *Hypothesis of Light*, which was addressed to the criticism that his famous, groundbreaking optical paper, the *New Theory about Light and Colours*, received on its presentation to the Royal Society in 1672. The ethereal mechanisms that Newton postulates in the *Hypothesis of Light* and in later texts are prefigured in a striking way in *Of Natures obvious laws* and linked explicitly to his understanding of the cosmology purveyed by contemporary alchemy.

Humores minerales: The Subterranean Generation and Degeneration of Metals

Humores minerales is essentially a list of postulates and queries provided with their own corresponding answers. Newton begins with the common observation that metals and minerals can be corroded by acids or sometimes even by water into aqueous solutions. In the subterranean realm, these

[9]Newton, *CPQ*, see 3–325.

liquids are carried downward toward the center of the earth, whence they are sometimes found by miners collecting in subterranean passageways. This is an old alchemical commonplace that already appears in the high medieval *Summa perfectionis* of Geber, where the dripping down of alums and vitriols (sulfates) is attributed to the action of "pontic" (corrosive) waters on them.[10] That observation had been updated and expanded in the sixteenth century by the German chymist Johann Grasseus, whom we encountered in chapter four. Grasseus argued that the subterranean corrosive waters, which also contain a hidden mercury, sink downward within the earth, where they encounter the sulfurous vapors that are always rising up from the earth's core. As Grasseus would say, "if the salty waters are pure and clear, and the sulfurous vapors pure, and they embrace one another upon meeting, a pure metal is thence generated."[11]

Humores minerales draws on the work of Grasseus, but modifies it in important ways. Like Sendivogius, the German alchemist had focused on the growth and formation of new metals, whereas Newton, here following the tradition of Basilius Valentinus and his source Nicolaus Solea, emphasizes the cyclical character of metallic generation, which presents a parallel decomposition of metals within the earth. According to Newton, the down-flowing, mineralized waters are in turn vaporized and driven back up to the surface by the earth's subterranean heat, leading to a perpetual circulation. But here he notes a problem. If we dissolve metals by means of acids in our laboratories and then distill off the solution, the metals typically remain behind in the bottom of the flask, and the liquid distills over independently. This easily verified fact would have been known to the generality of early modern chymists. Newton continues by pointing out that if anyone should argue the subterranean menstrua to be of a different order, and "sharp" enough to dissolve the metals into a volatile form, that person would have to answer a number of further questions. Why is it that metallurgists and vulgar chymists have been unable to replicate anything like this ultrapowerful corrosive in their laboratories? And if it indeed existed, would it not actually destroy the metals entirely? How would they return to their old form after having been reduced to such volatility? Since the replies to all of these questions lead to the conclusion that the metals cannot be distilled, Newton answers thus:

> Hence with the metals continually drawn downwards, never ascending ^so long as they remain metals it would be necessary that in a few years the greatest part would have vanished from the upper earth, unless they are conceded to be generated ^there de novo.[12]

It is important not to misunderstand Newton's reasoning at this juncture. Although "vulgar chymists" cannot make a menstruum so powerful that it

---

[10] William R. Newman, *The Summa perfectionis of Pseudo-Geber* (Leiden: Brill, 1991), 731.

[11] Johannes Grasseus, *Arca arcani artificiosissimi de summis naturae mysteriis*, in *Theatrum chemicum* (Strasbourg: Haeredes Eberhardi Zetzneri, 1661), 6: 294–381, see 305–7.

[12] Dibner 1031B, 6v: "Hinc metallis continuò deorsū delatis et ~~in metallica~~ nunquam ^dum sunt metalla ascendentibus ~~nisi metallica forma destructa in~~ <~~illeg.~~> ~~volatilitatem deducantur~~: necesse esset ut intra paucos annos maxima pars a superiori terrâ evanesceret, nisi ^ibiq̣ e novo generari concedantur."

can volatilize a metal and destroy its form, it does not follow that this is impossible for the adepts. To the contrary, paraphrasing a passage from book 6 of the *Aeneid*, where the Sybil explains that it is easy to descend to Hades but quite another matter to reascend, Newton adds, "hic labor hoc opus est" (this is the labor, this is the work). Like Aeneas's return from Hades to the land of light, the volatilization of metals that allows them to rise up as spirits is difficult to reproduce in a flask, but not beyond the realm of possibility, at least not for those who are true adepts in alchemy.[13] Newton's point is that if such a volatilization of the metals is indeed possible, it can only occur after a particular metal has lost its metallic nature. The spirit or vapor into which the metal is reduced by this alteration will bear the same relationship to the metal as spirit of vitriol (sulfuric acid) does to the vitriol (iron or copper sulfate) out of which it is distilled.[14] Although Newton's meaning here is a bit obscure, he is probably suggesting that the normal, nonvolatile form of the metal contains a hidden volatile component in the same way that vitriol was thought by chymists to hold an acid spirit within itself. Only in their altered form as extremely attenuated vapors or fumes can the metals return to the surface of the earth after having been carried down in corrosive solutions.

Having established that the metallic fumes must rise back up to the higher levels within the earth, Newton now develops a theory as to how their regeneration occurs. The theory is complicated, and it requires both active and passive participants, being guided by a biological heuristic. Newton's idea is that the rising metallic spirits or fumes encounter other metals that have been corroded into their dissolved state by the mineral waters. Thus we have metals in two different states of dissolution: those that have merely been corroded into relatively gross metallic particles are sinking down in liquid form, while the much more volatile spirits consisting of subtle corpuscles, which have lost their metallic nature as a result of their extreme attenuation, are rising up. When the two parties meet, the much more active spirits mix with the dissolved metals in solution, and both enter into a "state of motion and vegetation." At this point, the spirits "putrefy and destroy the metallic form and convert it into spirits similar to themselves." After having been putrefied and subtilized, the metals carried down in the acid solution lose their metallic nature in turn and also rise up, having themselves been converted to spirits. This accounts for the formation of the volatile metallic spirits in the first place; they have been volatilized by other metallic spirits. Thus there is

---

[13] It is tempting to think that Newton meant to draw an explicit comparison between the sinking down of metallic solutions and Aeneas's descent into Hades. One must be cautious here, however, for the Vergilian expression "hic labor hoc opus" appears in the works of other contemporary chymists as well. Newton's employment of the passage may well stem from contemporary chymists rather than from Vergil himself. One finds it, for example, on page 110 of the Latin edition of Basilius Valentinus's *Triumphwagen* published by Pierre Jean Fabre as *Currus triumphalis antimonii fratris Basilii Valentini* (Toulouse: Petrus Bosc, 1646). Newton copied the expression in his early manuscript Keynes 64, 4v. The Vergilian formula also appears in Hadrianus à Mynsicht, *Thesaurus et armamentarium medico-chymicum* (Lyon: Ioan. Antonius Huguetanus, 1645), 6. Newton extracted it from Mynsicht in his manuscript Keynes 41, 1v.

[14] Dibner 1031B, 6v: "Idem esset inter hunc spiritum et metallū destructum ac inter vitriolū et spiritum ejus."

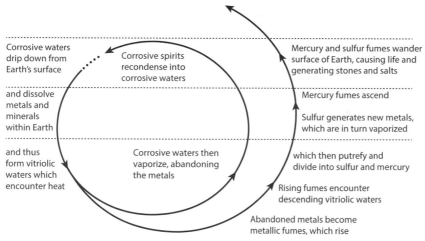

Corrosive waters drip down from Earth's surface

Corrosive spirits recondense into corrosive waters

Mercury and sulfur fumes wander surface of Earth, causing life and generating stones and salts

and dissolve metals and minerals within Earth

Mercury fumes ascend

Sulfur generates new metals, which are in turn vaporized

and thus form vitriolic waters which encounter heat

Corrosive waters then vaporize, abandoning the metals

which then putrefy and divide into sulfur and mercury

Rising fumes encounter descending vitriolic waters

Abandoned metals become metallic fumes, which rise

FIGURE 7.1. Figure illustrating Newton's theory of metallic generation and regeneration, based on *Humores minerales*.

a perpetual cycle of the more active form of the metal converting the passive version into a state of activity like its own (figure 7.1).

Although Newton does not use the Latin word for leaven or yeast here, there can be little doubt that when he employs the word "putrefy," he is thinking in terms of fermentation, a concept that repeatedly surfaces in *Of Natures obvious laws*. In that text he goes so far as to equate putrefaction with fermentation in the following words:

> nature ever begins with putrefaction or fermentation whereby there is an intimate union & exertion of S^{pts} $\wedge$& purgation of impuritys.[15]

A good entry into Newton's reasoning about fermentation can be gained from a brief consideration of modern sourdough bread. Just as sourdough starter can be passed on from one batch of dough to another, "subtilizing" the flour paste by giving it the fluffy texture of bread dough, so Newton's metallic spirits act on the dissolved metals to turn them into vapors like the spirits themselves. Once this transformation has occurred, the newly formed metallic spirits, like sourdough starter transported from the newly formed dough to fresh flour paste, can transform yet other metals into a state like their own. Hence the circulatory process in Newton's subterranean world continues its revolutions indefinitely once the initial metallic spirits have been formed. But there is more to *Humores minerales* than this alone. After describing the circulation of dissolved metals and metallic spirits, Newton then brings in the traditional alchemical theory that the metals are composed of sulfur and mercury.

The destruction of the metals in solution by metallic spirits is accompanied by a separation of the dissolved metals into the two principles, sulfur and mercury. Building on traditional alchemical theory, Newton asserts that the mercurial principle is unfixed or volatile, while the sulfur principle possesses more fixity. He then explicitly employs the chymical language of

[15] Dibner 1031B, 1r.

FIGURE 7.2. Saturn as a lame old man watering his garden. From Michael Maier, *Symbola aureae mensae duodecim nationum* (Frankfurt: Lucas Jennis, 1617).

Sendivogius, saying that the mercury is the *magnes* or magnet to the sulfur, which in turn acts as *chalybs* or steel; hence, although they share a mutual attraction, they can be separated. Then adding an echo from Grasseus, Newton states that the mercury is lame or "wounded in the foot" (*pede læsus*) like the titan Saturn; traditionally viewed as a personification of time and its effects, Saturn was often pictured as a lame old man, often with a scythe, and sometimes with an hourglass (figure 7.2).[16] Because it is "lame," the mercury cannot therefore mature into a fully metallic form "for a very long time." The dissociated sulfur, on the other hand, sinks back down toward the center of the earth and is cooked into a metal. It can either fall into a commodious, clean passageway beneath the earth that allows it to be decocted by the hot, metallic spirits into gold, or if it encounters impurities and insufficient heat, it can become iron. In either case, given sufficient time, the sulfur or metal can again be rendered volatile and rise up as a metallic spirit. Newton finally terminates *Humores minerales* by generalizing his theory even further, for the sulfur and mercury that have formed the object of his discussion are not just the progenitors of metals alone:

> These two spirits above all wander over the earth and bestow life on animals and vegetables. And they make stones, salts, and so forth.[17]

[16] At Keynes 48, 20r, Newton again uses the expression "pede laesus" for Saturn, which he attributes to Grasseus's *Arca arcani*. I have located the passage in the *Theatrum chemicum*, 6: 326. For a classic discussion of this *topos*, see Raymond Klibansky, Erwin Panofsky, and Fritz Saxl, *Saturn and Melancholy: Studies in the History of Natural Philosophy, Religion, and Art* (New York: Basic Books, 1964).

[17] Dibner 1031B, 6r: "Et hi duo præsertim spiritus extravagantur terram et animalibus et vegetabilibus vitam largitentur. faciuntꝗ lapides salia &c."

At this point it becomes entirely clear that Newton is restating Sendivogius's influential theory that the interior of the earth as well as the atmosphere are permeated by an aerial niter that acts as a principle of life and growth. Sendivogius had argued that the niter or *sal nitrum* existed both in an active and a depleted form. Its life-giving power had to be recharged by a periodic volatilization that allowed it to be carried up into the outer reaches and replenished. When it returned to the surface of the planet in this active form, it joined to a fixed niter within the earth; the combination of the two produced "flowers and all things." Hence the association between the more volatile form of the aerial niter and mercury on the one hand, and the more fixed variety of it and sulfur on the other, was an obvious move to make.

Two features of paramount importance stand out in Newton's account of the generation of metals. The first is the fact that his metallic spirits and dissolved metals are really the same thing, though in different states of activity and subtlety. Just as sourdough starter and unleavened flour paste are both essentially dough, so the dissolved and spiritual metals both consist of metallic material. When the active form works on the passive one, this activity can therefore be seen as the working of one thing on itself rather than as a change dependent on the adventitious importation of heterogeneous ingredients. The same thing was true of Sendivogius's niter in its aerial and earthy forms; it was an agent acting on itself. There is no cause for surprise in this structural similarity between Newton's theory and that of the Polish alchemist, for the likeness of their theories goes further than the mere fact that Newton had read Sendivogius. Both authors are reflecting on an alchemical desideratum that extends backward to the origins of the art in late antiquity and emerges in a medieval source that they both knew well: the *Emerald Tablet* of Hermes Trismegistus. This cryptic document, supposedly inscribed on a massive slab of emerald and found between the hands of its entombed and eponymous owner, became a favorite subject of alchemical interpretation among the Latin alchemists of the High Middle Ages and their heirs. Probably stemming from Neoplatonic sources as well as hermetic ones, the text uses enigmatic language to describe a parallelism between the tellurian and cosmic spheres.[18] In a later translation that Newton himself made of the *Emerald Tablet*, the Egyptian sage intones that all particular materials are actually made from "one thing" acting by means of itself alone:

> That wch is below is like that wch is above & that wch is above is like y^t wch is below to do ye miracles of one only thing. And as all things have been & arose from one by ye mediation of one: so all things have their birth from this one thing by adaptation.[19]

Through the vague and oracular language of Hermes one can make out the claim that the multiplicity of the world arose from one miraculous matter,

---

[18] Ursula Weisser, *Das "Buch über das Geheimnis der Schöpfung" von Pseudo-Apollonius von Tyana* (Berlin: Walter de Gruyter, 1980). See also the classic study by Julius Ruska, *Tabula Smaragdina: Ein Beitrag zur Geschichte der hermetischen Literatur* (Heidelberg: Carl Winter's Universitätsbuchhandlung, 1926).

[19] Keynes 28, 2r.

and that this "one thing" generated such a plurality by acting on itself. Medieval and early modern alchemists, intent on modeling their chrysopoetic efforts on the generative processes of nature, would therefore frequently assert that the overall process for making the philosophers' stone should employ only one essential ingredient, avoiding the addition of heterogeneous materials. Geber's famous *Summa perfectionis*, for example, stated that "there is one stone and one medicine in which the magistery consists, to which we neither add anything extraneous nor remove anything."[20] Employing the far more biologically oriented alchemy of the early modern period, Newton interpreted this ability of a unitary material to act on itself as a process involving fermentation. As we shall see, this had important consequences for his laboratory practice.

A second key feature of *Humores minerales* lies in the opusculum's insistence on activity in the vapor state. The "spirits" to which Newton alludes are merely materials that have been divided into such a state of subtlety that they become active and penetrating. The belief that metals form within the earth from the recirculation or cohobation and gradual condensation of subterranean mercury and sulfur fumes extends well into the Middle Ages, of course, and is already found well developed among the Islamic alchemists.[21] Early modern chymists such as Sendivogius and Grasseus were building on these ideas with the additional help of information from miners and metallurgists. As we saw in chapter four, numerous chymists even went so far as to argue that metals were not only formed within the earth, but that they also underwent degeneration there. This idea, already present in the *Sarepta oder Bergpostill* of Johann Mathesius, the Lutheran pastor of Joachimsthal, was adopted by various sixteenth- and seventeenth-century alchemists, including the highly influential Basilius Valentinus and Michael Maier. But Newton pushed the idea much further than his sources by making the paired degeneration and regeneration of minerals a cyclical process required in order to explain the presence of metals in the earth's upper crust.

In *Of Natures obvious laws* Newton would also advance beyond his sources in the insistence that he would place on the "intimate union & exertion of Spirits" taking place during mineral fermentation. Newton's term "intimate union" implies a microstructural rearrangement of particles, while "exertion," in his seventeenth-century usage of the term, refers to the discharge or emission of spirits.[22] What Newton had in mind was a profound transformation of matter divided into such minute particles that they could act on one another in an intimate fashion; in order to attain the required degree of attenuation, the materials in question had to be in a vaporous, or perhaps even gaseous, state. In that condition of extremely minute division,

<hr/>

[20] Newman, *Summa perfectionis*, 639.

[21] See, for example, the *De aluminibus et salibus* of pseudo-Rhazes, in Robert Steele, "Practical Chemistry in the Twelfth Century," *Isis* 12 (1929): 27: "Scias quod corpora mineralia sunt vapors qui inspissantur et coagulantur secundum mensuram servitutis nature in spatio longa."

[22] Smithsonian Institution, Dibner 1031B, 1r: "nature ever begins with putrefaction or fermentation whereby there is an intimate union & exertion of S^{pts} $\wedge$& purgation of impuritys." See "exert" and "exertion" in *OED*, consulted June 1, 2016.

the spirits could ferment or "vegetate" and transform into something radically different from their origin. The relationship of these mechanistic and hylozoist concepts is in fact the central problem addressed in *Of Natures obvious laws*, a natural progression of ideas from the purely alchemical cosmology presented in *Humores minerales*.

Conclusion: *Humores minerales* and *Of Natures obvious laws* as the Basis of Chymical Practice

Already in the Middles Ages, alchemists were arguing that chrysopoeia could only succeed if the practitioner mimicked or even replicated the processes by which nature itself creates the precious metals. As chymistry and mining technology underwent their rapid and closely intertwined development in early modern Europe and the Americas, a new and deeper relationship between miners and alchemists emerged, making it increasingly possible to model the practice of chrysopoeia on geochemical concepts. Newton's description of the cyclical generation and corruption of minerals and metals beneath the earth reflects a further elaboration of ideas drawn directly, and sometimes indirectly, from authors such as Paracelsus and his pseudepigraphers, as well as Johann Mathesius, Nicolaus Solea, Johann Grasseus, Michael Maier, and the school of writers who wrote under the pseudonym of Basilius Valentinus, in addition to the ubiquitous Sendivogius and Philalethes. And yet despite his reliance on these and other sources, Newton reworked this material into a novel and striking theory that combined the fermentation and putrefaction of metallic materials with concepts drawn from the mechanical philosophy. As our next chapter will elaborate, his view was that behind the phenomena presented to our senses by the action of gross corpuscles undergoing mechanical combination, analysis, and transposition, lay another world forever inaccessible to our sensory organs because of its extreme minuteness. This was not the universe of the classical atomists or of Cartesian mechanism, but a kingdom in which nature exercised its true intent by secretly guiding the processes of generation and corruption.

It was imperative both for the natural philosopher and for the chrysopoetic chymist to see behind the spectacle presented by macrolevel appearances and their immediate mechanical causes, to penetrate into the invisible realm of the metallic "seeds" or *semina*. The natural philosophy implicit in this view would develop further in Newton's published scientific works, especially in the growing list of queries that accompanied successive editions of his *Opticks* from 1704 to 1717. The practical chymistry, on the other hand, would resurface in the decades of laboratory work reflected in Newton's experimental notebooks belonging to the Portsmouth collection at Cambridge, and in the systematic studies that he made of myriad alchemical texts during those same years. Newton's cyclical theory of metallic generation purveyed in *Humores minerales* and *Of Natures obvious laws* provided him with a strong basis for the alchemical practice that would emerge in his laboratory notebooks. In the notebooks we see an extreme emphasis on reiterated sublimation and dissolution of materials that is entirely sui generis when compared to the practice of other seventeenth-century chymists. It is clear that Newton was focusing on reactions, as we would now say, that took place in the state of vapors or even gases, in order to produce substances of increasing volatility and reactivity. Despite the decades-long attempt to base his own laboratory practice on contemporary models, especially the method for making the sophic mercury provided by Eirenaeus Philalethes in his influential *Secrets Reveal'd* of 1669, Newton's unrelenting emphasis on vaporous reactions would lead him in a very different direction. In reality, his creative reading of Philalethes and other chymists allowed Newton to find his own idiosyncratic methods cryptically buried within their texts. The reason for this is not hard to appreciate when one understands the theoretical basis supplied by the early opuscula found in Dibner 1031B. *Of Natures obvious laws* would carry this process even further than *Humores minerales*, going so far as to use a famous experiment of Boyle, the "transmutation" of water into earth, to advise that the gross concatenations of matter could be "unraveled" and stripped away by mechanical means to free the latent spirit trapped within.[23] Laboratory techniques such as sublimation and dissolution in strong menstrua, even if they employed mechanical means, could liberate the active *semina* and make them able to interact with one another and ferment. In this fashion the "mineral spirit" might be made to "receive metallick life" and restore "the pristine metalline forme" to other minerals and metals. Under the controlled conditions of the laboratory, where heat and moisture could be regulated and the tender, mineral spirits incubated free from impurities in their own vessels, who could say what wonders might occur?

In reality, Newton never achieved the goal of freeing the metallic spirit and fermenting it with less mature minerals in order to produce the philosophers' stone. His laboratory notebooks show that even in the late 1690s he was still at the stage of making preparatory reagents rather than forming "our Embryo," as Philalethes described the nascent philosophers' stone gestating in its sealed incubating flask. Yet the multiple substances that Newton did produce, such as "the net of Vulcan," the "sophic sal ammoniac," "volatile Venus," and "Diana," were all either products of vaporous reactions or else materials that he would employ in subsequent sublimations (or in some cases both). Although we will have to wait for a later section of the present book in order to examine these practices and products in detail, it is beyond doubt that the ideas developed and recounted in *Humores minerales* and *Of Natures obvious laws* provided Newton with the theoretical background underlying his persistent and enduring pattern of laboratory experimentation in chymistry. In the next chapter I will examine *Of Natures obvious laws* in detail in order to bring to light his multiple yet simultaneous objectives for chymistry. Newton's aim of unraveling the secrets of the adepts cannot be disentangled from his goal of understanding the inner workings of the earth,

---

[23] Dibner 1031B, 3r, modeled on Robert Boyle, *The origine of formes and qualities* (Oxford: Richard Davis, 1666), 399–400.

and neither end can be separated from his desire to confute the billiard-ball mechanism of Descartes by framing a comprehensive new natural philosophy. While *Humores minerales* purveys Newton's response to the early modern fusion of alchemy and mining that led to a cyclical view of mineral generation and corruption, *Of Natures obvious laws* represents a further stage, a sort of chymical revery on the world. In a certain sense, *Of Natures obvious laws* provides the master key to Newton's long alchemical adventure, and for this reason it requires its own chapter.

EIGHT

Toward a General Theory of Vegetability and Mechanism

Written around the same time as *Humores minerales* or possibly a bit later, *Of Natures obvious laws & processes in vegetation* covers an astounding range of topics. The little treatise begins with a detailed consideration of the similarities and differences between mineral generation and that of animals and vegetables, then passes to a quite original theory of the different methods by which nature produces two common products, sea salt and niter, incidentally invoking the aerial niter theory of Sendivogius. After this, Newton presents his view that the earth is itself a living creature and uses its respiration to account for gravity, leading him into an intricate discussion of different "airs" as well as the relationship of even more subtle materials, namely, ether and the "body" of light, to one another. From here he launches into a discussion of God and provides several attempts to improve on the proofs that René Descartes had supplied for the existence of the divinity. In the final paragraphs of the text, Newton then returns to the theme of generation and employs the principle of vegetability, which he attributes to the action of tiny alchemical *semina*, to distinguish between the growth and activity imparted by nature from the more superficial processes of mechanism. Here too *Of Natures obvious laws* contains an implicit critique of the Cartesian mechanical philosophy based on principles derived from Newton's readings in chymistry.

A general comment is in order before we examine this fascinating material in detail. Newton clearly composed *Of Natures obvious laws* after his acquisition of Eirenaeus Philalethes's influential *Secrets Reveal'd* (1669), for he supplements his reading of Sendivogius, Grasseus, and other chymical authors with important elements taken from the "American philosopher." This use of Philalethes is particularly noticeable in Newton's repeated references to the multiple "regimens" or stages of the process leading to the formation of the philosophers' stone in a sealed vessel after the sophic mercury and a quantity of gold have been sealed up and heated for an extended period of time. As we discussed in chapter two, each of the regimens was supposed to display a particular color in the transparent, heated flask, and in *Secrets*

Reveal'd, Philalethes claims that each stage corresponds to a different planet in the traditional set of qualities attributed to each celestial body. Philalethes says the initial stage, corresponding to Mercury, is variegated, while the subsequent regimen of Saturn is primarily black, followed by Jupiter's multicolored hues, Luna's white, Venus's green, Mars's orange, and Sol's red. According to *Secrets Reveal'd*, each of the stages takes about thirty to fifty days to reach its completion.[1] *Of Natures obvious laws* accepts the veracity of the Philalethan regimens unconditionally, and Newton even goes so far as to build elaborate theories of generation and corruption on the account given of them by *Secrets Reveal'd*. This interaction between alchemical practice and Newtonian theory is far from accidental. When reading *Of Natures obvious laws* as well as *Humores minerales*, one should not forget that Newton's lucubrations not only share a chymical inspiration for these theoretical developments; they also reveal a common practical goal of fabricating the summum bonum of early modern alchemy, the philosophers' stone.

The Divisions of the Text: Section One

Although *Of Natures obvious laws* betrays a strong conceptual similarity to *Humores minerales*, the structure is significantly different. Whereas *Humores minerales* was a sustained set of questions and answers devoted to the single topic of the generation, decay, and regeneration of metals and minerals, *Of Natures obvious laws* appears to be a fragment of a commonplace book dealing with a much broader set of issues. The opusculum is structurally similar to Newton's student notebook, *Certain Philosophical Questions* (Questiones quaedam philosophicae), which he compiled for the most part while an undergraduate at Trinity College. Like *Certain Philosophical Questions*, it consists primarily of topical entries preceded by "Of," such as "Of ye actions & passions of grosser matter," which is found on the first folio of Dibner 1031B. These topics range from heat, light, fire, and cold to the nature of God, which are all topics covered by *Certain Philosophical Questions* as well.[2] Similarly, both *Of Natures obvious laws* and *Certain Philosophical Questions* contain numbered subtopics that divide a broader issue into parts that could be explored independently.[3] Both texts were conceived as tools of contemplation and discovery rather than finished treatises and reflect an active mind in the midst of working out a host of different questions. Perhaps the best way to view *Of Natures obvious laws* is as a continuation of *Certain Philosophical Questions* in which chymistry has come to play a central role that it lacked in the earlier text.

We can divide *Of Natures obvious laws* into five rough parts, providing a structure that will allow us to see the working out of Newton's thoughts.

---

[1] Philalethes, *SR*, 90–109.

[2] See Newton, *CPQ*, 330–35 for Newton's own index of the topics covered in the notebook.

[3] See the section headed "Magnetic attraction" on pp. 377–79 of Newton, *CPQ*, where Newton introduces six subtopics under the general entry.

The two sides of the first folio (1r–1v) use ideas drawn from chymistry to explore issues of "vegetation" across the three realms of the animal, vegetable, and mineral genera. In accordance with its Latin etymology (from *vegetatio*), Newton uses "vegetation" primarily to mean "growth" and "animation." In seventeenth-century English, the term did not have the strict association with the plant world that it typically does today. This section presents twelve numbered heads first followed by a section titled "Notes of agreement," where Newton systematically compares the similarities between animal and mineral generation, then passing to another section labeled "Dissimilitudes," where he points out the differences between them. A second conceptual division can be detected with Newton's next topical entry, "Of ye production of ye upper region from mineralls" beginning on 1v. Here we find material similar to that in *Humores minerales*: the same emphasis on mineral fumes appears in more developed form. But there is a highly significant difference in that *Of Natures obvious laws* does not deal with the generation of metals at this point, but rather with the production of salts. Our third main division appears at 2v, where Newton passes from the formation of salts in the upper crust of the earth to the subterranean generation of air. This leads into a fascinating discussion of the role of air and another thin, material substance, the ether, in the production of gravity. The topic of ether in turn leads him to consider the relationship between that subtle medium and an even less physically tangible "substance," namely, light itself. This section on air, ether, gravity, and light is followed by a fourth division—a succession of short, heterogeneous entries beginning after a blank section with "Of Heat" on 4r and continuing to "Of God" on 4v. These headings are set off by considerable empty space, especially after the entry on God, which is followed by two-thirds of a page with no writing. Finally, folios 5r through most of 6r make up a fifth section that is nominally devoted to a discussion of putrefaction and its role in generation and vegetation. Although the main subject is generation, corruption, and growth, these concluding folios actually provide the basis for a detailed discussion of the relationship between vegetability and mechanism, and the means of testing their boundaries.

One feature belonging to *Of Natures obvious laws* is worthy of immediate comment. Despite its many similarities to *Certain Philosophical Questions*, the text is markedly different in its treatment of chymistry. Newton's student notebook already makes frequent appeal to the works of Robert Boyle, especially in its consideration of colored bodies and of color changes in reactions, confirming the fact that even in the mid-1660s Newton had an acquaintance, albeit bookish, with chymistry.[4] The use of Boyle is also found in *Of Natures obvious laws*, but here Newton builds on hints and allusions that the older chymist had made to possible chrysopoetic implications in his work. Even more strikingly, Newton takes the enunciations of self-styled adepts like Sendivogius and Philalethes as accepted matters of fact, using them to support the general theory of generation and corruption purveyed by *Of Natures obvious laws*. The work is emphatically not a critical analysis

---

[4] See Newton, *CPQ*, 452–63.

of alchemical theories of growth and decay, but rather an excursus on those theories that uses their data as starting points for further speculation. This indiscriminate acceptance of chrysopoetic dicta as fact provides further support for the view that Newton had already enlisted himself among the ranks of the adepts, even if he realized that he was at this point a pupil rather than a master. His private allegiance to the sons of art would prevail throughout his remaining years in Cambridge, and for all we know, he may never have given up the dream of joining their number.

One can already see this feature of Newton's thought in *Of Natures obvious laws*' first paragraph. The twelve numbered heads begin with a discussion of the dendrites grown within flasks by early modern chymists. Like Sendivogius and Grasseus, among many others, Newton believed that metals and minerals grew within the earth, with the subterranean mineral veins corresponding to the branches of terrestrial trees. The artificial production of mineral dendrites provided a laboratory-based support to this theory and promised to give further clues about the invisible processes of metallic generation and multiplication within the earth. If Newton had performed experiments with dendritic formations at this early point in his career, and there is no evidence that he had done so, the products would probably have been such commonplace metallic vegetations as the "tree of Diana" made by placing a silver-mercury amalgam in a dilute solution of additional silver and mercury dissolved in nitric acid, or perhaps a silica garden made with "oil of glass" (potassium silicate) and ferric chloride. The tree of Diana was already a feature in chymical textbooks, and the alchemical entrepreneur of Amsterdam, Johann Rudolph Glauber, had popularized silica gardens a generation before Newton's efforts.[5] At the same time, Newton's major early source for the preparation of the philosophers' stone, Eirenaeus Philalethes, stated that the ingredients of the stone, when sealed up in a flask and heated, appeared "sometimes like to a pure silver Tree shining with branches and leaves."[6] Employing the parallelism and tension between art and nature that characterizes alchemical thought in general, Newton concludes that such factitious growths produced in the chymist's flask give evidence for the way that metals and minerals grow beneath the earth.[7]

But Newton is not content merely to argue for an identity between the subterranean mineral trees and those grown in a flask. His goal is to universalize. Thus he states that vegetation is "ye sole effect of a latent spt & that this spt is ye same in all things," differing only in its degree of maturity. This latent, or rather hidden, spirit often acts across a wide range of individuals, producing the mass fermentation of wines in autumn as well as the spread of "putrefaction," possibly referring here to disease ("ye contagiousnes of putrefact<ion>"). The loss of the latent spirit results in debilitation, as in the

---

[5] For a roughly contemporary description of the tree of Diana, see Nicolas Lemery, *A Course of Chymistry* (London: Walter Kettilby, 1677), 42–45. For the silica tree in Glauber's work, see Johann Rudolph Glauber, *A Description of New Philosophical Furnaces* (London: Tho. Williams, 1651), 20–21, 160–61.

[6] Philalethes, *SR*, 101. See also 105.

[7] For the pervasive occurrence of the art-nature dichotomy in alchemy and the traditional, premodern use of the aurific art as a focal point for discussion of the natural and the artificial, see my *Promethean Ambitions: Alchemy and the Quest to Perfect Nature* (Chicago: University of Chicago Press, 2004).

emission of seed during copulation, and here Newton is probably thinking of sexual generation across the three kingdoms of nature. He concludes the paragraph with a final numbered topic that is clearly a point to be explored later: "Why the two Elixirs are the most nourishing amicable & universall medicine to all beings what ever." This appeal, which expresses no doubt whatsoever as to the reality of the "two elixirs," refers to the philosophers' stone in its white and red forms. Traditionally, the white or lunar stone was thought to be made in an earlier stage of the process that would lead in due course to the solar, or red form. The lunar stone could either transmute base metals into silver or heal human diseases according to a common view, while the red stone was the chrysopoetic agent par excellence. The point of Newton's comment is that these marvelous products were "medicines" for both the bodies of humans and of metals; hence, their curative operation was universal. He takes the existence of the two elixirs as a matter of fact and links their beneficent effect to the operation of the hidden spirit.

These initial points are followed by an additional twenty-six numbered "Notes of agreement," consisting primarily of comparisons between mineral and animal generation, growth, and corruption, though with occasional references to the plant world as well. Here too Newton mines the literature of the aurific art, drawing evidence for generation in general from the differently colored regimens leading to the philosophers' stone after the sophic mercury has been sealed up in a flask with gold. Only when we keep this in mind do Newton's remarkable subsequent comments make sense. He is using the regimens as a generational model in part because the sequential stages of development as revealed by their changing colors in a sealed glass flask are visible to the eye of the chymist, at least according to his sources (among whom Philalethes no doubt figures prominently). The passage is vaguely reminiscent of Aristotle's famous use of developing eggs in his *Historia animalium* and *De generatione animalium*. In the same way that Aristotle dissected chicken eggs at various stages of their gestation in order to reveal the conversion of embryo to chick, so Newton employs reports of the developing philosophers' stone within its transparent vessel during successive regimens to arrive at greater knowledge of assimilation and growth across the realms of nature.[8] This is the purport of the following entries, particularly numbers fifteen and eighteen:

> 15 after conjunction the matter is apt to grow into all figures & colours though transitory because ye motion is not yet terminated. 16 In ye same Oare severall metalls are found all wch vegetate distinctly. 17 That salt cheifly excites to vegetation 18 That in ye first days of ye stone green is ye only permanent colour & so in ye least mature vegetables.[9]

The "conjunction" to which Newton refers is the assimilation of the gold by the Philalethan sophic mercury that occurs at the beginning of the regimens leading to the philosophers' stone. In *Secrets Reveal'd*, Philalethes had said

---

[8] Aristotle, *Hist. an.* VI 3 561a4–562a21; *De gen. an.* III 2 753b17–754a15.
[9] Dibner 1031B, 1r.

that the two materials will produce "divers colours" for the first twenty days in their heated flask, to be followed by "a most amiable greenness" lasting another ten days; only after this green stage does the matter within putrefy "like unto a coal in blackness" to be followed in succession by other colors.[10] Echoing this language, Newton says that the colors are initially "transitory," and then followed by green, "ye only permanent colour" in the first regimen leading to the philosophers' stone.

Unlike Philalethes, however, Newton draws lessons from this "experiment" that can be applied to human generation and the assimilation of nourishment. The "fact" that the elixir in both its white and red forms unites with our body and conserves it indicates that metals share a common principle of vegetation with our flesh. As in *Humores minerales*, the metallic vapors and fumes are constantly rising up out of the earth and interacting with the terrestrial biome. For further evidence of this interaction, Newton turns to "healthfull & sickly yeares, the barronnes of grownd over ₘᵢₙₑₛ &c." In other words, mineral exhalations lead to mass health or disease, and account for the absence of plant life in the vicinity of mining excavations. The ability of metals to combine with our bodies and participate in their growth leads Newton to conclude as follows: "therefore o^r bodys vegetate as they doe in a glas." Behind this seemingly offhand comment is a justification of Newton's application of the philosophers' stone to the study of human generation. Because our bodies grow as metals do, we can learn the secrets of human biology by observing the gestation of the sophic mercury as it passes from an inchoate amalgam to the perfection of the philosophers' stone.

Despite this strong affirmation of the analogous relationship between biological and mineral vegetation, however, Newton notes in the following passage marked "Dissimilitudes" that there are also significant differences. Metals have no stable shape, unlike our bodies. Also, they are "augmented in vertue" during their growth, probably referring to the path from putrefaction to the philosophers' stone again. Additionally, our growth does not involve this total putrefaction, unless one thinks that this occurs when the male seed is deposited in the female body. They attain a "supereminent fixity," moreover, and grow without need of air, unlike us. And in a rather obscure comment that could refer to the philosophers' stone and its putative ability to transmute many times its own weight of metal into gold, Newton says, "They can convert ^2 or 3 nay 10 ^or more times their owne weight of nourishment at once." Finally, he concludes with the observation that in metals, every part is "sperme," whereas in animals it is only a tiny portion of our bodies.

The Generation of Salts in *Of Natures obvious laws*: Section Two

This rather intricate discussion then passes to a consideration of "the production of the upper region from mineralls" in our second division of the text. Newton begins this section in a way reminiscent of *Humores minerales*, saying

[10] Philalethes, *SR*, 81.

that "mettalls dissolve in divers liquors to a saline or vitriolate substance." In- stead of launching into a consideration of the putrefaction induced by fumes meeting these vitriolic liquors as he did in *Humores minerales*, however, New- ton now takes the discussion in a different direction. He jumps into an appar- ently quite original treatment of the formation of sea salt and niter by means of a putative interaction between water and the metallic fumes that rise up from the earth's depths. Although Newton's ideas about sea salt and niter are his own, it will be necessary here to introduce another contemporary source that may have been the immediate pretext for his theory of saline generation.

I refer to Bernhard Varenius's well-known *Geographia generalis*, a work that Newton himself edited and published in Cambridge in 1672. It is im- portant to examine Newton's debt to this influential author, for not only does it illuminate the immediate context that led him to write *Of Natures obvious laws*, it also reveals the degree to which Newton was combining ideas from very different scientific genres. Varenius is known primarily as a geographer, but the *Geographia generalis* contains scattered comments on the generation of metals and minerals, as well as statements about the char- acter of the ocean's salinity and on salts more generally, which Newton took quite seriously. Varenius often acknowledges his source in chymical matters to have been one "Thurnheuserus," who may well have been the prolific al- chemical writer Leonhard Thurneisser zum Thurn (born before 1531–died 1596). Among Thurneisser's bewildering array of publications was one on the nature of mineral waters, which could well have been known to Var- enius.[11] At any rate, the author of the *Geographia* displays a keen interest in the formation of mineral waters, which leads Varenius into the related area of metallogenesis as well. Like Grasseus, Varenius proposes that underground water can dissolve salts and vitriol—he adds sulfur as well—and this water is thereby impregnated with such minerals. Having a pronounced atomist streak, Varenius says that such mineral waters can in turn dissolve the metal- lic granules that they encounter into atoms, which they then unite with. As a result, "corporeal mineral waters" are formed, "which contain solid particles of minerals (*fossilia*), but so small, minute, and thoroughly mixed that they cannot be made out by sight," although they can settle in due time, like "the chymical waters in which metals <are> dissolved."[12]

The meaning of "corporeal mineral waters" becomes clearer when Var- enius passes to the generation of metals proper. The metals are generated beneath the earth when "vapors and fumes are condensed on the protruding angles of the rocks, to which they adhere; first they come together into a soft substance, and then they are condensed." Hence, although Varenius does not use the miners' and chymists' term "Gur," he too thinks that the metal- lic vapors can pass through a soft, immature stage on their way to becoming full-blown metals.[13] Moreover, Varenius adds that waters can penetrate into

<hr />

[11] Leonhard Thurneisser, *Pison: Das erst Theil, Von Kalten, Warmen Minerischen und Metallischen Wassern* (Frankfurt an der Oder: Johan Eichorn, 1572).

[12] Bernhardus Varenius, *Geographia generalis* (Cambridge: Henricus Dickinson, 1672), 189.

[13] Varenius, *Geographia* (1672), 190.

the areas that contain these immature metals and metallic fumes, with the result that "they are impregnated by them, and they thus become spiritual, metallic, mineral waters." In other words, a volatile, metallic component derived from the still imperfect metals can penetrate into the water to produce "spiritual," metallic solutions. In contradistinction to the "corporeal mineral waters" formed by acid dissolution, the "spiritual mineral waters" are fully volatile—they do not leave behind a fixed residue when evaporated. In sum, Varenius allows for two modes by which metallic waters can be generated—either by direct solution of the immature metallic fumes in water or by solution of the fully formed metals in subterranean acidic solutions. In the former case, a totally volatile solution is formed, while in the latter, metallic waters of a fixed nature arise.

It is likely that Newton's introductory lines about saline generation are loosely inspired by Varenius's discussion of sea salt, for both Newton and the author of the *Geographia generalis* make the seemingly odd claim that while the sea is saltier in the tropics thanks to the higher volume of fresh water evaporated off by the sun there, seawater cannot be rid of all its salt by means of distillation.[14] Indeed, Newton's words betray the direct influence of Varenius's assertion that seawater contains both a fixed salt that is left behind in distillation and a volatile salt that evaporates with the water: "Because the sea is perpetually replenished wth fresh vapours it cannot bee freed from a salin tast by destillation, that salt arising wth ye water wch is not yet ~~indurated~~ concreted to a grosser body."[15] This passage surely recapitulates the following words of Varenius:

> The Learned Chymists, or true Naturalists, have hitherto laboured in vain, that they might find out an Art by which they might distill and abstract fresh water from the water of the Ocean, which would be of great advantage; but as yet their Labours have proved fruitless: for although, as well in the decoction as distillation, Salt may be left in the bottom of the Vessel, yet the water separated by decoction as well as distillation, is yet found salt, and not fit for men to drink, which seemeth wonderful unto those that are ignorant of the cause. Yet Chymistry, that is, true Philosophy, hath taught the reason; for by the benefit of this we know that there is a twofold salt in Bodies, or two kinds of salt, which although they agree in tast, yet they much differ in other qualities: one of these Artists term fixed, the other volatile salt. The fixed salt, by reason of its gravity, is not elevated in distillation, but remaineth in the bottom of the Vessel; but the volatile salt is full of spirit, and indeed is nothing else but a most subtile spirit that is elevated by a very light fire, and therefore in the distillation ascendeth with the fresh water, and is more firmly united by reason of the subtilty of the Attoms.[16]

[14] For these two claims in Varenius, see Varenius, *Geographia* (1672), 109 and 112.

[15] Dibner 1031b, 1v.

[16] [Bernhard Varenius and] Nicolas Sanson, *Cosmography and geography in two parts, the first, containing the general and absolute part of cosmography and geography, being a translation from that eminent and much*

Varenius is employing much the same line of reasoning that he used in his discussion of volatile "spiritual metallic waters" and "corporeal metallic waters." In nature, tiny atoms of light weight are found mingled in with larger, heavier ones; distillation merely separates the two types of particles by raising the smaller and leaving the bigger behind. Hence it is possible for the smaller atoms of the volatile salt to ascend while the larger, fixed ones remain behind, just as the spiritual metallic waters could be completely distilled while the corporeal ones left a residue on their distillation. The same ideas linking subtlety to volatility and grossness to fixity pervade Newton's reasoning as well, and indeed are standard features of medieval and early modern alchemy.

But Newton differs markedly from Varenius in bringing niter into his discussion of salts. Probably stimulated in a general way by Varenius's claim that sea salt contains components of varying volatility, Newton asserts that niter is a looser, less fixed salt than sea salt, and that the difference between the two salts arises not from a chymical diversity between their ingredients but rather from the fact that the niter is made when the metallic fumes combine with "subtile invisible" water vapor, whereas sea salt originates from the combination of the volatilized metals with liquid water or mist. A preponderance of water causes the fumes to be "overwhelmed & drowned," which kills their fermentative activity and results in the immediate formation of sea salt.

Newton lays out his justification for this interesting theory of niter and sea salt by invoking evidence drawn from the laboratory and from the world at large. First, he asserts that "the fixt salt left in ignition returns to <niter> by dissolution." This is surely a reference to Robert Boyle's famous analysis and synthesis of saltpeter, first described in his 1661 *Certain Physiological Essays*, and then elaborated in the 1666 *Origin of Forms and Qualities*. Boyle's experiment, as we already discussed in its relation to Newton's early optical discoveries, worked by injecting a red-hot coal into saltpeter and thereby igniting it. The product of this ignition, which we now call potassium carbonate, was then dissolved in nitric acid to produce further saltpeter. Boyle recognized that the initial and final products were the same, and therefore called the process a "redintegration" (resynthesis) of niter. Interestingly, Newton here seems to focus solely on the physical features of the experiment—the fact that the fixed salt left by ignition is "dissolved" into saltpeter, without considering the chemical fact that the solvent has to be nitric acid. This omission on Newton's part is a calculated move intended to bring the experiment into conformity with his theory, whereby the looser, more subtle niter is formed by mere mechanical "dissolution" of the more fixed and impassible potassium carbonate. Newton then launches into a detailed comparison of niter and sea salt in the world at large in order to confirm his idea that sea salt is merely a more fixed version of niter:

---

esteemed geographer Varenius (London: Richard Blome, 1693), 79. For the Latin of this passage, see Varenius, *Geographia* (1672), 112.

Hence also little or no ☉ is in ye sea for its <illeg.> same <illeg.> insensible quantitys there may be in it <illeg.> becaus ye grosse water stifles all or ye far greatest p̱t of the exhalation the aire indeed is replenished wth this exhalatiō from neibourig regions & so may impregnate rain water wth wth <sic> niter & so it may replenish receive niter from rivers but <illeg.> ye proportion is inconsiderable compared to all those vapors y^t arise into it. And all this will appear more then conjectur by considering 1 fums do arise ^plentifully, 2 and fumes they will abide wth water in a pellucid form & 3 therefore appear in evaporation of a saline forme. 4 they must therefore produce something like salt copiously 5 there are noe such p̱oducts but ⊖ & ☉ generally found: 6 These are generally washd down by ye descent of water hence ☉ is most copious in houses & dry places, hence also the sea is salter y^n the earth 7 these salts would therefore soone <illeg.> vanish if they were not constantly new generated & this is further confirmed by their bee plentifully produced in places where there was none before & where they could not bee had but out of y^e <illeg.> vaporous air nay that it descends wth rain yet in that saline form <illeg.> ascend wth it $^{it descends}$ is two gros to ascend wth it [that tis noe stranger for it to præcipitate out of vapors upon rock y^n out of water upon the sides of a vessell.] They ^are therefore constantly generated & that out of a most subtil vapor y^t ascends wth as little heat as water.[17]

The upshot of this passage is once again that the metallic fumes permeate water or water vapor to produce either sea salt or niter respectively. But here Newton buttresses this claim with the argument that these two soluble chemicals would soon be washed down into the depths of the earth and therefore disappear from its surface unless they were in fact regenerated constantly. This counterfactual passage is quite reminiscent of the ideas already discussed in *Humores minerales*, where Newton argued that metals and their ores were constantly being washed downward and were too fixed to resublime as such—they therefore had to be regenerated from the interaction of fumes and metalliferous waters just as niter and sea salt are generated from water vapor and water descending in the form of rain.

The main thrust of Newton's argument so far has been "the production of the upper region from minerals," as his heading announced, and in the following lines he pursues this topic, arguing that salts "concrete" or crystallize into rocks, precious stones, and sand beneath the earth. These in turn are gradually crushed by nature's action into clay, which is nothing but powdered stone. Hence all of these natural products originate ultimately from the same metallic fumes that "wander over the earth," as Newton said in *Humores minerales*, and produce "stones, salts, and so forth." But there is more to this section than a mere discussion of generation for its own sake. In reality, Newton is already thinking of the distinction between vegetation and mechanism, and the way in which these two fundamental causal agencies interact with each other. The underlying concept guiding Newton's

[17] Dibner 1031B, 2r–2v.

view of vegetation is principally the aerial niter theory of Sendivogius, a fact that the English natural philosopher reveals in a canceled passage before his treatment of redintegration. In speaking of the more open texture of salts produced from metallic fumes that associate slowly with water vapor as opposed to being "drowned" in the liquid, Newton refers explicitly to Niter as a "ferment" or agency: "y^t spt being the ferment of fire & blood &c." This is undoubtedly a reference to the Sendivogian aerial niter, which was supposed to provide heat and life to the body as well as serving as the "food" of fire. Although these words are struck through, Newton returns to the same ideas a few lines later in a passage whose meaning would be obscure without the canceled section. Here he refers to the relationship between the aerial niter and ordinary niter or saltpeter, in terms of "ye affinity of y^t spt wth niter." He then glosses "that spirit" with the following insertion:

^y^t ^{is} y^e <*illeg.*> ferment of fire & all vegetables ye other most apt to take fire & most <*illeg.*> promoting vegetation of all salts.[18]

In his abbreviated way of working out these ideas, Newton is contrasting the ferment of fire and all vegetables, namely, the Sendivogian aerial niter, with "the other," that is, ordinary saltpeter, which as an essential ingredient of gunpowder is indeed apt to take fire, and as a source of nitrates is also able to promote vegetation in the modern sense of the word. Behind his discussion of the mechanical difference in texture that produces sea salt from saltpeter lies a deeper concern with the nontextural features of niter that allow it to act as an agent of vegetation.

These ideas of Sendivogian inspiration are followed by additional nitrous ruminations. Newton argues in a heading labeled "Of sal ✳. Sal gemmæ &c. alume &c" that salts actually impede generation until they are "incited" to it by some other agent. Summoning up another element often used in support of the aerial niter theory, Newton then adds that this is why salts are used to preserve meat, as in the cases of salt pork and corned beef. And yet despite this preservative quality, under the proper circumstances the same salts can themselves be made to decay. Thus when salts are made to putrefy in the upper crust of the globe, they generate "a sort of blackish rotten substance," in other words common humus. When plants grow out of fat earth, they are actually assimilating this putrefied salt, and when they die, they return to it. In support of his claim that ordinary soil is for the most part putrefied salt, Newton again adduces the phenomena described in the regimens leading to the philosophers' stone, which typically included putrefaction at an early or initial stage: "Nay since metalls may putrefy into a black fat rotten stinking substan<ce> why not earth $_{salts}$ also." The black, fat, rotten, stinking material is the sophic mercury after it has begun acting on gold and digesting it. As Philalethes says in *Secrets Reveal'd*, the early stages in making the philosophers' stone are accompanied by "a most black colour, and a most stinking odour."[19] Again we witness Newton's reliance on Philalethes for the regimens of the alchemical process.

[18] Dibner 1031B, 2r.
[19] Philalethes, *SR*, 22.

Newton then steps back from his theory of subterranean generation and points out that despite the fact that salt, stones, earth, and even water ultimately descend from the "metalline nature," their physical textures have been so changed and wrought on that they "hinder or destroy ye work," if mixed with the sophic mercury in the generation of the philosophers' stone. The reason for their alienation from the underlying metallic substance is obvious: they have lost their original subtlety and become thick, material bodies not on account of vegetation, but because of "a gros mechanicall transposition of pts." As a result of this confusion and agglomeration of the corpuscles making them up, such earthly bodies must be "reduced back to their first order & frame" before they can enter into the work of transmutation. This reduction, moreover, cannot be performed by vegetative means, since the original conversion of the metallic fumes took place mechanically. It is helpful to recall Newton's theory of the generation of sea salt and niter here. They were not formed by vegetation, but by the mechanical interaction of the metallic fumes with water, either in its vapor state to yield niter, or in its liquid form to generate sea salt. Because of their origin, they must be treated first by mechanical means if we wish to reduce them back to their primitive state.

At this point, Robert Boyle reenters the discussion, in even more remarkable guise than he did in the passage where Newton mustered the redintegration of niter as a support for his transformation of niter into sea salt. Although Newton does not mention Boyle by name, the following passage depends tacitly on the experimental section in the older chymist's *Origin of Forms and Qualities*:

> Yet ye reduction of these <*illeg.*> is possible to bee performed by <*illeg.*> mechanicall ways unravelling their production. Water by ye suns heat & by assention & descention will yeild earth as hath been tryed by distilling it often, Also ^standing water w will putrefy by the suns heat, corrupt & let fall a fæculent earth & that <*illeg.*> successively w^{th}out period. Out of these earths may be extracted a salt. This salt may be brought to putrefy & ye minerall sp^t thereby set loose from ye water wth wch it was concreted & so returnes to ye same state it had when at <*illeg.*> its first ascent out of ye earth y^t is to ye nearest metalline matter & (though debilitated by these changes) yet <*illeg.*> if pervading ye earth where other metalls vegetate might enter them ^receive metallick life & by degrees recover ^their pristine <*illeg.*> metalline forme.[20]

In this extraordinary passage, Newton enlists one of Boyle's most famous demonstrations of the mutability of substances, namely, "the Transmuting of Water into Earth," which appears as "Experiment IX" of the *Origin of Forms and Qualities*. There Boyle discusses how he produced a fine, fixed, white earth by reiterated distillation of rainwater. Although he was aware of the possibility that the earth might simply consist of glass powder that had been "fretted" by the hot water, consultation with an unnamed colleague who had

[20]Dibner 1031B, 3r.

performed a similar experiment led Boyle to conclude that the water had genuinely transformed into earth.[21] Because the experiment seemed to open a door to the possibility of transmutation in general, *The Origin of Forms and Qualities* refers to it as a potential "*Magnale* in Nature" and explicitly states that it might make "the Alchymists hopes of turning other Metals into Gold, appear less wild."[22] After all, Boyle adds, two bodies of entirely different degrees of fixity and specific weight have been shown by the experiment to be mutually transmutable. Why should it be impossible, then, for a light, volatile metal such as tin to become heavy, permanent gold?

Newton takes these rather cautious but suggestive comments by Boyle and builds an alchemical revery on them. It is not merely the "assention & descention" of water by distillation that can lead to earth, but also mere sitting and stagnation. And this earth, whether produced by distilling or sedimentation, can then be mechanically unraveled to reduce it back to its original metalline nature. A salt may be extracted from the earth, presumably by lixiviation, which can then be induced to putrefy by means that Newton leaves unstated. This will free the metalline fumes from the water with which they were originally "concreted" into a salt in the first place, and the spirit will thereby be released. Newton then seems to suggest that the whole operation may even have a practical goal rather than serving as a mere illustration. The "debilitated" metallic spirit freed by mechanical unraveling might regain its primitive potency if it were added to the "earth where other metalls vegetate." In other words, if the liberated metallic spirit is added to ores or minerals that have not yet touched the refiner's fire, a fermentation might take place. The result, Newton says, would be that the weakened spirit extracted from the artificial earth would receive "metallick life" from the still-living metals, and as a result "by degrees recover ^their pristine *<illeg.>* metalline forme." The use of the plural "their" in the final phrase reveals that it is the ores or minerals that will regain their original, pristine, metalline condition. The subject of the transitive verb "recover" (meaning "bring back") is the mineral spirit that has been extracted from the salt, and the object is the pristine, metalline form of the "other metals" already growing within the earth.[23] What Newton is suggesting is that the liberated spirit will act on the living ores in much the same way that the ascending metallic fumes in *Humores minerales* attacked descending metallic solutions and converted them to yet more spirits.

Despite the practical overtones of Newton's description, it seems unlikely that he is advising himself first to transform water into earth and then to reverse the process in order to free the hidden metallic spirit. After all, why would he have gone to the trouble of making a generic earth by Boylean means when it is a material found wherever humans make their habitation? More probably, he considered the sequence of operations leading from water

---

[21] Boyle, *The Origin of Forms and Qualities*, in *Works*, 5: 432–33; 1666, 399–400.

[22] Boyle, *The Origin of Forms and Qualities*, in *Works*, 5: 438; 1666, 417.

[23] For other examples of this archaic, transferred sense of the verb, see *OED*, s.v. "recover," at 6.a: "To regain or get back for another person; to bring back or restore to (also rarely unto, into) a person, country, etc."

to metallic spirit as a lesson in the mutability of matter, as Boyle intended it. Nevertheless, Newton's comments show that he viewed the extraction of a salt from earth (or from a particular "earth" or mineral), followed by the liberation of a spirit from that salt, and then by the use of that spirit as a ferment to return other minerals to spirits (their "pristine, metalline forme") as a feasible set of operations. As we will see in later chapters of the present book, this was precisely the modus operandi that Newton pursued in trying to arrive at the *arcana majora*, the higher secrets of the aurific art. His sublimations upon sublimations of salts and metals leading to materials of ever greater volatility, and his explicit references to fermentation as a goal in multiple experiments, surely reflect the attempt of a chymist trying to replicate the subterranean processes of geochemistry in order to arrive at the very tools that nature uses in making and transmuting metallic material. Here I must stress that the almost obsessive focus on extractions and sublimations followed by dissolutions, precipitations, and yet more sublimations, was not the normal operating procedure of seventeenth-century alchemy, despite the seemingly endless repetition of these processes in Newton's experimental notebooks. Even a cursory comparison with the laboratory practices of other chymists, especially those of Newton's favorites Boyle and Starkey, shows that they did not follow this path.[24] Newton's alchemical experimentation, though heavily informed by his knowledge of contemporary chymistry, is in fact sui generis. It represents a highly idiosyncratic interpretation of Newton's sources as seen through the heuristic spectacles summarized in *Humores minerales* and *Of Natures obvious laws*.

Air, Ether, and the Earth-Vegetable: Section Three

After terminating his section on the production of salts, Newton passes to a new topic in *Of Natures obvious laws*' third division. Here he is concerned with the generation of "air," meaning what we would today call diverse gases. He makes it clear at once that he means actual gases rather than vapors: the latter condense into "water" when cold, while the former do not. This observation is found at the head of a numbered list consisting of five entries that describe different means by which air can be generated or set free from a material matrix in which it is trapped. The first method by which air is liberated is, "out of water by freezing it." This probably refers to a set of experiments that Newton carried out in January 1670 (New Style) and then wrote up in his long experimental notebook, CU Add. 3975.[25] He took a glass vial partially filled with water, inserted a tube into it, and froze it with a mixture of snow and salt. As the water froze to ice, it rose up in the tube to a height that Newton marked. When he thawed the frozen water again, he noted

---

[24] For Starkey's experimental work, see Newman and Principe, *LNC*. For Boyle's alchemy, see Principe, *AA*.
[25] CU Add. 3975, 20v–21v. Newton says on 20v that he performed the experiments on "Christmas Ian 28, 29, & 30, 1669." These must be Old Style dates where January is included in the same year as the previous December. Given the fact that Old Style dating was also eleven days behind New Style in Newton's day, all the dates given would fall in New Style January 1670.

that bubbles were released, and that these consisted of "permanent aire." Repeating the experiment in a more quantitative manner convinced Newton that the air had been generated by the act of freezing; it had not merely been absorbed and then released, as we would say today.

Entries two through four are of chymical character and indeed concern phenomena known to all chymists in Newton's day. They refer respectively to the generation of gases by solutions mixed together, such as occurs when acids react with alkalies, the production of gaseous fumes by destructive distillation of sulfates, nitrates, and chlorides to make the mineral acids, and the release of gas that often occurs in the dissolution of metals in acids. Finally, Newton adds a fifth entry to account for "airs" produced by fermentation. His examples are the "flying" of bottled beer, presumably meaning its carbonation, and "swelling after a stroake," perhaps simply referring to swollen flesh after an injury. In all five of these cases, Newton says that the "constringed air" is liberated because the internal "parts" or corpuscles of a body are set to working among themselves. The fact that it is so easy to make tangible bodies release air provides evidence, he says, of the fact that most terrestrial materials are "but Æthereall concretions," a topic that he will develop at length in the next paragraph. Before passing to that argument, however, Newton cannot resist bringing the philosophers' stone back into his discussion. After listing his five methods of producing or liberating air, he then considers the reverse operation, where air is reduced to a "gross body" by combination with another material. Almost certainly drawing again on Philalethes, Newton says that this occurs in the early regimens:

> I know but one instance & y^t in ye stone wher during ye firs solution much air is generated, enough to burst a weak glas <*illeg.*> wch yet returns after to ye stone againe.[26]

This putative release and return of air is described several times in the recounting of the regimens given by *Secrets Reveal'd*. Philalethes urges first that the glass flask must be strong in order to contain the winds released "in the forming of our *Embryo*" and later claims that directly before the stage of blackness and putrefaction, "the Winds are ceased."[27] Not unreasonably, Newton has interpreted these passages to refer to a release and reabsorption of gas within the alchemist's sealed vessel.

Having established that gross matter can both release and combine with air (and ether), Newton then passes to one of the most remarkable passages in *Of Natures obvious laws*. Here we find a melding of alchemy with Newton's early thoughts about the mechanical origin of gravity. The fact that the juvenile theory of gravity presented in an alchemical context by *Of Natures obvious laws* relies on the impact of invisibly small corpuscles rather than employing force at a distance belies the claim of earlier scholars, such as Westfall and Dobbs, that Newton's concept of gravitational attraction

---

[26] Dibner 1031B, 3v.
[27] Philalethes, *SR*, 61–62, and 83.

stemmed from his alchemical readings.[28] Already in *Certain Philosophical Questions*, Newton had mined and built on the *Two Treatises of Body and the Soul* (1644) by the virtuoso and cavalier Kenelm Digby in order to explain the falling of bodies.[29] The abbreviated theory that Newton presents in *Certain Philosophical Questions* argues that bodies are carried down toward the center of the earth by an unidentified matter that passes through their pores in its descent and in the process pushes them down. Because of its extreme subtlety and rapid speed, this matter does not stop its passage at the surface of the earth, but penetrates to a point deep within the planet. In a way similar to Newton's counterfactual arguments about the necessity of regeneration in the case of salts and metals, *Certain Philosophical Questions* points out that this putative material underlying gravitation would "swell the earth" as a result of its collection within the planet if it did not return back its surface. It must therefore reascend after its descent, but unless it were somehow changed in form, it would have the same force in its reascent as it did in its fall, and there would be no gravity. Thus Newton speculates that the corpuscles of the reascending matter are of a "grosser consistency" than when they fell and rise at a lower speed, meaning that they can no longer penetrate the pores of the bodies that they carried down. Thus in rising up again, when they meet the stream of finer, descending particles and bodies, the ascending corpuscles are merely pushed aside and do not inhibit their fall.[30]

Some of these ideas resurface in *Of Natures obvious laws*, but here they are given an explicitly chymical vesture. Thus the rising matter that exhales from the earth is now attributed to "minerall dissolutions & fermentations," in accordance with the experimental evidence that Newton adduced in the previous paragraph. No doubt alluding to the *Witterungen* described by Basilius Valentinus and other early modern chymists, Newton says that these mineral exhalations are "very sensible in mines." The air produced in this fashion rises up into the atmosphere until it loses its gravity in the ethereal regions. The ether is "comprest thereby & so forced continually to descend," whereupon it serves the same role of carrying down bodies that it encounters as the subtle matter did in *Certain Philosophical Questions*. Interestingly, Newton supplies a rough estimate of the amount of air being generated. Supposing that the air rises to the height of a mile in three or four days, "wch it may doe wth a very gentle ~~motion~~ & insensible motion," Newton states that this would be equivalent to a layer of water five feet deep around the globe; apparently at this time he viewed the density of air as being about a thousandth of water's. His point is that this massive release of air and its entry into the zone

[28] Richard Westfall, "Newton and the Hermetic Tradition," in *Science, Medicine, and Society in the Renaissance*, ed. A. G. Debus (New York: Science History Publications, 1972), 2: 183–98, see especially 193–94. For the view of Dobbs on alchemy and gravity, see her *Foundations of Newton's Alchemy* (Cambridge: Cambridge University Press, 1975), 211–12. Dobbs later backed away somewhat from her claim that alchemy was responsible for Newton's move to an immaterial gravitational force. See *JFG*, 15 (for the "Clavis") and 207–8, where she admits that "the story no longer seems quite so straightforward."

[29] Newton, *CPQ*, 288–93, 393. For Digby's theory of gravitation, see his *Two treatises in the one of which the nature of bodies, in the other, the nature of mans soule is looked into in way of discovery of the immortality of reasonable soules* (Paris: Gilles Blaizot, 1644), 76–85.

[30] McGuire and Tamny, *Certain Philosophical Questions*, 363–65, and 427.

of the ether must compress the ether and force it back down to the earth. But just like the subtle matter of *Certain Philosophical Questions*, Newton's ether penetrates deep below the earth's surface. The full integration that he has made between his mechanical view of gravity and the geochemistry of Sendivogius becomes apparent in the following passage, which gives cause for thought, or rather for astonishment in its treatment of the earth as a living organism. The alchemical aerial niter has become the ether, or at least a vehicle for it:

> there tis gradually condensed & interwoven wth bodys it meets there ^& promotes their actions beeing a tender fermet. but in its descent it endeavours to beare along w^t bodys it passeth through, that is makes them heavy & this action is promoted by the tenacious elastick constituon whereby it takes ye greater hold on things in its way; & by its vast swiftness. Soe much Æther ought to descend as air & exhalations ascend, & therefore y^e Æther being by many degres more thin & rare then air (as air is y^n wather) it must descend soe much the swifter & consequently have soe much more efficacy to drive bodys downward then air hath to drive them up. And this is very agreeable to natures proceedings to make a circulation of all things. Thus this Earth resembles a great animall ^or rather inanimate vegetable, draws in æthereall breath for its dayly refreshment ^& vitall ferment & transpires again wth gross <illeg.> exhalations, And according to ~~natures~~ the condition of all other things living ought to have its times of beginning youth old age & perishing.[31]

Just as Sendivogius's aerial niter penetrated the earth, collected in subterranean pockets, and combined with the tellurian materials located there to form different metals and minerals, so Newton's ether is condensed and "interwoven" with the terrestrial substances that it encounters in its underground passage. Unlike the alchemical aerial niter, however, Newton's ether has acquired a quite different function, namely, the mechanical impulsion of bodies that forces them to descend. Without question we are witnessing Newton's hybridization of two distinct traditions—one of them originating in the ancient atomists' attempts to explain the fall of bodies as a rain of atoms, a school of thought that reached its full maturation under the aegis of the seventeenth-century mechanical philosophy, and the other a distinct alchemical school of thought descending from the cosmic circulatory system described in enigmatic terms by the *Emerald Tablet* of Hermes and filled out by early modern chymists such as Sendivogius and his followers.

Yet Newton was not the first to graft these two traditions into a unified cosmic system. I have already pointed to the influence of Digby's *Two Treatises of Body and the Soul*, a work that alludes to the chymists' theory that there is a central sun within the earth that may "raise up vapours, and boyle an ayre out of them, and divide grosse bodies into atomes."[32] Here and else-

---

[31] Dibner 1031B, 3v.
[32] Digby, *Two Treatises*, 89.

where Digby combines his protomechanical philosophy with ideas taken from chymistry. In his later discourse on the powder of sympathy (1658), Digby would even explicitly invoke the aerial niter theory and quote its author to the effect that air contains a "hidden food of life" (*occultus vitae cibus*).[33] Another English contemporary whose use of the aerial niter theory is well known was Newton's famous adversary in the 1670s, the mechanical philosopher Robert Hooke.[34] In his 1665 *Micrographia*, Hooke argued at length that the reason why combustion takes place is because of a chymical dissolution of bodies. As he puts it, "Air is the *menstruum*, or universal dissolvent of all *Sulphureous* bodies," so that their destruction by fire is literally a solution like that of metals in an acid. According to the *Micrographia*, it is not the whole body of the air that acts as this menstruum, but rather a component mixed in with it, "that is like, if not the very same, with that which is fixt in Salt-peter."[35] Clearly we are back in the realm of the aerial niter. At other points in his *Micrographia*, Hooke hinted at an ethereal explanation of gravity and even went so far as to suggest that vapors endowed with an elastic quality generated within the center of the earth are the cause of mountains and earthquakes.[36] Newton was keenly interested in the *Micrographia*, as copious notes of his from around 1665 testify, and it may well have helped to stimulate his early interest in chymistry. But of course Hooke, unlike Sendivogius and Philalethes, was not an adept.[37]

Neither Hooke nor Digby, despite their use of the aerial niter theory, show Newton's commitment to the Sendivogian model of the organic earth, which he unforgettably describes as a living being whose respiration accounts both for the renewal of the atmosphere and for the gravitation of falling bodies. This giant organism is an "inanimate vegetable" rather than an animal, presumably because it lacks the sensitive and motive faculties associated traditionally with animals but retains the nutritive capacity of plants. One must take "inanimate" in its most literal, technical sense here, meaning that which lacks an Aristotelian animal soul. There can be no doubt that Newton means to say that the planetary globe is alive, since it is continually breathing. Its inhalation of the subtle ethereal material accounts for gravitation and the maintenance of life, while its exhalation of "gross," depleted matter provides the second half of the circulatory process that allows its continued existence. Nonetheless, it ages as any living creature does, and will eventually meet an end.

From the earth and its respiration Newton turns to a closer description of the material that it breaths in, the ether. Here too one can detect the powerful influence exercised by his alchemical reading:

[33] Kenelme Digby, *A late discourse made in a solemne assembly of nobles and learned men at Montpellier in France touching the cure of wounds by the powder of sympathy* (London: R. Lownes and T. Davies, 1658), 36.

[34] For Hooke and niter, see Robert Frank, *Harvey and the Oxford Physiologists* (Berkeley: University of California Press, 1980).

[35] Robert Hooke, *Micrographia; or, Some physiological descriptions of minute bodies made by magnifying glasses with observations and inquiries thereupon* (London: Royal Society, 1665), 103–4.

[36] Hooke, *Micrographia*, 22, 244–45.

[37] Alan Shapiro, *The Optical Papers of Isaac Newton* (Cambridge: Cambridge Univerity Press, 1984), 8–9.

This is the subtil spirit wch searches ye most hiden recesses of all grosser matter which enters their smallest pores & divides them more subtly then any other ^materiall power w^t ever. (not after ye way of common menstruums by rending them ^violently assunder &c) this is Natures universall agent, her secret fire, y^e ~~materiall soule of all matter~~, y^e ~~sole~~ onely ferment & principle of ^all vegetation. The material soule of all matter wch being constantly in- spired from above pervades & concretes wth it into one form & then if incited by a gentle heat actuates ~~it & makes it vegetate~~ & enlivens it but so tender & subtile is it w^thall as to <illeg.> vanish at ye least excess and (having once begun to act) to cease acting for ever & congeale in ye matter <illeg.> at ye defect of heat; ^unless it receive new life from a fresh ferment. And thus perhaps <illeg.> a great pt if not all the moles of sensible matter is nothing but Æther congealed & interwoven into various textures whose life depends on that <illeg.> pt of it wch is in a middl state, not wholy distinct & lose from it like y^e Æther in wch it swims as in a fluid nor wholly joyned & com- pacted together wth it under one forme but in som degree ^condensed united to it & yet remaining of a much more rare ^tender & subtile disposition & so this seems to bee the principle of its acting to resolve ye body & bee mutually condensed by it & so mix under one form ^being of one root & grow together <illeg.> till ~~they attain~~ the compositū attain ye same state wch ye body had before solution.[38]

Newton here alludes to two types of material division, one involving a subtle spirit that dissects bodies at a profound level of their structure, and the other employing "common menstruums" like the mineral acids to rend them asun- der in a more superficial fashion. In large measure, Newton is following an old alchemical tradition that made a distinction between "sophistical," or specious transmutation involving the mere interchange of gross corpuscles, and genuine metallic transmutation, which was thought to involve the infil- tration and substitution of extremely minute particles subsisting within the larger ones. Alchemists had long realized that the mineral acids, for all their corrosive hissing and bubbling, do not penetrate to the deep recesses of mat- ter and induce irreversible transformations.[39] The sophic mercury of Philale- thes was an attempt to circumvent the problem of superficial dissolution by cleansing quicksilver of its impurities and reducing it to a state of ultrapure subtlety that could penetrate into the smallest pores of base metals and radi- cally transmute them. Hence in *Secrets Reveal'd*, the "American Philosopher" dismisses vulgar menstrua—the mineral acids—as mere "external Agents, after the manner of fire, though somewhat different."[40]

Newton's expression "secret fire" for the universal spirit penetrating and dividing matter is also borrowed from his alchemical sources. Phila- lethes, paraphrasing Sendivogius, says that the Chalybs or metallic sulfur is "a Spirit, very pure beyond others," and "an infernal Fire, secret in its

[38] Dibner 1031B, 3v–4r.
[39] On this topic, see my *GF*, 92–169.
[40] Philalethes, *SR*, 25–26.

kind."[41] Like his chymical mentors, Newton thinks that the invisible, fiery agency within matter is stirred up and animated during fermentation; his ether, like the aerial niter, acts as a "ferment" or yeast-like material that resolves and actuates the otherwise passive matter and unites with it. Under the proper conditions, the ether can induce the passive matter to develop further, but with the least "defect of heat" it congeals, just as Newton's metalline fumes congealed on their exposure to water and its vapors to produce sea salt and saltpeter. Thus Newton is able to argue that the entire globe of the earth is probably "nothing but Æther congealed & interwoven into various textures." And yet this transformation of living ether into sensible matter does not mean that the stuff out of which the world is made has "died." To the contrary, the ether is not merely congealed into an inanimate condition but also inhabits brute matter in a "middle state" where it is intermixed with it and acts on it. Newton's next paragraph reveals what he has in mind, and strongly suggests that this rumination also stems from his reading of chrysopoetic authors.

After arguing that the ether congeals to form sensible matter and yet lives on within it to act as a guiding agency, Newton qualifies his theory with the following comment:

> Note that tis more probable y^e æther is but a vehicle to some <*illeg.*> more active spt. & ye bodys may bee concreted of both together, they may imbibe æther as well as air in genetion <*sic*> & in y^t æther ye spt is intangled. This spt perhaps is ye body of light.[42]

Thus the gradation of subtlety that exists when we compare air to ether does not end with these two substances. The ether contains a still more active, almost immaterial spirit, and this spirit may even be "the body of light." These words cannot help but bring to mind the work of Sendivogius again, who claimed in various passages that every material body contains a *semen* or seed that acts as its principle of activity. The Polish chymist even goes so far as to say that this seed is a "spark" (*scintilla*) or point of light occupying 1/8,200 of "whatever body" it inhabits.[43] At the center of its material *emboîtement*, the Sendivogian spark of light is protected from excesses of heat and cold, and is free to act on its "container." Given the heavy influence of Sendivogius throughout *Of Natures obvious laws*, it is likely that the Polish alchemist provided Newton with the immediate pretext for this claim that the ether is a vehicle for the "body of light." Just as Newton adopted the Sendivogian aerial niter and converted it into his own vital ether, it was an obvious move for him to transform the animating spark of light at the center of "whatever body" into an almost immaterial spirit entangled with grosser matter and acting as a corporeal envelope for light.

It is fascinating to see how these Sendivogian notions blend with Newton's earlier speculations from *Certain Philosophical Questions*, where his

---

[41] Philalethes, *SR*, 7.

[42] Dibner 1031B, 4r.

[43] Michael Sendivogius, *Novum lumen chemicum*, in Nathan Albineus, *Bibliotheca chemica contracta* (Geneva: Jean Antoine and Samuel des Tournes, 1654), 11 and 115. In the latter passage, "8200" has been printed incorrectly as "280."

reading of Descartes stimulated him to ask, "Why does air moved by light cause heat or why does light cause heat?"[44] *Of Natures obvious laws* picks up this line of questioning when Newton provides a list of corroborating evidence to support the claim that the vegetative spirit is the body of light. In the numbered list that follows, he points out among other things that both the spirit and light are sources of prodigious activity, that "all things may be made to emit light by heat," and that heat excites both light and "the vegetable principle." One cannot refrain from thinking here of the much later ruminations that fill the queries in the successive editions of Newton's *Opticks*, terminating with his famous *Query 30* (1717), where Newton says that interchangeability of bodies and light "is very conformable to the Course of Nature, which seems delighted with Transmutations."[45] I will return to the *Opticks* in due course, but for now it is enough to see that already by the early 1670s Newton had committed himself to the view that light lies hidden within the deepest recesses of matter, where it serves as a principle of action. As he would put it more cautiously in 1717, "may not Bodies receive much of their activity from the Particles of Light which enter their Composition?"

From Heat to God: Section Four

Newton's ruminations on the earth-vegetable and its ethereal, gravity-inducing nourishment terminate in a discussion of the relationship between sensible matter, light, and heat. These considerations seem to have stimulated a chain of subsequent thoughts that are represented by short jottings in *Of Natures obvious laws*; with these abbreviated entries we begin our fourth division of the text. Heat, light, fire, cold and freezing, fluidity, hardness, and volatility and fixity follow one another in rapid succession, some of these words appearing as mere headings. It is interesting to note that these topics are precisely the sort of material that forms the subject of Boyle's numerous experimental histories. They are in fact the bread and butter of the mechanical philosophy in its midcentury, Baconian form. Newton's goal is not that of recounting the Boylean version of the mechanical philosophy, however, but of interpreting it in the light of his adherence to the newly elaborated Sendivogian aerial niter *cum* ether. Heat and light, for example, result from the ether's rapid entrance between corpuscles ("parts") or its sudden extrusion; fluidity is maintained by the agitation of the ether among corpuscles; hardness and union of corpuscles stem from their roughness and from the exit of the ether.

At least one of these entries employs chymistry in the goal of responding to Cartesian questions that Newton had already raised in *Certain Philosophical Questions*. In his student notebook, Newton had asked: "Whither things congeale for want of agitation from yᵉ ethereal maters," and then tagged this

---

[44] McGuire and Tamny, *Certain Philosophical Questions*, 360–61.
[45] Newton, *Opticks* (London: W. and J. Innys, 1718), 349.

"Cartes."[46] Chymistry provided him with the answer. Observing that various liquids can be supercooled without freezing, and that even under normal circumstances freezing point is not proportional to the viscosity of the material, *Of Natures obvious laws* distinguishes cold from freezing and argues that congelation requires more than Descartes's simple absence of motion. Appealing to a phenomenon first mentioned by medieval alchemists and affirmed by Boyle, Newton says, "Cold is only rest, freezing is by an agent as fumes of lead coagulate ☿." This old experiment already appears in the *De aluminibus et salibus* of pseudo-Rhazes, an Arabic text that exercised a huge influence on medieval Europe once translated into Latin.[47] Boyle also reports occasional success in using this method to solidify quicksilver by tying it up in a rag and inserting the packet into a cavity made in fused lead while it is cooling, though the English experimenter says that the operation will not always succeed.[48] Probably relying on Boyle for this "matter of fact," Newton generalizes from the experiment of congealing mercury with lead vapors to say that freezing occurs when "any agent" settles on the cooled corpuscles and thus makes their surface rough, or rather "adhæres to their out side & acquiesces by cold."

From these very general physical phenomena Newton then passes to the more particular problems posed by living creatures and their interaction with nature. His thoughts tumble forth in a quick sequence of combined headings: "Of yᵉ contrivance of vegetables & animalls. of sensible qualitys. Of yᵉ soules union." Clearly these topics were food for future research in Newton's ongoing attempt to respond to the excessive mechanism of Descartes and to augment or rectify the less rigid version of the mechanical philosophy purveyed by Boyle. That most Cartesian of topics, the soul's union with the body, leads Newton then to a heading that actually does receive more than three or four lines of text: "Of God." Newton's treatment of God at this point has been the cause of serious misinterpretation on the part of at least one historian and must therefore receive our consideration here, despite the fact that it appears to be something of a digression. Notwithstanding her careful scholarship, Dobbs's overall attempt to portray Newton's alchemy as having a theocentric purpose misled her into thinking that his entry "Of God" was intended to make "room for the nonmechanical laws of vegetation." Her view is that Newton was "reminding himself of God's unlimited power to institute *any* series of causes," whether mechanical or not, in order to accommodate his alchemical concept of vegetation.[49] In the interpretation of Dobbs, this passage belonged to Newton's overall employment of alchemy to demonstrate "divine activity in the world." Unfortunately, Dobbs failed to see that Newton's entry "Of God" is actually not inspired by alchemy, but consists rather of a very specific response to Descartes's proofs for the existence of the divinity.

---

[46] Newton, *CPQ,* 360.
[47] Steele, "Practical Chemistry," 26.
[48] Boyle "The History of Fluidity and Firmness," in *Works,* 2: 180; 1661, 218.
[49] Dobbs, *JFG,* 115.

The heading "Of God" is followed by a dense web of logical premises, thought experiments, and conclusions to be drawn from them. The first two paragraphs run as follows:

> Of God. what ever I can conceive w^th^out a contradiction, either is or may ~~effected~~ ^bee made^ by something that is: I can conceive all my owne powers (knowledge, *<illeg.>* activating matter, &c). wi^th^out assigning them any limits Therefore such powers either are or may bee made to bee.
> Example. ^All the dimensions imaginable are possible^. A body by accelerated motion may ~~becom infinitely long or~~ trancend all ~~space~~ distance in any finite tim assigned ^also it may becom infinitely long^. This if thou denyest tis because thou apprehendest a contradictiō in the notion & if thou apprehendest none thou wilt grant it *<illeg.>* ^to the^ pour of things.[50]

In reality there is nothing alchemical about this passage, nor is it intended to clarify the realm of divine activity. It follows on responses to Descartes in *Of Natures obvious laws*, and is much closer to the Cartesian-inspired jottings found in *Certain Philosophical Questions* than it is to Newton's alchemical sources. If we examine a related passage found in Newton's student notebook at the end of his notes on Descartes's *Meditations* and *Responses*, the Cartesian inspiration of Newton's comments becomes quite clear:

> Ax: ~~That thing~~ Tis a contradiction to say, that thing doth not exist, ~~wch may bee conceived~~ whose existence implys no contradiction, & being supposed to exist must necessarily exist. The reason is y^t^ an immediate cause and effect must be in ye same time & there fore ye præexistence of a thing ~~must~~ ^can^ bee no cause of its post existence (as also because y^e^ ~~former~~ ^after^ time depends not on ye former time). Tis onely from the essence of it that a thing can ~~by it owne~~ perpetuate its existence w^th^out extrinsicall helpe. W^ch^ essence being sufficient to continue it must bee sufficient to cause it there being ye like reason of boath.[51]

According to the editors of *Certain Philosophical Questions*, this is a Newtonian gloss on the ontological proof for God's existence in Descartes's "Fifth Meditation." But Newton may well have been thinking of other portions of the *Meditations* as well, and especially of the "Second Set of Objections," where the following criticism is raised against the ontological proof: "From this it follows not that God really exists, but only that he ought to exist if his nature is something possible or non-contradictory."[52] It is in the light of this

---

[50] Dibner 1031B, 4v.

[51] See Newton, *CPQ*, 464. As McGuire and Tamny point out, the passage is inspired by Descartes's fifth *Meditation*. I have compared the transcription to the digital scan posted by the Cambridge University Library (http://cudl.lib.cam.ac.uk/view/MS-ADD-03996/170, consulted June 2, 2016). The term "post existence," altered by McGuire and Tamny to "past existence" in their normalized version of the text, is not a slip of the pen on Newton's part. The point is that if existence is implied by essence, as in the Cartesian ontological proof for God's existence, then cause and effect must be simultaneous.

[52] I owe this reference to an extended discussion with Roger Ariew. The translation is from René Descartes, *Meditations, Objections, and Replies*, ed. and trans. Roger Ariew and Donald Cress (Indianapolis: Hackett, 2006), 74. Gideon Manning has also found echoes of the third Meditation in Newton's comments, a fact that he has kindly related to me in a personal exchange.

criticism of Descartes that one should approach Newton's emphasis on non-contradiction. The concerns expressed in *Certain Philosophical Questions* are an outgrowth of the criticisms of the ontological proof found in the *Opera philosophica* of Descartes that the young Cantabrigian studied as a student.[53] Similarly, Newton's passage "Of God" in *Of Natures obvious laws* testifies to his encounter with Descartes's ruminations on the existence and nature of God: it is not the affirmation of nonmechanism that Dobbs asserts.

The third and final paragraph of Newton's entry "Of God" is also an attempt to improve upon the Cartesian proof of God's existence. The argument runs as follows. The universe might have been otherwise than it is; since God is unconstrained in his ability to create, he could well have created other worlds quite different from our own. To phrase this another way, God's decision to make our world as we know it was "noe necessary but a voluntary & free determination." According to Dobbs, Newton is here justifying "to himself his empirical investigations into the laws of vegetation" by expanding the range of God's creative powers in accordance with arguments for voluntarist theology.[54] But in fact this is not in itself an argument for theological voluntarism, *pace* Dobbs. Rather, the absolute free will of God is assumed, from which Newton draws the following conclusion: "And such a voluntary [cause must bee a God.] determination implys a God." In other words, the fact that our world exists with all of its particularities was a willed decision, and the act of willing implies a being that is capable of making "a voluntary & free determination," that is to say, a creator god. In this fashion, the young Newton thinks that he has succeeded in proving the existence of God where Descartes failed.

What then is this entry "Of God" doing in the midst of a text that is otherwise preoccupied with natural philosophy and alchemy? One might just as well ask the same question about the entries on God in *Certain Philosophical Questions*, and indeed, the same answer would apply. We must remind ourselves that *Of Natures obvious laws* is a commonplace book, albeit fragmentary, organized around topical entries that need not be closely related. *Certain Philosophical Questions* was an earlier exercise in the same genre. The passage "Of God" in *Of Natures obvious laws* looks more like a digression stemming from Newton's responses to Descartes than a series of thoughts that grew integrally out of Newton's concern with alchemical vegetation. One cannot claim the entry "Of God" as an illustration of an integral affinity between alchemy and religion in Newton's mind. It is instead an example of his desire to subvert and supplant Descartes.

---

[53] Roger Ariew has kindly pointed out to me that Leibniz made great use of a "contradiction clause" quite similar to Newton's. In his *Monadology*, for example, Leibniz says, "Thus God alone (or the necessary being) has the privilege, that he must exist if he is possible. And since nothing can prevent the possibility of what is without limits, without negation, and consequently without contradiction, this by itself is sufficient for us to know the existence of God *a priori*"; translation by Roger Ariew and Daniel Garber in G. W. Leibniz, *Philosophical Essays* (Indianapolis: Hackett, 1989), 218. Moreover, the main elements of this argument already appear as early as 1676 in Leibniz's *De summa rerum*. Ariew has also provided me with the references for these: see G. W. Leibniz, *De summa rerum* (New Haven, CT: Yale University Press, 1992), 47–49, 63, 91–107.

[54] Dobbs, *JFG*, 114–15.

Determining the Limits of Mechanism and Vegetability: Section Five

Newton's profound engagement with Cartesian and Boylean mechanism in *Of Natures obvious laws* leads him, in the fifth and final division of the text, to return to the vexed problem of distinguishing between purely mechanical processes and those that he links to a principle of vegetation. The content at this point is closely related to that found earlier in the manuscript, especially at the end of the second section, where in the context of putrefaction, Newton first distinguishes mechanism from vegetation. The text of section five is taken up mostly with further attempts to refine and test the distinction between mechanical change and vegetation. To those unfamiliar with the mental habits of early modern natural philosophers and chymists, however, the early paragraphs of section five may seem to be a distraction from this goal. After some initial "rules" that spell out the necessity and modus operandi of a preceding putrefaction when a thing is about to undergo transmutation "from w^t it is" into a radically different substance, Newton passes to a discussion of the traditional distinction between nature and art. Why should this hoary birfurcation, redolent of Aristotelianism and adopted by Renaissance painters, sculptors, and literati as a means of glorifying the verisimilitude of their artistic endeavors, detain Newton in his discussion of mechanism and vegetation?

The answer is twofold. On the one hand, the art-nature distinction was traditionally a favorite subject of alchemical writers, who had for centuries been justifying the legitimacy of their manufactured gold by arguing that seemingly "artificial" human fabrications could be as "natural" as those made by nature itself. Employing a rather elastic version of the bifurcation between the artificial and the natural, alchemical texts typically made the argument that the aurific art "aided nature" to perfect itself, in the same way that medicine acted as the handmaid of nature in curing debilitated patients. By this logic, alchemical products, which could involve such radical transmutations of "species" as the conversion of base metals into gold, might be seen as being no more unnatural than the patient cured by medicine. Because of this medial position between art and nature, alchemy had been adopted as the focus of a broader discussion on the boundaries of the artificial and the natural by the scholastics of the High Middle Ages, and this discussion was still very much alive in the seventeenth century when Newton was composing *Of Natures obvious laws*.[55] It was therefore an obvious move for Newton to incorporate some of this traditional discussion into his heavily alchemical text.

In addition to the natural tendency to bring the art-nature discussion into an alchemically flavored text such as *Of Natures obvious laws*, however, there is a more pressing reason for the appearance of the dichotomy there. The same scholastic discussion that debated the ability of alchemists to "aid nature" in the laboratory rather than acting in a purely artificial way, typically distinguished in a somewhat rigid fashion between "perfective arts"

[55] For the central role of alchemy in the art-nature debate, see Newman, *Promethean Ambitions* (Chicago: University of Chicago Press, 2004), especially chapter two.

that helped nature such as medicine and alchemy and those that worked "extrinsically" without engaging the Aristotelian internal principle of change (Aristotle's *archē kineseōs*) that allowed natural things to grow and progenerate.[56] A paradigm example of such starkly artificial arts lay in the realm of machines, which employed mechanical principles such as the law of the lever to produce effects that were sometimes stupendous, but which did not involve the natural processes of generation, growth, or transmutation. Hence the more rigid form of the distinction between art and nature was in large part coextensive with the bifurcation between the operation of machines and the working of nature's agency, and the latter dichotomy mapped closely onto Newton's separation of mechanism from vegetability. It is largely for this reason that Newton occupies himself with the art-nature division in *Of Natures obvious laws*' section five.

Newton begins his section on art and nature like his alchemical forebears, with the claim that human art or technology is not restricted to mere mechanical change. Instead, art may "promote" or encourage nature's fermentative or putrefying action so that genuine transmutation may be effected at the will of the human operator. He puts these ideas in the following words when discussing putrefaction:

> Nature only works in moyst substances
> And wth a gentle heat
> Art may set nature on work & <*illeg.*> promote her working in y^e
>   production of any thing what ever. Nor is ye product less naturall then
>   if nature had produced it alone.[57]

In the following lines, Newton illustrates this traditionally alchemical perspective with examples such as a child born from a mother who "took physic," a tree grown and watered in a garden, and insects produced by artificially induced spontaneous generation from a carcass kept in a heated flask. None of these products are rendered "artificial" merely because they have received the benefit of human care, and in a like manner the products of alchemical intervention, as long as they employ subtle processes such as fermentation, are still natural products. In order to effect such radical changes as alchemy lays claim to, the chymist must employ "a more subtile secret & noble way of working" than mere mechanical transposition. Instead of working in the gross fashion of mechanism, the alchemist must make use of a "vegetable spirit" diffused in the form of "seeds or seminall vessels" throughout the mass of the matter that it inhabits. Here, once again, we are in the realm of the Sendivogian aerial niter theory, which had postulated the existence of a tiny, seedlike spark at the center of bodies that emanates a virtue or power responsible for qualitative difference and change. As we have seen throughout *Of Natures obvious laws*, Newton is elaborating on ideas that he had inherited from early modern chymistry.

---

[56] For Aristotle's *archē kineseōs*, see above all *Physics*, book 2. The presence of this innate agency was what made a thing "natural"; its absence implied artificiality.

[57] Dibner 1031B, 5r.

But the final paragraphs in *Of Natures obvious laws* also advance Newton's agenda by employing chymistry to facilitate the identification of products of mechanism more generally with the domain of the artifactual or adventitious. As he puts it earlier in the text, "vegetation is yᵉ only naturall work of metals."[58] This is not to say that mechanism is unnatural in a categorical sense, since in Newton's universe, "Natures actions are either vegetable or purely mechanicall." For him, mechanical activity belongs to the domain of the natural world along with vegetability, of course, but it need not reflect the guided activity that characterizes vegetation. To Newton, as to Sendivogius, Philalethes, and many other early modern chymists, vegetation implied a goal-directed process governed by the tiny *semina* or "seeds" implanted deep within matter. The traces of Aristotelian teleology are clearly discernable here, just as they are in Newton's overarching distinction between art and nature. But since Newton admitted an extensive role for mechanical action in the natural world, he was confronted by a dilemma. How do we know when vegetation as opposed to mere mechanism lies behind a natural process? Chymistry would provide him with a way to solve this problem.

As we have already seen, *Of Nature's obvious laws'* second section conveys a quite original treatment of the formation of sea salt and niter by means of a putative interaction between water and the metallic fumes that rise up from the earth's depths. There Newton employs Boyle's redintegration of saltpeter in order to justify his claim that the substance consists of a looser, less tightly compacted structure than that of sea salt. In section five of the text, Newton again employs redintegration but in a more general sense as a test case for determining whether a given material has been made by mechanical or rather by vegetative processes. In short, materials that can be analyzed and resynthesized fit Newton's criterion for mechanical products, whereas substances produced by vegetation are not fit products for resynthesis or, to use the Boylean term, redintegration.

The processes of salt production that we examined earlier in this chapter are manifestly not instances of vegetation, since they involve only a mechanical change in texture brought on by corpuscular interaction between metallic fumes and water. Newton places these changes in the class of such purely mechanical operations as the mixing of differently colored powders to produce new colors (as when jumbled blue and yellow granules give the appearance of green), the dissolution of metals in mineral acids, and the separation of cream into butter, curds, and whey by churning. As for vegetation, Newton defines it here in the following terms:

> Natures actions are either seminall ^vegetable or ^purely mechanicall (grav. flux. meteors. vulg. ₍Chymistry₎ <)>
> The principles of her vegetable actions are noe other then the ~~seeds~~ seeds or seminall vessels of things those are her onely agents, her fire, her soule, her life,
> The seede of things that is all that substance in them that is attained to the ~~full~~ fullest degree of maturity that ~~that~~ is in that thing <*illeg.*> so that

[58] Dibner 1031B, 3r.

there being nothing more mature to act upon them they acquiesce. Vegetation is nothing else but ye acting of w^t is most maturated or specificate upon that wch is <*illeg.*> less specificate or mature to make it as mature as it selfe And in that degree of maturity nature ever rests.[59]

In drawing this sharp distinction between mechanical and vegetative processes, Newton had to confront an obvious potential objection. Although the artificial operations employed by a laboratory technician in cases of "vulgar chymistry" might be purely mechanical, there are many instances where a hidden, indwelling nature may actually be driving operations that seem to our senses to be mere mechanism. This seminal "vegetable substance," acting as a latent "invisible inhabitant," may direct grosser particles to take on the structure of bones, flesh, wood, fruit, and other materials subject to growth. As Newton clarifies a few lines later:

> So far therefore as ye same changes may bee wrought by the slight mutation of the textures of bodys in common chymistry & such like experiments ~~may~~ may judg that ~~there is noe other cause that will~~ such changes made by nature are done ye same way that is by ye sleighty transpositions of ye grosser corpuscles, for upon their disposition only sensible qualitys depend. But so far as by ~~generation~~ ^vegetation such changes are wrought as cannot bee done wthout it wee must have recourse to som further cause And this difference ~~is seen clearest in fossile substances~~ is vast <*illeg.*> & fundamental because nothing could ever yet bee made wthout vegetation wch nature useth to produce by it. [note ye instance of turning Irō into copper. &c.][60]

The point of this passage is that even seemingly mechanical operations in nature can be directed by the hidden, seedlike entities that occupy an "unimaginably small" portion of matter and are immeasurably smaller than the "gross" corpuscles involved in mechanical change. Once again, we are in the realm of Sendivogius's spark-like *semina* that dwell in the heart of matter and direct its actions. But since the changes that we can perceive by means of our senses only involve the larger corpuscles as opposed to the tiny particles existing at the deepest recesses of bodies, how then can we distinguish between the purely mechanical operations of ethereal gravitation, fusion, meteorology, and vulgar chymistry and the vegetative processes employed by nature? Newton responds by asserting that any laboratory process that allows one to retrieve the initial ingredients from what we would today call a "chemical compound" or recreates the compound from its analyzed ingredients reveals that the compound in question was a mere mechanical combination rather than a product of vegetation. A similar ideology underlay Newton's experimental analysis and synthesis of white light, and the use of decompounding followed by recompounding as an index of mere mechanical change in *Of*

---

[59] Dibner 1031B, 5r.
[60] Dibner 1031B, 5v.

Natures obvious laws probably also had its sources in Boyle's work. Newton puts it thus in *Of Natures obvious laws*:

> Thus acid two pouders mixed each to a 3^d colour, $^{ye\ unctuous\ pts}$ in milk by a little agitation concret into one mass of butter Nay all y^e operations in vulgar chemistry (many of wch to sense are as strange transmutations as those of nature) are but mechanicall coalitions $^{\wedge or\ seperations}$ of particles as may appear in that they returne into their former natures if reconjoned or (when unequally volatile) dissevered, & y^t wthout any vegetation.^{61}

In other words, all the ordinary reactions that Newton groups within the realm of "vulgar chemistry" are mere mechanical interactions like the mixing of blue and yellow powders to produce the color green, and this is demonstrated by the retrievability of their unaltered ingredients by analysis or their recombination by synthesis. As we have already seen, Newton used the redintegration of niter as a paradigmatic case of such purely mechanical change earlier in *Of Natures obvious laws*. The fact that the "fixed salt" made by igniting niter could be returned to ordinary saltpeter merely "by dissolution" meant that the change was a mechanical one of "texture & constitution" rather than a transmutation induced by vegetation. It is likely that Newton has the same process in mind here, though the reference to unequal volatility suggests that he has broadened his scope to include compounds that can be separated by mere sublimation or distillation rather than combustion. Like earlier alchemists, Newton viewed such separations and recombinations as a sort of change that took place between "the grosser corpuscles" of bodies. Real transmutation, which Newton has in mind when he speaks of vegetation, had long been thought of in alchemy as something that occurs at a deeper microstructural level of matter.

To the Newton of the early 1670s, who had not yet embraced the principle of long-range action at a distance that marked his mature *Principia*, the phenomena exhibited by falling bodies, melting materials, changes in the atmosphere, and inorganic chemical reactions were all explicable by means of microlevel particles acting mechanically on one another. Vegetation, on the other hand, is a goal-directed process whereby a more mature seed leads a less mature material into a state of maturity equivalent to its own. In other words, vegetation is the procedure whereby generation and growth occur in the natural world. In Newton's mind, it is clearly the operation by which nature retains and replenishes the species of the world around us. Even if the phenomenal world may appear at first face to operate by purely mechanical means, nature employs vegetative processes at a deeper level to drive the corpuscular interactions that result in generation and growth. Hence in reiterating the distinction between mere mechanism and vegetation, Newton says the distinction is "vast and fundamental" in that "nothing could ever yet bee made wthout vegetation wch nature useth to produce by it."

It is remarkable that this sweeping claim for the role of vegetation is supported only by the seemingly mundane phenomenon found in Newton's

---

^{61} Dibner 1031B, 5v.

following comment—"note ye instance of turning Irō into copper. &c." Here Newton is referring to the cementation process of refining copper out of vitriolate solutions by adding iron to them; this was a standard piece of the empirical evidence many chymists used to support the reality of transmutation. Newton had displayed a keen interest in this topic, which he derived from his reading of the chymist Maier, in his 1669 letter to Francis Aston, composed only a short time before *Of Natures obvious laws*. Along with various other secrets such as extraction of gold from river water by means of mercury, the possible use of pendulum clocks to determine longitude at sea, and the tricks that the Dutch might employ to combat shipworms, Newton mentions there a technique by which the miners at Schemnitz "change Iron into Copper." The simultaneous deposition of copper and dissolution of iron in vitriol springs, which would eventually serve him as an example of "elective affinity," appears here as a vegetative process driven by tiny semina. The phenomenon resurfaces in Newton's alchemical florilegia of the 1690s, and it forms a significant part of the discussion in the famous *Query 31* of the 1717 *Opticks*, where he discusses elective affinities at length and uses them as evidence that matter is endowed with immaterial forces.

Despite Newton's later understanding that this was merely a process of copper plating onto iron that was itself dissolving, his blanket acceptance of the accounts of the philosophers' stone and transmutation given by Philalethes and Sendivogius suggests that at the time of writing *Of Natures obvious laws* he probably still accepted Maier's view that a genuine transmutation had occurred rather than a mere deposition of copper already present in the vitriol.[62] We have seen that throughout *Of Natures obvious laws*, Newton builds on the "facts" supplied by chrysopoetic authors and uses them as support for his discussion of putrefaction, fermentation, and vegetability. These topics formed a tightly knit cluster in Newton's mind and would resurface in combination throughout his scientific career. We will address the integration of these concepts in Newton's later published science in the final chapters of this book, particularly in relation to his mature work in optics.

To conclude, the alchemical cosmology purveyed by *Of Natures obvious laws* reflects the same wide-ranging attempt to integrate different scientific topics and methods that one encounters in Newton's undergraduate notebook, *Certain Philosophical Questions*. The two documents share a concern with phenomena such as gravity, heat and cold, freezing, and other topics dear to the mechanical philosophy, but the appearance of vegetability and *semina* in *Of Natures obvious laws* represents a radically new departure for Newton. Mere mechanism now seemed inadequate to him, and even phenomena that appeared to the eye to be strictly mechanical could actually be driven by vegetative agencies at a deeper level. Indeed, processes that resulted in products which could not be disassembled into their original parts or ingredients, must of necessity have stemmed from the vegetative action

---

[62] Maier's insistence on a genuine transmutation of iron into copper is unequivocal. His words are "Hic ferrum transmutatur revera in ipsum per aquas vitriolatas naturales." See Michael Maier, *Symbola aureae mensae duodecim nationum* (Frankfurt: Lucas Jennis, 1617), 525.

of *semina*. It is important to stress that Newton's position dovetailed closely with the much older view of many alchemists that terrestrial materials consisted of agglomerated particles (*grossae partes*) that were themselves composed of smaller particles (*subtiles partes* or *minimae partes*) that retained their full being within the larger corpuscles. It had been an easy matter for seventeenth-century chymists such as Eirenaeus Philalethes to equate the *minimae partes* of earlier alchemy with the living seeds of the Sendivogian and Helmontian traditions. Newton's corpuscularian reading of alchemy, though heavily conditioned by his knowledge of Cartesian and Boylean mechanism, also found a strong foundation in the works of the alchemists and even in the geographical work of Bernhard Varenius, who had himself benefited from the sons of Hermes.

But despite the fascination of its alchemical world picture, we must not forget that *Of Natures obvious laws* is more than a treatise of natural philosophy. Underlying its seductive theorizing and its recurrent attempts to distinguish between mechanical and vegetative agencies, *Of Natures obvious laws* intentionally sets forth a path to chymical practice. From Newton's rewriting of Boyle's "transmutation" of water into earth in *The Origin of Forms and Qualities* as a story about "unravelling" the structure of matter and then using the liberated material to galvanize ores into their "pristine metalline forme" one detects the former's practical goals. The same thing is even more obvious in Newton's emphasis on ever more subtle media, ranging from invisible water vapor and heavier mist to air, ether, and the "body of light," where the decreasing particle size of the respective media corresponds to the increasing activity of the medium in question. The passage from mechanism to vegetability mapped closely onto the analysis of matter by means of increasingly powerful menstrua, which could in Newton's view liberate the minute, hidden *semina* trapped within the confines of coarser matter. But these dissolvents could not act in the violent fashion of the ordinary mineral acids; instead, their affinity with the solvenda allowed them gently to coax forth the secret, active particles within by a process involving fermentation. In the following chapters, we will see how Newton developed these ideas in the context of practice and used them to decipher the enigmas transmitted to him by the adepts. The culmination of this effort may be found in his laboratory notebooks, where the twin desiderata of increasing subtlety and affinity between menstruum and solute, or sublimandum and adjuvant, emerge as the dominant themes in a remarkably sophisticated research project extending over more than a generation.

NINE

The Doves of Diana

FIRST ATTEMPTS

The immediately previous chapters examined Newton's early integration of alchemical theory and natural philosophy in his two short treatises from the period between 1670 and 1674, *Humores minerales* and *Of Natures obvious laws & processes in vegetation*. Around the same time as he was composing those documents, or perhaps a little earlier, Newton had also begun the process of extracting practical, chrysopoetic information from his alchemical readings. His earliest attempts probably focused on Basilius Valentinus, as we saw in chapter five, but he soon launched into a much more ambitious attempt to decipher the difficult texts of Michael Sendivogius.[1] A close examination of Newton's very early efforts at decoding the allegorical writings of the alchemists presents remarkable surprises. It has long been known that Newton had a keen interest in antimony, for example, but no historian until now has realized that the young adept-in-training had an equal if not greater fascination with the metal lead. Indeed, lead and its ores would play a major role throughout Newton's career as a chymist, though in increasingly complex ways. Fortunately, Newton's earliest forays into the decipherment of chrysopoetic literature are quite straightforward, however, and the writings of his youth reveal clearly the methodical fashion through which he attempted to extract practical sense from these often bewildering documents.

As we will see, Newton's analytical technique was pragmatic and flexible. Believing that his sources were authentic adepts rather than mere vulgar "tyros," Newton assumed that they would not use the common *Decknamen*

---

[1] Newton's exposition of Basilius Valentinus in his very early manuscript Keynes 64 must have been composed before the Sendivogian *Collectiones ex Novo Lumine Chymico quæ ad Praxin Spectant* found in Keynes 19. The text in Keynes 19 regurgitates a passage that Newton took from Jean Pierre Fabre's translation of the Basilian *Triumphwagen* and other works in Basilius Valentinus, *Currus triumphalis antimonii fratris Basilii Valentini* (Toulouse: Petrus Bosc., 1646), 117. The Basilian passage, found in Keynes 64 on 4v, is reproduced without statement of source in Keynes 19 on 3r: "Antimonium enim apud veteres dicebatur Aries Quioniam <sic> Aries est primū Signum Zodiaci in quo Sol incipit exaltari & Aurum maxime exaltatur in Antimonio." The absence of further Basilian borrowings in Keynes 19 merely reflects Newton's awareness of the fact that the texts of Philalethes, Sendivogius, and Jean d'Espagnet are closely linked to one another, whereas the Basilian corpus is not.

of the day in a rigid, standardized way. Hence, a given allegorical term, such as *Luna* (moon), did not have to refer to its traditional alchemical referent, silver, but might conceal a multiplicity of substances whose true meaning could change in different contexts and whose sense could only be discovered by skillful reading. We have already given an overview of Newton's alchemical sources; although a number of them will figure in the present chapter, the three that stand out most in his early readings are Michael Sendivogius, Eirenaeus Philalethes, and Johann Grasseus. Newton knew at least some of the works by the first two of these authors before acquiring the bulky *Theatrum chemicum* in 1669, and they form the backbone of his earliest attempts to render the riddling language of the adepts into actual laboratory practice.[2]

Two manuscripts are of paramount importance here. The first, Babson 925, now held at the Huntington Library, was probably composed before Newton's 1669 acquisition of the *Theatrum chemicum*. It uses only texts that were in print before Newton bought the *Theatrum*, even when those works were also present in the multivolume collection.[3] The other manuscript, Keynes 19, is probably slightly later than Babson 925, since it contains the same ideas in more developed form, but still probably dates from around 1669.[4] Both of these manuscripts contain expositions of Sendivogius's 1604 *Novum lumen chemicum*, one of the most enigmatic and immensely popular chrysopoetic writings of the seventeenth century. Let us begin here with Babson 925, since a description of this short document will take little time. Babson 925's Sendivogius commentary bears the title *Loca difficilia in Novo Lumine Chymico explicata* (Difficult Places in the *New Chymical Light* Explained), and its tiny, careful script covers only one side

---

[2] For Newton's acquisition of the *Theatrum chemicum* in April 1669, see Harrison, *Library*, 7–8.

[3] See Karin Figala, John Harrison, and Ulrich Petzoldt, "*De scriptoribus chemicis:* Sources for the Establishment of Isaac Newton's (Al)chemical Library," in *The Investigation of Difficult Things,* ed. P. M. Harmon and Alan Shapiro (Cambridge: Cambridge University Press, 1992), 135–79. On p. 153, Figala, Harrison, and Petzoldt argue that Newton acquired the *Artis auriferae, quam chemicam vocant,* an important collection of alchemical texts, by the early 1670s. The *Artis auriferae* contains two versions of the famous *Turba philosophorum,* a medieval translation from Arabic. The *Turba* is also found in volume 5 of the *Theatrum chemicum* in its 1622 and 1660 editions. Yet on 2v of Babson 925, which consists of an index of terms drawn without acknowledgment from the *Turba philosophorum,* Newton has only used the first version of the *Turba,* printed in volume one of the *Artis auriferae* (Harrison no. 90, Wren Tr/NQ.16.121). This is easy to demonstrate since Newton gives not only page numbers in Babson 925, but in many instances line numbers as well. Although Figala, Harrison, and Petzoldt do not point this out, it is therefore likely that Newton composed Babson 925 before purchasing the *Theatrum chemicum,* and that his acquisition of the *Artis auriferae* predated his ownership of the other collection. Figala, Harrison, and Petzoldt are probably correct in their claim that Babson 925 made use of Nathan Albineus's *Bibliotheca chemica contracta* for Sendivogius and d'Espagnet. See "*De scriptoribus chemicis,*" 159–60n87. Since there is a 1654 edition of Albineus's book, Newton's use of this text does not rule out a pre-1670 date for Babson 925. As the authors also point out, the fact that Newton later owned the 1673 edition of Albineus's collection (Harrison no. 220) is inconclusive, since Newton could have been using someone else's copy before he acquired his own.

[4] I rely here on Dobbs's dating of Keynes 19, which seems sound. As she says in *FNA,* 152, the manuscript contains a number of Newton's early uncrossed Saturn symbols and lacks any crossed ones, and it omits any references to texts found in the *Theatrum chemicum*; the former fact indicates that the manuscript is early, and the latter points to a date of composition before or not long after April 1669. As Dobbs also points out, the manuscript refers to Philalethes's *SR,* which was published in 1669. This fact, in combination with the absence of the *Theatrum chemicum,* indicates that the manuscript was likely to have been composed in that year.

of a small folio. The text consists of recondite passages taken verbatim from the *Novum lumen chemicum* with the occasional spare remark by Newton in his customary square brackets. As we saw in analyzing *Of Natures obvious laws & processes in vegetation* (Dibner 1031B), Newton was fascinated with the cosmic system of circulation between the heavens and the earth described by Sendivogius. Basing himself in part on the *Emerald Tablet* ascribed to Hermes Trismegistus, Sendivogius had argued that the earth is replenished by a thin, material spirit, namely, the *sal nitrum* or aerial niter, which not only animates life on earth but also sinks into the pores of the globe and serves as the first matter of metals and minerals. It was this theory that provided Newton with the primary basis for his "theory of everything" in the early to mid-1670s, and Sendivogius's assurance in the ultimate simplicity of nature lying behind the bewildering world of appearances is stated multiple times in the *Novum lumen chemicum*. As the Polish adept puts it, "I say that Nature, which God made before time and in which He placed a spirit, is One, True, Simple, and Integral."[5]

Sendivogius's insistence on the simplicity of nature had implications extending well beyond the realm of pure theory, for like many alchemists, he viewed the practice of chrysopoeia as an attempt to imitate nature's operations on a smaller scale. The idea that "ars imitatur naturam" (art imitates nature) was an engrained habit of mind for early modern thinkers in general and for alchemists in particular. A stock feature of Aristotelian philosophy and Galenic medicine, the belief that technology and artisanal practice should mimic nature had roots that extended back into the origins of Greek representational art.[6] If one took the idea seriously, as Sendivogius clearly did, then it was necessary to understand the workings of nature in detail so that one could model the laboratory operations of alchemy on them. For the reader of the *Novum lumen chemicum*, this could have another implication, namely, that the theoretical portions of the text describing nature's actions in chymical terms were meant to serve primarily as allusive descriptions of alchemical processes. This is in fact the way that Newton read Sendivogius, as the following paragraph from Babson 925 illustrates. Newton first reproduces a line from the end of Sendivogius's treatise where the Polish chymist is paraphrasing the *Emerald Tablet*. Then the young Cantabrigian introduces his commentary with a bracket (which he neglects to close):

> Conclusion: Its father is Sol, its mother Luna, the wind has borne this in its womb, its nurse is earth, etc. [it is sublimed mercury (or its seed, called sal nitrum on account of its vegetability, sal alkali on account of the attraction of its masculine seed (which is called the central water), and sal ammoniac on account of its volatility); for its father it has antimony, and

---

[5] Michael Sendivogius, *Novum lumen chemicum*, in Nathan Albineus, *Bibliotheca chemica contracta* (Geneva: Jean Antoine and Samuel des Tournes, 1654), 4.

[6] See my *Promethean Ambitions* (Chicago: University of Chicago Press, 2004), which argues that alchemy provided a focal point for the art-nature debate of the Middle Ages and early modern period.

for its mother mercury of Saturn, and the wind or air—that is, the impure mercurial water—has borne it in its center.[7]

Here we see Newton's attempt to arrive at the meaning behind Hermes's description of the mysterious *una res*—the "one thing" that is born of the sun and moon, carried in the belly of the wind, and nursed by the earth in the *Emerald Tablet*. Sendivogius had built his theory that there is an aerial niter circulating between the earth and the heavens on this idea. Unlike his exposition in *Of Natures obvious laws*, however, Newton here interprets Hermes (in the version given by Sendivogius) as describing a laboratory process. On the principle that art imitates nature, this was a permissible and even obvious interpretive move to make. Thus Babson 925 tells us that the "one thing" is either sublimed mercury or its seed, and that this in turn is concealed under other *Decknamen*. The first cover name, *sal nitrum* or saltpeter, refers to the ability of the seed to "vegetate"—to change and grow. The second, *sal alkali*, is intended to connote attraction. Newton is thinking here of another passage in the *Novum lumen chemicum*, where Sendivogius compares his aerial niter to the hygroscopic, alkaline material salt of tartar, which absorbs water out of the atmosphere.[8] Finally, Newton says that Sendivogius employs the *Deckname* sal ammoniac in reference to the volatility of the aerial niter, an appropriate designation since sal ammoniac sublimes at a relatively low 338°C.

So far, Newton's interpretation of Sendivogius could apply uniquely to the cosmic circulation of the aerial niter. But his subsequent comments reveal unequivocally that he has much more than cosmology in mind. When Newton says that the hermetic "one thing" has antimony for its father and mercury of Saturn for its mother, and that it is found in the center of an impure, mercurial water, he has moved into the world of laboratory practice. This is assured by a slightly later passage in Babson 925, where Newton explicitly links antimony and Saturn to the making of the "elixir," that is, the philosophers' stone:

> Man (the elixir) is created from earth (Saturn and antimony), and he lives by means of the air (the seminal metallic seed), which we call dew by night and rarefied water by day, whose invisible, congealed spirit is better than the entire world.[9]

To Newton, the creation of man described in the *Novum lumen chemicum* is actually a creation of the philosophers' stone or elixir. Although Sendivogius says nothing about antimony, Saturn, or the seminal metallic seed in

[7] Babson 925, 1r: "Conclus: Pater ejus est Sol, Mater Luna, portavit illud ventus in ventre suo, & nutrix est Terra &c [id est Mercurius sublimatus (vel semen ejus <*illeg.*> dictum Sal<*illeg.*> nitri propter vegetebilitatem, Sal Alcali propter attractionem seminis masculini (quod aqua dicitur centralis) & Sal Armoniacum propter volitabibitatem <*sic*>) habet patrem ♃ antim: matrem ☿us ♄i & portavit illud ventus sive Aer id est aqua mercurialis in rer impura in centre suo."

[8] Sendivogius, *Novum lumen chemicum*, in Albineus, *Bibliotheca chemica contracta*, 52.

[9] Huntington Library, Babson 925, 1r: "Creatus est homo (elixir) de terra (♄ & Antim̄) & ex aere (vapore metalico seminali) vivit quem nos rorem de nocte de die aquam vocamus rarefactatem, cujus spiritus invisibilis congelatus melior est quam terra universa."

the passage that served as Newton's immediate source, the young chymist has decoded his Polish forebear's allusive text into a recipe for practice.[10] Yet beyond the fact that this practice seems to involve antimony, some substance hidden behind the planetary *Deckname* "Saturn," and a metallic seed or impure, mercurial water, Babson 925 tells us very little. In order to learn Newton's detailed interpretation of Sendivogian alchemy as he had decoded it in the earliest phases of his alchemical endeavor, we must turn to another manuscript, Keynes 19.

Unlike Babson 925, Keynes 19 contains a sustained and ambitious attempt to extract the practical sense of the *Novum lumen chemicum*. The primary text in the manuscript consists of three folios (front and back) written in double columns, and is titled above the left column of the first folio *Collectiones ex Novo Lumine Chymico quæ ad Praxin Spectant* (Summaries from the *New Chymical Light* Which Pertain to Practice). The right column is titled *Collectionum Explicationes* (Explanations of the Summaries). This is followed on 3v by a shorter exposition on the French chymist Jean d'Espagnet (labeled *Arcanum Hermeticæ Philosophiæ Opus* by Newton) with its own parallel *Explicationes*, and a still shorter commentary on Sendivogius's *Dialogus Mercurij Naturæ & Alchymistæ* (4v), which is found after the twelve tractates in the *Novum lumen chemicum* in the edition of Albineus that Newton was using. Although the entire manuscript consists of only four folios, it is remarkably rich and gives a powerful sense of the mental energy that Newton had already invested in dissecting his alchemical sources by the second half of the 1660s. In the following exposition of Keynes 19, I follow the order of Newton's presentation for the most part, since he begins by identifying the principles of the art and then builds his practice on those principles. He is not interested in providing a commentary to the entire *Novum lumen chemicum*; instead, he is trying to determine the nature of the laboratory processes cloaked behind the "noble Polonian's" mystification. As we will see, the interpretation here is the same as the one in Babson 925, though in Keynes 19 we encounter both the reasons behind Newton's interpretation and his suggestions for putting them into practice.

The Lead Process and Sendivogius

In Keynes 19, Newton begins his summary and explanation of the *Novum lumen chemicum* with the fourth tractate. One can hardly improve on the clarity of his prose:

> In the fourth Tract. There is a single seed of the metals, the same in lead and in gold, the same in silver as in iron. And a bit later he says, if the

---

[10] See Sendivogius, *Novum lumen chemicum*, in Albineus, *Bibliotheca chemica contracta*, 48: "Creatus homo de terra, ex aëre viuit: est enim in aëre occultus vitae cibus, quem nos rorem de nocte, de die aquam vocamus rarefactam; cuius spiritus inuisibilis congelatus melior est quam terra universa."

fatness comes to pure places by subliming it becomes gold, but if that fatness comes to impure places, it becomes lead.[11]

The *pinguedo* or fatness here refers to the principle of sulfur, which since the Middle Ages had been viewed as the second component of metals, along with mercury. This corresponds rather closely to the corresponding passage of Sendivogius's *Novum lumen chemicum*, although Sendivogius's point is that *all* metals come from a single seed, whereas Newton has put the emphasis on two lineages, one for lead and gold, the other for iron and silver.[12] What is remarkable about this is that by emphasizing these two genealogies, Newton manages to privilege lead, as one can immediately see by consulting the corresponding passage in his *Explicationes*, where he comments on the foregoing text:

> Whence if the impurity contracted from <an impure> place can be separated from lead, you have the matter from which gold is made by digesting.[13]

It is entirely characteristic of Newton to deduce alchemical practice from the putative generation and growth of metals beneath the surface of the earth, in conformity with the principle that art should imitate nature. Hence he concludes from the fact that lead is merely unripe gold whose maturation has been blocked by "impurity," that the removal of this impurity from the lead will lead to gold. In the next few extracts, Newton confirms his interpretation of Sendivogius. Where the Polish alchemist says that there is one metal that consumes all of the others except gold and silver, and is virtually their water and their mother, Newton says that this must again be a reference to lead, which was commonly used by assayers and chymists to separate the base metals from gold and silver by means of cupellation. This is a quite sensible reading on Newton's part, since crude antimony or stibnite, the other obvious candidate, would have consumed silver along with the base metals, whereas lead would not attack the silver.

So what is one supposed to do with lead, once it has been chosen as the starting point of transmutation? Newton's answer is remarkably straightforward and simple. Since the lead needs to be purified, why not employ antimony, the very material that refiners used to refine gold, and that Eirenaeus Philalethes had used to purify his sophic mercury? As we will shortly see, Philalethes enters into Newton's understanding of Sendivogius in a major way, and *Secrets Reveal'd* was probably the basis for the assumption that antimony, along with lead, is one of Sendivogius's fundamental materials. What other substances did Newton think Sendivogius employed in making the philosophers' stone at the time of writing Keynes 19? The

[11] Keynes 19, 1r, at note "a": "Tractatu quarto. Vnicum est semen [metallarum <*sic*>] idem in ♄o quod in ☉e invenitur, idem in Luna quod in marte. Et paulo post ait, Si pinguedo venit sublimando ad loca pura fit Sol, si vero pinguedo illa venit ad loca impura, frigida, fit Saturnus. A."

[12] Sendivogius, *Novum lumen chemicum*, in Albineus, *Bibliotheca chemica contracta*, 15.

[13] Cambridge University, King's College, Keynes 19, 1r, at note "a": "Vnde si impuritas a loco contracta potest a saturno separari, habes materiam ex qua ☉ fit, digerendo."

answer, remarkably, is none at all. Despite the fact that Babson 925 may have seemed to locate three materials in Sendivogius's practice—antimony, Saturn (which we now understand to mean lead), and an "impure mercurial water," Keynes 19 is quite insistent in its claim that the true work of alchemy employs only two material starting points. As Newton puts it without equivocation—"There are no more than two starting materials: lead and antimony" (*nec plura sunt quam duo principia, Plumbum & Antimonium*).[14] By examining another of the passages where Newton insists on the need for only two materials, we will now see that despite first appearances, this is no contradiction between Babson 925 and Keynes 19. In commenting on a passage where Sendivogius says to take eleven grains of "our gold" and one grain of "our Luna," Keynes 19 glosses "our Luna" in the following fashion:

> That is, the mercury of lead. For at the end of the Conclusion of the Tractates, he says to work on Sol and our Luna which is overspread with the sphere of Saturn (i.e. on Sol and the mercury which is coagulated by fetid sulfur into the dark form of lead) and not in a third material. But there will be three materials unless earth and Luna are held to be the same thing. Moreover, reason persuades that some part of the lead should be reduced into mercury so that a digestion take place.[15]

Newton is quite correct that Sendivogius advised taking only two starting materials, although the studied vagueness of the *Novum lumen chemicum* makes it very unclear what those substances were.[16] But this raises the problem again, that Sendivogius often speaks of the inclusion of mercury as though it were a third substance to be added. In order to defuse this problem, Newton interprets Sendivogius's "Luna" as the mercury of lead, which has been coagulated by a "fetid sulfur" to form the dark metal; had its sulfur been more pure and subtle, the product would have been a more noble metal. As for the "earth" that Newton alludes to in this passage, it is merely the metallic lead from whence the mercury can be drawn; the mercury of lead and the metal itself are really one thing, which happens to have been corrupted and coagulated by a "fetid sulfur."[17] Hence the seeming contradiction with Babson 925 (and with Sendivogius himself) has been resolved. The "impure mercurial water" of Babson 925 is not common quicksilver or any other substance beyond the initial crude antimony and lead; it is the internal mercury of the lead itself.

But this still leaves the problem of the "Sol" that Sendivogius claimed to combine with "our Luna which is overspread with the sphere of Saturn." We

---

[14] Keynes 19, 1r, at note "f."

[15] Keynes 19, 1v, at note "o": "Hoc est mercurij ex plumbo. Nam in fine Conclusionis Tractatuum, ait, operare in sole & luna nostra quæ obducta est Sphæra saturni (id est sole & mercurio qui coagulatur sulphure fœtida in obscuram formam plumbi) & non in tertiâ materiâ. Tres autem erunt materiæ nisi Terra & Luna pro eadem habeantur. Imo ratio suadet ut aliqua saltem pars plumbi in merium <sic> redigatur ut digestio fiat."

[16] Sendivogius, *Novum lumen chemicum*, in Albineus, *Bibliotheca chemica contracta*, 54.

[17] As Newton says on 1v of Keynes 19, at note "l": "Terra ista est plumbum nam in Tract 9no, Plumbum est cum quo aurum undecies coit."

now know that Saturn is lead and Luna the mercury of that metal, and we know that Newton's lead process employed antimony to purify the lead. If we are to avoid the inclusion of more than two materials in the process, then "Sol" cannot refer to its traditional alchemical referent, gold. And in fact, Newton explicitly tells us that Sendivogius's "Sol" or gold is really the very antimony with which his Luna combines, in a passage that describes his lead process in detail:

> If gold (that is, antimony, which fills the place of gold) is digested eleven times with lead (which, since it was practically referred to as the water or mother of the metals, is held as feminine with regard to its mercury), when the operation is repeated with the addition of new lead and the gold dissolved thereby, its sulfur and that of the lead will float upon the mercury; if it is again conjoined with that mercury, it purges the mercury by means of the dregs sinking down, and the mercury of the philosophers is produced.[18]

Here the pervasive hylozoism of Sendivogius enters at full force. The "gold," or rather crude antimony, is male, and the lead, or its mercury, female. Following Sendivogius, Newton says that this "gold" should be refined with lead eleven times, and then it will release its male seed or sulfur. Since Sendivogius's "gold" is actually stibnite in Newton's view, this refers to a process of digesting crude molten antimony with lead eleven times. According to Newton, this should bring about a separation of the sulfur in the stibnite and the sulfur of lead, both of which will float as a slag on the surface of the "mercury," which remains in the bottom of the crucible. It seems clear that Newton, like various other seventeenth-century alchemists, is here thinking of the regulus of purified metallic antimony and unreacted lead that would be found at the bottom of the crucible as a so-called mercury. He then says that the supernatant sulfur should be recombined with the regulus or mercury multiple times, which will result in further cleansing, leading eventually to a fluid sophic mercury.

Following Sendivogius, Newton also employs the terms "Magnes" (magnet) and "Chalybs" (steel) for the male and female ingredients that go into the sophic mercury. As though this multiple terminology for the same ingredients were not confusing enough, Sendivogius says in his epilogue to the *Novum lumen chemicum* that he has now reversed the two terms, so that the previous "Chalybs" is now the "Magnes" and vice versa.[19] Again, faced with Sendivogius's masterful equivocation, Newton reduces the Pole's complex allegory into a practical recipe with only two players—lead and antimony:

---

[18] Keynes 19, 1r, at note "d": "Si aurum (i Antimonium quod vices auri ^& masculi supplet) undecies digeratur cum plumbo (quod, cùm ferè ut aqua vel mater metallorum dictum fuit, ~~ejus mercurius~~ respectu nempe ☿ⁱʲ sui, pro feminâ habeatur) operatione scilicet repetita per additionem novi plumbi, auro sic dissoluto, sulphur ejus & plumbi supernatabit mercurio, quod cum isto ☿° rursus conjunctum purgat illum per fæces decidentes & producitur ☿ Philosophorum."

[19] Sendivogius, *Novum lumen chemicum*, in Albineus, *Bibliotheca chemica contracta*, 49: "Magnes est noster, quem in praecedentibus chalybem esse dixi."

In the Epilogue he says "our Magnes is that which I called 'Chalybs' in the preceding tractates." But they call lead the magnet because its mercury attracts the semen of antimony just as a magnet attracts steel.[20]

The sexual imagery of the male and female, sulfurous and mercurial seeds found respectively in antimony and lead, is fully compatible with the Sendivogian terms "Magnes" and Chalybs." As Newton explains a little later in the text, the female seed of the lead has a magnetic attraction that opens the pores of the antimony and draws out its male seed. It is the "copulation" of these two materials that renders the masculine seed of the antimony, which Newton (following Sendivogius) also calls the "radical humidity" of the metalloid, fertile.[21] The multiple digestions that Newton describes in Keynes 19 are not mere reheatings of dead matter—they are repeated copulations of the living seed within antimony and lead in order to produce a fruit of a higher order, the sophic mercury in its fully graduated form.

Reducing all of Sendivogius's mysterious language to the interaction of lead and antimony may sound deceptively simple, but in fact Newton tries to build a detailed practice on the allusive hints strewn hither and yon in the *Novum lumen chemicum*. At times, for example, Sendivogius seems to be alluding to two successive processes that require different glassware.[22] Following these hints, Newton says that the elevenfold digestion of lead and antimony, which leads to the release of the mercury of lead, must be carried out two times. But in the second iteration, ten parts of antimony are used instead of eleven.[23] Similarly, Newton argues that once the mercury of lead has been extracted by these processes, it must be recombined with the sulfur of the antimony and digested twice, in different proportions. These extractions or purgations of the mercury of lead, followed by recombinations and digestions with lead and antimony, are intended to "graduate" and fertilize the mercury of the lead, so that it eventually becomes the sophic mercury of the alchemists. Sublimation of the mercury is also involved, though not in a way that is entirely clear. The mercury should be divided into two portions, and the second of these should again be split into six or eight parts, "depending on whether you want to sublime the mercury seven or nine times."[24]

This level of specificity is manifestly lacking in the *Novum lumen chemicum*, a text that steers resolutely clear of the details involved in actual laboratory practice. In fact, Newton has imported this sevenfold or ninefold sublimation of the sophic mercury into the alchemy of Sendivogius from

---

[20]Keynes 19, 1v, at note "e": In Epilogo ait, Magnes est noster quem in p<*illeg.*>ræcedentibus Chalybem esse dixi Plumbum autem dicunt magnetem quia ᵍᵘˢ ᵉʲᵘˢ attrahit semen Antimonij sicut magnes ferrum Chalybem."

[21]Keynes 19, 1v, at note "t." For the medieval theory of the radical humidity, see the works by Michael McVaugh, Chiara Crisciani, and Giovanna Ferrari cited in chapter four of this volume.

[22]Sendivogius, *Novum lumen chemicum*, in Albineus, *Bibliotheca chemica contracta*, 37–38, 54, and 63.

[23]Keynes 19, 2v, at note "k": "Si dissolvas plusquam undecim partes Saturni Plumbi in opere primo ^vel conjungis plusquam decem in secundo, medicina non inde melioratur propter nimiam abundantiam Mercurij Saturnialis respectu Sulphuris Antimonialis."

[24]Keynes 19, 2r, at note "b": "Reliqua autem aqua concipiatur dividi in sex vel octo partes æquales prout velis mercurium septies vel novies sublimare."

a much more explicit source—Philalethes's *Secrets Reveal'd*. In making his sophic mercury, Philalethes had recommended "seven or nine" eagles, that is, sublimations.[25] Newton's early commentary on Sendivogius is, in reality, a hybrid of Philalethan and Sendivogian beliefs, with the balance tilted decisively toward Philalethes in the area of practice. For this reason, *Collectiones ex Novo Lumine Chymico quæ ad Praxin Spectant* has as much to tell us about Newton's interpretation of Philalethes as it does about his reading of Sendivogius. Let us now examine the comments in the text that explicitly derive from *Secrets Reveal'd*, which may well have been the only Philalethan text that Newton knew at the time of writing his *Collectiones*.

As we have seen, Newton's early belief that lead was the proper matter from which the philosophical mercury should be extracted derived from Sendivogius's *Novum lumen chemicum*. Basing himself on Sendivogius's claim that lead would become gold beneath the surface of the earth if the impediments of its impurity were removed and its seed were allowed fully to mature, Newton straightforwardly devised a method of purging them away. His chosen technique employed antimony, which was the assayers' agent for refining gold, and Eirenaeus Philalethes's agent for purifying quicksilver and converting it to the sophic mercury. In essence, Newton believed that Sendivogius and Philalethes were describing different aspects or portions of the same process, namely, the purification of lead by antimony and the concomitant interaction of the male, sulfurous seed found in the antimony with the feminine mercury of the lead. A process of repeatedly purging followed by recombination of the sulfur and mercury were supposed to lead, eventually, to a liquid sophic mercury. It is obvious, then, that the young Newton had an idiosyncratic reading of *Secrets Reveal'd* in which the metal lead played a major role. To what degree was Newton committed to this Sendivogian reading of Philalethes, and how did he justify it? In order to answer these questions, let us return briefly to Keynes 19. In considering another *Deckname* for the mercury of lead, Newton glosses the words of his sources with his own comments in square brackets:

> The menstruum of the world [that is, the water whence all things have taken their origin] in the sphere of ☽ [that is, in the form of living Luna or mercurial water; thus Sendivogius says in the conclusion to the Treatises, our Luna is overspread with the sphere of Saturn, that is, in the form of lead] is extracted from our earth or lead with the foresaid water of *sal nitrum*; for Eyrenæus says in his fourth chapter, "Our magnet <*illeg.*> (i.e. Lead) hath an occult center abounding wth salt (which is here called *sal nitrum*) wch salt is ye menstruum (or, ye vertue in the menstruum) in the sphære of ye Moone wch knows how to calcine ☉."[26]

[25] Philalethes, *SR*, 56.

[26] Keynes 19, 2r, at note "c": "Menstruum mundi [ie. aqua unde omnia in mundo duxerunt originem] in sphærâ ☽* [id est in forma ~~mercurij~~ lunæ vivæ vel aquæ Mercurialis. sic in conclusione Tractatuum ait, Luna nostra obducta est sphærâ Saturni i.e. ~~observata~~ formâ plumbi] est aquâ ^salis nitri p^rdicta de terrâ nostrâ vel plumbo extracta; dicit enim Eyrenæus Capite 4to, Our magnet <*illeg.*> (i.e. Lead) hath an occult center abounding wth salt (quod hic dicitur sal nitri) wch salt is ye menstruum (or, ye vertue in the menstruum) in the sphære of ye Moone wch knows how to calcine ☉."

Here we see Newton repurposing the Sendivogian passage about "our Luna"—the mercury of lead—being overspread with the visage of the dark, Saturnian metal. His point here is to equate the mercury of lead with another Sendivogian *Deckname*—"menstruum of the world in the sphere of Luna." But in order to certify that his interpretation is correct, Newton turns to "Eyrenaeus," that is, Philalethes, whose words he quotes verbatim from *Secrets Reveal'd*. Newton's careful reading of that text showed that Philalethes was building on the noble Pole in his own fourth chapter, where the American author uses the Sendivogian expression "*Menstruum* in the Sphere of the Moon."[27] Given the obvious fact that Philalethes was acting as a commentator of the *Novum lumen chemicum*, it made complete sense for Newton to assume that the "American philosopher" corroborated his view that lead was the starting point for the sophic mercury. One can imagine Newton's excitement at finding these shared passages among his favorite authors; an old alchemical maxim had it that "liber librum aperit" (one book opens another), and the aspiring chymist had found two authors whose complementarity cast a mutual beam on each. The noble Pole and his American successor were signaling that the young Cantabrigian was on the right track.

Newton and the Doves of Diana

Since we have shown that Newton's readings of Sendivogius and Philalethes were highly interdependent, it will now be useful to back up a bit and refresh our understanding of Philalethes's actual process for making the sophic mercury. We will also have to examine the mythological language in which Philalethes and his alchemical forebears clothed this desideratum and the ingredients required to make it. Unlike Newton in the late 1660s, we have the great advantage of knowing exactly what Philalethes—or the man behind that name, George Starkey—was describing in *Secrets Reveal'd*. Thanks to Starkey's "Key into Antimony," part of the surviving letter that the young New Englander wrote to Robert Boyle in 1651, we know that the Philalethan process began by smelting stibnite with iron in order to arrive at the star regulus of antimony. Starkey then combined the metallic antimony with two parts of refined silver in order to produce an alloy that would easily amalgamate with quicksilver. After separately purifying his quicksilver with vinegar and salt, Starkey sublimed it multiple times (seven to nine) from the antimony-silver alloy, which eventually resulted in the sophic mercury. As one can see, the two parts of refined silver, which are called "the two doves of Diana" in *Secrets Reveal'd*, played a key role in the process.[28]

Newton, however, did not have Starkey's "Key into Antimony" at the time of composing Babson 925 and Keynes 19, although he did acquire a

---

[27] Philalethes, *SR*, 8. This closely paraphrases Sendivogius's words on page 37 of Albineus's edition of the *Novum lumen chemicum*: "Aqua verò Illa debet esse menstruum mundi, ex sphaera Lunae, toties rectificatum quod possit calcinare Solem."

[28] William R. Newman, "Newton's *Clavis* as Starkey's 'Key,'" *Isis* 78 (1987): 564–74.

Latin translation of the text (the *Clavis*) later in his career. But by that time he was wedded to his own, entirely different interpretation of the doves of Diana, which he never entirely abandoned even after acquiring Starkey's letter. Unaware of the need for silver in Starkey's process for the sophic mercury in 1669, Newton had already come to his own conclusion about the material referent behind the term "doves of Diana" by the time of writing Keynes 19 if not before. Even in this very early manuscript we see Newton committing himself to the view that the doves of Diana signify a material entirely distinct from silver. Before we describe that material, however, we must come to terms with the origin of the expression "doves of Diana," since Newton does not mention that *Deckname* in his *Explicationes* to Sendivogius, but rather in the commentary to Jean d'Espagnet's *Arcanum philosophiae hermeticae* that immediately follows in Keynes 19. Like many alchemists of the sixteenth and seventeenth centuries, Newton's source d'Espagnet builds on the old belief that ancient mythology was actually for the most part encoded alchemy. It is easy to see how myths such as that of Jason and the golden fleece could be read as thinly veiled allegories of the chrysopoetic quest. In order to acquire the golden fleece, Jason had to defeat its sleepless guardian, the dragon of Colchis, and use fire-breathing, bronze bulls to plow a field sown with the teeth of the dragon. Dragons and fire had always been popular among alchemists, and the decoding of classical myth into precepts of the aurific art extends well into the Middle Ages.

Another favorite source for chymists was Book 6 of Vergil's *Aeneid*, where Aeneas descends into Hades in order to confer with the shade of his father Anchises. In Vergil's recounting, Aeneas had to enlist the aid of the Sybil of Cumae in southern Italy in order to carry out this feat. On entering her customary trance, the prophetess informed him that he must find and tear off a golden bough deep in an obscure grove and carry it to the goddess of the underworld, Proserpina. Saddened by news of this seemingly impossible task, Aeneas prayed to his mother, the goddess Venus, for guidance. His prayer was answered in the form of two doves who descended from the sky, led him to the tree bearing the elusive golden bough, and landed on it. This myth resurfaces in stanza fifteen of d'Espagnet's *Arcanum*, where the French alchemist skillfully blends his commentary with the italicized text of the *Aeneid*. For d'Espagnet, the golden bough is a symbol for the philosophers' stone, which the alchemists obscure behind dark words:

> In nothing do they strive so bitterly as in hiding their golden bough, *which the whole grove covers and shadows hide in dark valleys*; nor does it yield to just any powers but it easily and willingly will follow him *who knows the maternal birds; and twin doves come flying from the sky, as it happens, beneath his very eyes.*[29]

---

[29] Jean d'Espagnet, *Arcanum philosophiae hermeticae*, in Albineus, *Bibliotheca chemica contracta*, p. 9, stanza 15: "In nullo tam acriter contendunt quam in celando ramo ipsorum aureo, *quem tetigit omnis lucus nec ullis cedit viribus & obscuris claudunt convallibus umbrae*: sed facilis volensque sequetur eum qui *maternas agnoscit aves et geminæ cui forte columbæ, ipsa sub ora viri venere volantes.*"

Hence it is the two doves of Venus who reveal to Aeneas the tribute that he must bear to the underworld. By implication, these doves must also serve in the preparation of the philosophers' stone.

The doves occur again in another passage of d'Espagnet's *Arcanum*, where he is discussing Jason's quest for the golden fleece.[30] There d'Espagnet says that the two doves of Venus and the *insignia* or "emblems" of Diana can be used to circumvent the ferocious beasts guarding the entrance to the alchemical art like the fire-breathing cattle and terrifying Colchian dragon that Jason had to defeat in order to acquire his golden prize. Both the alchemist and Jason achieve this difficult task not by force and threats, but by mollification. They soothe or "soften up" their adversaries (*mulcebunt* in Latin) by gentle means rather than attacking them directly. Apparently d'Espagnet is playing on the story that Jason had been given the means to defeat these monstrous beasts by Medea, enchantress and daughter of the King of Colchis, who had fallen in love with him. In one alchemical reworking of the story, Medea bequeaths four gifts on Jason: an ointment with which he could anoint his body and protect himself against venom and fire; a soporific drug that would put an end to the dragon's eternal wakefulness; a limpid water that would extinguish the fire of the bulls; and perhaps most importantly, an "image" or medallion of the sun and moon that Jason should wear around his neck to assure that everything turned out successfully.[31] Thus Jason was able to reduce direct confrontation with the dragon by using the soporific drug to narcotize the beast, and once the dragon was sleeping, the leader of the Argonauts, wearing his protective talisman of the two celestial luminaries, extracted the teeth that he would subsequently sow with the help of the fiery bulls. Perhaps d'Espagnet had this magical medallion in mind when he spoke of the emblems or *insignia* of Diana that the alchemist would need, alongside the doves of Venus. The hunter goddess was typically viewed as a deity of the moon, so the presence of Luna on Jason's talisman would allow for mention of Diana as well.

Obviously, d'Espagnet's discussion mentions two doves of Venus rather than two doves of Diana. Nonetheless, by bringing the "emblems of Diana" into the discussion, he opened the door to a substitution of the goddess of the hunt for the goddess of beauty. The fluid use of tropes is as much a part of early modern alchemy as it is of seventeenth-century poetry, and to expect a rigid, one-to-one association of alchemical images would only lead us astray. Whatever d'Espagnet's precise meaning may have been, Newton well understood this principle of substitution, as he reveals in the following comments to d'Espagnet's *Arcanum*: "the emblems of Diana and the doves of Venus are the same thing."[32] If these emblems could become doves, then in the literary

---

[30] Jean d'Espagnet, *Arcanum philosophiae hermeticae*, in Albineus, *Bibliotheca chemica contracta*, 23.

[31] These four gifts of Medea to Jason are already discussed in an alchemical context in Michael Maier, *Arcana arcanissima* (s.l.: 1614), 64–65. Later in the *Arcana arcanissima* Maier says that the image of the sun and moon was a "pentaculum" or talisman, and that this refers to the dissolution of "Sol" and "Luna" into the prime matter. See Maier, *Arcana*, 73–74.

[32] Cambridge University, King's College, Keynes 19, 3v, at note "c": "Dianæ insignia & veneris columbæ idem sunt, nempe Sulphur aquæ mercuriali supernatans."

world of alchemy their owner should also be capable of such permutation; why not then substitute "doves of Diana" for "doves of Venus"? In fact, Newton had already been beaten to this game by no other than his perennial stalwart—Philalethes. Just as d'Espagnet advised, Philalethes urges the alchemist in *Secrets Reveal'd* to use gentle methods to tame the wild beasts of alchemy. And just as in d'Espagnet, the mollifying methods include two doves, but now they are the Philalethan doves of Diana rather than those of Venus. This is reprised in another passage of Newton's early commentary to d'Espagnet:

> Eyrenæus says in chapter 2, "yᵉ Doves vanquish yᵉ Lyon by asswaging him." That is, the sulfur penetrates the mercurial water and expels its impurity not by hostile force but by amicable insinuation.[33]

For the early Newton, the Philalethan doves of Diana are identical to d'Espagnet's Venereal doves, but unlike d'Espagnet, Newton gives us a clear material referent. From the passage, it is obvious that the doves of Diana refer to a certain sulfur that penetrates and purifies a particular mercury by gentle means. But what is this sulfur, and what this mercury? If we return to the earlier passage where Newton stated unequivocally that the Venereal doves and Diana's emblems are identical, the answer is revealed:

> the emblems of Diana and the doves of Venus are the same thing, that is to say the sulfur floating on the mercurial water.[34]

Suddenly we are transported from the ethereal realm of classical poetry to the fiery furnaces of Newton's laboratory. We are now back in the world of the sulfurous slag that floats on the mercurial regulus formed by reducing antimony from its ore, stibnite, by means of lead. The doves, be they d'Espagnet's Venereal version or the Dianic variety of Philalethes, both refer to the "sulfur" that first appears as a slag and is then recombined with the lead-antimony regulus multiple times in Newton's lead process. But then, one might reasonably ask Newton, why do Philalethes and d'Espagnet speak of two doves instead of one? The aspiring chymist would respond that they are merely providing another clue. One need only recall what Newton said early in his commentary on the *Novum lumen chemicum*—that during the repeated addition of lead to crude antimony in fusion, both "its sulfur and that of the lead will float upon the mercury."[35] In short, the two doves of Diana refer to the two respective sulfurs, that of antimony and that of lead, which are released in the production of Newton's sophic mercury. In another passage of his commentary to d'Espagnet, Newton nails down this point

[33] Keynes 19, 3v, at note "d": "Eyrenæus cap 2° ait, yᵉ Doves vanquish yᵉ Lyon by asswaging him. Nempe Sulphur non hostili vi sed amicabili insinuatione penetrat mercurialem aquam & expellit ejus impuritatem."

[34] Keynes 19, 3v at note "c": "Dianæ insignia & veneris columbæ idem sunt, nempe Sulphur aquæ mercuriali supernatans."

[35] Keynes 19, 1r, at note "d": "Si aurum (i Antimonium quod vices auri ^& masculi supplet) undecies digeratur cum plumbo (quod, cùm ferè ut aqua vel mater metallorum dictum fuit, ~~ejus mercurius~~ respectu nempe ☿ⁱʲ sui, pro feminâ habeatur) operatione scilicet repetita per additionem novi plumbi, auro sic dissoluto, sulphur ejus & plumbi supernatabit mercurio, quod cum isto ☿° rursus conjunctum purgat illum per fæces decidentes & producitur ☿ Philosophorum."

while adding yet another *Deckname*—"The twin sulfurs, whether doves or crows, float above the mercurial water."[36] In fact d'Espagnet does speak at one point of crows that metamorphose, becoming doves, but let us keep our attention focused on the bird of peace rather than on that of carrion.[37]

By continuing our analysis of Diana's doves, we will see the extraordinary attention that Newton paid to every detail of the text serving as the source of his commentary. One might be inclined to call his interpretation "literal," even to the point of absurdity, but that would entirely miss the point, since the extreme liberties of his interpretation belie the usual sense of the term. There is nothing in d'Espagnet's *Arcanum* to make a naive reader think of antimonial slag. For Newton, every word of his alchemical sources was pregnant with meaning, and that meaning could only be extracted by focusing on the exact nature of the images and syntax employed by the author in question. The idea of extracting the "general sense" from a source and moving on was completely alien to his consciousness. The riddles of the adepts were too precise for that. One can get a clear sense of this emphasis on precision by considering Newton's further comments on the doves of Diana.

As we now know, for the young Newton the two doves are the sulfurs from the stibnite and from the lead, which he earlier said should float above the molten mercurial regulus in the refining crucible. But he does not stop at identifying the supernatant sulfurs with Diana's doves. As is often the case, the excuse for Newton's interpretation here lies with Philalethes. In *Secrets Reveal'd*, the "American philosopher" had said to "Learn what Diana's Doves are, which do vanquish the Lion by asswaging him," and in another passage of the same text, Philalethes returns to the Doves, saying:

> In this, let *Diana* be propitious unto thee, who knows how to tame the wild Beasts, whose two *Doves* shall temperate the malignity of the *Air* with their feathers.[38]

Why does Philalethes discuss not only the doves of Diana but also their feathers? To Newton, the answer was obvious. For him even the feathers of Diana's doves have the sense of specific *Decknamen*. He makes this entirely clear in his commentary to the *Novum lumen chemicum*:

> The water is abstracted seven times from the sulfur so that it may be rectified until a white powder which is here called "cinders" is separated. Or according to Eyrenaeus it is called "the feathers of the Doves," & to scatter the cinders into water or the feathers into the air mean the same thing.[39]

Here Newton alludes to the seven (or nine) sublimations that were required for the sophic mercury in *Secrets Reveal'd*. He now refers to the molten mercurial regulus from the stibnite-lead mixture as a "water," which is repeatedly

---

[36]Keynes 19, 4r, at note "w": "Sulphura sive Columbæ sive corva supernatabant aquæ mercuriali."

[37]D'Espagnet, *Arcanum philosophiae hermeticae*, in Albineus, *Bibliotheca chemica contracta*, p. 40, stanza 69.

[38]Philalethes, *SR*, 13.

[39]Keynes 19, 2v, at note "h": "Aqua, ut rectificetur, septiës abstrahitur a Sulphure donec relinquitur pulvis albus quod cineres hic dicuntur, vel apud ~~alies authores~~ Eyreneum plumes columbarum & cineres in aquam vel plumes in aerem spargere idem sonant."

recombined with the supernatant sulfur, namely, Diana's doves, until a white powder is released. It is this white powder, which d'Espagnet says the alchemist should recombine with the mercurial water and Philalethes says to scatter in the air, that receives the name "feathers of the Doves."

To summarize all of this, Newton's Philalethan interpretation of Sendivogius's process began with lead, which was supposed to be purified with stibnite eleven times and then another ten times in a "second operation." The resulting regulus or mercury would be covered by a floating slag containing the sulfur of the stibnite and that of the lead, which was for Newton the secret meaning behind Diana's doves. The mercury of lead then had to be recombined with the two doves multiple times in order to receive the full purification that would result in the sophic mercury. In Newton's interpretation, based on Philalethes's multiple sublimations of the sophic mercury, this purification should be carried out by repeated sublimations of the antimony-lead regulus from its sulfur. Eventually, this was supposed to lead to the separation of a white powder—the "feathers" of the doves, which would serve as the basis of still other operations. Like the mercury of lead and the two sulfurs, the feathers of the doves were supposed to be a derived material produced from the initial ingredients, metallic lead and crude antimony, during the long set of operations described in Newton's commentaries.

It is impossible to overstress the fact that Newton's early interpretation of Sendivogius—although indeed based on Philalethes—represents the young Cantabrigian's idiosyncratic interpretation of *Secrets Reveal'd*. We know from Starkey's 1651 "Key into Antimony," written for his friend and patron Robert Boyle, that there is actually no lead process lurking behind the extravagant cover names of the Philalethan text. Moreover, Starkey's process really did employ quicksilver, by which I mean the mercury of the modern periodic table, not a putative "mercury of lead," along with stibnite, iron, and silver. The iron and stibnite were employed in making the star regulus of antimony, which was then fused with silver to make an alloy that would easily amalgamate with the quicksilver. By reducing the range of ingredients to a mere two—lead and stibnite—Newton was making a drastic oversimplification. And yet this parsimony seemed to be exactly what Sendivogius was calling for when he argued that the great work consisted of only two things. And after all, the noble Pole had insisted time and time again that Nature is simple, and that the alchemist should mimic her by following her simple path. By restricting himself to lead and antimony, Newton thought he was doing precisely that. Hindsight reveals that applying the Sendivogian stricture on more than two ingredients could only lead to a misinterpretation of Philalethes. And yet Newton was far from unique in misunderstanding the "American philosopher," and the pervasive, though not universal, belief that the adepts were all really discussing a single process under their colorful menagerie of doves, lions, dragons, and crows inevitably led to a sort of unintentional homogenization. As we will now see, Newton soon found himself questioning his own preliminary interpretation of Philalethes, which was only the beginning of his career-long struggle to interpret the secrets of the adepts.

Second Thoughts about Lead

At some point after writing the *Loca difficilia in Novo Lumine Chymico explicata* found in Keynes 19, probably within a year or two, Newton encountered a text that would force him to reconsider his early interpretation. To judge by two sequential synopses found in a very early manuscript now kept in Jerusalem (MS Var. 259, National Library of Israel), Newton devoured *The Marrow of Alchemy* soon after acquiring it, probably in the first years of the 1670s.[40] *The Marrow of Alchemy* is another text written by George Starkey under the pseudonym of Eirenaeus Philalethes, published in two parts in 1654 and 1655. Although the text is similar in many ways to *Secrets Reveal'd*, the *Marrow* also differs from that text in a number of significant points. First, the *Marrow* is an English poem written in stanzas of quatrains and couplets; *Secrets Reveal'd*, to the contrary, was a Latin treatise that had been translated into English (from the *Introitus apertus ad occlusum regis palatium*, written by Starkey under the pseudonym "Anonymous Philalethes"). Hence unlike *Secrets Reveal'd*, the *Marrow* did not come to Newton through the filter of a translation. Another obvious difference between *Secrets Reveal'd* and the *Marrow* lies in the clarity with which the later text describes the starting point of the process for making the sophic mercury, namely, the antimony ore stibnite and its reduction by means of iron. As Philalethes says in Newton's paraphrase:

> Saturns child is sable coloured wth argent veines, all volatile, in its native crudity it purgeth ☉^s superfluity, its o^r dragon wch Cadmus assailed ^in vain[41]

Mineral stibnite is indeed black or "sable," and it can have shining, silvery streaks in it. The ore releases visible fumes at temperatures attainable in a charcoal-burning furnace, thus distinguishing it unequivocally from lead, and if these clues were not enough, Philalethes then alludes to the use of crude antimony in refining gold. The subsequent reference to Cadmus—the Greek hero who founded Thebes—attacking Saturn's child or "our Dragon" in vain simply means that when iron is used to make the star regulus, the iron disintegrates in the process of reducing the antimony, most of it disappearing into the slag (though some can also remain in the regulus).

That Newton did not fail to observe these broad hints appears from his parenthetical insertions in Var. 259. Commenting on Part 2, Book 1 of the *Marrow*, he says "O<u>r Dragon Saturns child (Antimony) conquers Cadmus (♂)," making it certain that at this point Newton interpreted "Saturn's

---

[40] Var. 259 is a composite manuscript made up of twelve different parts composed at different times. Var. 259.7, the part containing Newton's synopses of *The Marrow of Alchemy*, is no doubt early, since it contains multiple occurrences of the unbarred Saturn symbol and no versions of the barred version. Additionally, Newton's title originally ran (7.2r) "The Marrow of Alchymy a fals Poem," but he subsequently struck through "a fals Poem," presumably after reading it. Given Newton's use of the *Marrow* as an authoritative source everywhere else that he mentions it in his alchemical *Nachlass*, this suggests an early misimpression of the *Marrow* that he immediately corrected. Additionally, the second synopsis of the *Marrow* in Var. 259 (7.3r) begins "At a 2^d reveiw Booke 1." If we take Newton at his word, this means that these notes reflect his second reading of the text.

[41] Var. 259.7.3v.

child" as antimony and Cadmus as the iron necessary for its reduction to the star regulus. In the next paragraph, Newton cements this identification with another parenthetical gloss: "Stanza 21 &c. Salt of Nature (Antimony) must bee joyed <sic> wᵗʰ ♃ in ye house of ♈ (♃ of ♂)." Philalethes had said in the commented passage that vulgar quicksilver needs to be "acuated" or sharpened by a salt that is found in the belly of "*Saturns* off-spring" before it can become the sophic mercury.[42] Thus Newton's comment refers once again to Saturn's child, here under the guise of a salt, which he equates unproblematically with antimony.

What is perhaps more surprising is Newton's identification here of the sulfur in the zodiacal house of Aries with the sulfur of iron. This directly contradicts his earlier interpretation of Sendivogius in Keynes 19. There Newton had paraphrased the noble Pole as saying "our water is drunk up in marvelous ways, but the best is that which is drunk up by means of our Chalybs found in the belly of the Ram."[43] Rather predictably, Newton had replied to Sendivogius in Keynes 19 by saying that "our water" meant lead, or rather the mercury of lead, which he elsewhere calls the Magnes or magnet in Sendivogian language. Since Magnes meant lead in the parsimonious interpretation of Keynes 19, the Chalybs must mean the other ingredient of Newton's very early lead process, namely, antimony. And on the strength of the association between the Chalybs and Aries, the zodiacal Ram must mean antimony too, a fact that Keynes 19 spells out in the following interesting words:

> For antimony among the old was called Aries, since the Ram is the first sign of the zodiac in which the sun starts to be exalted, and gold is exalted above all else in antimony.[44]

Despite his evident satisfaction with this exercise in creative word origins, Newton was forced to abandon his interpretation of antimony as Aries in Var. 259 because the *Marrow* had made it transparently obvious that crude antimony was Saturn's child, and Saturn's child had to be joined to the martial sulfur in Aries in order to become the star regulus. The sulfur in the Ram's belly could no longer be crude antimony, since it was now the means by which that material is converted to its regulus.

The fact that Newton now understood the need for iron in making the star regulus means, of course, that he could no longer insist on a process that involved only the reduction of stibnite by means of metallic lead. Does this mean that lead simply dropped out of his interpretation of Philalethes now, to be replaced by iron? Interestingly, it does not. In glossing a particularly difficult passage of the *Marrow*, Newton says the following:

> ♄ child & ♂ united yᵉ fæces are purged, the pure sinks downe & powered forth shows a starr. Soe yᵉ souls of ♂ & ♄ are inseperably mixed till mars

---

[42] Philalethes, *Marrow*, part 2, book 1, stanza 23, p. 6.

[43] Keynes 19, 3r, at note "f": "Aqua nostra hauritur miris modis, sed ista est optima quæ hauritur vi Chalybis nostri qui invenitur in ventre Arietis."

[44] Keynes 19, 3r, at note "f": "Antimonium enim apud veteres dicebatur Aries Quioniam <sic> Aries est primū Signum Zodiaci in quo Sol incipit exaltari & Aurum maxime exaltatur in Antimonio."

soul bee fixed, y^{n} it leaves ♄ & in tryall is found perfect ☉. But this is done by ♀ (♄^{s}) mediation for by ♀ association Diana seperates them.[45]

The first part of this section paraphrases Philalethes's relatively open description of the star regulus. The mixing of the "souls of Mars and Saturn" refers to the combining of the invisible sulfur found in iron with the mercury of antimony, which purges the stibnite of its dregs and leaves the regulus pure. The "souls" or sulfurs are inseparably mixed in that both are carried up when the regulus is sublimed over a heat source. When Philalethes says that the soul of Mars can be fixed, that is, rendered nonvolatile, he is referring to a process that Starkey was developing in the early 1650s for creating "antimonial metals" by distilling the sophic mercury from alloys of metallic antimony, silver, copper, lead, or tin. We know from Starkey's surviving notebooks that these experiments ultimately resulted in failure, but when he was writing *Secrets Reveal'd* and the *Marrow*, he was confident of their success and presented them in positive, albeit obscure, terms.[46] What is important for our present purpose, however, is that Newton glosses Philalethes's reference to "Venus" as "Saturn" with the parenthetical comment "by ♀ (♄^{s}) mediation." Newton clearly realized that in the Philalethan system of *Decknamen*, Venus and Saturn could both mean antimony. This fact is underscored by his subsequent comment:

Stanza 52 &c. The Burning 🜍 being rightly seperated from ye purer parts of Antimony (wch is o^{r} Venus uniting ☿ & 🜍 of ♂ together) & the dregs removed there appeas <*sic*> a nut ^like at mettall but very brittle & easily fusible. Cause ♂ to embrace this o^{r} ♀ & both shall bee purged & thou <*illeg.*> shalt see a starr [& ♀ shall mediate twixt ♂ & Diana].[47]

Newton correctly read this passage as a description of the production of the star regulus of antimony by means of iron fused with stibnite. The brittle, fusible "nut" is the star regulus, which has been denuded of the excessive sulfur found in crude antimony. Newton explicitly equates the "purer parts" of antimony, that is, the regulus lying hidden within stibnite, with Venus. The regulus contains both the mercury of antimony and the congealing sulfur of iron, which makes it a solid. As he says again two lines later, the iron must embrace "this o^{r} ♀" so that both it and the stibnite can be purged of their undesirable sulfur. Only after this purgation has been achieved can the star regulus emerge.

Newton's sensitivity to the polysemic character of Philalethes's language had its limitations, however, at least at the time when he was composing Var. 259.7. There is nothing in his abstract to make one believe that he understood that in the *Marrow*, Venus could also mean copper in addition to denoting antimony. Starkey had actually written the *Marrow* after his own process for making the sophic mercury had evolved away from the one he described in *Secrets Reveal'd*. Hence, although the operations described in

45 Var. 259.7.2v.

46 See Newman and Principe, *LNC*, 212–16.

47 Var. 259.7.2v.

the *Marrow* are very similar to those of *Secrets Reveal'd*, there are also important operational differences between the two works. Since Philalethes was Newton's major guide to practice, at least in these early years, this presented a problem for the Cantabrigian alchemist. In a word, Starkey had discovered between 1651 and the publication of the *Marrow* that the two doves of Diana, the two parts of refined silver to be added to star regulus of antimony in order to make an alloy capable of uniting with quicksilver, were unnecessary. In reality, it was possible to accomplish the same amalgamation with the much cheaper metal copper. Thus the *Marrow* would explicitly demote the doves of Diana in favor of the new technique employing copper as a mediator between the star regulus and quicksilver:

> Some use *Dianæs* doves for to prepare
> The water, which a tedious Labor is,
> And for to hit it right, an Artist rare
> May twice for once unfortunately miss:
>> The other way (which is most secret) we
>> Commend to all that Artists mean to be.[48]

From Newton's perspective, Philalethes could not simply have changed his mind about the proper way of arriving at the philosophers' stone. A perfect adept like Philalethes, who claimed in *Secrets Reveal'd* to have already acquired the philosophers' stone by the age of twenty-three, could not really be stumbling along and happening on new and better processes (as Starkey actually was doing in the 1650s). If the processes described in *Secrets Reveal'd* were only work in progress, how could the self-professed adept Philalethes have been speaking honestly? Since Newton had no doubts that Philalethes was a genuine adept, there had to be another answer. In the very early notes composing Var. 259.7, Newton therefore had good reason to overlook the allusions that Starkey made to his new copper process, and to interpret Venus in a rigid sense as antimony alone.

Second Thoughts about Antimony

Although Newton realized that the Philalethan *Marrow of Alchemy* had the production of the star regulus of antimony by means of iron at its core, he was still not willing to abandon a role for lead. His earlier confidence in the central importance of the heavy metal, as revealed by Keynes 19, was not easily shaken. Thus lead would reemerge as a topic for research in a slightly later set of notes. The manuscript in question, Yale University, Mellon 79, appears to have been composed well after Newton's early synopses found in Var. 259, and probably dates from the mid-1670s.[49] Like his earlier abridgements of

---

[48] Philalethes, *Marrow*, part 2, book 1, stanza 66, p. 16.

[49] Mellon 79 refers multiple times to the *Metallographia* of John Webster, a book that was first published in 1671, thus providing a *terminus post quem*. It is harder to establish a *terminus ante quem*, but the mixed use of unbarred Saturn symbols (five occurrences) and barred ones (four occurrences) suggests that the text was no later than the middle of the decade. Additionally, Alan Shapiro notes in his groundbreaking study of

the *Marrow*, Mellon 79 imparts important hints and reveals the progress of Newton's thoughts. The text consists of extracts from eight authors with occasional parenthetical comments by Newton. Of the eight pages making up the manuscript, six deal with lead or lead ore in one way or another. The instigation for this attempt to arrive at new knowledge about lead clearly lies with Newton's ongoing attempt to extract the sense of Philalethes's *Marrow*. The first passage in Mellon 79 dives into the nature of "Saturn," and subsequent comments make it clear that Newton was now inclined to read the planet as meaning lead:

> Saturn though vile & base to see, is of o^r secrets all ye ground In ♄ is hid an immortal soul. Untie its fetters wch do it forbid to sight for to appear then shall arise a vapour shining like pearl orient. To Saturn Mars wth bonds of love is tied who is by him devourd of mighty force whose spirit divides saturns body & from both combined flow a wondrous bright water in wch y^e Sun doth set & loos its light. Venus a most shining star is embrac't by ♂. Their influences must be united for she is y^e only mean between y^e Sun & o^r true argent vive to unite them inseparably. Marrow of Alchemy <*illeg.*>$^{p. 1}$. lib. 3.[50]

As Newton indicates, these passages stem from part one, book three, of the *Marrow*. The obvious problem that this paragraph presented was whether to interpret Saturn as lead or as crude antimony. In Keynes 19, Newton had read "Saturn" unambiguously as lead. But as we have just seen, the *Marrow of Alchemy* is more forthcoming in its description of the reduction of the crude antimony with iron than is *Secrets Reveal'd*. In fact, the passage just cited from the *Marrow* is relatively unambiguous in its description of Saturn's (stibnite's) conversion into a "wondrous bright water" (molten regulus) after consuming Mars (iron), and Newton picked up on these hints in Var. 259. But if Saturn stands for antimony, then where is the lead that Newton thought to be essential to Philalethes's sophic mercury in Keynes 19? The role of lead, I believe, is the fundamental problem that Newton is grappling with in Mellon 79.

At the same time, the introductory paragraph from Mellon 79 contains the perplexing statement "Venus a most shining star is embrac't by ♂." The genuine meaning of this passage, which can be pieced together from Starkey's surviving laboratory notebooks and letters (recently edited), is that the metal copper (here Venus) combines with the putative sulfur in iron (Mars), which was previously carried over into the star regulus when the metallic antimony was reduced from its ore. Starkey said to carry out the reduction with iron horseshoe nails that would be added to the stibnite at high temperature

Newton's watermarks that the paper of Mellon 79 bears the same watermark as Newton's famous "Hypothesis of Light" sent to the Royal Society on December 7, 1675, and also that of a letter written to Hooke two weeks later. See Alan Shapiro, "Beyond the Dating Game: Watermark Clusters and the Composition of Newton's Opticks," in *The Investigation of Difficult Things*, ed. P. M. Harman and Alan E. Shapiro (Cambridge: Cambridge University Press, 1992), 181–227, see 195.

[50] Mellon 79, 1r.

along with some saltpeter as a flux.[51] Throughout *Secrets Reveal'd* Starkey (writing under the guise of Philalethes) refers to this sulfur of iron as "our Gold," or "our Sol," and much of his discussion focuses on this hypothetical ingredient of the iron.[52] When he speaks in the above passage of Venus uniting with Mars, this is therefore a veiled description of an alloy made with copper and metallic antimony in the form of the star regulus (again, the latter was thought to carry within it "our Sol," the sulfur of the iron).[53] This also allows us to decode Philalethes's claim that copper (Venus) is "yᵉ only mean between yᵉ Sun & oʳ true argent vive to unite them inseparably." The "Sun" here is not metallic gold, as in the traditional system of alchemical referents descending from medieval sources, but again, the invisible sulfur of iron in the regulus. The copper acts as a medium between the star regulus (containing "yᵉ Sun" or sulfur of iron) and the quicksilver that allows one to make them combine as the sophic mercury. In other words, the *Marrow* is here describing Starkey's new process for making the sophic mercury with copper instead of the silver denoted by the "two doves of Diana."

All of this interpretation is clear and straightforward if one possesses Starkey's letters and laboratory notebooks, but none of these were available to Newton at the time of composing Mellon 79. Hence Newton had to come to his own conclusions about the obvious differences between *Secrets Reveal'd* and the *Marrow of Alchemy*. That he did not choose in favor of the processes given by Starkey himself is revealed in the second paragraph of Mellon 79:

> Oʳ water flows from fourfold spring, wᶜʰ is but 3 & wᶜʰ but 2 & wᶜʰ but 1 [♄ ♂ ♀ ☿] Tis ♄'s ofspring who keeps a well in wᶜʰ drown ♂ & then ♄ behold his face in't wᶜʰ will seem fresh & young when yᵉ souls of both are blended together, for each need be amended by th' other. Then a star shall fall into yᵉ well. Let Venus add her influence for she is nurs of oʳ stone, yᵉ bond of Crystalline ☿. This is yᵉ spring in wᶜʰ oʳ Sun must dy. <illeg.> Saturn's child is yᵉ serpent wᶜʰ devours Cadmus wᵗʰ his companions. Though't be defiled yet thou shalt wᵗʰ a shour wash of its blackness till yᵉ ☽ appear shining most bright. Marrow of Alk. part 1. lib 4 st 59.[54]

Although it is not immediately clear what Starkey meant by a "fourfold spring," the process that he hid behind his allusive Philalethan language was once again the making of the sophic mercury. "Saturn's offspring" as usual refers to crude antimony, which drowns Mars by fusing it during the making of the star regulus. Venus is the "bond of Crystalline mercury," meaning that copper acts as a mediator between the star regulus and the quicksilver, allowing them to amalgamate. The sophic mercury, once completed, is the spring in which the Sun,

[51] Newman and Principe, *LNC*, 21–31.

[52] As at Philalethes, *SR*, 54, 57, 62, and throughout.

[53] For confirmation of this reading, compare the passage in question (Philalethes, *Marrow*, part 1, book 3, page 44) to part 2, book 1, pages 15–17. Stanzas 59, 60, 63, 65, and 69 incontestably use "Venus" or in the final instance "Aeneis" to mean copper, since Philalethes here says that the combination of Venus and the star regulus yields his purple antimony-copper alloy, the net. It is true, however, that immediately before these passages, on page 14, stanza 56, Philalethes refers to antimony as "our Venus."

[54] Mellon 79, 1r.

here meaning metallic gold, must die. This refers to the subsequent sealing up of metallic gold in the sophic mercury on which it dissolves and is supposed to undergo a series of color changes corresponding to the different planets, on the way to becoming the philosophers' stone. The reference to Saturn's child and the snake that ate Cadmus is just a repetition in mythological language of the reduction of metallic antimony by means of iron.

As we can see from Newton's bracketed comments, however, his reading differed strikingly from Starkey's original intent. To Newton, the four members of the spring are "[♄ ♂ ♀ ☿]"—in other words, lead, iron, copper, and quicksilver. He has now written antimony completely out of the picture! Only if one willfully ignored Starkey's many transparent clues about the antimonial nature of Saturn's offspring could this be a legitimate move. But such an eschewal of the passage's obvious sense could be justified on the assumption that Philalethes's seemingly unguarded language was actually a trap for the foolish. And this interpretive path fit with Newton's growing appreciation of the principles behind alchemical hermeneutics. Had Geber not intoned that the adepts were at their most deceptive when they seemed to be speaking openly? If so, then the expression "Saturn's offspring" might well refer to the underground product or "offspring" of the planet Saturn's rays, traditionally thought to be lead rather than antimony.

The absence of antimony becomes all the more striking as one proceeds through Mellon 79. After the passage just examined, Newton paraphrases another section from the *Marrow* that we have already examined, where Philalethes gives a seemingly unequivocal description of native stibnite, the ore of antimony:

> The substance wch we first take in hand is mineral, compound of ☿ & crude sulphur, Saturn's child, sable coloured wth argent veins, wholly volatile, most brittle, &c is o^r Dragon w^ch ♂ assailed in vain for a star shewd y^t Cadmus could not abide his force.[55]

As we have already seen, Newton was not blind to these obvious clues; in Var. 259 he had explicitly interpreted the mineral, sable substance with argent veins as crude antimony. The only plausible explanation for his new view that Saturn's offspring was lead is that he thought the "plain text" description given by the *Marrow* was so obvious that it must be a red herring thrown out by Philalethes to lead the unwary astray. At any rate, the rest of Mellon 79 makes it quite clear that Newton at this point in his career had decided that Philalethes was describing a lead process, and one that did not employ antimony.

The next author whom Newton quotes, for example, is "Zimon," one of the interlocutors in the *Turba philosophorum*, an important text of Arabic alchemy translated into Latin in the High or Late Middle Ages.[56] Since

---

[55] Mellon 79, 1r.

[56] For the *Turba philosophorum*, see Julius Ruska, *Turba philosophorum: Ein Beitrag zur Geschichte der Alchemie* (Berlin: Julius Springer, 1931). See also Martin Plessner, *Vorsokratische Philosophie und griechische Alchemie in arabisch-lateinischer Überlieferung: Studien zu Text und Inhalt der Turba philosophorum* (Wiesbaden: F. Steiner, 1975).

Zimon's process for making his equivalent of the sophic mercury involves only lead and copper, it is likely that Newton transcribed the passage here in order to corroborate his new opinion that antimony was not involved. The passage runs as follows:

> Take copper and put lead with it until it becomes thick [♀ & ♄]. This will be the lead of which the wisemen say "copper and lead become the precious stone."[57]

Zimon goes on to say that gold must then be added to this "precious stone," but antimony is entirely absent from his process. The key ingredients of the stone are lead and copper. With Zimon's passage, Newton ends the purely corroboratory part of Mellon 79, having apparently satisfied himself that antimony is not employed in the making of the sophic mercury.

The Subterranean Generation of Metals and Minerals: A New Turn

After Mellon 79's corroboration by ancient authority of the role of lead in chrysopoeia, Newton passes to gathering extracts from authors treating the generation and natural history of the metal. As we will see, Newton's growing concern with the subterranean development of minerals in nature marks an important phase in his development. Judging from its sources and use of graphic symbols, Mellon 79 was probably written within five years of the composition of Newton's theoretical texts *Humores minerales* and *Of Natures obvious laws & processes in vegetation* (Dibner 1031B), which rely heavily on alchemical sources for their cosmology and theory of metallogenesis. As we saw in examining the earlier documents Babson 925 and Keynes 19, neither of them concerns itself much with the natural history of the earth. Newton almost certainly composed these two documents before 1670, thus before tackling the subterranean generation of metals described in Dibner 1031B. Hence the pre-1670 manuscripts are naked attempts to extract the operational recipes from Sendivogius, d'Espagnet, and Philalethes without much thought about metallic generation. Mellon 79, on the other hand, falls into the pattern of Dibner 1031B, representing one of Newton's early attempts to grasp the tellurian processes leading to the natural generation of metals and to put this knowledge to practical use.

The first of the writers on metallogenesis in Mellon 79 is the pseudo-Paracelsian author who composed *De natura rerum*, an influential text containing a wealth of mineral knowledge. Newton begins with a discussion of the *Witterungen*, or "outbreathings" of mines, which form a major point of discussion in the pseudo-Paracelsian tradition and in the *Testament* of Basilius Valentinus, as we saw in chapter four. These colored "coruscations" were thought to indicate the presence of an unripe mineral, one that was still *in primo ente* (in its first essence) as Newton says, echoing pseudo-Paracelsus.

---

[57] Mellon 79, 1v: "Sumite æs & ponite plumbum cum eo donec spissum fiat [♀ & ♄]. Hoc erit plumbum de quo sapientes dixerunt Æs & plumbū lapis fiunt pretiosus."

Hence miners intent on locating new lodes would look for these exhalations, along with more obvious signs such as the presence of "a fat clayish earth" sometimes accompanied by bright colors. The presence of this passage in Newton's notes suggests that he was now turning his gaze toward minerals as opposed to fully developed metals, and the rest of the manuscript adds further support to that suspicion. This was a perfectly reasonable path for him to pursue, since Sendivogius (and other authors) had stressed that art should imitate nature, and that the successful alchemist was attuned to nature's processes and methods.

From pseudo-Paracelsus, Newton turns to a much more contemporary source, namely, the *Metallographia* of John Webster, published in 1671. This interesting and knowledgeable text must have only recently appeared when Newton wrote Mellon 79. Webster's *Metallographia* is a self-styled natural history of metals and minerals, but one that is highly dependent on early modern chymical texts, including the more esoteric among them. Newton jumps immediately from pseudo-Paracelsus's discussion of the signs that miners use in searching out rich lodes to Webster's more specific description of the discovery of lead ores:

> In digging for Lead in ye north parts of England, ye signs y^t most encourage them are grey or blewish stones flints or slates, red or yellow clay or earth or that wch appears of many colours; but especially to find some pieces of such ore as they call loos and <*illeg.*> shaken or some twiggs of a vein, &c Webster's History of metals chap: 6. p 103.[58]

We may surmise that Newton's interest in this passage lay not so much in the acquisition of lead per se, but rather in the bluish, red, yellow, and multicolored earths that Webster alludes to. These could all have been immature lead "in its first essence," as pseudo-Paracelsus said. This suspicion is strengthened by Newton's subsequent borrowings from Webster in Mellon 79. He reproduces more material on the minerals found with lead ore, such as a "fat & clammy earth" that is found with the lead mineral proper. It certainly sounds as if Newton is trying to fill out the information on embryonic minerals that he got from pseudo-Paracelsus. What follows is one of a handful of references in the entire alchemical *Nachlass* of Newton that come from the famous metallurgical writer Georg Agricola's *De re metallica*. Surprisingly, Newton uses this celebrated text merely to back up Webster on the issue of different colored earths indicating rich mineral lodes. Clearly, the developing chymist found his countryman to be the more stimulating author!

After citing Agricola, Newton soon returns to Webster's discussion of lead ores and associated minerals. The Lancashire or Yorkshire mine men call the yellowish red earth that accompanies the ore "the brown hen" and say that when the hen is present, her "blue chickens," the deposits of richer ore, are not far off. A similar relationship of proximal to distal is found when ores occur in mineral "strings" (stringers) that connect to veins, and these to major trunks. The organic language here is no accident. As we saw in chapter four on early

---

[58] Mellon 79, 1v. See John Webster, *Metallographia; or, A history of metals* (London: Walter Kettilby, 1671).

modern alchemy and mining, it was common belief among miners and alchemists of the period that metals grew underground like giant, subterranean trees. Newton is building on this idea when he extracts further information from Webster on "Cauk, bastar Cauk, black Chert, Wheat stone, Sheafe," all indicators of good lead. Similarly, various types of spar, which Webster calls the "rudiments of Gems," can point to the presence of good lead ore. The idea that the variously colored spars are immature gemstones once again reveals that Newton's interest here lies in the "first essences" of minerals, particularly that of lead, more than in the fully formed metal.

Newton then provides excerpts from several English alchemical poems that had appeared in Elias Ashmole's 1652 collection *Theatrum chemicum britannicum*.[59] Lead features prominently in these, and in the six-hundred-odd words that Newton has copied from the second poem, "Bloomfield's Blossoms," he has underlined only one: "our <u>Lead</u>." Newton's newfound interest in the *primum ens* or first essence of metals also reveals itself once more in striking fashion. The aspiring chymist recapitulates two pages from the *Instructio patris ad filium de arbore solari* (The Father's Teaching to His Son about the Golden Tree). This interesting treatise appears in volume six of the 1661 *Theatrum chemicum* and purports to have been translated from a French manuscript into Latin. Although written with the studied vagueness characteristic of many books of early modern alchemy, the *Instructio* bears striking similarities to the work of Sendivogius. Again on the principle that alchemy should mimic nature, the alchemist of the *Instructio* needs to acquire the starting material from which metals grow, which is their "first matter," like the *primum ens* of pseudo-Paracelsus.[60] The author develops an analogy between the first matter from which the sophic mercury should be extracted to the fatty earth that a farmer prepares with dung in order to grow wheat. Hence the *Instructio* refers to the alchemists' starting material as a *terra virginea*—a virgin earth that must be impregnated in its own fashion with a sulfurous *pinguedo* or fatness. All of this is standard alchemical language, but the *Instructio* takes on a seemingly Sendivogian tone when the author says, in a passage excerpted by Newton:

> The virgin earth is a material that is not found upon the earth of the living; it is a corporeal spirit or a spiritual earth, the niter of the wise <*Nitrum sapientum*>, a heavy fatness and a juicy earth: to be sure it is found in valleys, plains, fields, caves, mountains, and in your own house, but it must be taken before the sun regards it.[61]

All of this riddling language can be taken to refer to the aerial niter, the "niter of the wise," which is a fertilizing material found everywhere, responsible for maintaining life on our planet. Hence, like Sendivogius, the *Instructio* argues

<hr />

[59] Elias Ashmole, *Theatrum chemicum britannicum* (London: Printed by J. Grismond for Nath. Brooke, 1652).

[60] Anonymous, *Instructio Patris ad filium de Arbore Solari*, in *Theatrum chemicum*, 6: 175.

[61] Mellon 79, 4r: "Terra virginea est materia quæ super terra viventium non reperitur, Est corporalis spiritus aut spirituale corpus, Nitrum sapientum, pinguis gravis et succulenta terra Vti☿ in vallibus, planis, campis, terræ speluncis, montibus, in☿ tua domo reperitur, sed priusquam Sol eam intueatur capienda."

in another passage copied by Newton that the virgin earth is "not that earth upon which we walk, but that which is suspended above us," apparently the aerial niter.[62] But how do we extract and collect this marvelous material? Here a "magnet" is necessary, and this, the *Instructio* tells us, is one of the great secrets of the art. Like the echeneis or remora, a partly mythical creature without bones or blood that sticks to ships by a "magnetic" power, the virgin earth must be extracted from our sea with "the magnet of the wise." After describing the virgin earth, the *Instructio* goes on to say that the fructifying fatness necessary for its fertilization is found in metals, in the form of their sulfur.[63] The virgin earth is inseminated by the male, sulfurous seed, and from this the sophic mercury emerges if the alchemist employs the right series of operations.

Although Newton copied this material faithfully in Mellon 79, he unfortunately provided no comments of his own. Thus we cannot say with certainty how he was interpreting the *Instructio*'s riddling comments at this stage of his career. But the fact that these extracts appear in a manuscript that is overwhelmingly devoted to lead and the subterranean minerals associated with it suggests strongly that he believed the "first matter" of metals should be extracted from one of these materials. Although the *Instructio* hints that the virgin earth is the aerial niter of Sendivogius, that fact by no means eliminates the possibility that this material should be extracted from a mineral. Sendivogius himself had spoken throughout the *Novum lumen chemicum* of a laboratory practice involving "Saturn," as we have seen, and his treatise terminates with an elaborate mythological enigma in which the titan usually associated with lead plays a major role. All of this is understandable if one remembers that the Sendivogian aerial niter circulated beneath the surface of the earth and provided the material out of which minerals were formed. It is likely that the inclusion of these *Instructio* passages at the end of Mellon 79 were intended to fill out the material from pseudo-Paracelsus and John Webster earlier in the manuscript.

Our examination of Newton's alchemy from 1669 to the mid-1670s has revealed a significant peregrination on the part of the self-assured, not to say overweening, youthful natural philosopher. Initially confident of his ability to decode Sendivogius and Philalethes with the help of lead and crude antimony alone, Newton soon ran headlong into the evasions of alchemical polysemy. Confronted with the fact that the Philalethes of the *Marrow of Alchemy* had employed the additional materials of copper and iron, Newton was forced to adjust. But with an unshakable confidence similar to the self-assurance that accompanied his critique of Descartes in *Certain Philosophical Questions*, the budding scientist refused to abandon his earlier interpretation in its entirety. His initial conclusions could not be absolutely wrong, but they must be modified to fit his evolving understanding of the masters. There must be a role for lead, as he had already intuited in 1669. Hence Newton turned to the subterranean processes that produced

---

[62] Mellon 79, 4r: Est terra sed non illa cui inambulamus verum illa quæ supra nos est suspensa."
[63] Anonymous, *Instructio Patris ad filium de Arbore Solari*, in *Theatrum chemicum*, 6: 179.

not only metallic lead but also its ores, gangue, and accompanying minerals. Incorporating the ideas of pseudo-Paracelsus, Webster, and other authors, Newton determined that lead and its associated minerals contained the first matter of metals in a particularly rich and accessible form. His pondering of the Philalethan *Marrow of Alchemy* had led him to turn his attention first from lead to antimony, and then back to lead, but now he had acquired a new emphasis on the subterranean materials that accompanied the generation of that metal. We will soon see how Newton fleshed out these early thoughts by consulting a number of other alchemical authors on the subject of metallic generation.

TEN

Flowers of Lead

NEWTON AND THE ALCHEMICAL FLORILEGIUM

Newton's bold attempt in Keynes 19 to decipher Sendivogius on the basis of two ingredients alone was rapidly discarded after his exposure to *The Marrow of Alchemy*, as we have seen. The greater clarity of that text made it impossible for him to maintain his parsimonious reading of Philalethes and Sendivogius, in which only lead and antimony had figured. And yet Newton's willingness to forsake antimony in Mellon 79 shows that the Sendivogian insistence on nature's simplicity still exercised a powerful hold on his mind. Our examination of Mellon 79 revealed that Newton was much more concerned with probing the mysteries of lead than antimony at this point, and that seems to have remained the case for some time. If we turn to Keynes 35, a large manuscript of around seventeen thousand words, the terms "antimony" and its Latin equivalent, as well as the symbol "♂," appear only five times. Comparing this to occurrences of "lead" and its Latin forms, as well as the symbol "♄," I find that lead and its Latin forms appear thirty-eight times in the same manuscript. This is not conclusive in itself, since expressions like "our lead" could be read as antimony, but it certainly justifies a closer look at Keynes 35. The manuscript is important for several other reasons as well.

Keynes 35 leads us into the middle period of Newton's long alchemical endeavor. Interestingly, this chronological interval corresponds roughly to the span that Westfall labeled "years of silence," the time between Newton's disgust and withdrawal from public science resulting from the contentiousness of Hooke and the foreign critics of his optical theory, which came to a head early in 1676, and the famous visit from Edmund Halley in August 1684 that eventually led Newton to compose the *Principia*.[1] As his obligations to the Royal Society and other outside distractions waned, Newton's commitment to alchemical decipherment swelled, like the documents resulting from it. Unlike any of the manuscripts that we have examined up to now, Keynes 35 cites almost the full panoply of printed works by Philalethes. Two of these works are decidedly more recent than *Secrets Reveal'd* or *The Marrow*

---

[1] Westfall, *NAR*, 335–401.

of Alchemy, namely, the English collection of Philalethan commentaries on the fifteenth-century alchemist George Ripley published by William Cooper in 1678 as *Ripley Reviv'd*, and the *Opus tripartitum*, a Latin collection of treatises also mostly written by Starkey under the Philalethes pseudonym and similarly published by Cooper in 1678. Hence the year 1678 will serve as an indisputable *terminus post quem* for the composition of Keynes 35. Finding the latest possible date at which the manuscript could have been written is a more uncertain task. It is striking, however, that Keynes 35 cites none of the authors about whom Newton grew excited in the period from the mid-1680s to the 1690s, including the French authors whom he would begin reading around the time of his alchemical collaboration with Nicolas Fatio de Duillier in the early 1690s. Nor for that matter does Keynes 35 refer to "Mundanus," the alchemical adept whose work forms the basis of the physician Edmund Dickinson's 1686 *Epistola ad Theodorum Mundanum*, a text that occupied Newton's full attention soon after its publication. Additionally, there are only a handful of references to pseudo-Ramon Lull in Keynes 35; I count only two occurrences of "Lullius," and these are derived secondhand from other authors.[2] Since Newton began to read Lull's pseudonymous alchemical works seriously only after his exposure to the elusive adept Mundanus, this also leads one to suspect that Keynes 35 was written before Newton's acquisition of Dickinson's work. All of this points to a rough *terminus ante quem* of 1686 for Keynes 35, giving a likely window of 1678–86 for its composition. More exactitude than that will have to await an exhaustive study of the physical clues provided by Newton's manuscript corpus.

Newton's immersion in chymistry during his "years of silence" not only led him to devour any and all written material that he could acquire on the subject, but even to adopt the literary style and genres of his sources. A particularly notable feature of Keynes 35 lies in the literary form of the document. It is one of the earlier instances of Newton's explicit compilation of authoritative snippets into a sort of topical anthology, complete with his own chapter headings. Although we saw him taking something like this approach in Mellon 79, that manuscript lacked topical headings or chapters. With Keynes 35, Newton was now adopting one of the favorite forms of literary exposition late medieval alchemists employed—the florilegium. This is particularly appropriate since one of the texts that Newton excerpts in Keynes 35 is the anonymous *Rosarium philosophorum* (usually identified by its beginning words, "Qui desiderant artis philosophicae scientiae"), which is precisely such a florilegium. The many *Rosaria* and other florilegia of late medieval alchemy typically wove a variety of authoritative *dicta* (sayings) by the masters into a meandering tapestry of quotation and paraphrase. Like the *Rosarium philosophorum* and other texts in the same genre, Keynes 35 collects passages from authors that are supposed to shed light on one another. It is clear from a surviving table of contents that is now found in

---

[2] Keynes 35, 8v and 16v. The first reference comes from Johannes Grasseus, *Arca arcani*, in *Theatrum chemicum* (1661), 6: 318. The second is taken from an anonymous *Rosarium philosophorum*, in *Artis auriferae* (1610), 2: 238.

Newton's famous alchemical concordance, the *Index chemicus*, that he had serious plans for the text now found in Keynes 35. The *Index chemicus* manuscript (Keynes 30/1) lists twelve chapter headings for this florilegium, of which only the first three and the final two survive in Keynes 35.[3] Whether the manuscript was never completed we cannot say, but its burned edges and missing fascicles argue compellingly that it is no longer as complete as it once was.[4]

Newton's more or less explicit adoption of the florilegium genre in Keynes 35 marks an important turning point in his alchemical career. He clearly recognized now that the approach of Keynes 19, where one author acted as a key to all the others, as Philalethes had to Sendivogius and d'Espagnet, was too simplistic. Philalethes himself had taught Newton the error of this early approach in the *Marrow of Alchemy*. As Newton's alchemical reading grew, his need to organize the huge mass of seemingly conflicting authors became an overriding concern, and one that would eventually lead him to compose the four successive drafts of the *Index chemicus* (in addition to two supplements). In the meantime, his adoption of the time-honored genre of the florilegium owed more to the literary habits of late medieval alchemists than it did to early modern innovations such as the commonplace book. Unfortunately for the modern scholar, Newton's florilegia provide many of the frustrations that one encounters in the medieval compilations themselves. In trying to extract the "flowers" of alchemical wisdom from the "thorns" thrown out to delude the unworthy (a prominent title among such florilegia was *Lily among Thorns*), alchemical compilers tended to submerge their own authorial identity behind the words of other authors. This is precisely the approach that became Newton's favored literary form as his alchemy progressed. We consider ourselves fortunate when he interjects even a symbol or two enclosed in his customary square brackets to reveal what he may have been thinking about his sources. Reading these texts is an acquired skill (not to mention an acquired taste), but in fact Newton's choice of which authors and passages to include and which to ignore can tell us a great deal about his goals. Keynes 35 provides an excellent illustration of such a use of other authors' words to tell Newton's own narrative.

In the following we will examine only the first three chapters of Keynes 35, the ones that deal with finding and extracting the ingredients of the philosophers' stone. The final two chapters, "On the Doubled Mercury" (De ☿io duplato) and "On Extracting the Living Gold and Conjoining it in the Hour of its Birth" (De auro vivo extrahendo & in hora nativitatis conjungendo), carry Newton into the higher reaches of alchemical speculation and have less

---

[3] The first three headings listed in the *Index chemicus* all appear in Keynes 30/1 on 1r. They are "1. Quomodo Metalla generantur ^et corrumpuntur^ in venis terræ. Arca Arcani p 305. Sniders Pharm Cath. p Sendivog. p 33 1," "2. De semine spermate et corpore mineralium," and "3. De mineralibus ^et metallis^ ex quibus lapis desumitur." The two final headings, "De ☿io duplato" and "De auro vivo extrahendo & in hora nativitatis conjungendo," are also found on 1r of Keynes 30/1. This wording is very close, though not identical, to that of the existing chapter headings in Keynes 35, at 1r, 3r, 5r, 19r (again at 23r), and 24r. There is also one heading in Keynes 35 that is lacking in the "table of contents" from Keynes 30/1, namely, "De Projectione" on 23v.

[4] See the physical description of Keynes 35 given in its online edition in *CIN* (under "Manuscript Information").

intersection with his actual laboratory practice. Keynes 35 begins with the chapter heading "How metals are generated and corrupted in the veins of the earth" (Quomodo metalla generantur & corrumpuntur in venis terrae). This chapter starts by paraphrasing the metallogenesis theory of Johannes Grasseus, taken from the German lawyer's *Arca arcani* in the *Theatrum chemicum*. Grasseus's theory of metallic generation is worth a brief review. As we saw in chapter four, the alchemist and counselor to the Archbishop of Cologne Ernst von Bayern promoted a combination of traditional ideas inherited from medieval alchemy and newer, biologically tinged views stemming from early modern mining literature. He speaks of salty, mercury-rich waters that drip down from mines and subterranean deposits, encountering sulfurous fumes that rise up from the center of the earth. In some cases, these materials are pure enough that their encounter produces metals directly, but in many others they are impeded by an intermixture of adventitious matter. In these instances, nature slowly cooks them into a mucilaginous slime that can be spread like butter (though presumably not on toast). Relying on the Lutheran preacher and mining expert Johannes Mathesius, Grasseus calls this immature metallic precursor "Gur." The Gur in turn is gradually incubated by the underground heat to become a "leady matter" (*materia plumbea*) in which a grain of silver or gold is always found. All of this material is reprised in the first chapter of Keynes 35, and what follows immediately are the competing *sal nitrum* theory of Sendivogius, a short, enigmatic paragraph by an author who will soon play a large role in our account—Johann de Monte-Snyder—and an even shorter paragraph by the pseudonymous alchemist Bernardus Trevisanus. The entire chapter runs for only a folio and a half; one gets the sense that Newton is eager to get to more practical matters.

The second chapter in Keynes 35 (though erroneously labeled "Cap 3" by Newton), "On the root, seed, sperm, and body of minerals" (De ^radice semine spermate & corpore mineralium), confirms this impression. Again Newton begins with Grasseus, but the text subtly shifts away from theory and toward operation. In the words of the German alchemist, "the subject of the Elixir must be extracted and its shell removed (*enucleari*) ... for it can only be obtained from that material out of which all metals arise."[5] This is the same underlying principle that drives the alchemy of Sendivogius; the ingredients of the philosophers' stone must come from the *primum ens* or first matter of the metals themselves. Hence it is vitally important for the practitioner to understand the generation of metals. Employing Grasseus's theory of metallogenesis, then, Newton comes to the following conclusion:

> Grasseus says these things about the root of minerals, then describes in detail how the stone should be taken from the first of the metals that arise from this root, namely lead.[6]

[5] Keynes 35, 3r: "Hic igitur sequitur subjectum Elixiris ex rebus simplicibus (quæ ex fontibus & scaturiginibus primorum mineralium originem ducunt) extrahi et enucleari oportere Neꝗ enim ex ulla re totius mundi, quam ex sola materia ex qua omnia metalla oriuntur hoc subjectum elici potest."

[6] Keynes 35, 3r: "hæc Grasseus de radice minerarum, postea describit prolixe quomodo lapis desumi possit ex primo metallorum quæ ex hac radice oriuntur nempe plumbo."

Thus we are clearly back in the realm of an alchemy for which lead is the foundation. Nonetheless, things could not be so simple, as Grasseus immediately lets the reader know. In the subsequent passages that Newton abridges from the *Arca arcani*, the German alchemist describes how even after he understood the rudiments of the art, he was still uncertain where to start, as he did not know the most proximal subject on which to begin his labors. Grasseus thus decided to make a voyage in the hope of finding the proper material. On the path between two mountains he encountered an old, country fellow wearing a long, gray cloak. His cap bore a black band (*velum*), and he had a white cloth around his neck; his belt was yellow and his leggings red. In his hand the old man held two lilies, one red and one white. The old man smiled when Grasseus made a comment about his appearance and replied that most people underestimated him and did not see him as the origin of the metals. He then added that the two flowers in his grasp were highly toxic unless distilled with other materials that were, all the same, of the same nature as the flowers. Their juice must be expressed and then combined in a certain proportion. When Grasseus pressed the old man for further information, he replied, "you wish to know many things without investigating them yourself," and promptly disappeared. After considerable labors, Grasseus then managed to determine the proper "proximal material" and drive off its venomous vapors, at which point the two flowers appeared, first the white and then the red. The alchemist sealed them up in a flask, and further favorable signs appeared, on which the old man, now identified as Saturn, reappeared and revealed additional marvels. At this point the second chapter ends in Keynes 35 without comment by Newton.

The introduction of the red and white flowers and the multiple colors of old Saturn's clothing suggest strongly that more than mere lead is involved in the making of the philosophers' stone. Hence Newton explores the possibilities of the number and nature of the necessary ingredients in the next chapter, which is appropriately titled "On the Minerals from which the Stone is drawn" (De Mineralibus ex quibus lapis desumitur).[7] Looking at this chapter will give us an excellent taste of the florilegium style and its highly fluid form. In examining such texts, the reader must keep a sharp eye out for the thematic factors that unify the successive extracts; it is not always obvious why Newton jumps from one passage to the next, but with a bit of practice it is usually possible to divine his goals. The first extract, taken again from Grasseus, immediately signals that Newton's interest lies in the number of necessary first ingredients: "There are three things necessary for the perfection of the Stone" (Tria sunt ad Lapidis perfectionem necessaria). After announcing the need for three things, Grasseus then goes on to describe this plurality in his usual cryptic style. The first is a "stone of the sun" that comprehends within itself a red lion or red, incombustible sulfur. The second is a white sulfur in a lunar subject, which contains "our mercury." The third is a stone that is medial between the first two and holds their natures within itself.

---

[7] Keynes 35, 5r.

Grasseus then reassuringly adds that it is necessary to hide these "metallic natures" from the ignorant and unworthy.

It would be foolhardy for us to attempt a decipherment of Grasseus's *Decknamen*, even though some of his successors had definite ideas about his meaning. Anonymous early modern reports view the initial driving off of venomous vapors that Grasseus describes as the roasting of argentiferous lead ore to free it with sulfur, in the way that sulfide-rich ores are typically treated. Similarly, his followers saw the sweet, crystalline compound of lead *saccharum saturni* (sugar of lead) lurking behind the term "white lily" and the red oxide of lead, minium, hiding beneath the *Deckname* "red lily." According to this interpretation, Grasseus derived both lead compounds from the roasted ore, not from the refined metal, first by extracting the sugar of lead with vinegar, and then by oxidizing that product in a furnace to arrive at the minium. It has also been argued that Saturn's colorful clothing represents the various color changes that the first matter undergoes during its stages or regimens, beginning with the black stage of putrefaction and ending with the red of the philosophers' stone.[8] For us the precise material referents of these terms are less important than what Newton made of them; at this stage of his chapter "On the Minerals from which the Stone is drawn," he seems mainly bent on determining the number of the initial ingredients rather than pinning down their precise character.

Newton's emphasis on ascertaining the number of materials required is born out in the next three extracts, which stem from Philalethes. This set of passages, all derived from Philalethes's English commentaries on the work of George Ripley, begin respectively, "In the first laborious prep<aration> o^r crude sperm flows from three substances," followed by "There is nothing wch can exalt tinctures bu<t our di>solving water, wch I told you flows from three spri<ngs>," and then, "There are in o^r Mercury three <mer>curial substances."[9] In the traditional style of the alchemical florilegia, Newton is piling up supporting passages from the sages so that he can compare them to one another and see if their similarities and differences produce any new information. The process is really a sort of preliminary stage of induction not entirely unlike the Baconian winnowing of particulars in the *Novum organum*, though Newton is getting this approach from alchemy rather than from the famous Lord Chancellor. Newton continues weaving together alchemical authorities for a total of thirteen folios in this chapter. Although the bulk of the extracts support the view that the alchemist should begin with three materials (5r–v, 6r–v, 9r, 11r), there are also passages that insist on two (9r–10r), four (6r, 13v, 15v, 16r–v), seven (6v), and even one that suggests "all the metals" (6v). Clearly the assiduous chymist had his work cut out for him.

It would be erroneous to claim that the only goal of Newton's third chapter was that of determining the number of initial ingredients, however. Since

---

[8] Thomas Lederer, "Der Kölner Kurfürst Ernst von Bayern (1554–1612) und sein Rat Johann Grasse (um 1560–1618) als Alchemiker der frühen Neuzeit" (Inaugural dissertation, Ruprecht-Karls-Universität, Heidelberg, 1992), 80–83.

[9] Keynes 35, 5r–v.

many of the extracted passages not only detail the number of substances to be employed but also go on at length about their nature, it would have been excessively artificial for Newton to ignore their detailed descriptions of materials. Another long extract from the *Arca arcani*, filling over two folios front and back (6v–9r), deals both with the number and nature of the required substances. Grasseus begins this passage with the claim that the philosophers' stone can be made from all the metals, as long as they "lie in their minerals," in other words as long as their ores have not been refined. But as soon as they touch the fire, their tinging spirits depart, leaving a dead body behind. Despite the fact that this spirit exists in all metallic ores, Grasseus continues, it is easier to extract it from those that are of easier solution, and "in which the *primum ens* and generative, multiplicative power is still present."[10] What materials satisfy this condition? In principle they might include the *minora et media mineralia* (lesser and intermediary minerals) such as salts and marchasites, but those lack the necessary metallic principles. Thus the philosophers affirm that we should take up a mineral on which Nature has begun her efforts to create metallic splendor but that she has left incomplete. Concluding his thought, Grasseus states, "Indeed, Nature first creates the metal lead" (Natura vero primo minerale plumbum creat) but has left it imperfect, and this should therefore be the metallic root. Grasseus then adduces the authority of the *Rosarius magnus* and Nicolas Flamel to bolster his claim that the *primum ens* or mercury of the metals is found in the "mineral of Saturn." Moreover, Grasseus says, Bernardus Trevisanus tells us that an immature ore is like an apple on a tree; as soon as it is plucked or falls to the ground, it ceases its passage of ripening to the more mature metal, silver or gold. Fusing or smelting an ore before its full maturity is like picking the unripe apple; it interrupts the path to full maturity. Once again, Grasseus's point is that refined metals are dead and inert, unlike the living, developing ores that populate the underground regions.

After continuing this line of thought with more examples and authorities, Grasseus then explicitly says on the authority of an obscure Master Degenhard that his lead is not *plumbum vulgare*, the ordinary metal, but rather *plumbum æris*, which can be translated either "lead of copper" or "lead of ore." The puzzling expression *plumbum æris* is not a term Grasseus invented, but one that appears in that hoary ark of medieval alchemical wisdom, the *Turba philosophorum*. There one Philotis advises that the whole mystery of the art lies in a substance that goes by many names but is called *plumbum æris* when in its crude form.[11] Newton knew the *Turba philosophorum* well and regarded it highly, so it is likely to have appeared to him that Grasseus had here deciphered a secret of considerable antiquity. Master Degenhard's views on *plumbum æris* would serve as one of Newton's foundational *dicta* to

---

[10] Keynes 35, 7r, paraphrasing Grasseus, *Arca arcani*, in *Theatrum chemicum*, 6: 309.

[11] Anonymous, *Turba philosophorum*, in *Artis auriferae* (1610), 1: 41. The term *plumbum æris* also appears in the version of the *Turba* printed in the 1660 edition of the *Theatrum chemicum*, 5: 50–51. For the history of the *Turba*, see Ruska, *Turba philosophorum*, and Plessner, *Vorsokratische Philosophie*, as cited in the previous chapter of this volume.

be repeated throughout his subsequent attempts to solve the riddles of the sages, and must therefore be examined now in some detail.

Diana's Doves and *Plumbum æris*

Describing the first ingredient of the philosophers' stone, Grasseus's Degenhard (as paraphrased by Newton) gives a characteristically circumspect description of *plumbum æris*:

> In that itself lies the mystery of the wise, and this is the lead of the philosophers, which they call *plumbum æris,* in which a splendid white dove is present, and this is called the salt of the metals in which the mastery of the art consists. But when he says a white dove is in lead that means it lies in metals that have not been exposed to the fire. For all the philosophers agree that that which is of easier solution exists in Saturn. The kernel must be extracted and the shell discarded.[12]

The main point here is that *plumbum æris* is the initial ingredient of the philosophers' stone, and therefore the key to the art. But equally important, at least to Newton, was Master Degenhard's claim that the *plumbum æris* contains a splendid white dove (*splendida alba columba*), and that this dove was equivalent to the "salt of the metals." To the assiduous Newton, bent on using one book to open another, this was a highly charged statement. We need only recall that Newton's first extant attempt to subject alchemical texts to serious scrutiny, Keynes 19, had already arrived at an understanding of doves in the context of chrysopoeia. There the fledgling alchemist had deduced that the two doves of Diana referred to the twin sulfurs of lead and antimony released in the slag of their refining. He had even gone so far as to argue that the feathers of the doves had a determinate concrete referent, namely, a white powder that would emerge during their subsequent processes of purification. We know that by the time of writing Keynes 35, Newton had abandoned the binary approach of Keynes 19, where lead and antimony alone were thought sufficient to produce the sophic mercury. It does not follow, however, that he now rejected every feature of this early interpretation. One can easily imagine his excitement on reading the passage from Master Degenhard recounted by Grasseus, and seeing that his earlier interpretation of Diana's doves was at least partially confirmed. It looked to him as though he was on the right track all along; the adepts were searching after a white, powdery substance, which could certainly describe a salt, found in lead. The fact that the *Arca arcani* speaks of one dove rather than two was of little consequence since, after all, Newton's earlier interpretation

---

[12] Keynes 35, 8r: "In ipso latet mysterium sapientum et hoc est, Plumbum Philosophorum quod plumbum æris appellant: in quo splendida alba columba inest quæ sal metallorum vocatur in quo magisterium operis consistit. Cum autem dicit columbam albam in plumbo id est in metallis ignem non expertis jacere. Omnes Philosophi consentiunt quod in saturno sit qui solutu facilior est. Nucleus igitur eximendus & folliculus abjiciendus."

had seen only one of Diana's doves in the material extracted from lead; the other came from the antimony used in the refining process.

There was one obvious difference between the *Arca arcani*'s "splendid, white dove" and Newton's early interpretation of the doves of Diana, however. There is no hint at all in Keynes 19 that the lead to be employed is anything other than the ordinary metal. And as we have just seen, Grasseus emphasizes throughout the *Arca arcani* that his lead is not the vulgar sort, but rather an "earth" or ore. Grasseus never tires of repeating that fused metals are dead, and that the skillful chymist will rely on the still living minerals fresh from their mines. Hence the white dove of Degenhard "lies in metals that have not been exposed to the fire," and apparently its most accessible source is the mineral that he refers to as *plumbum æris*. In the long segment of the *Arca arcani* quoted by Newton here, Grasseus goes on to cite other authorities who support his view that a "white lead" or white material drawn from lead is required for making the philosophers' stone, adding that this material is crystalline in character. As the section draws to a close, however, Grasseus returns to a subject that had already appeared in chapter one of Newton's florilegium. The subject is that elusive precursor to the metallic ores that exercised a perennial fascination on early modern chymists, namely, Gur.

Newton was already deeply interested in the *primum ens* or "first rudiments" of metals by the time he composed Mellon 79, as we saw from his paraphrases of pseudo-Paracelsus and the anonymous *Instructio patris ad filium de arbore solari*. Thus it was a foregone conclusion that when Grasseus introduced a discussion of Gur into the midst of the practical instructions in the *Arca arcani*, Newton would take up the passage. It is important to examine it here, since it raises interesting issues about the relationship to Gur, which was supposed to be a metallic precursor, and the recognized ores of the metals:

> Degenhard, Lull, and Mathesius write that the matter before it is congealed into a metallic form is like the coagulum of the butter of milk, which is separated as butter is, which they call Gur; I have found this in the mines in which Nature has made lead. If such a matter can be prepared on the surface of the earth, it is a sign not only that the true matter has been obtained, but that the true path has been found out. I can produce this to hand, by the grace of God, and in the space of an hour it putrefies and becomes first black, then ruddy, and finally darkly red. The philosophers call this "milk of the virgin." If a little of our metallic salt is put into our water, it will become like milk; if much is put in, it thickens like butter and can be spread like fat. I have said these things so that you may have no doubt of the matter. They call this leady material litharge and plumbum æris.[13]

[13] Keynes 35, 8v: "Degenhardus Lullius & Mathesius scribunt materiam priusquam in metallicam formam congelatur esse instar lactis coaguli butyri; quæ ut butyrum diducitur quam Gur vocat, quam ego quoꝗ in fodinis in quibus natura plumbum paravit inveni. Si talis materi <*sic*> supra terram præparari potest, signum est quod non solum vera materia habetur, sed veram quoꝗ viam observari. Hanc ego per Dei gratiam in manu parare possum: quæ etiam unius horæ spatio in calore in putrefactionem abit & primum nigra dein rubicunda & tandem fusco rubra fit. Hanc Philosophi lac virginis vocant. Si aliquantulum Salis metallici in aquam nostram inditur, lactis instar fiet, si multum imponitur crassescit ut butyrum et instar pinguedinis illiniri potest.

Newton's close paraphrase from the *Arca arcani* repeats some of Grasseus's earlier discussion of Gur, such as the fact that it can be spread like butter, but it also adds new information. Here Grasseus explicitly says that Gur is found in lead mines, and he introduces a new topic—the replication of Nature's underground preparation of Gur on the face of the earth, carried out by the alchemist in his laboratory. Precisely what Grasseus has in mind is very difficult to make out, but the fact that the alchemist can prepare or use Gur indicates that he has found both the true matter and the correct path forward. What he means by the "true path," evidently, is the initial putrefaction of the Gur leading to blackness, followed by varying shades of redness. Finally, Grasseus concludes the section by providing various alternative names for Gur, one of which is the by now familiar *plumbum æris*.

Because of the identification between Gur and *plumbum æris*, we are led to wonder about the exact relationship between this precursor to lead and the typical ore of lead, galena. Galena is a relatively soft mineral, but it cleaves on impact and is by no means capable of being spread like butter. The shiny or dark gray material is often very rich in silver, and Grasseus immediately makes a major point of that fact in the remaining text that Newton excerpted. Perhaps Grasseus thinks of the Gur as a sort of "yeast" hidden in the ore and making it ferment and develop; this would fit with the etymology of the word, since its origin, *Gärung*, means "fermentation" in German. At any rate, he says that the quality of the *plumbum æris* is directly related to its silver content. Although good examples are found in the mines of Villach in Austria, and Meißen and Annaberg in (modern) Germany, the *plumbum æris* from Joachimsthal (modern Jáchymov in the Czech Republic) is deficient in silver. Grasseus in fact mentions numerous other mines in the *Arca arcani*, particularly ones in Hungary and Poland, but Newton did not include them in his excerpt, perhaps believing that he could acquire the ores from German-speaking lands more easily. In the end, Grasseus's discussion leaves one unsatisfied as to the exact relationship between Gur, *plumbum æris*, and ordinary galena or lead ore. Were more than one of these required in the making of the philosophers' stone? And what exactly is the relationship between *plumbum æris* and copper? Should we read the *aes* in its name as the red metal, or as a generic ore, knowing that either meaning is possible in Latin? The answers to these questions, were, I believe, as unclear to Newton as they are to us.

The reader may be surprised to learn that of the more than ten thousand words in the third chapter of Keynes 35, less than 125 are actually Newton's own (I count 121). He has rather skillfully extracted the passages that make his authors say what he wants, and culled out the rest; this was ever the way of the florilegium. But there is at least one highly significant Newtonian comment buried in the midst of all the other authors. Directly after a long passage from Thomas Norton's fifteenth-century *Ordinall of Alchimy*, in which the author says that the philosophers' stone must be found in two things,

---

Hoc declaravi ut de materia nullum dubium habeatis. Hanc materiam plumbaginem Lithargyrium et plumbum æris nominant."

"*Magnetia and Litharge*," Newton refers to a section where Norton identifies "Magnetia" with *res æris*, literally "a thing of copper." The expression "*res æris*" is actually part of an elaborate wordplay in Norton's *Ordinall* where he is treating the word "Magnetia" as though it were an acronym, and then deriving further words from several of the letters. During all of this riddling play, the author makes it quite clear that he is referring to the extended sense of the Latin *æs* or copper to mean "money" (as in the expression "a few coppers").[14] Newton, however, was struck by the similarity between the expression *res æris* and the term *plumbum æris* from Grasseus. For Newton, Norton's language was a hint that actual copper was somehow involved. The comments that Newton makes are deeply suggestive of his goals, which will reemerge more fully in our subsequent chapters. Let us quote his conclusions here:

> Note y^t as Norton calls it res æris so others call it plumbum æris & Snyders saith: Neptune & Venus make make <*sic*> to fly ye snak wch els beneath must lye. Also the Marrow of Alkimy: Venus hath a central salt ye key of all secrets. And ~~Snyders~~ Eyrenæus: Dianas Doves are enfolded in ye everlasting arms of Venus. And again: This work Diana knows to perform if she be enfolded in y^e <*illeg.*> arms of Venus.[15]

The profusion of references to Venus in this passage shows that Newton was pulling his sources together and isolating their common ingredients—lead and copper. This was not the case for Norton and Grasseus alone but also for Philalethes (Eyrenæus) and Johann de Monte-Snyders as well. The quotation from Snyders is taken from the very end of his *Metamorphosis of the Planets*, an extended alchemical allegory that would occupy Newton profoundly in the 1680s and 1690s; we will discuss it at length in a subsequent chapter. What are most interesting for the moment, however, are the passages from Philalethes. Once again, Diana's doves emerge as a point of discussion. Newton has brought together passages from *Secrets Reveal'd* (chapters fourteen and fifteen) that speak of enfolding the doves of Diana in the arms of Venus. If we build on our previous analysis suggesting that "Diana's doves" (or at least one of them) was for Newton a salt of lead, all of this begins to come into focus. Just as *plumbum æris* seems to have meant "lead of copper" for Newton, so the embrace of the doves by Venus suggested the union of a lead salt acquired from an unrefined mineral with either copper or an unspecified copper compound, presumably the "Central salt" of Venus referred to here. The massive collocation and culling that we have witnessed in Keynes 35 was intended to arrive at just this sort of result. The riddles of the adepts could, and must, be reduced to practice. Exactly what form this practice would take will form the subject of much of the remaining book.

But before we pass to that, it will be useful to cement our understanding of the doves of Diana as a salt of lead by glancing at Newton's colossal

---

[14] Thomas Norton, *The Ordinall of Alchimy*, in Ashmole, *Theatrum chemicum britannicum*, 42–43. For the passage in a modern edition of the *Ordinal*, see John Reidy, *Thomas Norton's "Ordinal of Alchemy"* (London: Oxford University Press, 1975), 38.

[15] Keynes 35, 10r.

Index chemicus in the final form that it took in the 1690s. Under the entry for *Columba*, the singular "Dove," Newton has begun with the following words: "The most white dove whiter than snow, drawn from the black crow, sacred to Venus and friend of the peacock, is the white salt of nature." After a few further comments, he cites Grasseus to the effect that this white dove is found above all in Saturn that has not touched the fire. As we have seen, this is a reference to *plumbum æris*, the Gur of lead. If we now turn to the *Index chemicus*'s immediately following entry for *Columbae Veneris* (Doves of Venus), the association with copper that we have already encountered in Keynes 35 reemerges: "The doves of Venus are the doves of Diana conjoined with Venus." What follows is a profusion of snippets taken mostly from Philalethes and Grasseus to the effect that the doves are a salt or salts, and finally a passage that confirms beyond reasonable doubt the identification that Newton has made between the doves of Diana and the splendid white dove of Grasseus. Referring to the *Arca arcani*, Newton now says the twin doves of Diana "are extracted from the not yet fused minerals. Grasseus, p. 298, 309" (Extrahuntur autem ex mineris nondum fusis. Grass. p. 298, 309).[16] Without question, then, Newton interprets the Philalethan doves of Diana to be identical to Grasseus's splendid white dove extracted from *plumbum æris*.

Conclusion: Interpretation and Experiment

The last two chapters have been devoted exclusively to Newton's literary analysis of alchemical riddles as verbal enigmas, albeit with the help of chymical theories concerning the generation of metals and ores. To judge by the remaining records of Newton's laboratory practice, incomplete as they are, he was unable to begin testing the fruits of his early decipherment until some point in the early 1670s. Even in his mature reading notes, however, Newton operated at two levels, the first of which treated the conundrums of the adepts as purely verbal entanglements that required decoding on their own terms before they could form the subject of laboratory experiment. The adepts were consummate tricksters, and a double meaning missed by their would-be interpreter would inevitably lead to failure. Only after the painstaking process of linguistic analysis could the hopeful practitioner pass to the laboratory. Nonetheless, it appears that at some point Newton may have put Grasseus's belief in a salt drawn from lead to the test. In a tantalizing fragment appended to a recently discovered manuscript copy of Philalethes's *Experimenta de praeparatione mercurii sophici* (Experiments for the Preparation of the Sophic Mercury), Newton has written the following comments (I have tentatively translated the Latin parts):

> White & blew ▽ in Lead mines ℔ iiij lead oare veines sublimate ℔ j Distill from a glass retort over an open fire and a liquor will go forth and it

[16] Keynes 30/1, 25r.

will be sublimed. Mix four pounds of this sublimate or fresh, leady earth with one pound of lead ore. Reiterate this till all comes over in a liquor then rectifye it leave noe fæces.[17]

Although this short paragraph presents some ambiguities, its unguarded language describes a practice in which the white and blue earth or clay found with lead ore is mixed with a previously prepared material that has been sublimed from the ore itself.[18] Once these two substances have been mixed together, they are distilled and the liquid collected; apparently, the heating is continued until they are dry, and then there is a sublimation of the dry materials in the retort. Alternatively, it may be that the distillation product is meant to be boiled to dryness separately and its *caput mortuum* then sublimed. Either way, the sublimate that is produced by this process (or possibly the initial sublimate) is then mixed with a pound of lead ore, and the distillation repeated; it seems that one can also use "fresh, leady earth" instead of the sublimate. Although we have no proof that Newton invented this process, and indeed, the symbol used for a retort in the recipe is not characteristically his, the processes described here are close to what one finds in his laboratory notebooks, which we will examine in a later chapter. The multiple sublimations and distillations, as well as the emphasis on ores and "earths" that have not yet been exposed to the refiner's fire, all resonate strongly with our knowledge of Newton's laboratory work. It is very plausible that these notes reflect either Newton's own thoughts or those of an alchemical collaborator. It has long been known that Newton traded secrets with his friends Nicolas Fatio de Duillier and the shadowy William Yworth; there were no doubt others as well. At any rate, the recipe is clearly an attempt to extract the "spirit" of lead ore by using minerals that accompany it in the mine as extractive agencies. Quite possibly Newton hoped that the product, either of the first sublimation or of the process as a whole, would be Diana's doves.

Newton's surprising interpretation of the Philalethan doves of Diana as a salt (or salts) of lead began as a purely literary exercise in deciphering texts and eventually transformed itself into actual laboratory practice. Whether the little passage on distilling and subliming lead ore and associated earths

[17] Sotheby Lot 75, recently acquired by the Science History Institute, final folio, verso: "White & blew ♄ in Lead mines ℔ iiij lead oare veines *<unclear>* sublimate ℔ j dist ex ☾ vitrea *<? with an odd tail>* igne nudo et prodibit liquor et sublimabitur. hoc sublim: vel recentem terram plumbeam ℔ iiij misce cum ℔ j Lead oare. hoc reitera till all comes over in a liquor then rectifye it leave noe fæces."

[18] One such ambiguity exists in the phrase ending with "hoc sublim": It is not clear whether this should be read as part of the next phrase (as I have translated it), or as the end of a sentence. In the latter case, the passage would run "White & blew ♄ in Lead mines ℔ iiij lead oare veines sublimate ℔ j Distill from a glass retort over an open fire and a liquor will go forth and this sublimate will be sublimed. Or mix four pounds of fresh, leady earth with one pound of lead ore. Reiterate this till all comes over in a liquor then rectifye it leave noe fæces." According to this alternative translation, the sublimate produced by the first set of operations would also be used in the second round of distillations, and the "vel" (or) would apply only to the use of "fresh, leady earth" instead of "White & blew" earth. In either translation there is also some question as to whether "hoc sublim" (this sublimate) refers to the initial sublimate that was prepared before the start of the described operations or rather to the sublimate that is produced by subliming the blue and white earth with the lead ore. Such seeming minutiae could make a major difference in actual laboratory practice.

reflects Newton's own ideas or not, an examination of his laboratory notebooks will show that distillates, sublimates, and extracts of lead ore played a major role in his mature experimental alchemy. Even though he abandoned his early, rigid adherence to the *dicta* of Sendivogius as executed in the binary practice of Keynes 19, Newton retained his belief that the doves of Diana were to be found in lead or its minerals. It is remarkable that he continued to build on an interpretation arrived at in his earliest attempts to come to terms with the riddles of alchemy, already evident in Babson 925, throughout his alchemical career, as shown by the entries for *Columba* and *Columbae Veneris* in the fully mature *Index chemicus*. One cannot avoid drawing a comparison with his better known discoveries in physics and mathematics, where one can find significant elements of Newton's mature science in his notebooks from the 1660s. The self-assurance that marked Newton's earliest discoveries in these better known areas was the same confidence that led him to his binary interpretation of Sendivogius and Philalethes in Keynes 19. It is true that Newton failed to divine the full sense of Philalethes's chrysopoetic allegories, but this takes away nothing from the records of his own remarkable chymical practice, which in turn fed into such influential "public" texts as *Query 31* of the *Opticks*. Before we can consider Newton's laboratory notebooks, however, we must first examine the evolution of his interpretation of alchemical literature up to its full maturity, which will form the subject of the next three chapters.

ELEVEN

Johann de Monte-Snyders
in Newton's Alchemy

Introduction: The Life of a Wandering Adept

The main characters who featured in the previous chapter were Michael Sendivogius, Eirenaeus Philalethes, and Johann Grasseus, all chrysopoetic writers whose work would fascinate Newton throughout his chymical career. Another self-proclaimed adept who increasingly influenced Newton from the mid-1680s onward was the elusive follower of Basilius Valentinus, Johann de Monte-Snyders or "Snyders," as Newton usually calls him. Given the major influence wrought by Snyders on Newton, this obscure though intriguing figure must be examined on his own terms. More than any of the foregoing figures, Snyders fits the picture of a wandering adept who would drift into town, perform a transmutation or two, and then mysteriously disappear. The seventeenth century collected accounts of these chrysopoetic performances in "transmutation histories," often stuffed with dates, names, and places in order to bolster the appearance of authenticity.[1] The transmutation histories were a powerful tool in convincing the learned world that chrysopoeia was a genuine phenomenon; Robert Boyle and Benedict Spinoza are just two of the many who followed up on such accounts in order to determine the truth behind them. By looking at several printed transmutation histories in conjunction with manuscript evidence and genealogical material, one can begin to piece together a picture of Snyders and his activities, though much remains to be learned. Snyders appears to have been active in the Rhineland, mainly in the 1660s, and then to have traveled to Vienna, where we have a reliable report of his performances.[2] Additionally, he wrote at least two influential texts that Newton pored over for decades. These were his *Tractatus de medicina universali*, which despite its Latin title was published in German in 1662, then translated into Latin in England and published in the Netherlands in 1666, and his *Metamorphosis planetarum*,

---

[1] For the genre of transmutation histories, see Newman, *GF*, 3–13, and Principe, *AA*, 93–98, 108–11.

[2] The conclusion to Snyders's *Tractatus de medicina universali* announces that his preparations are available for purchase "zu Cöllen bei besagtem Jacopo Hanßen." Given the proximity of Cologne (Köln) to Aachen (where we know Snyders was active), about fifty-three miles, it seems likely that the reference is to the Rhineland city rather than to Cölln in Prussia. See Johann de Monte-Snyders, *Tractatus de medicina universali* (Frankfurt am Main: Thomas Matthias Götsen, 1662), 122–23. There is also further evidence linking Snyders to Cologne, some of which we discuss below.

also written in German and published in 1663. We will treat these remarkable treatises at length in due course, but for the moment, let us examine the reports of Snyders as an itinerant adept.

A number of accounts of Snyders's transmutations emerged within a generation of the events themselves; we will consider only two of the most compelling here. The first appears in *De Goude Leeuw* (The Golden Lion), a work by the Dutch chymist Goosen van Vreeswyk published in Amsterdam in 1675. Van Vreeswyk, who styles himself "berg-meester" (mine master), transmits a report of Snyders's activities given to him by "Mr. Guilliaem," a mintmaster of Aachen, when Van Vreeswyk visited that city on October 29, 1670. In 1667, "Mr. Snyders," whom Guilliaem already knew, appeared out of nowhere after a twelve-year absence. Showing up on the mintmaster's doorstep, Snyders took a ring off of his finger and challenged Guilliaem to determine its composition. The ring appeared to be gold, but when struck with a hammer, it shattered rather than displaying the malleability of the noble metal. The mintmaster then assayed it with stibnite in the usual fashion, no doubt expecting the supposed "gold" to be severely corroded by the test. To his astonishment, the metal not only remained undamaged, but it also "ate up" the crude antimony that had been added for the test. Astonished, Guilliaem moved on to other methods of assaying, using "saltpeter and sulfur." The result was only that the metal improved its appearance: it was now "the most beautiful gold in the world." After all this effort, the befuddled mintmaster and the alchemist retired to a local drinking establishment to relax, and the adept promised to return the next morning with yet further evidence of his transmutational prowess. This he did, commanding Guilliaem to fuse twenty-eight *Loot* of lead the following day—probably equivalent to about fourteen ounces—with about a quarter ounce of copper.[3] To the molten mass Snyders had Guilliaem add three and a half grains of a powder wrapped in paper, that is, a little over two hundred milligrams. When cool, the alloy was gray and glassy, but six successive fusions with blasting gradually converted it to a golden color. An assay of the final product revealed that it was "like the most beautiful ducat-gold" and that it weighed fifteen *loot*. Hence the material had lost some thirteen and a quarter *loot* in the repeated meltings—almost half its weight. After this apparently successful trial, Snyders promised to meet Guilliaem again after dinner, but following a pattern typical of wandering adepts, he failed to make the appointment, and in fact disappeared with the freshly manufactured gold.[4]

Our second report of Snyders occurs in the midst of the weighty history of the Duchy of Carniola or Krain (in modern Slovenia) written by Johann Weichard von Valvasor, an aristocrat of that province. Valvasor's account, though published in 1689, recounts a firsthand experience that he had with Snyders in Vienna during October 1666. While staying at the "Arnoldisch

---

[3] Assuming that the "loot" in Van Vreeswyk's account is the German "loth." See Jacob Grimm and Wilhelm Grimm, *Das Deutsche Wörterbuch*, at http://dwb.uni-trier.de/de/; consulted July 4, 2016.

[4] Goosen van Vreeswyk, *De Goude Leeuw, of den Asijn der Wysen* (Amsterdam: Johannes Janssonius van Waesberge, 1675), 6–12.

Haus" near the Red Tower with one "Herr Meintzer," Valvasor met "Herr Johann de Monte Sniders" and his wife. Evidently Valvasor struck up a friendship with the seeming adept and was rewarded with a demonstration of chrysopoeia. As in the case of the mintmaster Guilliaem, Snyders did not actually perform any transmutation himself, but apparently provided Valvasor with an elixir or powder of projection that allowed Valvasor to perform the feat. Thus Valvasor says that without any chicanery he himself "tinged a pound of lead into precious gold with a grain of tincture."[5] Although he was convinced of the reality of this transmutation, the level-headed historian adds the important caveat that the tincture could only convert as much lead to gold as the gold that was employed in the fabrication of the tincture itself. Hence there was no net gain in this process, and the tincture was only a "particular," unlike the philosophers' stone per se, which would have been a "universal." This is hard to understand, given that one grain of the elixir (about sixty-five milligrams) supposedly transmuted an entire pound of lead to gold. But Valvasor explains that the tincture was a "concentrated extract" (*eine concentrirte Extraction*) of gold, requiring a considerable amount of the precious metal to make. Hence it was unprofitable, and in the end Snyders was unable to make good on promises that he had presented to his backers in Vienna. Fearing for their lives, the alchemist and his wife escaped in secret, though not before paying their respects to both Valvasor and their landlord. Surprisingly, Valvasor adds that he and Snyders stayed in contact after this abrupt departure and continued to correspond for the rest of the alchemist's life.[6]

How Snyders managed to perform these tricks remains a matter of conjecture, but for us the important thing is that he was able to convince both assayers and critically minded intellectuals such as Valvasor of his transmutational ability, even if this proved to be his undoing. The fascination that Snyders's work exercised on Newton is therefore no great surprise. Already by the 1660s the German chymist's fame had reached as far as England, for the anonymous editor responsible for the Latin version of Snyders's *Tractatus de medicina universali* says that he performed the task of translation in London, and that he presented the manuscript to a group of learned men there, mostly doctors, in order to determine whether it merited printing.[7] Additional evidence of Snyders's English reputation is found in Kenelm Digby's posthumous *Chymical Secrets and Rare Experiments* (1682), where one of the recipes bears the title "Snyders's Secret, as he gave it me himself the 22 of July, 1664."[8]

---

[5] Johann Weichard Freyherr Valvasor, *Die Ehre deß Hertzogthums Crain* (Nuremberg: Wolfgang Moritz Endter, 1689), 415: "So habe ich gleichfalls / im October 1666 Jahrs / zu Wien nahe beim Roten Thurn / in dem Arnoldischen Hause / (woselbst ich damals bei dem Herrn Meintzer / eben in selbigem Hause in der Kost war / mit meinen eigenen Händen / ohn einigen Betrug / ein Pfund Bley / mit einer Gran Tinctur / ins kostlichste Gold tingert."

[6] Valvasor, *Ehre deß Hertzogthums Crain*, 416.

[7] Anonymous, *Chymica vannus* (Amsterdam: Joannes Janssonius à Waesberge and Elizeus Weyerstraet, 1666), 7. The information is confirmed on the title page of the *De pharmaco catholico*, the translation of Snyders's *De medicina universali*, which is found separately paginated at the end of the *Chymica vannus*.

[8] George Hartman, *Chymical secrets and rare experiments in physick & philosophy* (London: George Hartman, 1682), 16.

Newton's Developing Work on Snyders

Newton had known Snyders's *Commentatio de pharmaco catholico*, the Latin translation of the German *Tractatus de medicina universali* published in 1666, since the 1670s, but fell under the spell of another Snyderian text, the *Metamorphosis of the Planets*, after he had already begun digesting the *Commentatio*. The baroque and at first face incomprehensible allegory presented by the *Metamorphosis* consists of multiple rendezvous and battles between the seven planets, pictured in traditional fashion as the Olympian gods. Somehow Newton managed to obtain an unpublished English translation of the text, and the fastidiously copied manuscript in his hand, complete with an illustration and several pages of preceding notes, shows the care that he invested in it (figure 11.1). As it happens, Snyders is one of the most difficult early modern chrysopoetic writers to make sense of. Although I do not claim to have fathomed the depths of his convoluted imagery, there are several early modern accounts—one by an alchemist who claims to have worked with the adept—that have escaped the attention of previous scholars and are worth examining in this chapter.

The reader of Newton's long and complicated florilegia from the second half of the 1680s and onward cannot fail to be impressed by the frequent appearance in them of Snyders. In fact, Newton had been honing his understanding of the *Commentatio de pharmaco catholico* for some time, and this fed into the vigor with which he then pursued the *Metamorphosis of the Planets*. Newton's mature florilegia make use of both the *Metamorphosis* and the *Commentatio*, so we will have to treat both texts here. For the *Commentatio* we are better stocked with materials in Newton's hand. First, there is a synopsis of the text that Newton probably wrote in the mid-1670s, which now forms part of the National Library of Israel's composite manuscript Var. 259.[9] That manuscript also contains Newton's very early synopsis of the Philalethan *Marrow of Alchemy* that we described in chapter nine, but the two abridgements date from different periods. This is not surprising, since Var. 259 as a whole consists of twelve small manuscripts of different dates bundled together at a later period by Newton and given a single table of contents in his hand.[10] Second, Newton wrote a short text called "A Key to Snyders" (part of Sotheby Lot 103) at an undetermined date, but almost surely after the aforementioned synopsis.[11] Although this manuscript is currently in private hands, I provide a transcript and translation as appendix two and will give an account of it after treating the synopsis in Var. 259. The two manuscripts mentioned so far deal wholly or for the most part with Snyders's *Commentatio*. That is obviously not the case for the third manuscript

---

[9] Like Mellon 79, Var. 259.10 contains a mixture of barred and unbarred Saturn symbols (♄ at 10.2v and 10.5r, ♄ at 10.1r, 10.2r, and 10.2v). This suggests that the two manuscripts may have been composed around the same time.

[10] Var. 259.0.1r.

[11] The "Key to Snyders" also displays the barred Saturn symbol characteristic of the post-1674 period. See folio 1v of the text in Appendix II. In terms of content, the "Key" represents a more advanced understanding of Snyders than the one given in Var. 259.10.

FIGURE 11.1. Drawing by Newton after the title page of Johann de Monte-Snyders's *Metamorphosis planetarum*. Reproduced from Cushing Medical Library MS at Yale University.

that we will consider in this chapter, Newton's transcript of the *Metamorphosis of the Planets*, found in the Harvey Cushing Medical Library at Yale University. Not only does this include Newton's copy of the Snyderian work in English translation, it also contains five dense pages of reading notes in which Newton repeatedly tries to come to terms with the text. These notes are particularly important for their linkages with Newton's own laboratory practice as revealed in his experimental laboratory notebooks, Cambridge University Portsmouth Collection, Additional MSS 3973 and 3975.

Finally, in order to give the reader a sense of what Snyders may actually have been doing in his laboratory, we will look at the anonymous "Secret of the Author of the *Metamorphosis of the Planets*," a manuscript that was given to Robert Boyle, and which is found in the library of the Royal Society (Boyle Papers 30, pp. 415–18). This text represents a version of what may be called the "standard" interpretation of Snyders among early modern chymists. The "Secret" varies considerably from most of Newton's notes on the German alchemist and helps to underscore the idiosyncrasy of the interpretation given by the Cambridge savant. Just as Newton molded the chrysopoeia of Philalethes to fit his own purposes, so it seems he refashioned the work of Snyders in his own image. An examination of these four documents will therefore provide an excellent Ariadne's thread to help guide us through the twisting passageways of Newton's mature florilegia.

A brief look at Snyders's *Commentatio* is enough to engender a sense of despair in the modern reader. After several pages spent dispensing the usual invocations to divine benignity and other platitudes common to many early modern chrysopoetic texts, Snyders launches into his main business—the revelation and concealment of his practical operations. The problem is that the concealment far outweighs the revelation. The text, which is divided into chapters with numbered paragraphs, to the modern eye reads like a succession of extended riddles. This is precisely how Newton read it too, and in accordance with the time-honored alchemical writing practice of dispersion of knowledge (*dispersa intentio*), he tried to reassemble the dissociated parts of Snyders's process. In order to give a sense of Snyders's style I will translate an early paragraph from his first chapter here, in Newton's paraphrase synopsis. This paragraph forms the beginning of Newton's early abstract of the *Commentatio* found in Var. 259.10:

> There are two solutions; the first when the matter is reduced into prime matter through prime matter; namely into the principles, as it were into a certain dry water which is not only called mercury, but sulfur. The dry solution comes about through the magical elements in an open fire; but the other solution through an astral seed which is a dry liquor that liquefies and flows like wax. And here we see separation because the menstruum takes up for itself only the noblest soul from the already acquired metallic sulfur; there is left a remainder of other dregs.[12]

What Newton learned from this is that the initial ingredient of the philosophers' stone, the unidentified "matter," must be subjected to two "solutions," one apparently over an open flame that reduces it to its principles (evidently the mercury and sulfur here), and another that requires a menstruum or dissolvent that is dry at room temperature but fluid when heated. This menstruum dissolves and absorbs the "sulfur," but leaves the rest as dregs or perhaps slag. Although it would be unbearably tedious to present a line-by-line analysis of Newton's synopsis, there is at least one other passage from

---

[12] Var. 259.10.1r: "Duæ sunt solutiones; prima cùm materia per materiam primam in materiam primam redigitur viz: in principia tanquam in siccam quandam aquam quæ non solum ☿ sed et ♄ dicitur. Solutio sicca per Magica fit Elementa igne aperto; <u>Alia autem</u> per astrale semen quod est siccus liquor qui cæræ <*sic*> fluit instar et liquescit. Et hic plerumꝗ deprehendimus separationem quia menstruum sumit ^sibi solummodo ^nobilissimam animam e sulphure metallico jam priùs adepto idꝗ cum remanentia aliquarum fæcum."

Snyders that we will encounter repeatedly, as it is a foundation stone of his doctrine in the *Commentatio*. I refer to his guideline that the work requires three "mysterious fires" in addition to the two solutions outlined above. In Newton's paraphrase, taken from Snyders's third chapter, the introduction of these fires appears thus:

> The solution cannot be performed without these three fires, of which the first forces the metal to fuse, the second has sympathy with the metallic fire and is double, or consisting of two contrary natures forced into a state of friendship; hence I will hold it for simple. This fire kindles the metallic sulfur and augments the element of fire in a metallic body. The third fire is the cold, metallic fire, practically like mercury. For it spreads through the body as if a spirit, aids the sympathetic fire to penetrate the whole, ignites the soul, renders the body porous, and is both beginning and end of the work: for it is a vehicle to the sympathetic fire and by that it is corrupted, due to sympathy and antipathy.[13]

It goes without saying, perhaps, that these three fires are actually substances that enter into the work. The first merely causes fusion, the second, called "the sympathetic fire" on account of its union of two opposites, penetrates the metals and ignites their innate sulfur; the third fire, which is cold, makes the metals porous and aids the penetration of the second fire, while also kindling their hidden "souls." What could Snyders be talking about?

Newton's "Key to Snyders"

Confronted with Snyders's obscurity, Newton turned to a second tool in his panoply of interpretive methods: after compiling his paraphrase synopsis, he composed a "Key to Snyders" (see appendix two where the text is reproduced). What this consisted of, despite its rather grandiose title, was merely a collection of the most suggestive passages from Snyders brought into an order that Newton thought appropriate, and accompanied by his own halting and tentative interpretations. Newton composed "keys" to other texts as well, such as one for the *Turba philosophorum* (Sotheby Lot 60, in private possession), which is similar in spirit. A look at the first paragraph in the *Key to Snyders* reveals both the tenor of the little text and some illuminating content. Here Newton provides comments both on Snyders's two solutions and on his three fires:

> The wet solution comes about through an astral seed which is a dry liquor flowing like wax. This liquor is the first fire, by which the metal is forced

---

[13] Var. 259.10.1r: "Solutio non potest fieri sine his 3 ignibus, Quorum primus metallum in fluxum adigere debet, secundus sympathiam habet cum igne metallico est♀ duplex sive ex duabus contrarijs naturis ^in amicitiam redactis compositus. ~~etq~~ Itaque pro simplici sumam. Hic accendit ♃ metallicum et elementum ignis in metallico corpore augmentat. Tertius est frigidus metallicus ignis, mercurio ferè similis. Nam metallum tanquam spiritus pervadit, procurat sympatheticum ignem in toto penetrare, animam incendit, metallum efficit porosum, principium est et finis et fundamentum operis: Nam est vehiculum ignis sympathetici et ab hoc corrumpitur ex sympathia et antipathia."

into flux. (p. 10). The second fire is a salt prepared from ♀ and horned Diana without the seed of gold. The third fire is the spirit of ☿ or rather the philosophical Venus ~~running like mercury~~ ^because it is referred to as practically like ☿. But perhaps the dry water ^impregnated with ♂ and ♀ because it is called metallic.

The *Key to Snyders'* first paragraph announces at once that Newton is considering the "wet solution" of the *Commentatio*: this is what Snyders called the "other solution" (*Alia*) at the very beginning of Var. 259.10 and contrasted to the dry solution that takes place "through the magical elements in an open fire." The *Key to Snyders* then jumps into a discussion of the three Snyderian fires, all of which pertain here to the wet solution. This is an important passage, for Newton reveals that the three fires are respectively the dry liquor flowing like wax, a salt made from "♀ and horned Diana," and "spirit of ☿ or rather the philosophical Venus." Although one might be tempted to accuse Newton of an ambiguity rivaling that of Snyders himself, there are significant clues buried in this, particularly in Newton's description of the second fire. The symbol "♀" can be taken straightforwardly to mean copper or a copper compound, since a few lines later Newton distinguishes it from "the philosophical Venus." If the ♀ of the second fire is not "philosophical," then its most likely referent is actual copper as in the traditional system of planetary *Decknamen*, or else one of its compounds. As for "horned Diana," namely, the goddess wearing her crown of a horn-like crescent moon, every Newtonian text that we have examined up to now suggests that this refers either to lead, a mineral of lead, or a lead salt, possibly in combination with antimony. Already in the very early commentary that Newton wrote to Jean d'Espagnet's *Arcanum hermeticae philosophiae* in Keynes 19, he was arguing that "horned Diana" referred to the two sulfurs that floated above a regulus of lead and antimony, and these in turn were equivalent to the Philalethan doves of Diana.[14] It appears, then, that the second fire consists of free or compounded copper, some compound(s) of lead, and possibly antimony, at the least. In the case of the third fire, Newton was clearly torn between conflicting interpretations. On the one hand it might refer to a "spirit of mercury" also called "Philosophical Venus"; on the other, it might be the first fire or dry water impregnated with iron (♂) and copper (♀), "because it is called metallic."

The second paragraph of the *Key to Snyders* continues Newton's attempt to pin down the third, metallic fire:

The metallic and mineral fire is the first matter, which is found in the mineral of Saturn as in its universal house. It must withdraw from this house on account of fear of the fiery, flying dragon who ignites the home of cold Saturn so that he is forced to die in it and his spirit is forced to exhale. If you can capture this spirit in a receiver, you have the universal menstruum, the astral fire, which has the likeness of dry water yet at the same time wet, which wets nothing but metals. It is light and heavy in weight beyond all other things. It is the true separator of the metallic ^impurities of sulfurs. It

---

[14] Keynes 19, 4r, at note "w": "Bina Sulphura sive Columbæ sive corva supernatabant aquæ mercuriali."

is similar to the double mercurial water. And it is called acid spirit and double corrosive. The soul of the King is reducible into oil by this alone.

Here we learn that the third, or metallic, fire is the first matter out of which metals are made. It must be extracted from "the mineral of Saturn," which in the language of Newton's sources could mean either an ore of lead or the primary mineral of antimony, stibnite. That Newton chose here to interpret it as an ore of lead seems evident from a later passage in the *Key to Snyders* where he explicitly distinguishes between "Luna [♂]" (unambiguously stibnite) and "that cold, Arietine ♄." If the cold Saturn is not stibnite, then it is probably a lead ore.[15] From this one must extract a spirit or vapor (which is also a dry water) by means of a fiery, flying dragon, and capture it in a receiver. In other words, the volatile component of a lead mineral must first be "loosened up" or released by another volatile material, and then distilled into the receiver of the retort or alembic. The extracted material will be a universal menstruum capable of dissolving "the soul of the King" (probably here metallic gold) into an oil.

The *Key to Snyders* follows this operation with ten additional Latin paragraphs, of which nine contain references either to the "mercury of Saturn," the "mineral of Saturn," "cold, Arietine ♄" or "my cold dragon." Only the last of these four requires explanation at this point. In the fifth paragraph of the *Key to Snyders*, Newton glosses his source with his usual square brackets: "Whoever understands my cold dragon [the third fire]" needs only to join it with its brother, the fiery, flying dragon, in order to make the ultimate goal of the art, which Snyders calls "the most general Universal" (*generalissimum Universale*). Newton has therefore equated the third fire, which we now know must be extracted from a certain mineral Saturn, with the cold dragon. From all this it is clear that the primary focus of the *Key to Snyders* lies in Newton's attempt to arrive at the means of extracting Snyders's third fire from "a certain mineral Saturn, not yet fused." For the reasons that we have just laid out, that mineral Saturn was for Newton almost certainly an ore of lead or another mineral found in lead mines.

There is yet more to be gleaned from the *Key to Snyders*. In a passage to which I have already alluded, Newton introduces antimony under the *Deckname* "Luna" and what we now believe to be a lead mineral under the term "cold, Arietine ♄." The same passage decodes yet another Snyderian cover name, "Solar Venus," as vitriol:

> From Luna [♂] and likewise from that cold, Arietine ♄ a mercury can be made: just as a certain ☿ which is gifted with a solar ♀ can be made from the mineral of Solar Venus [⊕], whence I have called it the ☿ of the ☉, since it must be taken for the generation of Sol. From these [at least from the two latter], the most general Universal is made. p. 72.

[15] The Snyderian term "Arietine" may refer to the fact that Saturn contains the first matter from which all metals are made, just as Aries is the first of the zodiacal signs.

The association that Newton makes between "the mineral of Solar Venus" and "$\oplus$" suggests strongly that the vitriol is a salt of copper, though not necessarily the copper sulfate that commonly went under the name of "blue vitriol" in seventeenth-century England. When we examine Newton's laboratory notebooks, it will become obvious that for him, Venereal vitriol could refer to other copper compounds besides the sulfate. Newton also adds a bracketed comment to the effect that Snyders's "most general Universal" is made "from the two latter," apparently meaning the "metallic fire" extracted from the cold, Arietine Saturn, and the vitriol just mentioned.

When all of this new textual material is assembled, it appears that Newton believed the wet solution of Snyders's *Commentatio* to require a "dry water" or volatile material drawn from a lead mineral with the aid of one or more copper compounds and heat. Antimony evidently also played a role, though exactly what at this point is unclear; quicksilver may also have been involved. These very materials resurface in Newton's later alchemy time and time again; in particular they form the focus of an important set of letters between Newton and his friend Fatio de Duillier in August 1693, as we will see in the later chapters of the present book. In the meantime, it is possible that Newton viewed the last paragraph that we examined as a mere restatement of the introductory one on the second and third fires, but for the moment it is better to suspend our judgment. In order to proceed further, it will be necessary to examine what Newton made of Snyders's baroque masterpiece, the *Metamorphosis of the Planets*.

Newton and the *Metamorphosis of the Planets*

The *Metamorphosis of the Planets* was without doubt composed after the *Commentatio*, since it refers to the *Commentatio* numerous times (usually as "my little treatise on the magical elements" or a variant of this expression). It is important to state at the outset that the *Metamorphosis* lays out a succession of veiled processes that appear to be somewhat different from those of the earlier text. While much remains the same, the *Metamorphosis* explicitly downplays the role of lead or lead ore (Saturn), despite the great emphasis that the *Commentatio* put on that material. It is entirely possible that the author of the *Metamorphosis*—for the sake of simplicity let us assume that he really is the author who composed the *Commentatio*—changed his mind about the starting ingredients of the philosophers' stone after writing the earlier text. We already saw in chapter nine that George Starkey made just such a radical change between writing *Secrets Reveal'd* and the *Marrow of Alchemy*, when he substituted Venus (copper) for the two doves of Diana (two parts of silver). Snyders may well have made a similar move when it came to the role of lead and its ores. At any rate, the evidence for a demotion of "vulgar" Saturn in the *Metamorphosis* is unambiguous, and indeed, the common interpretation of Snyders's processes among seventeenth-century chymists held no special place for lead or its minerals. In Newton's English manuscript of the text, we encounter the following passage, which gives a good idea of Snyders's new approach to lead:

> This my doctrine will at first to many seem exceeding wonderfull, but if they give good heed to my writing they will not think it so very strange, although this in respect of the Lunary birth appear repugnant to my first treatise concerning the Magical Elements. ffor there I mentioned that the earthly Saturn or Lead conteined & produced the seed of Silver & Quicksilver, whence in the Lead Mines in the very deepest of them much unripe Silver is found &c. This is true and is not now gainsayed by me, but serveth here for an elucidation of the former, That ye common decrepid Saturn as a bastard of ye true Saturn participated such a nature from ye feminine Lunar Child of ye world, & if so be the defective Saturns salt were thereto disposed, the Lead Mines which hold Silver might produce greater advantage. Now it may well be concluded whence it proceeds y^t those who place their hopes upon the mercury of this Saturn, are deceived in their opinion.

Snyders is manifestly worried here that readers of his two books will accuse him of inconsistency. The "earthly Saturn" explicitly refers to lead here, and as Snyders says, he now wants to rid his readers of the hope that the "mercury of Saturn" can lead to the philosophers' stone. It was precisely the mercury of Saturn that Snyders touted in the *Commentatio*, as we saw from Newton's careful collation of the many passages that he gathered in the *Key to Snyders*. Here in the *Metamorphosis* Snyders tries to extricate himself from his earlier claims by saying that it is indeed true that argentiferous lead might in principle be matured to a point where it yielded more silver than it does, but that it is in practice impeded by its defective, internal salt. Hence it is better to avoid this "decrepid Saturn as a bastard of ye true Saturn" and to turn one's attention to the "true Saturn" itself. What could this true Saturn be? Whatever Snyders himself may have meant, Newton would understand it to be antimony.[16]

Newton wrote four short, successive sets of notes on the *Metamorphosis* that now form part of a manuscript kept by the Cushing Medical Library at Yale University, which he titled *On ye Metamorphosis of ye Planets*. In order to keep these commentaries distinct, I will number them one through four.[17] The Cushing manuscript also contains Newton's transcript of the *Metamorphosis* itself, along with a copy of the famous illustration that appears on the title page of the printed book (figure 11.2). Unlike anything that we have considered so far, Newton's *On ye Metamorphosis of ye Planets* strongly resonates with his laboratory practice as recorded in his own experimental notebooks. These notebooks, Cambridge University Library Portsmouth Additional 3973 and 3975, form the basis of three subsequent chapters, so I

---

[16] As at Cushing 3v, where he explicitly associates "Saturn" with "Antimony" and "the double nature"— "∧The double nature p 11 l 12 p 9 l 31 Antimony p 25 l 9. Saturn p 26 lin 1"

[17] Henceforth I refer to folios 2r–2v of the Cushing manuscript as *On y^e Metamorphosis of y^e Planets 1*; 3v as *On y^e Metamorphosis of y^e Planets 2*; 4r as *On y^e Metamorphosis of y^e Planets 3*; and 5r as *On y^e Metamorphosis of y^e Planets 4*.

FIGURE 11.2. Title page of Johann de Monte-Snyders, *Metamorphosis planetarum* (Amsterdam: Johan Jansson, 1663). This image, or a copy of it, served as the model for Newton's drawing in the Yale Cushing MS.

will not consider them deeply here; nonetheless it may prove useful to refer to them occasionally. The comparison between "the bastard Saturn" or lead and antimony is one of the first things that one encounters on examining *On yᵉ Metamorphosis of yᵉ Planets 1*. There Newton provides a brief synopsis of Snyders's "demotion" of lead that we just discussed:

> "c. 3 p. 8 l 24 NB. The ☿ of <*illeg.*> ♄ is much inferior to ye ☿ of ♂. That is only Lunary, this solary: this is ye true matter not that. p. 9. l 21 This is Cerberus, the three headed dragon &c."

Obviously, Newton got the point of Snyders's newfound downgrading of lead. The mercury of ♄ (clearly lead here) is inferior to that of antimony because it pertains only to the moon (silver), and not to the sun (gold); as Snyders had said in the *Metamorphosis*, this was made evident by the fact that lead ore is often argentiferous. The "true matter," which contains the seed of gold, is the mercury of antimony, also signified by terms such as Cerberus and the three-headed dragon. How then does Newton understand the term "mercury of antimony" here? Following the same practice that he had already employed in his early Sendivogius commentary in Keynes 19, Newton thinks of the metallic regulus of antimony as its "mercury." This emerges clearly from a passage in *On yᵉ Metamorphosis of yᵉ Planets 4*, where Newton adds his usual bracketed comments to a passage from Snyders: "Mercury [♂] by humbling himself [into <*illeg.*> Reg.] is exalted." In other words, the internal mercury of the stibnite is "humbled," that is, it sinks down as a regulus within the crucible or iron cone in which the ore is smelted while the slag floats to the top; to Snyders, this is like a subject bowing to a ruler. Yet at the same time, the internal mercury is exalted, because the term "regulus" means "little king" in Latin: hence the conversion to regulus entails its ennoblement. In order to understand Newton's interpretation of the *Metamorphosis* as a whole, we must now take a quick look at Snyders's plot line.

Most of the baroque allegory of the *Metamorphosis* takes place in heaven, which is ruled by a monarch who also goes by the name "double nature" because he is a hermaphrodite.[18] A number of the Olympian gods and goddesses (Saturn, Mars, Venus, Jupiter, Luna, and Mercury, all planetary names, of course) are successively exalted to become his companions or consorts, and then each of them is in turn cast out from the kingdom on high to be humiliated. In the meantime, the monarch himself falls in love with Venus, but she rejects him and is subsequently raped by a dragon. After wandering disconsolate in the desert, she is rescued and carried up to heaven by an eagle, where for a time she sits on high. Then Mars becomes infatuated with Venus, but instead of responding in kind she has a tryst with the sun god Phoebus, infuriating both her would-be lover Mars and her husband, Vulcan. In an important section of chapter fifteen, Vulcan burns Venus and Phoebus to cinders by means of a firework made of "an uninkindled fire, of a fiery Air, & of a Vegetable Salt."[19] He then collects some of the ashes, dissolves them

[18] Cushing, 11r–11v.
[19] Cushing, 20v.

in common well water, and gives them to Mars to drink, which the warrior god refuses to do. Vulcan then adds white wine to the liquid, whereon the solution changes to "a thick & most beautifull red essence." Mars, now regretting the loss of the goddess, begs Vulcan for the red "blood," and when he is granted his wish, he gathers the remainder of the ashes, lixiviates and filters them, and adds vinegar to the solution. By these and other means, Mars and Vulcan eventually recombine Venus's (and Phoebus's) spirit, soul, and ashes to carry out a palingenesis and bring her back to life. In the course of this operation, Venus undergoes a series of color changes that correspond to the alchemical regimens—the stages of maturation that the sophic mercury was thought to undergo in becoming the philosophers' stone. In the end she is absorbed into the monarch himself, who is now identified as King Solomon.[20]

Bewildering as this story may be, it only recounts a fraction of the trysts, rivalries, battles, monstrous births, and fiery deaths that fill Snyders's *Metamorphosis*. Any number of chymical operations and materials could lie hidden behind these lurid *fabliaux*, but Newton chose his own, distinctive path through the maze. In essence, Newton interprets the rise and fall of each planetary god as its sublimation and precipitation, such as it might undergo those operations in a laboratory. Of course most metals in their normal state are not easy to sublime even when molten. Hence Newton interprets Snyders to be saying that they must be sublimed by using regulus of antimony, which volatilizes on its conversion to antimony trioxide in the heat of a charcoal-burning furnace. The practice of subliming various metals with antimonial mixtures and compounds would form the backbone of Newton's laboratory protocols from the first half of the 1670s onward, so it is of great interest that he finds it in the *Metamorphosis*. He spells out the sublimation of antimony regulus in *On y* Metamorphosis of y* Planets 3* when glossing Snyders's oration consoling the hermaphroditic monarch on his rejection at the hands of Venus:

> In textu. lin 3 O thou honourable [Regulus] astral [idem] earthly [idem] salt, moist [sal] dry, light he [in sublimation] heavy [in y* metallick form] & chosen electrum.

The refined antimony is "honourable" because again its name, *regulus*, simply means "little king" in Latin. It is "astral" because it can form the crystalline pattern of the star regulus during its cooling, and "earthly" when still in the form of its ore. Newton's term "sal" or salt may again refer to the crystalline property of the star regulus. Another term that Snyders uses for the hermaphroditic monarch is "double nature," which Newton highlights by glossing "light" with "in sublimation," and "heavy" with "in y* metallick form." There can be no doubt that Newton is taking Snyders to be referring to the sublimation of antimony regulus. But what of antimony's role in subliming other metals? Immediately before this passage in *On y* Metamorphosis*

<hr/>

[20] Cushing, 23r.

of ye Planets 3 there is another one where Newton spells out the use of antimony regulus in helping other metals to sublime:

> Cap 4. In titulo. The double nature ~~fals~~ harmonizeth wth ♀ in ye sublimation & is forsaken by her after sublimation. ♀ <illeg.> is ⊕.[21]

In other words, antimony regulus first sublimes copper or one of its compounds and is then forced to separate from it after the sublimation has taken place; the separation, as Newton spells out elsewhere in his commentary, occurs by means of precipitation. Moreover, we learn that this is not ordinary copper as such, but rather a Venereal vitriol (⊕) that is being sublimed with the regulus. Again, this conforms closely to Newton's own practice as recorded in the Portsmouth manuscripts containing his laboratory notebooks, as we will see in due course. Rather than going into further detail with Newton's process on copper vitriol here, which I will cover exhaustively in a later chapter, let us consider the role of the other metals as he interprets them in Snyders's text. Another revealing passage from the same commentary throws further light on antimony regulus and its uses. Newton is here glossing the same chapter as above, where Snyders addresses the hermaphroditic monarch after his failure to win the heart of Venus:

> l 19 It is not enough for thee to be a little King (Regulus wch carries up ♀) but thou strippest thy self of thy royal ornaments & thy purple diadem (Venus wch in —— is purple) thou bestowest on ~~them ye no~~ others & therewith cloathest ye naked (metals divested of their feculent natural cloathing) &settest them (by making them regulus's & subliming them) in thy kingdom. Thou purgest ye leapers (impure metals by making them regulus's).

After again referring to the sublimation of Venereal vitriol, Newton interprets Snyders's comments about the monarch clothing the naked gods and setting them into his kingdom as references to the making of reguli with the different metals, followed by their sublimation. As we saw in our analysis of Newton's early commentaries in Keynes 19, he had long ago adopted the Philalethan idea that mercury could be cleaned and "acuated" by antimonial regulus, and he had transferred those properties to lead. The attempt to purge lead of its impurities by means of antimony formed the basis there of Newton's interpretation of Sendivogius and d'Espagnet. It is not surprising, then, that the same idea would reemerge in greater complexity in Newton's much later interpretation of Snyders. His view now is that the metals must be first cast into reguli in order to purify them, and then sublimed. This is another practice that we find described at length in Newton's laboratory notebooks. In addition, Newton finds a justification here for his view that multiple metals should be fused together into the same regulus before their sublimation. The pretext for this comment appears in chapter 5 of the *Metamorphosis*, where the planetary gods assemble and vie with one another to

---

[21] Cushing, 4r.

show their support for the monarch. Mercury, who bows his head in submission and "humbles" himself, is accepted first. Newton interprets the passage thus:

> Chap 5. In textu. l. 2. Venus Mars & Mercury come <*illeg.*> in yᵉ Regulus wᶜʰ Iupiter discovered by a Signat Star. Mercury stept in bare-foot (naked below or in yᵉ Regulus) hubling <*sic*> himself (sinking down) in hope to be exalted (by sublimation<)>.

When Newton claims that Venus, Mars, and Mercury "come in yᵉ Regulus," this is an unambiguous reference to a compound regulus made of copper, iron, and mercury (perhaps again the "internal" mercury of antimony). The reference to Jupiter and the signate star is an obvious allusion to the addition of tin to the composite regulus and the star formation that Newton supposed would form on the surface of the cooling alloy. The final lines of Newton's gloss indicate once again that the making of the regulus should be followed by its sublimation.

So far, we find Newton interpreting Snyders to mean that reguli of the metals should first be made—probably both individual and compound—and that these, or at least some of them, should be sublimed from a vitriol made with copper. Not surprisingly, things get more complicated as we progress into the final set of Newton's notes. *On yᵉ Metamorphosis of yᵉ Planets 4* is particularly rich for its resonance with what we know of Newton's laboratory practice. Here Newton has cobbled together snippets from seven pages of Snyders (some widely separated in the original text), giving line numbers after each page. This is a classic example of reassembling a text that one believes to have undergone "dispersion of knowledge" (*dispersa intentio*) at the hands of the original author:

> Mercury [☿] by humbling himself [into <*illeg.*> Reg.] is exalted [p. 11,5 ^¹¹ by the Venereal property & evaporated Neptune ^& thereby becomes an eagle (11.36 & 16.10. ^For He hath a metallic body & was originally born of yᵉ same mother with yᵉ Monarch, namely of yᵉ great world & is yᵉ eartly <*sic*> black eagle & was washed wᵗʰ yᵉ the <*sic*> corrosive of Neptune & by yᵉ Venereal property exalted into a most beautifull crystalline weighty essence (13. 6,7,8,9) ffor Mercury had exceeding large wings wᶜʰ by the Venereal property & hardened lye of yᵉ briny ocean was exalted into a very beautiful white colour. p 42. l. 25, 26, 27. For ☿ is ♃ & ☿ & this essentiallized ☿ is the eagle the air & yᵉ salarmoniack of Phers. p. 19. l. 7, 8, 9 This ☿ is not malleable like ♃ but brittle (p. 12. l. 16, 20.) & speaks in yᵉ Hungarian language (p. 13. l. 1. & 14.9) & is yᵉ double nature p. 14. l 19 & <*illeg.*> the only means of enlivening dead metals. p. 14 l 15, 33.²²

There is an initial making of a regulus here, followed by its sublimation with "the Venereal property," probably alluding again to a type of copper vitriol. What then is "evaporated Neptune" and the "corrosive of Neptune" a few lines later? These expressions occur in the same form in Snyders's text, but

---

²²Cushing, 5r.

in Newton's hands they contain a relatively straightforward allusion to the material that would form his standard acid "menstruum," which he referred to throughout his experimental notebooks as "liquor of antimony" (or alternatively as "vinegar," "spirit," or "salt" thereof). We will deal with this substance at length in a later chapter; for now it is enough to say that Newton would typically dissolve a material in it, then filter, wash, and evaporate the product, following this with sublimation using an antimony compound or regulus in combination with sal ammoniac.[23] One such mixture of a white antimony compound and ammonium chloride is precisely what Newton referred to very excitedly in his laboratory entries of 1681 as "sophic sal ammoniac."[24] Hence it is no surprise here to see him say "♂ is ♁ & ☿ & this essentiallized ☿ is the eagle the air & yᵉ salarmoniack of Phers." In decoding Snyders's sal ammoniac of the philosophers as antimony (or more likely a mixture or compound of antimony), Newton was finding his own laboratory practice in the text of the master. The passage concludes, finally, with Newton's reassertion that when Snyders says "mercury," he really means antimony; antimony is indeed brittle in its metallic form, and the fact that Snyders has the planetary god Mercury speak "yᵉ Hungarian language," is for Newton an allusion to the common seventeenth-century view that the best antimony came from Hungary.[25]

So far, Snyders seems to have kept true to his newfound disenchantment with lead. Among the base metals, copper, iron, and tin have all appeared in the operations mentioned up to now, but so far lead is absent. Is it actually the case that Snyders has utterly abandoned "the old Saturn?" In truth, despite his protestations, the German alchemist could not entirely foreswear the heavy metal. The *Metamorphosis* contains an entertaining episode in which the gods are vying with one another to claim kinship with the monarch. The first god to approach the monarch is Mercury, who humbles himself appropriately and is accepted. Next follows Jupiter, who "mounted upon the wings of his nimble Eagle" (the eagle's name is "Bismuth") and flew off to the royal palace to present himself. Having at first been left out of the assembly, Saturn then stumbles into court to plead his case. As Snyders puts it, the lame god "scrabled with his hands, & with his stump-foot could not so much as rais himself up," but he still manages to make his plea. Old Saturn argues that the other gods have unfairly left him out, and that any decisions require his presence "as a prover of them all." This is an obvious reference to the use of lead in the time-honored assaying test of cupellation, which was employed to free gold and silver from base metals and other impurities. Prompted by this, Jupiter interrupts in his native language, English (an allusion to Cornish tin) and begins a pompous oration in favor of his own

---

[23] Although he uses the term "Neptune" for his menstruum several times in CU Add. 3973, this term may indicate that he had dissolved an additional material in the liquor of antimony before going through the remainder of his protocols. See CU Add. 3973, 21v.

[24] See CU Add. 3973, 13r, CU Add. 3975, 62r.

[25] See for example Basil Valentine, *Last Will and Testament* (London: Edward Brewster, 1672), 105. This text was well known to Newton. He copies out the claim that Hungarian antimony and vitriol are the best in British Library MS Additional 44888, 1v.

"spiritual angelical & altogether divine" qualities. But the god Mercury intercedes on Saturn's behalf, pointing out that Jupiter is weak and that common fire burns up his residence, meaning that the metal tin, the traditional referent of the planet Jupiter in alchemy, cannot withstand the fire of calcination. Jupiter's arrogance and Mercury's list of his shortcomings enrage the monarch, and as a result he casts Jupiter down to dwell with Saturn. This, however, has the result of elevating Saturn himself, who is now transformed into "ye Saturn of Philosophers."

What did Newton make of these confusing developments? In *On ye Metamorphosis of ye Planets 1*, Newton announces that this episode refers to "a collateral work wth ♃"; in other words, it is not part of the sequence of operations involving the internal mercury of antimony, Venus, and Neptune that we already analyzed. It is only with *On ye Metamorphosis of ye Planets 4*, however, that Newton gives a detailed synopsis of the Jupiter-Saturn parable. Much of this consists of a recapitulation of the various obvious clues cast forth by Snyders: Jupiter is "a malleable metal" (tin), his eagle's name is Bismuth, his native language is English (again tin), and he "thunders by his crackling quality," a reference to the well-known "cry" of tin, which makes a creaking sound when bent. Then Newton repeats that Jupiter is denied a place with the monarch and "commanded down to Saturn." It is not until the end of Newton's paraphrase that we learn something radically new:

> Venus is the Green Lyon [or most High] who wth her hot fiery volatil salt spirit educeth by the help of the lunary little world a fiery mercurial spirit out of the cold Dragon p. 19. 12, 13, 14. Neptune & Venus cause Dragon to fly.[26]

With one important exception, almost all of these words come verbatim out of chapter eight of the *Metamorphosis*, which follows Saturn's apotheosis. We must briefly consider this section now. As soon as the titan's conversion to the philosophical Saturn has been achieved, a company of philosophers bursts on the scene, headed by the venerable alchemists Geber and Hermes. Each philosopher bears a heraldic emblem and a *dictum*: for Hermes it is a Phoenix surrounded by the famous words from the *Emerald Tablet*—"That wch is above is like that wch is beneath, & on the contrary." At the end of this troupe is found "a most envied man but yet a most true Philosopher," the humble Benedictine monk, Basil Valentine. Because of his humility, Basil is called on to judge the assembly of the philosophers and arrive at the common nugget of truth in their sayings. The result is actually a pastiche of snippets that Snyders has pulled mostly from the large and heterogeneous corpus attributed to Basil Valentine. The last of these snippets, which Newton has carefully collated with multiple related pages (and line numbers), is precisely the passage that we just quoted, beginning "Venus is the Green Lyon."[27] But consulting Snyders's text reveals that Newton has made one very significant

[26] Cushing, 5r.
[27] Cushing, 16r.

change—in his usual brackets he has added the modifier "[or most High]."
For Newton, Venus in her Green Lyon role has become the monarch!

At first this might seem a passing enthusiasm or even a slip of the pen on Newton's part, but in fact it is much more significant than that. In a slightly earlier part of *On yᵉ Metamorphosis of yᵉ Planets 4*, Newton begins a passage on the double nature that sounds as though it should refer to regulus of antimony, as indeed it did in the previous three commentaries to the *Metamorphosis*. Here, however, he concludes very differently, saying:

> This Monarch is yᵉ double nature (p. 11. l. 12) ^the Dragon (p. 16.14) the son of yᵉ old Dragon p. 11. l. 21 & p. 13. l. 7 <*illeg.*> & hath a metallic♀ body (p. 13. l. 6.<)> This Monarch is therefore ⊕ vol.

All of the terms that we had come to associate with regulus of antimony in Newton's reading—monarch, double nature, Dragon, and so forth, now relate instead to "⊕ vol." This is an abbreviation for "volatile vitriol," and it refers to the Venereal vitriol that Newton already said should be sublimed from regulus of antimony in *On yᵉ Metamorphosis of yᵉ Planets 3*. The difference is that now the volatile vitriol has become the monarch. Remarkably, the expression "⊕ vol." can also be found in Newton's experimental notebooks, where we see him working out the details of his interpretation of Snyders. In an entry dated July 18, 1682, we find Newton imbibing an alloy of copper and antimony regulus with his proprietary vinegar of antimony in which volatile vitriol has been dissolved; he then sublimes the resulting mixture with "✶," probably meaning sophic sal ammoniac.[28] Newton's laboratory records reveal unequivocally that "volatile vitriol," "volatile Venus," and "our Venus," which all meant the same thing in the end, acquired increasing importance to him as his alchemical project developed. It became, in fact, his favorite tool for subliming metals and other materials—hence his granting of the epithet "most High" to the volatile vitriol in *On yᵉ Metamorphosis of yᵉ Planets 4* is no accident.

As we have just seen, Newton's four commentaries to Snyders's *Metamorphosis of the Planets* link much more closely to the surviving portions of his laboratory records than do any of the other texts that we have examined so far. Before proceeding into the details of this linkage, however, it is important to determine the degree to which Newton's interpretation of Snyders was peculiar to himself. This is not a matter of idle curiosity but rather a pressing concern, for a resolution of this problem will help to answer a further question. Was Newton's decoding of the *Metamorphosis* and the *Commentatio* a simple matter of transferring allegory and *Decknamen* into laboratory practice, or was Newton instead justifying his existing protocols by reference to these authoritative texts? The history of alchemy provides abundant examples of both practices. To give but one illustration, Newton's avatar George Starkey went to great lengths under his Philalethan disguise to find an authoritative pedigree for his antimonial practice in the works

---

[28] CU Add. 3975, 68r.

of the fifteenth-century alchemist George Ripley.[29] But in reality, Ripley's alchemy used antimony hardly at all, and depended heavily on the use of lead compounds instead, which he hid under the *Deckname* "sericon."[30] Whether Starkey realized it or not, his use of Ripley was principally one of demonstrating that "Philalethes" belonged to the same lineage as the great masters of the art. On the other hand, Starkey's laboratory work really was heavily influenced by another chymical writer, the sixteenth-century Paracelsian Alexander von Suchten, whose name and reputation lacked the numinous quality of Ripley's.[31]

When Newton decoded Snyders, was he, as in Starkey's use of Ripley, merely finding authoritative *dicta* on which to pin his own, independently arrived at processes? The remarkable detail that Newton extracted out of Snyders and the endless hours that he spent at this endeavor militate against such an interpretation. There are also numerous instances where one can show Newton choosing a particular operational path at a crossroad on the basis of his textual interpretation. Moreover, a large number of his chymical experiments recorded in the two Portsmouth laboratory notebooks (CU Add. 3973 and 3975) are attempts to test processes derived from sources such as Snyders, Philalethes, and Sendivogius. But at the same time, one must admit that the hints and allusions given by these authors would on their own be a poor guide indeed to actual laboratory practice, which had to be learned elsewhere. Newton was an extremely skilled chymist for his day, as even a cursory glance at his experimental notebooks reveals. How much of his own interpretation, based on personal experience at the bench, was he bringing to these recalcitrant texts? This question can be answered best, perhaps, by taking a quick glance at other early modern interpretations of Newton's sources, particularly of Snyders. We are fortunate in this endeavor to have a document that claims to have been written by an associate or operant of Snyders himself.

The Secret of the Author of the *Metamorphosis of the Planets*

Among the many chrysopoetic writings found among the heterogeneous papers collected by Robert Boyle there is one in a continental hand with the Latin title *Arcanum authoris Metamorphosis planetarum* (Secret of the Author of the *Metamorphosis of the Planets*).[32] This document, which has received no previous scholarly attention, claims to have been written by an anonymous coworker of Snyders. At the end of the first set of operations, he explicitly says, "I have done this with the author" (sic feci cum authore),

---

[29] Newman, *GF*, 117–25.

[30] Jenny Rampling, "Transmuting Sericon: Alchemy as 'Practical Exegesis' in Early Modern England," in *Chemical Knowledge in the Early Modern World*, ed. Matthew Daniel Eddy, Seymour H. Mauskopf, and William R. Newman, *Osiris* 29 (2014): 19–34.

[31] Newman, *GF*, 135–41.

[32] Royal Society, Boyle Papers 30, pp. 415–17. I thank Michael Hunter for providing me with a scan of this document.

a statement that is repeated again in other language later in the text. Such a statement might at first be met with a jaundiced eye, but there is at least one other piece of evidence suggesting that whoever wrote this little tract had firsthand knowledge of Snyders and his work. After he finishes recounting the processes that come from "the author of the *Metamorphosis of the Planets*," the composer of the *Secret* gives another, similar set of operations introduced by the phrase "The Archbishop of Cologne has proceeded thus" (hoc modo procedebat Arch. Col.). Although far from guaranteeing authenticity, this attribution to the Archbishop of Cologne is important for two reasons. First, we know from Snyders's *Commentatio* that he was in fact operating in the city of Cologne for a time, probably in the early 1660s.[33] Moreover, at that time, the Archbishop of Cologne was Maximilian Heinrich of Bavaria, a famous aficionado of alchemy who is said to have even traveled incognito to Amsterdam in order to acquire secrets from the local adepts.[34] These facts give some degree of credence to the author of the *Secret*, although the matter requires further research that need not form part of the present book.

What is important for us is the fact that the *Secret* gives an entirely different interpretation of Snyders's work from what we have found in Newton. The *Secret*'s recipe consists of several stages. The first is to make butter of antimony by distilling corrosive sublimate (our mercuric chloride) with martial regulus of antimony, that is, antimony that has been reduced from stibnite by means of iron. Once the "the martial, antimonial butter" has been made, it is allowed to absorb water from the air, which the substance will in fact readily do. Multiple distillations and exposures to the damp atmosphere eventually liquefy all the butter, which is then distilled to remove the lighter fraction, and the rest is preserved. The author assures us that this is the chosen menstruum, which he has made with Snyders himself. The *Secret* then passes to the next stage, which consists of grinding a mixture of "four ounces of crude tartar, eight ounces of vulgar sulfur, and one pound of saltpeter," and then deflagrating it. This is in fact a variation on the explosive known today as "fulminating powder" or "yellow powder," although the modern composition normally uses a different proportion of these ingredients.[35] Like gunpowder, it employs potassium nitrate and sulfur, but it substitutes potassium carbonate (the salt of tartar) for charcoal. According to the *Secret*, this deflagration yields an "inflaming salt," which is subsequently melted

---

[33] The first German edition of the *Commentatio*, published in Frankfurt am Main in 1662, ends with an advertisement for Snyders's medicaments (122–23), which he says can be bought from a certain Jacob Hanßen in "Cölln," i.e., Köln. Adam Gotlob Berlich, the editor of the 1678 German edition of Snyders's *Commentatio* (*De medicina universali*) states that Snyders seems to have been "of the Reformed religion," and therefore could not have lived in Catholic Köln. Hence Berlich argues that "Cölln" must mean the Prussian city of Cölln am Spree. But this is a specious argument, since no evidence is given for Snyders's religious affiliation, and other Reformed chymists, such as Johann Moriän, definitely did live in Köln a few years earlier. For Moriän and his Cologne associates, see John Young, *Faith, Medical Alchemy, and Natural Philosophy: Johann Moriaen, Reformed Intelligencer, and the Hartlib Circle* (Aldershot: Ashgate, 1998), 3–25. For Berlich's report see Johann de Monte-Snyders, *Tractatus de medicina universali* (Frankfurt: Thomas Matthias Götzen, 1678), 15.

[34] See online *Deutsche Biographie*, under "Maximilian Heinrich, Herzog von Bayern" (http://www.deutsche-biographie.de/sfz59377.html, accessed March 4, 2016).

[35] Tenney L. Davis, *The Chemistry of Powder and Explosives* (New York: John Wiley and Sons, 1941–43), 30–31.

with finely laminated gold. The result, according to the *Secret*, will be a purple powder that should then be boiled in water to become "the solar lixivium." The text then advises to dissolve this material in spirit of salt (hydrochloric acid) with the following result:

> a very subtle skin of gold will be precipitated, which the author has called sulfur of gold. Wash it, edulcorate it, dry it, and the ferment is then ready: we have done this.

Once the menstruum and the ferment made from the subtle skin of gold have been produced, they must be combined, and this is the third step of the *Secret*'s process. Four ounces of the menstruum and one ounce of the ferment are placed in a long-necked matrass, and another vessel of the same type is inverted over its mouth. They are sealed together with lute and heated in a sand bath, with the result that they will undergo the stereotypical set of color changes that alchemists believed necessary for the production of the philosophers' stone—black, iridescent (the peacock's tail stage), green, white, yellow, and finally red. Interestingly, however, the *Secret* does not make exalted claims for the final product; it does not have the power of universal transmutation traditionally associated with the philosophers' stone. Instead, two ounces of the red product must be layered in a crucible with an ounce of gold in leaves. The two materials are then put in a reverberatory furnace for three days at high temperature, in order to fix the powder. At the end of the process, the author says he had a "particular" (*particulare*)—not a "universal" like the philosophers' stone. His final comments are worth quoting:

> This powder is quite heavy and red. This is injected into fused silver and it tinges it. I have injected two grains into a pound and a half of lead swirling about in a cupel and I had two ounces and one drachm of the purest, most precious gold.

Even if the red powder is not the philosophers' stone, this is still not a bad result, since the two grains of transmuting agent would only amount to about 130 milligrams in modern measurements! Perhaps this result is not as impressive as the one witnessed by Valvasor in 1666, but the *Secret*'s recipe is otherwise similar in that it produces a "particular," like the Slovenian nobleman's "concentrated extract" of gold, rather than the universal elixir.

Although there are obvious technical problems with the *Secret*'s account, the claim that something like its processes lie at the heart of the *Metamorphosis* is probably not unfounded. The reader may recall that chapter fifteen of the *Metamorphosis* contained a sequence in which Vulcan and Mars burned Venus and Phoebus to cinders with a strange firework composed of "an uninkindled fire, of a fiery Air, & of a Vegetable Salt."[36] These are very likely *Decknamen* for native sulfur, niter, and salt of tartar, precisely the ingredients of the *Secret*'s "inflaming salt." The absence of copper (Venus) in the *Secret*'s recipe could simply mean that it is a variant form of the operation described in the *Metamorphosis*. Venus does at any rate enter into the *Secret*'s

---

[36] Cushing, 20v.

recipe when the author adds an addendum explaining that the corrosive sublimate used to synthesize the butter of antimony menstruum should be made with quicksilver, salt, and vitriol. As he says, the sublimate is thereby "impregnated with the soul of copper."

Other interpreters of the *Metamorphosis* describe processes that are far more similar to that of the *Secret* than they are to Newton's notes on Snyders. The famous cavalier and chymist Kenelm Digby, for example, claimed in his posthumous *Chymical Secrets* to have met Snyders on July 22, 1664, and to have received a recipe from him. Digby's process proceeds with exactly the same proportions of niter, sulfur, and salt of tartar as the *Secret*'s, but Digby first fuses his gold with regulus of antimony, then adds the powder. He repeats this multiple times until all of the regulus is converted to slag, then runs through a succession of processes culminating in the extraction of a salt with distilled vinegar. It is this salt, finally, that is combined with butter of antimony, just as the *Secret*'s ferment was combined with the antimonial menstruum, and like the *Secret*'s process, Digby's leads to the *nigredo* or black stage resulting, on further heating, in a fixed powder. In short, the operations given in Digby's *Chymical Secrets* appear to be a variant form of the same basic scheme related by "The Secret of the Author of the *Metamorphosis of the Planets*." Was Newton entirely unaware of this interpretation, with its emphasis on the deflagrating yellow powder made of niter, sulfur, and salt of tartar? In fact he was not, as a fascinating little document in Newton's hand now kept at Columbia University reveals. The text, titled "Three Mysterious Fires," was a late product of collaboration between Newton and a circle of London alchemists. I will provide a full study of this document in chapter seventeen, but first it is imperative to show how Newton merged the chryso-poeia of Snyders with that of Philalethes and other authors to arrive at his own distinctive blueprint for the production of the philosophers' stone. The basic model for practice that Newton had established by the final quarter of the seventeenth century would serve as his exemplar for the remainder of his career as an alchemist. Let us therefore pass to the next chapter, in which we will consider another text where Newton's full and idiosyncratic use of Snyders emerges in all its details.

TWELVE

Attempts at a Unified Practice

We have now examined several of Newton's interpretations of Johann de Monte-Snyders alongside the standard interpretation of the German author's work. I have made the point already that Newton's commentaries to Snyders link more closely to the surviving portions of his laboratory records than do any of the other texts that we have examined so far. This does not mean, however, that Snyders was the only alchemist who supplied Newton with workable processes. To the contrary, he derived fully operational procedures from Philalethes and other alchemists as well and tried to piece together operations from multiple authors. To Newton, no single alchemist had revealed the entire set of processes necessary to acquire the philosophers' stone. On the alchemical principle that "one book opens another" (*liber librum aperit*), it was necessary to assemble the full set of stages from multiple authors in order to arrive at success. In order to get a sense of Newton's mature melding of Snyderian motifs with those drawn from other chymists such as Philalethes and Sendivogius, we need to examine at least one other manuscript from the King's College collection, Keynes MS 58. This fascinating text, though heavily Snyderian in orientation, brings in elements from Philalethes and Sendivogius as well in an attempt to arrive at foundational elements of a "master process" in Newton's overall chrysopoetic quest.

Keynes 58 is an unusual manuscript in certain respects. First, as B.J.T. Dobbs already recognized, the document provides an uncharacteristically clear window into Newton's transfer of elaborate, metaphorical texts such as Snyders's *Metamorphosis* into actual laboratory practice. This makes it a valuable resource indeed. But Keynes 58 is unusual in another respect for a text of Newton's maturity. It adopts the two-column method of exposition that we encountered in one of Newton's earliest alchemical expositions, the commentaries on Sendivogius and d'Espagnet in Keynes 19. Like that document, Keynes 58 provides text in the left column keyed to notes in the right one. The similarity in form might at first suggest that Newton wrote Keynes 58 at a time not far removed from that of Keynes 19, and indeed Dobbs placed the composition of Keynes 58 in the 1670s, tentatively proposing the middle of that decade.[1] But there are excellent reasons for doubting such an early date for this manuscript.

First, the text of Keynes 58 clearly exhibits a stage in which Newton had completely absorbed Snyders's *Metamorphosis of the Planets*, had combined its contents with those of the German alchemist's *Commentatio de pharmaco catholico*, and was even in the process of bringing in other alchemical authors to fill in the holes in Snyders's presentation of the great work. Hence the text is surely of a later date than Newton's four sets of notes on the *Metamorphosis* in the Cushing Library of Yale University. While this in itself does not rule out the second half of the 1670s, there are other clues that suggest a significantly later date. In his demonstrably early manuscripts such as Keynes 19 and Var. 259.7, Newton refers to Eirenaeus Philalethes simply as "Eyrenæus," "Eyreneus," or by other variants of the putative adept's first name.[2] The substitution of "Ey" for the "Ei" diphthong, if not the preference for the adept's first name, reflects the sources that Newton was using. The earliest Philalethan works available to Newton, such as *Secrets Reveal'd*, refer to the American chymist as "Eyræneus Philaletha" or even employ the bizarre spelling "Æyrenæus" for his first name.[3] Only in works that postdate Newton's "Lullian turn" at the hands of Dickinson and Mundanus do we find him consistently referring to Eirenaeus Philalethes merely as "Philaletha."[4] Hence the presence of "Philaletha" in Keynes 58 is a clue that the manuscript dates from the mid-1680s at the earliest, and could even be later.[5]

Disregarding an earlier and later section of Keynes 58 where Newton is apparently copying from other authors, as well as a mathematical figure related to Book 2 of the *Principia*, the work consists (between folios 2r and 4r) of his attempt to work out several key processes that depend largely on Snyders.[6] In reality, these folios contain three successive drafts of the same text, though the fact that the first draft is in English (2r–3r) and the two following ones in Latin (3r and 4r respectively) might at first obscure their relationship. Newton numbers parts of these processes and keys them to their

[1] Dobbs, *FNA*, 167. Dobbs's dating is based on the following observations—"MS 58 is in the handwriting of the 1670s and may perhaps be dated more precisely around the middle of the decade since some of the symbols for lead are crossed and some are uncrossed." But her observation about the hand is entirely impressionistic, and by my count the manuscript contains twenty-six occurrences of the crossed Saturn symbol to only two of the uncrossed version. This makes the uncrossed Saturn symbols look like simple slips of the pen rather than indicating a medial period where Newton was moving from one form to the other.

[2] Var. 259.7.2r; Keynes 19, ff. 2r–4v.

[3] Philalethes, *SR*, 120, where the seemingly illiterate authorial form "Æyrenæus" is used.

[4] The following manuscripts all employ the form "Philaletha" and also refer to "Mundanus" (meaning that they are dependent on Edmund Dickinson's *Epistola ad Theodorum Mundanum* of 1686): Keynes 21, 38, 41, 48, 54, 56, and 57; Dibner 1032B and 1070A; and Babson 421. Another manuscript, Dibner 1041B, mentions "Philaletha" but not Mundanus, yet its heavily Lullian character testifies to the influence of Dickinson's *Epistola*. Newton's adoption of the form "Philaletha" in his maturity probably reflects his growing use of continental sources, which often employ that version of the adept's name instead of "Philalethes."

[5] Keynes 58, 3r and 4r.

[6] The illustration pertains to the efflux problem in Book 2, Section 7 of the first edition of the *Principia*. I owe this information to the kind help of George E. Smith. See his article "Fluid Resistance: Why did Newton Change His Mind?" in *The Foundations of Newtonian Scholarship*, ed. Richard H. Dalitz and Michael Nauenberg (Singapore: World Scientific Publishing, 2000), 105–36. Unfortunately, Newton's habit of reusing old paper renders the presence of this diagram in Keynes 58 of little use for dating purposes.

notes with Latin letters; thus in order to avoid confusion, I will call them draft α (2r–3r), draft β (3r), and draft γ (4r). Fortunately, draft γ gives headings to the different processes that occur in all three drafts. They are: 1. "Dry Water" (*Aqua Sicca*); 2. "Saturnia;" 3. Eagle and Sceptre of Jove (Aquila & sceptrū ♃is Iovis); and 4. Jupiter. No text is given after draft γ's second heading "Saturnia," however, suggesting that Newton thought better of allocating this product its own entry.

It is important at the outset to understand what Newton is doing in Keynes 58. The operations that he describes are prescriptive; they are not attempts to describe actual experiments that he has carried out in the laboratory. They belong to the genre that George Starkey called "conjectural processes," the blueprints for practice that remained to be tested at the bench. For that very reason, they are exceptionally interesting for what they can teach us about the underlying modus operandi represented in Newton's experimental notebooks, where he actually put such procedures to the test. But before we can make that examination, we must first consider Keynes 58 from a literary point of view, extracting Newton's often obscure meaning from the text itself. In order to do that, I will focus on draft α and its accompanying notes, occasionally referring to the successive drafts insofar as they throw light on the nature of the processes described. In order to follow Newton's complicated procedures it may be helpful for the reader to have the edited version of Keynes 58 open here, in the diplomatic version found in the *Chymistry of Isaac Newton* project.[7]

How to Make Dry Water

As we know from draft γ, the first set of operations in the three renditions is intended to produce a "dry water." Draft α begins with a paragraph introduced by a number one with the following claim, which, thanks to its idiosyncratic symbols, may at first seem utterly incomprehensible. The passage and its symbols have in fact resisted decipherment up to the present moment:

> 1. ℞ Io⊖ ᵃ + Co⊖ <*illeg.*> + Lv, ana. dissolve & digest them in yᵉ blood of yᵉ green Lyon till they be dry. Then imbibe them again wᵗʰ ye ᵇ blood mixed wᵗʰ yᵉ double sᵖᵗ. & digest till various colours appear & yᵉ clouds vanish. <*illeg.*>

The command to take (℞ stands for "Recipe" or "Take" in the Latin imperative) "Io⊖ ᵃ + Co⊖" can be decoded by referring to note "a" (referenced in superscript after "Io⊖"), which is found in the manuscript on the right column of the page. The note says "Io + Co are yᵉ two breasts of Venus male & female ♂ & ♀. Ex ♂ & ♀ fit ⊕ effectualis." Having already examined Snyders's *Metamorphosis of the Planets*, we are in a surprisingly good position to make sense of this. The reference to the two breasts of Venus tells us that Newton is thinking of chapter twenty-one of the *Metamorphosis*, where Cupid drinks

[7] *CIN* at www.chymistry.org.

from the two breasts of his mother Venus and then transforms into a Swan. In that shape he manages to seduce or rape Diana, who subsequently gives birth to a radiant, solar child. But the important thing for us is Newton's clue that Io and Co refer to "♂ & ♀" and that from them is made an effectual vitriol (Ex ♂ & ♀ fit ⊕ effectualis). On the basis of these hints I propose that "Io" and "Co" are simply abbreviations of "iron ore" and "copper ore." Newton often abbreviates "ore" as "o," even incorporating the letter into his proprietary system of alchemical symbols, so that ♂ᵒ means iron ore and ♀ₒ means copper ore. The appearance of the standard symbols for iron and copper, "♂ & ♀," in the accompanying note suggests that he is doing much the same thing here. If that is so, then the additional symbol for salt, ⊖, merely means that Newton is saying to take a salt extracted from iron ore and a salt extracted from copper ore, rather than one from the refined metals. This accords very well indeed with Newton's general preference for ores over refined metals, as we have seen in the foregoing chapters of this book. As for the claim that an "effectual vitriol" can be made from these two salts, this could easily refer to a mixture of the two salts themselves, which could well be described in the language of seventeenth-century chymistry as vitriols.

What then are we to make of the mysterious letters "Lv"? Since Newton has just used an acronym for the two ores of iron and copper, we can expect an acronym here too. His adjacent note, unfortunately, is not hugely helpful here: "Lv <illegible deletion> Cupid or Venus & Cupid together." Although Newton's On yᵉ Metamorphosis of yᵉ Planets in the Cushing manuscript give a host of synonyms for Venus and Cupid, one term stands out as a possible expansion of "Lv," namely, "Leo viridis," which is Latin for "Green Lion." Newton's laboratory notebooks suggest strongly that for him (as opposed to some of his sources), the Green Lion was a copper compound, though not copper vitriol. What precise copper compound Newton had in mind here is not certain, although we will have opportunity to return to this question later.

Finally, there is the equally cryptic expression "yᵉ blood of yᵉ green Lyon." It might at first face seem unlikely that Newton would initially tell us to take the Green Lion and then to dissolve it, along with the effectual vitriol, in its own blood. In fact, however, the entire process outlined so far involves reiterated use of the same ingredients. The parallel text given by draft γ is particularly helpful in making this clear, so I will quote it here in English translation:

> Take equal amounts of the double vitriol extracted by the first menstruum <apparently the term "equal amounts" refers to the two salts making up the double vitriol>. Dissolve it in the same menstruum. Add green ♀ or its salt together with Cupid, and the ~~putrefaction~~ operation will begin. When the matter appears dry, imbibe it again with the same menstruum mixed with the double, solar spirit, and when the colors have disappeared and the vapors gone away, add once and again the salt of Saturn extracted by the same menstruum.[8]

---

[8] Keynes 58, 4r: ℞ ⊕ duplex menstruo primo extractum ana. Dissolve in eodem menstruo. Adde ♀ viridem vel salem ejus una cum cupidine, & incipiet ~~putrefactio~~ operatio. Quando materia apparet sicca imbibe

Draft γ tells us unequivocally that the initial salts of iron and copper are "extracted," or made by subjecting the ores to the corrosive action of a menstruum. The dried salts are then dissolved again in the same menstruum along with "green ♀ or its salt together with Cupid." This dissolution of the two salts or "double vitriol" with "green copper" or its salt and Cupid corresponds exactly to draft α's injunction to digest the salts of iron ore and copper ore and the Green Lion "in ye blood of ye green Lyon till they be dry." A simple process of textual substitution therefore allows us to see that the "green ♀ or its salt" can only be the Green Lion, and the blood of the Green Lion cannot be other than the same menstruum that was initially used to make the salts of the ores. What was this mysterious menstruum in plain terms? An examination of Newton's experimental notebooks will show that it was very likely aqua regia (made from aqua fortis with the addition of sal ammoniac) in which crude antimony had been dissolved.[9] As we will shortly see, it is also highly likely that throughout Keynes 58 Newton envisaged that the menstruum should also contain his copper compound "Green Lion," which would nicely account for the expression "blood of ye green Lyon."

A final point of uncertainty lies in draft α's expression "ye double s^{pt}," which becomes "the double, solar spirit" in draft γ. Fortunately, we can turn to draft β here, which gives an extremely helpful note to the term "double spirit." There Newton refers to "Snyders in chapter —— where he calls the double spirit the milk of natural Venus."[10] The milk of Venus refers back to *Metamorphosis* chapter twenty-one, where Cupid drinks from the twin breasts of Venus. And as we already saw, those two breasts allude to the salts of iron ore and copper ore that were combined to make up the effectual vitriol. As before, we can therefore substitute one term for another. The double spirit here is simply the effectual vitriol dissolved, once again, in the blood of the Green Lion, perhaps (though not necessarily) in distilled form. As for the claim in draft γ that the double spirit is "solar," this no doubt refers again to *Metamorphosis* chapter twenty-one, where Diana gives birth to a solar child after having been ravished by Cupid, whose transformation into a swan was effected by the milk, or "double spirit," of Venus. Since the double spirit led to the birth of a solar child, it must itself contain a solar seed.

After all this, one might expect to have arrived at Newton's "dry water," the aim of the whole exercise, but we are not quite there yet. Draft α follows the paragraph beginning "℞ IoΘ [a] + CoΘ" with a second section numbered "2." that explains how we should proceed to the dry water (*aqua sicca*):

> 2. Then for aqua sicca add [c] ♄ wth its menstrue once & again, & digest till it be a [d] black pouder. Thus ^by subliming this have you ye two saturns or Doves, ye Aqua sicca.[11]

iterum cum eodem menstruo, admisto etiam spiritu duplici solari, et ubi colores evanuerunt & vapores iterum deficiunt adde semel atq iterum salem saturni eodem menstruo extractum.

[9] For the pervasiveness of this substance in Newton's chymical experimentation, see our chapter fourteen.

[10] Keynes 58, 3r: "Sniders cap :ubi sp^{tû} duplicem vocat lac ♀is naturalis."

[11] Keynes 58, 2r.

Compared to the foregoing material, this seems relatively straightforward. In order to make the dry water, we must add lead that has been dissolved in "its menstrue" to the solution already prepared by digestion of the effectual vitriol, Green Lion, and Lion's Blood. Then everything must be digested together until a black powder forms, and when this is dry, it is sublimed. The product will be the dry water, which Newton now tells us is identical to "yᵉ two saturns or Doves." Remarkably, Newton still maintains that the Philalethan doves of Diana are salts or sublimation products that involve lead, building on the interpretation that he had arrived at years before in his Sendivogian commentary of Keynes 19! Draft γ adds another Philalethan note, asserting that the black powder is the product that the American adept described in the "porta prima," a section of his *Ripley Reviv'd*.[12] Finally, β and γ add that the lead must be dropped in in two stages, and that for the sublimation bole armeniack (a powdered clay used in high temperature sublimations and distillations) should be added, which reflects Newton's own experimental practice in his laboratory notebooks.[13] After that point, however, draft β sharply diverges by conflating the preparation of the dry water with that of the next product in the other two drafts, Jove's eagle. It will therefore be best to consider that variation in combination with draft α and γ's instructions for making the eagle of Jove.

Instructions for the Eagle and Scepter of Jove

As we proceed into Keynes 58, one can only be surprised at the remarkable degree of precision that Newton hoped to extract from Snyders's *Metamorphosis*. Here we will encounter Jove's eagle, his scepter, his "bolt" or thunderbolt, and even his hand, all of which refer to specific materials in Newton's interpretation. Like most of the chymical substances that Newton hoped to make in his laboratory, the dry water was a precursor to other factitious materials; in this case the just-mentioned products. In draft α, section two, Newton inserts a parenthetical operation in brackets, telling us to begin an operation with the black powder formed by digestion of dissolved lead with the effectual vitriol, Green Lion, and Blood of the Green Lion. We should take a small amount of this black powder and divide it into two equal portions. To one of these portions, one must add an equal amount of the ore of the Eagle of Jove. Fortunately, this material is easy to identify since Snyders explicitly said in the *Metamorphosis* that the eagle's name is "Bismutum"; thus we are to use the ore of bismuth for the process with the first half of the black powder.[14] As for the second portion of the black powder, Newton says to add in "some of yᵉ eagles & ♃'s mixed," meaning the ores of bismuth and

---

[12] Keynes 58, 4r: "Et materiâ tandem in pulverum nigrum & aridum conversa (quod <*illeg.*> est celebris illa calcinatio, porta prima) per sublimationem a triplo boli Armeni ad misti habebis aquam siccam."

[13] A related set of experiments using tobacco-pipe clay when subliming lead ore dissolved in liquor of antimony and other menstrua is found in CU Add. 3973 at 43v. These experiments come well after the last date in the manuscript, "Feb. 1695/6" on 30v.

[14] Cushing, 12r.

of tin.[15] These two halves must then be digested separately, whereupon each will again turn black.

At this point, Newton says to take "half" ("ye first" is struck through) of the newly produced black material and to digest this in turn with "♂ & saturnia," very possibly iron dissolved in Newton's perennial liquor of antimony. To this one is supposed to add "extracted calx of ye eagle," which signifies a salt dissolved out of the calx of bismuth, probably by means of liquor of antimony again. By this means, Newton says that the calx will be "mercurialized," and will distill over on heating. At this point Newton returns to the other sample, presumably the black material made by digesting the initial black powder with the combined ores of bismuth and tin. He tentatively identifies this with "Jove's scepter" and says that it should now be subjected to a further fermentation with Jove's bolt, again along with tin and bismuth ("ferment again wth his bolt and add ♃ & his eagle &c."). What is Jove's bolt? In an immediately preceding canceled paragraph, Newton identifies it with "ye last ~~black~~ calx fermented anew w^th ♂ & ye water." This black calx is probably the very same substance as the black material dissolved in "♂ & saturnia" that we discussed at the beginning of the present paragraph, and as in that case, Newton is very likely using ♂ & saturnia to mean iron dissolved in liquor of antimony.

So far, all of this complicated set of procedures could be viewed as deriving from Snyders's *Metamorphosis*, but at the end of section two in draft α Newton reveals that he has much bigger goals than a mere interpretation of this one text. In a word, the dry water, along with Jove's various appurtenances, all feed into one of the grand designs of Newton's chrysopoetic project—the making of Mercury's caduceus. I will treat this subject at greater length in another chapter, but we encounter an explicit if passing reference to it here:

> W^th this rod & ye two serpents (double sp^t, or rather ⊕ of ♂ & ♀ extracted wt^h ye juice of saturnia) ferment ☿ & cleans it.

This immensely revealing passage tells us that Newton believes the foregoing processes will yield the "rod" or staff of the hermetic caduceus along with the two snakes intertwined around it. The serpents are once again the "effectual vitriol" extracted from the ores of iron and copper by means of Newton's liquor of antimony. But the rod itself offers some ambiguity. It is possible that at the time of writing Keynes 58 the rod was identical in Newton's mind to the scepter of Jove, whose production involved the use of tin and bismuth.[16] Yet the phrasing of the passage opens the possibility that the rod was the salt of lead before having been fermented with the two serpents to produce the black powder. As we will see, the latter interpretation accords better with Newton's late manuscript *Praxis*, which we will examine in its place. Either way, the caduceus as a whole in Keynes 58 appears to be a combination of

---

[15] Keynes 58, 2r.

[16] The identification between the hermetic caduceus and the scepter of Jove is even more explicitly made in Dibner 1070A, at 8r: "Consimili sublimationum operatione Iupiter cum Aquila resolvitur in ☿^um per ☿^um vi sceptri sui quod et Mercurij virga est." A variant of the same passage is found on 20v.

the material produced in paragraphs one and two with Jove's scepter. The magical staff of Mercury is supposed to have the power of fermenting quicksilver, one of the more advanced stages of Newton's chrysopoetic endeavor. We can see, then, that these operations held tremendous significance for his chymistry. They were supposed to provide the keys that would unlock the deepest secrets of metals, and probably of matter in general.

Given the major significance that the "dry water" of draft α's paragraphs one and two held for Newton, it is no surprise that he would want to refine his procedures further. Additionally, it is quite clear that he was uncertain about many points of detail in his decoding of Snyders, even though he obviously believed himself to be generally on the right track. Thus we find Newton now adding a new paragraph, numbered "3" in draft α, where he lays out the proper technique for "mercurializing mercuriall bodies." Apparently what he has in mind is rendering metallic materials fluid and distillable. The particular substances that he has in mind are "ye two eagles <illeg.> <illeg.> Venus & ♃." An explanatory note clarifies the new plurality of the eagles: "This <illeg.> ♃^{vial} ♄ ∧sublimed is one eagle as ♄ alone is another." The first eagle here, sublimed Jovial Saturn, probably refers to a calx or mineral of bismuth that has been sublimed with the lead-based black powder of draft α, paragraph two. Newton then gives instructions for preparing these materials, which involve fermenting the already prepared dry water "wth a grain of ye old putrefied matter," as though he were leavening dough with yeast. The old, putrefied matter refers to the black powder that was described at the beginning of Newton's section two, where lead was digested with the effectual vitriol, Green Lion, and Lion's Blood. If one adds the calces of copper or bismuth to this, they should "mercurialize" and become distillable, according to Newton's present operation. The result will be that these materials are graduated or elaborated, so that they become "♀ ye daughter of Saturn & Iuno ye wife of ♃'s eagle." As one can see from the deleted reference to Juno, Newton was still actively working out the chymical referents of Snyders's allegorical figures, and had not yet arrived at certainty.

Newton's active decipherment is revealed even more clearly by a long explanatory note that he adds to draft α's paragraph three. He was uncertain, in the case of Jove's eagle, whether it should be mercurialized with "ye eagles salt, or ^{a} minera." In other words, should one add to the fermenting dry water (again, the product of paragraph two in draft α) a bit of the salt of bismuth, or rather some bismuth ore? His resolution to this problem is revealing, as it shows how Newton supplemented his interpretation of Snyders with mythological material that had already received an alchemical reworking at the hands of Newton's sources. Given the richness of this paragraph, it is worth quoting it in full:

> a And rather of ye minera because ♄ (not ye Lyons blood) ate a stone instead of ♃ & spewed it out again. Perhaps ♄ fermented will mercurialize ye stone wthout more ado because he spewed it up again after he had devoured it. Quære 1. whether ♄ must eat ye stone for ♃ <illeg.> so soon as ye sp^t has dissolved ♄ & is satiated but not yet grown to a dry black calx (as is

most <*illeg.*> probable ^becaus otherwise ♄ will distill over) or afterward yᵉ sublimation of ♄? Quære 3 whether this stone be crude or its calx. Both may be tryed to see wᶜʰ ☿ is best. Quære 2 whether yᵉ aqua sicca will not be^come more fixt by joyning wᵗʰ some bodies as ♂ or ♀ & so let yᵉ ☿ come over alone.[17]

Newton's way of resolving the question of whether to use salt or rather ore of bismuth probably betrays the influence of Michael Maier's *Atalanta fugiens*. The twelfth emblem in this learned attempt to reduce the mythology of the ancients to alchemical allegory vividly pictures Saturn in midair spewing out a stone bigger than his own head (figure 12.1).[18] Maier wove an elaborate allegory out of the myth that the titan Saturn ate his children, and that Jove tricked him by substituting a stone for himself. Saturn found the stone to be indigestible, and so vomited it up again. For Maier, Saturn represents the *nigredo* stage of the alchemical work, when the matter is dark like lead; after it has undergone successful operations it turns white and is represented by the stone that Saturn regurgitated. The same myth was subjected to alchemical interpretation by Philalethes as well, in the *Marrow of Alchemy*. Relying on a surprising mastery of obscure points in classical mythology, Starkey (under his Philalethes pseudonym) had even given a proper name to the stone that Saturn ate, "Abadir," while also rechristening Saturn as "old Aberipe" and Jupiter as "most noble Abrettane."[19]

Newton was therefore the heir of an established interpretive school when it came to the story of Jupiter's ingenious escape from Saturn. He resolves the problem of choosing between the salt and ore of bismuth by reference to the myth. But what does it mean when he bases his conclusion on the fact that "♄ (not yᵉ Lyons blood)" ate the stone and regurgitated it? How does this lead him to choose the ore over the salt? The answer must be that the salt of bismuth would have been produced by dissolving either the metalloid or its ore in a menstruum, and as we have seen, the blood of the Green Lion is precisely such a dissolvent. The fact that Jove's stone was not eaten by the Green Lion's blood means that the dry ore or metalloid was not dissolved in a menstruum. Hence no salt could be produced, leaving the other alternative, the ore, as the correct choice. Here we see Newton explicitly basing the details of an experimental investigation on the words of a myth, albeit one transmitted to him via alchemical sources. He is manifestly not merely fitting the myth to a predetermined experimental practice, he is deriving at least some parts of the practice directly from the myth.

But this does not mean that Newton's laboratory practice was a simple transfer of allegory into laboratory practice, as the three numbered queries that follow reveal. All three questions are to be resolved by experimentation, though it is not clear that Newton ever progressed far enough through the elaborate set of procedures that we have already described to test them.

[17] Keynes 58, 2v.

[18] Michael Maier, *Atalanta fugiens* (Oppenheim: Johann theodor de Bry, 1618), 57.

[19] Philalethes, *Marrow*, Part 1, Book 2, pp. 39–40. Starkey did not invent these odd names. They are all found, for example, in the widely diffused classical dictionary of Carolus Stephanus, *Dictionarium historicum, geographicum, poeticum* (s.l.: Jacob Stoer, 1609) under the headings "Abadir," "Aberides," and "Abretanus."

Figure 12.1. Saturn vomiting forth Jove in the form of a stone. From Michael Maier, *Atalanta fugiens* (Oppenheim: Johann Theodor de Bry, 1618).

The first asks whether the stone (bismuth ore) should be fed to Saturn (the dry water) before or after the dry water has been sublimed from the black powder out of which it is made. Newton equates the regurgitation of the stone with the mercurialization of the bismuth ore, which suggests that he is thinking that the ore, once liquefied and made distillable like quicksilver, should be separated from the dry water by distillation. But this raises the problem that the dry water is itself supposed to be volatile; hence, one must determine whether it is better to add the ore before the dry water has reached its maximum volatility (before being distilled from the black powder) or after the dry water has been distilled. Newton tentatively opts for the first choice. The second query (which oddly comes after the third) is related to the first. Again, Newton is clearly concerned with the problem of getting Saturn to regurgitate the stone, but he suggests a laboratory-based operation, in principle testable, to achieve that end. He now has the idea that by conjoining the dry water to "some bodies" (that is, metals) like iron or copper, he might induce the dry water to combine with them and thereby let go of the mercurialized ore (called "☿" here), allowing it to distill. Finally, the third query asks whether one should use the ore in crude form or calcine it first. The answer is again to be found in the laboratory—"Both may be tryed to see w^{ch} ☿ is best."

The fourth and final numbered section of Keynes 58, draft α, returns to the all-important topic of fabricating Jove's scepter, which Newton treated in passing in section two. Some of this rephrases the earlier material, but there are revealing alterations as well:

> 4 Ioves scepter probably is Salt of his eagle extracted out of y^e minera wth y^e Lyons blood. His bolt is y^e new ferment made of Aqua sicca ~~of Iove~~ $_\wedge$sale ~~aquilalis~~ ^{Iovialis} impregnata & y^e two serpents This he taks in his hand whether y^e ~~or~~ melted metal or ore or extracted salt.[20]

Unlike section two of draft α, where Newton tentatively suggested the black, Saturnine powder as a starting point for making Jove's scepter, he now thinks it more probable that the scepter derives from a salt that has been extracted from the ore of bismuth by means of the Lion's Blood. His interpretation of Jove's thunderbolt, on the other hand, is closer to the version that he proposed but deleted in section two. There he suggested fermenting a portion of the Saturnine, black powder with "♂ & ye water" (probably iron dissolved in Newton's liquor of antimony) and then fermenting ore of tin or Jove's scepter along with metallic tin and the eagle of Jove. In this earlier, deleted version, Jove's scepter was an already fabricated material used to make Jove's bolt. The new version also begins with the black powder but does not rely on the use of Jove's scepter. Instead, it says to impregnate the black powder with "Jovial salt," probably meaning a salt made of tin dissolved in liquor of antimony, along with the two serpents, which we know now to be the iron salt and copper salt(s) making up the effectual vitriol of section one. When Newton adds that Jove takes this bolt in his hand "whether y^e ~~or~~ melted metal or ore or extracted salt," this must mean that the thunderbolt can be combined with tin in any of these three forms—as a molten metal, while still in its unrefined ore, or as a salt.

Once Jove's scepter and thunderbolt have been synthesized, one can move on to the next succession of operations. Newton's goal here is to produce the final product listed in draft γ, "Jupiter," or as draft β has it, "prepared Jove."[21] First one must combine the scepter and bolt by adding either an amalgam of tin or its ore or salt, along with the scepter, to the bolt, which is already in a state of fermentation. Then one adds more tin amalgam to the foregoing mixture, along with Jove's eagle. Despite the multiple fermentations that have already taken place, Newton ponders whether one more may be necessary, and then adds the following lines, as though to reassure himself that he has understood his sources correctly:

> Perhaps a new ferment must beg<*illeg.*>un as at first <*illeg.*> ~~(though it may be not wthout ♄, experience will show)~~ Then ye two Ioviall salts

---

[20] Keynes 58, 2v. Presumably the supralinear "Iovialis" is a mistake by Newton for "Ioviali," which would then modify "sale" (giving "Jovial salt" in English).

[21] Keynes 58, 4v and 3r.

<illeg.> (male & female *<illeg.>* his hand & scepter) Then ♃ amalgamd wth his eagle. This is ye most natural way & will do if any will, unless ♄ be any where to be added.

This passage provides the usual interpretive difficulties, but I believe it is best read as a summary of the foregoing procedures. We already saw that Newton made an association with Jove's bolt and his hand when he said that the god would hold the bolt regardless of its molten, mineral, or saline state. It is likely, then, that we can substitute "bolt" for "hand" in the present passage. If so, Newton is merely saying that he has put the scepter in the hand of Jove, that is, combined the "two Jovial salts," and then added "♃ amalgamd wth his eagle," namely, the tin amalgam and bismuth that he described several lines above. It is reassuring to know that this is the "most natural way & will do if any will."

Conclusion

As I suggested above, much of Keynes 58 consists of Newton's transformation of his ideas about Snyders and other alchemists into the form of conjectural processes. It is highly unlikely that he actually performed many of the complex operations making up a series of well over thirty stages in Keynes 58 (see figure 2.3). Indeed, Newton himself confirms this fact indirectly in a short memorandum found directly after the final version of the text (draft γ) on folio 5v. The most remarkable feature of this short agenda is the plain language used for describing the initial ingredients that are necessary for carrying out the first and second paragraphs of Newton's instructions for a "Dry Water." The Snyderian and Philalethan *Decknamen* drop out for the most part, to be replaced by the language of "vulgar chymistry." Although this fact has been noticed by previous historians, it has escaped scholarly attention that Newton is restricting himself to the preparation of what might be called "precursor" materials, which enter into the three versions of his process, drafts α, β, and γ as fully formed substances. What Newton is proposing here is not to carry out the multiple operations that populate the three successive drafts supplied by α, β, and γ; rather, he is searching out the initial ingredients whose preparation is required before he can even begin.

The fact that Newton had not yet attempted even the preliminary procedures for making "Dry Water" reveals strikingly that the subsequent stages, "Saturnia," "Jove's Bolt," and "Jupiter," must also have been conjectural processes, since they represent later steps in the same series and are dependent on the completion of the earlier operations. I reproduce Keynes 58's "to do" list here, so that we may see how it confirms the explanation of the text given in the present chapter:

To be tried
1. Extract ♀ from green Lyon w^th ⅍ diluted & make ye menstrue of this
2. Try if y^t menstrū will dissolve Lead ore.
3. Get y^e ⊕^s & try ye ferment.

The list tells us that Newton intends to test ("try") his interpretation of the Dry Water process by carrying out three sequential experiments. First he plans to extract copper (♀) from "green Lyon" by means of diluted aqua fortis (𝆑). The "copper" here is almost certainly a "vitriol," that is, a crystalline salt made by dissolving a metal or metallic compound in an acid and then evaporating the solution. Since the "green Lyon" is the source of this vitriol, it must itself be a copper compound, and given Newton's general prediction for unrefined metals, there is good reason to suspect a native mineral of copper. From this extracted product he intends to make a "menstrue," that is, a corrosive for dissolving other materials. Although Newton does not say so, it is entirely possible that the aqua fortis mentioned here would have contained both sal ammoniac, to "sharpen" it and turn it into a form of aqua regia, and stibnite. The use of these three substances in making menstrua pervades his laboratory notebooks to such a degree that it would not have been necessary for him to spell out all three ingredients, and one can find examples of such truncated description throughout his experimental records.[22] But what of the linkage between the green lion mentioned here and the first paragraph of draft α? To make matters short, the "menstrue" described at item number one is identical to the "blood of the green Lyon" found in paragraph one of Newton's Dry Water instructions. This is the solvent in which "Io⊖ a + Co⊖ <illeg.> + Lv" must be dissolved in order to initiate the entire subsequent sequence of operations.

The second stage in Newton's agenda instructs him to test whether the freshly prepared menstruum will dissolve lead ore. As in the case of the menstruum made from the green lion, this operation must be carried out in order to make a required ingredient on which a host of later operations depend. The use of this product is spelled out in the Dry Water instructions at the beginning of section two (again in draft α). There Newton says, "Then for aqua sicca add ♄ wth its menstrue once & again." In order to combine the lead with the end product of section one, the metal must be dissolved in its own corrosive, whose fabrication is presupposed in the Dry Water instructions. What Newton is telling us in his "to do" list is that he is going to carry out experiments in order to determine what precisely that lead-dissolving menstruum may be. Finally, in the item labeled as number three, Newton is reminding himself to acquire "the vitriols" (y^e ⊕^{s}) and to test "the ferment." These vitriols can only refer to the salts of iron ore and copper ore (Io⊖ + Co⊖) that stand at the very head of draft α, the combination of which Newton called an "effectual vitriol." Along with the green lion and its blood, these two salts or vitriols make up the list of ingredients required for paragraph one of draft α to be carried out. The instructions to try the ferment, on

---

[22] See for example CU Add. 3973, 5r, where Newton speaks of using an old sample of "☿ once acted on by 𝆑." In the next paragraph he spells out in more detail how he made this product: "The last summer I had dissolved ☿ in about 4 times as much 𝆑 wth ✳." The use of sal ammoniac (✳) is only mentioned in the second paragraph but is implicit in the first. Further evidence for the use of sal ammoniac and stibnite here is also found in an important letter from Newton's friend Fatio de Duillier in August of 1693, which I will consider in a later chapter.

the other hand, cannot be pinned down with any certainty, since ferments abound throughout draft α and its successors alike.

Again, one cannot overstress the fact that Newton's brief "to do" list in Keynes 58 provides instructions for making and testing the preliminary ingredients on which the entire chain of following operations depends. This short agenda reveals that despite his astonishingly detailed lists of interlinked procedures decked out in the colorful language of alchemy, Newton by his own metric had only begun to penetrate the secrets of chrysopoeia in his maturity. As we shall see in due course, his laboratory notebooks also share this characteristic. They consist for the most part of short sequences of operations intended to test particular features of his interpretations of chymical writers, and on the rare occasions where Newton does exultantly announce a discovery, the finding usually consists of the successful decipherment of a particular alchemical reagent rather than a final product. Such products as "sophic sal ammoniac," "our Venus," and "Diana," all of which appear in CU Add. 3973 and 3975, represent preliminary tools for the making of the philosophers' stone, not advanced stages in its preparation. Even at the end of his three-decade preoccupation with alchemy, Newton had barely begun his apprenticeship in the shop of the adepts.

But he had made progress. The documents that we discussed in the previous chapters, along with the ones examined here, reveal a marked evolution in Newton's understanding of the adepts. His analysis of Sendivogius and Philalethes in Keynes 19 relied on the purification of lead by antimony to produce a regulus that could be reduced to the sophic mercury by a process involving multiple sublimations—the seven or nine eagles of the American adept. There is nothing about copper in Keynes 19, and indeed, the first, hesitant emphasis on the red metal appears in Newton's commentary to the *Marrow of Alchemy* found in Mellon 79, even though the major focus of that manuscript is lead. Both copper and lead emerge as research interests in Newton's early florilegium Keynes 35, where he was grappling with the mysterious *plumbum æris* of Johann Grasseus, but at this stage he was still unsure of the practical relationship between the two metals. Newton then turned to Johann de Monte-Snyders in order to provide further clarification. His *Key to Snyders* reveals what the German alchemist taught him: both metals were necessary for the alchemical magnum opus, along with antimony and quicksilver. Newton then employed this growing repertoire of ingredients in his interpretation of the Snyderian *Metamorphosis of the Planets*, where he also found such additional materials as iron, tin, and bismuth. But here we can see a further fact of paramount importance. The sublimations and distillations of antimony and mercury emphasized in the Philalethan text *Secrets Reveal'd* now merged with the ingredients emphasized by Snyders to develop into what one might call the "standard practice" of Newton's experimental laboratory notebooks, with their incessant emphasis on sublimation, dissolution, and resublimation. Finally, in Keynes 58 Newton put this information into the form of a linear series of operations. Beginning with copper and iron "vitriols" probably made by crystallizing a solution of the metals' ores dissolved by aqua regia containing antimony, Keynes 58 then instructs that a green salt of copper be added to this, probably for the sake of inducing a fermentation. After this, lead and its menstruum are added in order to arrive at a dry water that is equivalent, in Newton's interpretation (at least at this point in time), to the Philalethan doves of Diana. Further operations follow in which tin and bismuth, or perhaps their ores, are employed on the dry water with the goal of producing the "rod & ye two serpents," in other words, the caduceus of Hermes. This exalted goal recurs in many of Newton's late alchemical manuscripts, and would even appear in truncated form in his extraordinary exchange with Fatio de Duillier in late summer 1693. But the summary that I have just provided was not the end of Newton's alchemical evolution. At some point in the second half of the 1680s, Newton acquired a compelling interest in the alchemical works attributed to the late medieval Mallorcan philosopher Ramon Lull. His newfound interest in Lull would lead Newton to the belief that the magnum opus consisted of thirty or more subordinate *Opera* (works), which he would string together to form the links in a fantastic chain of operations. We will examine this development in the next chapter.

THIRTEEN

The Fortunes of Raymundus

NEWTON'S LATE FLORILEGIA

Introduction: Dickinson, Mundanus, and Newton's Lullian Turn

When we follow the development of Newton's alchemy into the period of its full maturity, from 1686 onward, the interpretive difficulties that we have encountered so far begin to seem relatively trivial. On the one hand, Newton's own understanding of alchemy had obviously changed. The Sendivogian simplicity of nature that drove his earliest, binary understanding of the alchemical work had given way to an almost unbelievably complicated series of operations required to produce and use the philosophers' stone. One manuscript from his mature period, Yale University Mellon 78, starts by listing thirty numbered operations called *Opera* (the Latin plural of *Opus* or "work") and then passes to explanations of them that end by introducing even more stages.[1] The first *Opus* begins in a way that seems strange, at least at first face. Newton calls it "The first manual preparation of the grapes" (*Uvarum praeparatio prima manuaalis <sic>*).[2] Since there has been no talk of wine making so far in this book, the reader may be justifiably surprised. But in fact the title tips us off to the source—by the second half of the 1680s, Newton had become fascinated with the work of pseudo-Ramon Lull, the late medieval author(s) whose extensive corpus includes important texts on the distillation of spirit of wine (ethanol).[3] Needless to say, the grapes are a *Deckname* for Newton, whose interpretation of pseudo-Lull's processes would have been unrecognizable to their original author. But the fact remains that Newton's new absorption of the Lullian alchemical corpus into

---

[1] See Mellon MS 78, 5v, where Newton has written two book titles upside down. The later of these is "Centrum naturæ concentratum. Or y⁰ salt of Nature regenerated. ffor y⁰ most part improperly called y⁰ Phērs stone. Written in Arabic by Alipili a Mauritanian born of Asiætick parents. Published in Low Dutch in 1694 & now done in English 1696. Price bound 1ˢ. Printed for John Harris at the Harrow in little Brittain." Hence 1696 is the terminus post quem for this section of the manuscript, as already noted by David Castillejo, *The Expanding Force in Newton's Cosmos* (Madrid: Ediciones de arte y bibliofilia, 1981), 20.

[2] Mellon MS 78, 1r.

[3] For pseudo-Lull, see Michela Pereira, *The Alchemical Corpus Attributed to Raymond Lull* (London: Warburg Institute, University of London, 1989). Pereira's articles on pseudo-Lull are too many to cite in their entirety here. Essential works include Pereira, *L'oro dei filosofi: Saggio sulle idee di un alchimista del Trecento* (Spoleto: Centro italiano di studi sull'alto Medioevo, 1992); Pereira, *Arcana sapienza: L'alchimia dalle origini a Jung* (Rome: Carocci, 2001); and, with Barbara Spaggiari, *Il Testamentum alchemico attribuito a Raimondo Lullo* (Florence: SISMEL, 1999).

his grand scheme for chrysopoeia was accompanied by an even more laby-rinthine form of florilegium than he had composed in the past.

It was during this period that Newton wrote long drafts of works consist-ing of multiple *Opera* like the ones whose titles are announced in Mellon 78. As in the manuscripts from the later 1670s and early to mid-1680s that we have already examined, these *Opera* consisted mostly of passages strung together from earlier authors with only occasional comments by Newton, often placed in brackets to set them off from the text. The focus of the in-troductory *Opera*, at least at the beginning of this undertaking, lay in the pseudo-Lullian corpus. The immediate instigation for this shift in Newton's reading interest stemmed mainly from the work of Edmund Dickinson, whose *Epistola ad Theodorum Mundanum* (Letter to Theodore Mundanus) was published with the imprimatur of the Vice Chancellor of Oxford Uni-versity in 1686. A fellow of the College of Physicians from 1677, Dickinson was a respected member of the Oxford and London medical community; he was also a fellow of the Royal Society from 1678 and had moved to London by the 1680s at the latest.[4] Dickinson's *Epistola*, as its title implies, consists mainly of an epistle or letter to the self-styled adept Theodore Mundanus, along with a comprehensive reply. About Mundanus, very little is known, al-though in his letter he indicates that he visited Dickinson twice, in 1662 and in 1678 or 1679; his letter is dated Paris, October 1684, and it is believed that he was French. We have already had occasion to examine Dickinson's views about the character of adepts and their writings earlier in the present book, but it will be necessary to consider his ideas about the proper starting materials of "the great work" more deeply.

Since it is necessary to have some grounding in the alchemy of Dickinson-Mundanus before descending into the complexities of Newton's mature flo-rilegia, I will begin this chapter with a brief consideration of the *Epistola ad Theodorum Mundanum* and Newton's synopsis of it before proceeding further. After this we will examine the cycle of *Opera* in their several com-plementary drafts, which will reveal the heavy influence exercised by Dick-inson, pseudo-Lull, Snyders, and other authors. The *Opera* may be dated conservatively to the period between the publication of Dickinson's *Epistola* in 1686 and a later stage in Newton's alchemy, namely, his intense collabora-tion with Nicolas Fatio de Duillier in the early 1690s. The collaboration led to Newton's use of a number of French alchemical authors that are not pres-ent in the *Opera*. The work with Fatio also contributed to Newton's produc-tion of a text that has received notice from other Newtonian scholars as in some sense the culmination of his alchemical endeavor, Huntington Library, Babson MS 420, otherwise known as *Praxis*. Thanks to the many difficulties raised by Newton's correspondence with Fatio and the *Praxis* manuscript, it will be better to save this material for a later chapter.

The subtitle of Dickinson's *Epistola* is "On the Quintessence of the Phi-losophers" (*De Quintessentia Philosophorum*), and this gives a strong hint

---

[4] *Oxford Dictionary of National Biography*, online version, under "Edmund Dickinson," consulted March 25, 2016.

of the approach that the Oxford doctor and his interlocutor would take. The term "quintessence" had been adopted in the fourteenth century by the alchemical Franciscan Jean de Roquetaillade (otherwise known as John of Rupescissa) as a term for the material that we now call ethanol or ethyl alcohol.[5] Although viniculture and brewing had been known since time immemorial, and sophisticated distillation apparatus had been invented by alchemists in the Roman Imperial period if not before, the isolation of ethanol by means of distillation was largely if not wholly unknown before the High Middle Ages. This mysterious "water of life" (*aqua vitae*) that burned with a blue flame, preserved dead flesh, extracted the active ingredients from numerous plants and minerals, and had the added benefit of imparting pleasure to those who imbibed it, was a surprisingly recent discovery in Roquetaillade's lifetime. The seeming incorruptibility of this "aqua ardens" (burning water) or aqua vitae made it a logical candidate for its own place alongside the traditional four elements, fire, air, water, and earth, which were consigned to the perennial cycle of corruption and decay that marks our sublunary world; thus Roquetaillade granted it the name "quintessence"—a sort of heavenly, fifth element in its own right. Roquetaillade's work was pillaged by the school of late medieval writers who adopted the name of the Mallorcan philosopher Ramon Lull, and under various titles, the most important of which was *De secretis naturae seu de quinta essentia* (On the Secrets of Nature or the Quintessence), his work on the quintessence acquired a new life as part of the pseudo-Lullian alchemical corpus.[6] The fact that this corpus had grown to well over a hundred works on chrysopoetic and medical alchemy by the seventeenth century made it only natural that Dickinson and Mundanus would associate the quintessence with Lull rather than Roquetaillade.

Needless to say, the qualifier "of the Philosophers" that Dickinson attached to "Quintessence" in his title was not without significance. As we saw earlier in this book, Dickinson regarded the adepts as fiendishly clever when it came to guarding their secrets from the hoi polloi. For him it was perfectly obvious that the Lullian school should use the term "quintessence" to mean something more than the alcoholic "branntwein" and "aquavite" that were by his time being sold on street corners by strong-water distillers along with the forerunners of modern-day gin and whiskey. The Lullian quintessence could no more belong among these common comestibles than the Catalonian sage himself could form part of the herd of vulgar distillers.

[5] For a recent study of Roquetaillade, see Leah DeVun, *Prophecy, Alchemy, and the End of Time: John of Rupescissa in the late Middle Ages* (New York : Columbia University Press, 2009). Still very valuable are Jeanne Bignami-Odier, *Études sur Jean de Roquetaillade* (Paris: Librairie Philosophique J. Vrin, 1952), and the pair of studies by Bignami-Odier and Robert Halleux, "Jean de Roquetaillade" and "Les ouvrages alchimiques de Jean de Rupescissa," in *Histoire littéraire de la France* (Paris, 1981), 40: 75–284. For the technical content of Roquetaillade's work, see Robert Multhauf, "John of Rupescissa and the Origin of Medical Chemistry," *Isis* 45 (1954): 359–67.

[6] For recent work on the influence of the pseudo-Lullian *De secretis naturae*, see Jennifer M. Rampling, "Analogy and the Role of the Physician in Medieval and Early Modern Alchemy," in *Alchemy and Medicine from Antiquity to the Enlightenment*, ed. Jennifer M. Rampling and Peter M. Jones (London: Routledge, forthcoming 2018).

And yet the full-blown quintessence as well as less exalted alcoholic spirits served very handily as model substances for describing the production of the philosophers' stone, or its precursor the sophic mercury. Since the beginnings of the aurific art in late antiquity alchemists had been stressing that the most perfect elixirs should be made from "one thing" acting on itself without extraneous additions.[7] Medieval texts such as the *Tabula smaragdina* of Hermes Trismegistus and the *Summa perfectionis* of Geber had further accentuated the necessity of working on a simple, unique material. Rectified spirit of wine could be seen as fulfilling this requirement; first the grapes had to ferment or "putrefy," which they could do on their own, then the alcohol-rich wine had to be distilled to separate the burning spirit from its watery phlegm. Finally, if one wanted to attain an even higher state of purity, the impure alcohol could be cohobated (circulated) with hygroscopic salt of tartar, which would attract more water, yielding an even stronger alcohol on distillation. And where did the salt of tartar come from? Since it was the product of calcining and leaching wine-lees or argol scraped from the inside of wine casks, it too obviously came from the wine itself.

Hence while addressing Mundanus, Dickinson argues that in order for metals to be led back to their *primum ens* or "primal crudity of their mercury"—a prerequisite for making the philosophers' stone—they must first be resolved by "your aqua vitae" (*vestra aqua vitae*) or "metallic wine." That Dickinson has something in mind other than our modern alcohol is assured by this and subsequent comments. He adds that he knows certain waters or menstrua that can extract the yellow color from gold and impart it to other metals, but the problem is that no more gold is produced by this than the amount destroyed in the process of extraction. The *aqua vitae* of Mundanus, on the other hand, slowly softens and dissolves gold, increasing and augmenting its tinctorial power so that it becomes ten thousand times more powerful in transmutation than gold dissolved in vulgar menstrua. What Dickinson and Mundanus have in mind is obviously something like the alchemical sophic mercury, and indeed the former refers to Mundanus's quintessence here as a "mercurial water" (*aqua mercurialis*).[8]

How then does pseudo-Lull, the famous promoter of the alchemical quintessence, figure in the exchange between Dickinson and Mundanus? Above all there is the matter of Lull's peerless authority; as one of the most famous *adepti*, he supposedly supported King Edward I in an unspecified "Holy War" by creating alchemical gold for him, while living in the church and hospital of Saint Katharine's by the Tower of London.[9] And as Mundanus also tells Dickinson, "Lull the Great" (*Lullius magnus*) managed to create an alchemical medicine that he employed when decrepit and near

[7] See the work of pseudo-Democritus and his commentator Synesius in Matteo Martelli, ed., *The Four Books of Pseudo-Democritus*, *Ambix* 60 (2013): supplement 1, S103, S127, S133.

[8] Edmund Dickinson, *Epistola ad Theodorum Mundanum* (Oxford, 1686), 72–73.

[9] Dickinson, *Epistola*, 151. See L. M. Principe, *The Secrets of Alchemy* (Chicago: University of Chicago Press, 2013), 73, for the legend that Ramon Lull helped the English cause in a Holy War usually said to have involved his coining gold "rose nobles" for Edward III, not Edward I.

death, thereby restoring himself to a state of youthful vigor.[10] Newton's own synopsis of Dickinson's *Epistola*, now found at the University of Texas, picks up on these cues and even recapitulates a process from the *Epistola* that is based on the pseudo-Lullian corpus. After recounting Mundanus's heavily veiled account of "our wine" or *aqua vitae*, Newton repeats the following discussion and recipe from the same source:

> This is our *aqua vitae*, our water of Diana or of Quicksilver through which we prepare the radical dissolution of metals and especially of gold, in this fashion: take good calx of gold prepared in the Lullian fashion and dissolve it in this *aqua vitae*. Then digest it for a while and distill the *aqua vitae* from it; make all the gold ascend with it by means of repeated affusions, digestions, and distillations of the spirit. The result is the true potable gold, the great medicine of metals and men, which can be made into an even higher medicine by reducing the dissolved gold into a true oil. You should do this by drawing off the solvent until it reaches the consistency of an oil, which is much more precious than pure gold.[11]

The *Epistola*'s rather vague directions for making an elixir modeled on pseudo-Lull's potable gold, along with the high praise directed at the Catalan sage by Mundanus, were enough to propel Newton into an intensive research project with Lullian alchemy at its center. From the second half of the 1680s onward, Newton became increasingly focused on collecting and interpreting the many alchemical treatises ascribed to Lull, a fact that has not escaped previous scholars. It is well known that various lists of Lullian alchemical works and purchasing desiderata for booksellers populate Newton's notes in the late 1680s and 1690s, though I believe that the important role of Dickinson's *Epistola* in this shift of interest has not been noticed previously.[12] And yet this is an important episode, not merely because it allows us a way into the labyrinthine complexity of Newton's late florilegia, but because it testifies to Newton's attunement to the particular setting that he inhabited. Like Philalethes and Yworth, Dickinson was a London phenomenon, and the fact that he did not attain the fame of Philalethes or the volume of publications flowing from Yworth's restless pen is all the more evidence that Newton's interest in Dickinson owed something to their shared environment. The work of pseudo-Lull was undergoing something of a revival in London during the last two decades of the century, as titles such as Johann Seger von Weidenfeld's *Secrets of the Adepts; or, Of the Use of Lully's Spirit of Wine* (1685), a text that Newton owned in Latin, testify.[13] It is no matter for surprise, then, that Newton would now turn his efforts to deciphering the riddles of the wise man of Mallorca.

---

[10] Dickinson, *Epistola*, 202, for the story of Lull's rejuvenation, and p. 170 for the epithet "Lullius magnus."

[11] University of Texas, Harry Ransom Humanities Center MS 129, 7r.

[12] Karin Figala, John Harrison, and Ulrich Petzoldt, "De scriptoribus chemicis: Sources for the Establishment of Isaac Newton's (Al)chemical Library," in *The Investigation of Difficult Things*, ed. P. M. Harmon and Alan Shapiro (Cambridge: Cambridge University Press, 1992), 135–79, see especially 145, 153n62, 155–56.

[13] Harrison, p. 260, no. 1719.

Newton's *Opera*: The Development of a Master Florilegium

At some point after digesting Dickinson's *Epistola*, Newton embarked on his attempt to organize the various operations and stages of his chrysopoetic project under the headings of *Opera* (Works). The linkage to Dickinson's *Epistola*, which Newton typically refers to simply as "Mundanus," is evident, for example, in the list of thirty numbered *Opera* in Yale University, Mellon 78.[14] But a more characteristic case of Newton's new interpretation of the Lullian quintessence is found in a group of closely related florilegia that emerged at various points after he had assimilated the *Epistola*. Some of these were already classed together by Dobbs, who rightly saw connections between two of the manuscripts at King's College, Cambridge, Keynes 40 and 41, and the Smithsonian Institution's Dibner MS 1070A.[15] These are in fact narrowly related drafts of the same text, to which one should add Huntington Library, Babson MS 417. They are not the only manuscripts in which Newton discusses the multiple alchemical *Opera* of course; less closely related texts include Babson 421, Keynes 21, and Keynes 23, among others, which need not form part of our present discussion. In the following, I will focus mainly on Dibner 1070A and Keynes 41, since these two manuscripts are tightly related to each other and Dibner 1070A contains perhaps the very earliest draft of Newton's *Opera*, though this is not certain.[16]

Like most of Newton's early drafts, Dibner 1070A is a jagged and saltatory collection of passages that often break off abruptly and pick up elsewhere in the manuscript. There appears to be a lacuna after folio 1v, followed abruptly on 2r by a fragment of what Newton calls "Opus quintum" (Fifth Work) in another draft. It is likely that Dibner 1070A originally had an "Opus tertium" (Third Work) and "Opus quartum" (Fourth Work)" that have dropped out here as a result of lost folios.[17] By combining Dibner 1070A with Keynes 41, however, one can arrive at a fairly complete idea of the *Opera* as Newton conceived of them soon after digesting Dickinson's *Epistola*. The combined text consists of nine *Opera*, but in Dibner 1070A they are not presented in consistent numerical order, nor are they entirely sequential. *Opus* one and two are followed by numbers nine, six, eight, a variant of six (which I call "Opus Sextum β"), and seven. At any rate, the online edition of Dibner 1070A (at www.chymistry.org) provides notes explaining Newton's numerous cross-references that link disparate portions of

---

[14] Mellon MS 78, 2r and 3v.

[15] Dobbs, *JFG*, 124n14.

[16] Because it is an early draft, Dibner 1070A contains Newton's thoughts in an unusually open and tentative form, in which they have not been subjected to subsequent editorial polishing that would tend to remove the traces of his original reasoning. See the deleted passages at Dibner 1070A, 7r, that begin with "Nonne," for example. Newton returns to these thoughts on 8r, but without the interrogative "Nonne."

[17] See Keynes MS 41, folio 7r, which begins a series of correspondences with Dibner 1070A, 2r. Folio 1v of Dibner 1070 ends abruptly with the heading "Idem aliter." If one compares this heading with the corresponding passages in Keynes 41, 3r, and Babson 421, 8r, it becomes quite clear that there is a lacuna at this point in Dibner 1070A. Probably a folded sheet (possibly more than one) was originally inserted after 1v in the manuscript that has subsequently been misplaced or lost. This may well account for the absence of "Opus tertium" and "Opus quartum" in Dibner 1070A as well.

the text, so in the following list of *Opera* we need only refer to the beginning folio of each section. Placing the *Opera* in sequential numeration produces the following list. I have added the headings for *Opera* three, four, and five as they are found in the more complete version provided by Keynes 41, since my analysis employs both manuscripts:

1r: Opus primum. Extractio spiritus (First Work. Extraction of the Spirit.)

1v: Opus secundum. Extractio Animæ. (Second Work. Extraction of the Soul.)

2r: <Opus tertium. Terræ calcinatio. (Third Work. Calcination of the Earth.)>

<Opus quartum. Salis imbibitio et sublimatio, in Terram albam foliatam, quæ est ☿ noster mineralis et primus motor. (Fourth Work. Imbibition and Sublimation of the Salt into the White, foliated Earth, which is Our Mineral Mercury and First Mover.)>

<Opus quintum. Acuatio spiritus rectificati cum rebus calidis sui generis per sublimationes, & conversio in mercurium vegetabilem ut et in Quintessentiam quæ cælum est Philosophorum, et Liquor Alkahest, quocum fit Aurum potabile. (Fifth Work. Sharpening of the Rectified Spirit with Hot Materials of its Own Genus through Sublimations, and Conversion into the Vegetable Mercury as also into the Quintessence which is the Heaven of the Philosophers, and into the Liquor Alkahest, with which it becomes Potable Gold.)>

5v: Opus Sextum. Purgatio et sublimatio Mercurij et Metallorum vulgi, Multiplicatio Mercurij nostri per dissolutionem & fermentationem infinitam Mercurij sublimati, et extractio auri vivi & saturni verissimi (Sixth Work. Purgation and Sublimation of Mercury and of the Metals of the Vulgar, Multiplication of Our Mercury through Dissolution and Infinite Fermentation of the Sublimed Mercury, and Extraction of the Living Gold and Very True Saturn)

19r: Opus Sextum β. Mercurij sublimatio septena et extractio ♃ (Sixth Work β. The Sevenfold Sublimation of Mercury and Extraction of Sulfur)

20r: Opus septimum (Seventh Work.)

15r: Opus octavum. Conjunctio Putrefactio & Regimen decoctionis Dispositio quam descripsimus præcedit putrefactionem. Conjunctio verò multiplex (Eighth Work. Conjunction, Putrefaction, and the Regimen of Decoction. The Disposition which we have Described Precedes Putrefaction. But the Conjunction is Multiple.)

3v: Opus 9. Ignis Pontani, et pondera sapientum et putrefactio. (Ninth Work. The Fire of Pontanus, the Weights of the Wise, and Putrefaction.)

Before analyzing the content of the *Opera* in Keynes 41 and Dibner 1070A, it is necessary to pose an essential preliminary question. What was

Newton actually trying to do with this material? Although it would be rash to argue that Newton acquired no new practical knowledge from the pseudo-Lullian corpus, it is safe to say that his main approach was one of accommodating the quintessence-based, largely medical alchemy of the Lullian *De secretis naturae* and associated texts to his own, far more metallurgically oriented chymistry. No doubt Newton hoped to extract additional practical secrets from his Lullian sources, but at the same time his *Opera*, particularly the first four of them, served de facto as assurances to Newton that he had interpreted the traditional secrets of alchemy correctly. Just as his own source Starkey had reinterpreted George Ripley's "sericon" as antimony, so Newton reinterpreted pseudo-Lull's alcoholic quintessence as a mineral product.[18] At the same time, however, Newton was adding preliminary stages to the operations that he had acquired from Philalethes and Snyders, either because he believed that those authors had left them out, or because he thought they were better described by others. It is sometimes said that Newton believed all alchemists to be saying the same thing in different words, but the *Opera* show that this is not true. Newton explicitly distinguishes the processes of Snyders from those of other chymists and points to the absence of important Philalethan processes in the German author.

The best way to think of Newton's *Opera* is as a succession of working notes rather than as a finished piece of work. Newton's goal was not to produce a final draft for dissemination to others but to combine related passages from the same author, from that author's commentators, and from others who might be saying the same thing in different languages, all for the purpose of arriving at the grand secret of chrysopoeia. His major interpretive techniques therefore lay first in the reassembling of passages that had been separated in accordance with the alchemical "dispersion of knowledge," then in the collecting of exegetical passages from other authors, and finally in the decipherment of *Decknamen*. Unfortunately, Newton's adoption of the florilegium style as his favored mode of expression means that his own voice can be difficult to discern among these diverse authors, and the additional fact that he was writing only for his own edification removed any need to clarify his thoughts for an audience. Nonetheless, the patient reader will find that an analysis of Newton's *Opera* provides much new material for understanding the development of his chrysopoetic endeavor. In the following, I will treat *Opera* one through four as a unit, since they represent a clear and sequential set of Lullian operations. *Opera* five and six require separate treatment, since they incorporate themes from Van Helmont, Snyders, Philalethes, Sendivogius, and a few other authors and weave them together with Lullian motifs into a dense and difficult fabric. We can dispense altogether with *Opera* seven, eight, and nine, since they are either too fragmentary to

---

[18] For sericon, see Jenny Rampling, "Transmuting Sericon: Alchemy as 'Practical Exegesis' in Early Modern England," in *Chemical Knowledge in the Early Modern World*, ed. Matthew Daniel Eddy, Seymour H. Mauskopf, and William R. Newman, *Osiris* 29 (2014): 19–34.

analyze in a reliable fashion, or else have been absorbed by Keynes 41 into *Opera* five and six.[19]

Opera One through Four—The Extraction of the Spirit and Soul, Calcination of the Residue, and Sublimation of the "Foliated Earth"

The "extraction of the spirit" serves as the subheading for *Opus* one, but the subject from which this material should be extracted is not immediately clear. Newton cites Mundanus to the effect that a certain chaos must be putrefied before "our wine" can be made from it, and this in turn must be distilled. Despite the vague character of these references to a preliminary "chaos," Newton drops several hints as to what he thought this material might be. The most important of these clues lies in a reference to a work by Adrian von Mynsicht, a German chymist of the early seventeenth century whose real surname may have been Seumenich.[20] Although Dibner 1070A contains only a brief reference to "Minschict" (Mynsicht) without further clarification, Keynes 41 expands on this, and its amplification makes Newton's interest in the German alchemist entirely understandable.[21] In the later draft, Newton closely paraphrases Mynsicht's instructions for operating on the chaos, as taken from the German chymist's *Thesaurus et armamentarium medico-chymicum* (Medico-Chymical Treasury and Armory). There Mynsicht says that a liquid should be separated from the chaos by first crushing the "ore of gold" (*minera auri*) into pieces the size of a hazelnut. These should then be distilled in a retort connected to a sealed receiver, all over low heat. A pound of the mineral will only yield one spoonful of the product, which is a "sweet and celestial water."[22] In Keynes 41, Newton first refers to this as a "blood," and then immediately identifies it as "our first, glorious mercury" and "our wine."[23]

Mynsicht's process may engender surprise in some readers, since native gold is usually found either as alluvial nuggets or embedded intact in other stones, such as quartz. The metal is famously resistant to attack by corrosives, whether natural or factitious. Nonetheless, gold compounds in mineral form do indeed exist. The best known of these today are gold tellurides, where the noble metal actually combines chemically with tellurium to produce a compound. But gold can also be found in some pyritic sulfide ores, where the metal, although free, is so finely dispersed as to be invisible and resistant

---

[19] In Dibner 1070A, *Opus* seven consists of three lines on folio 20r. *Opus* eight consists of a numbered list occupying half of folio 15r. Dibner 1070A's *Opus* nine (3v–4v), on the other hand, had been divided and absorbed into Keynes 41's *Opera* five and six (5r–20r).

[20] Rolf Gelius, "Mynsicht, Adrian von," *Neue Deutsche Biographie* 18 (1997), 671, online version; http://www.deutsche-biographie.de/pnd117624756.html, consulted March 25, 2016.

[21] Dibner 1070A, 1r.

[22] Hadrianus à Mynsicht, *Thesaurus et armamentarium medico-chymicum* (Lyon: Joannes Antonius Huguetan, 1645), 5.

[23] Keynes 41, 1r.

even to modern refining techniques involving cyanide.[24] These ores were a subject of sustained discussion in eighteenth-century mineralogical treatises such as the celebrated *Essay towards a System of Mineralogy* by the Swedish chemist Axel Frederic Cronstedt. The *Essay* describes multiple types of "mineralized gold," including the pyritic ores alongside auriferous cinnabar and a type of blende.[25] There is no reason then to think that Mynsicht's process was implausible *ab initio*, even if the identity of the volatile component that he claimed to distill out of his mineral cannot at present be identified.

But what did Newton make of Mynsicht's "chaos"? We know from Newton's earlier interpretations of alchemists such as Grasseus and Snyders in Keynes 35 and elsewhere that he had a strong predilection for operations that began with extractions from unrefined ores. This is also quite evident from his experimental laboratory records, where he even went so far as to devise special graphic symbols for the ores of the metals and metalloids known to him. Nowhere in those records do we find experiments with gold ores, however, which would be strange if Newton had actually interpreted the Lullian quintessence as a product deriving from such minerals. Additionally, we have no records of Newton trying to obtain gold ores from apothecaries or other purveyors of minerals, even though requests for crude antimony, copper ore, cinnabar and lapis lazuli, as well as the minerals of tin, iron, lead, bismuth, zinc, and cobalt have survived.[26] These facts suggest the possibility that Newton was interpreting Mynsicht's *minera auri* as a *Deckname*, which would have been a fairly natural move given the frequency with which alchemists spoke of hidden "gold" latent within other materials.

There is also another piece of evidence to suggest that Newton did not accept Mynsicht's report entirely at face value. In the reworked draft of the *Opera* found in Keynes 41, Newton adds a passage taken from Mundanus before the one from Mynsicht. Here too Newton is gathering information on the chaos from which the philosophical wine must be extracted, but this passage is particularly interesting because it contains the square brackets that characteristically contain Newton's attempts to decode *Decknamen* into workable materials. I therefore provide a translation of the Latin passage here:

> Our entire mercury is a salt from two saline substances which share the same root and lineage. Take the very acid sulfur and very oily mercury, remove all the feculency through sublimation or distillation [of the ♃], and render the mercury quite pure and subtle by means of common salt [of ♂] or vitriol [our volatile vitriol], or both. When they are purified in

---

[24] See the intelligent, short article by the mining engineer Charles Kubach, "Recovery of Gold in Pyritic Sulfide Ores," at http://mine-engineer.com/mining/minproc/gold-in-pyrite.htm, accessed July 29, 2017. For knowledge of gold ores in the Renaissance, see Robert W. Boyle, *Gold: History and Genesis of Deposits* (New York: Van Nostrand Reinhold, 1987), 51–64.

[25] Axel Frederic Cronstedt, *An Essay towards a System of Mineralogy* (London: Charles Dilly, 1788), 2: 524–27.

[26] See Babson MS 433, 1r, for one such list of desiderata. Nor are gold minerals mentioned among the actual purchases that Newton made from apothecaries as recorded on 174v of CU Add. 3975.

this fashion, resolve them and reduce them to one with the help of the distilled water. This is our Chaos (and grape), which after a proper fermentation and digestion, will give a clear and uniform liquor which is our wine. Mundanus, p. 182, 183, 197, 198.[27]

In this passage, Mundanus first equates his mercury with salt and then says that it is actually composed of two saline substances that are clearly to be identified with the acid sulfur and oily mercury that follow. Let us disregard the playful obfuscation by which Mundanus equates all of these materials and focus on Newton's interpretation. It is obvious that Newton thinks of the acid sulfur and oily mercury as composing one substance from which the corrupting sulfur must be sublimed or distilled off. The "mercury" that is left behind must then be purified and subtilized by means of "salt of antimony" or "our volatile vitriol," or by both together. In fact, "salt of antimony" and "volatile vitriol" are actual substances that Newton employed and described in his experimental records. "Salt of antimony" is the active ingredient in Newton's ubiquitous "liquor of antimony" produced by dissolving stibnite in aqua regia, and "our volatile vitriol" is a compound produced by first imbibing copper or a copper-bearing mineral with liquor of antimony and then subliming the product after several stages of purification, as I describe in a later chapter devoted to Newton's laboratory notebooks. Thanks to Newton's parenthetical notations, these materials are fairly unambiguous. But unfortunately he supplies no such pointers to the identity of the "chaos" of Mundanus and Mynsicht. What then are we to make of this mysterious material?

Although it is impossible to point with absolute certainty to the identity of the initial "chaos" in Newton's interpretation, he does provide us with several important clues. If one turns to "Opus quartum" in Keynes 41, for example, Newton describes a stage in the refinement of the chaos that involves subliming it. The result, he says, will be "the feathery alum of Basilius Valentinus."[28] This can only be a reference to a passage in the *Last Will and Testament* of Basilius that we already examined in chapter five of the present book. In the "Elucidarius" found there, the author advises to "ask counsel of god *Saturn*," who will provide "a deep glittering *Minera* for an offering, which in his Myne is grown of the first matter of all Metals."[29] As Newton already pointed out in the early manuscript commentary found today in Jerusalem (Var. 259), "this is Antimony."[30] In the same passage, Basilius then

[27] Keynes 41, 1r: Noster mercurius totus est sal ex duabus salinis substantijs quæ eandem radicem stirpemꝗ sortiuntur, productus.—Sume sulphur valde acidum et ☿um valde oleosum; fæculentiam omnem per sublimationem [🜍is] aut destillationem remove, mercuriumꝗ valde purum et subtilem communi sale [♁ij] aut vitriolo [n̄ro volatili] aut utroꝗ simul effice. Quando sic purificantur illa resolvas & in unum reducas ope aquæ destillatæ. Hoc est nostrum Chaos, (& vitis) quæ post debitam fermentationem & digestionem, clarum et uniformem liquorem dabit qui est vinum nostrum. Mundan, p. 182, 183, 197, 198.

[28] Keynes 41, 3v: "tunc in idoneum vas repone quod oblinere debes ad altitudinem ad quam ascendit materia, et quod a fæcibus exurget igne forti sublima per horas 24 (vel donec ascendat ad modum albissimi pulveris vel folior.ii Lunæ aut Talchi splendidi. Raym. Codicil. 211, vel Aluminis plumosi Basil. Valent."

[29] Basilius Valentinus, *Basilius Valentinus, Monk, of the Order of St. Bennet: His Last Will and Testament* (London: W. B., 1658), 127.

[30] Var. 259.11.7r.

says to prepare the "deep glittering" mineral and to sublime it with bole or tile meal, which will result in a noble sublimate "like little feathers, or *alumen plumosum*, which in due time dissolveth into strong and effectual water." This must have been what Newton had in mind when he referred to the feathery alum of Basilius in "Opus quartum," which provides us with a powerful clue that he thought the initial chaos was stibnite. A further clue emerges from Newton's late collaboration with the Dutch distiller William Yworth, which we will examine in chapter nineteen. To make matters short, Newton grafted a process developed by Yworth for distilling and subliming a variety of products from stibnite onto a Lullian discussion of the quintessence of wine. It certainly appears as if this late "Yworthian" compilation by Newton represents a further development of the ideas and practices already under development in Keynes 41 and the other *Opera* drafts.

If Newton did in fact interpret the chaos as referring to the ore of antimony, then how did he think the "common salt of antimony" or "our volatile vitriol" should be used on that substance? It is quite difficult to say, given the vagueness of the text and the absence of further parenthetical comments, but Newton's laboratory notes found in CU Add. 3973 and 3975 do offer some pointers. As I will show at length in due course, Newton spent years developing the subliming agents that he named "sophic sal ammoniac" and "volatile Venus." Both of these employed antimony in conjunction with other materials, and Newton was concerned that the sulfur in crude antimony would corrupt the products of the sublimations made with the help of these adjuvants. It is quite possible that the antimonial "chaos" of Keynes 41 was stibnite that Newton was trying to free from its sulfur by means of "common salt of antimony" and "our volatile vitriol" as a preliminary step toward making the Lullian philosophical wine. The fact that the antimony in these compounds or mixtures would be acting on the antimony in the stibnite was not a redundancy but an advantage, since "one thing" would be acting on itself in accordance with the old alchemical advice to avoid extraneous ingredients. To proceed further into speculation at this point, however, would be rash.

Once we understand that Newton wanted to extract the Lullian spirit from a mineral, and probably from stibnite, the rest of the four initial *Opera* fall into place. The absence of further bracketed comments with clear connections to Newton's laboratory practice intimates that this textual material, insofar as it extends beyond mere summary of pseudo-Lull and his commentators, consists largely of conjectural processes.[31] The remainder of *Opus* one, after the initial "wine" or spirit has been extracted, is taken up with a succession of processes for rectifying it. The "wine" must first be putrefied for thirty days over a gentle heat in order to separate the four elements, or else it can be subjected to shorter bouts of repeated putrefaction.[32] It is repeatedly

---

[31] In the first four *Opera* in Keynes 41, I find only nine bracketed comments, of which the first three are the references to sulfur, salt of antimony, and volatile vitriol that we already discussed. The rest consist of textual clarifications rather than references to Newton's own experimental products.

[32] Keynes 41, 1v. It may seem odd that the "wine" (instead of "grapes") would have to be putrefied and then distilled, but this conflation is an inevitable consequence of Newton's collocation of unrelated sources and assumption of mutual identity among them.

distilled as well, leaving a residual earth in the bottom of the vessel that looks like liver. After these repeated distillations have been carried out, the "wine" is further rectified by successively distilling ever smaller fractions of the distillate in order to purge it of watery phlegm. Citing pseudo-Lull's *De secretis naturae*, Newton says that one will know the process to have been perfected when a cloth impregnated in the solution can be lit like a wick.[33] All of this is obviously based on old processes for maximizing the strength of *aqua ardens* or the quintessence, namely, ethanol distilled from wine by isolating it from the water naturally found in it. Yet Newton is undeterred by the fact that the operations in the Lullian *De secretis naturae* pertain to real alcohol and real wine; in *Opus* one he continues to spin out the Lullian processes in even finer detail and reiterates additional stages with minute precision. Thus after distilling off the strongest part of the spirit, one must attach a different receiver to the still and collect the watery phlegm. After all the phlegm has been distilled off, an earth will remain in the bottom of the vessel like molten pitch or thick honey. These products will now provide the basis for *Opus* two.

It would be unduly tedious to present the next three *Opera* in all their detail. Their most striking feature is the reiterative character of their multiple digestions, putrefactions, sublimations, and distillations, intended not only to purify a single product but also to lead to a variety of salts, oils, and solutions. They may be summarized as follows. In *Opus* two (Extraction of the soul), the watery phlegm produced in the distillations of the first *Opus* is poured on the black, earthy residue, whereupon the residue immediately dissolves. The liquid is then distilled repeatedly over a low heat, leaving the "pitch" or tar-like residue harder than before. Then the "rectified ardent water"—namely, the stronger fraction of *aqua ardens* made in *Opus* one (not the phlegm) is divided into two parts. One part must be preserved apart for creating a menstruum or mercury with which "you sublime the earth of the stone."[34] As for the second part of the ardent water, it is putrefied on the pitch and distilled repeatedly, until the feces seem burned—then the soul has been extracted. *Opus* three (Calcination of the earth) advises to calcine the residue in a sealed vessel until it flows on a red-hot silver plate. Then one must extract a salt from the dregs with the phlegm or with rainwater. Finally, *Opus* four (Imbibition and Sublimation of the Salt into the Foliated Earth) states that an eighth part of the spirit is to be poured on the calcined earth or the salt, and it is allowed to soak and heat until it combines with the calcined earth and only water distills off; this is repeated seven more times. When the earth has drunk up its weight of spirit and turned white, one must sublime it. The sublimate will be the white sulfur of nature, a clear and resplendent crystalline mercury, and a salt that goes by many names, which Newton also

---

[33] Keynes 41, 2r: "Rectificatur autem donec ^flegma nullum faciat et tota comburendo evanescat & 2 linum in ea tinctum inflammet. Raym. Lib. secr. p. 34, 35." This corresponds to Raymond Lull, *Tractatus brevis et eruditus* (Strasbourg: Lazarus Zetzner, 1616), 34–35.

[34] Keynes 41, 2v: "unam partem serva pro creando menstruo seu ☿io ut cum eo terram lapidis sublimas et cum alia abstrahas animam a terra picea ut sequitur."

calls "white, foliated earth" (*terra alba foliata*).[35] But Newton is still not done. The foliated earth or fixed salt must now be elaborated into an oil by imbibing it repeatedly with the rectified spirit produced in *Opus* two (where Newton advised to divide it into two samples).[36] Only after this saline oil has been sublimed multiple times is Newton ready to pass to the fifth *Opus*.

Opus Five: Lullian Quintessence and Helmontian Alkahest

The elaborate title of the fifth *Opus* reveals the diverse themes that it covers—"Sharpening of the Rectified Spirit with Hot Materials of its Own Genus through Sublimations, and Conversion into the Vegetable Mercury as also into the Quintessence which is the Heaven of the Philosophers, and into the Liquor Alkahest, with which it becomes Potable Gold." As this suggests, Newton did not think that the first four *Opera* had arrived at the Lullian quintessence in its full, celestial glory. In order to achieve this end, one must first take three ounces of the initial sample of the rectified spirit produced in *Opus* two, namely, the portion that was not used to extract the soul from the residual earth at the bottom of the vessel, and add it to one pound of the white, foliated earth (possibly in its unctuous form as produced at the end of *Opus* four, though this is not clear). Multiple putrefactions and distillations with fresh spirit follow; to simplify Newton's presentation, the aggregate of these operations produces a *menstruum simplex*. This is digested from thirty to sixty days, and after still further operations, which include removing a sediment and dissolving ordinary gold in the quintessence, "it will be converted into a glorious and odoriferous Quintessence."[37] Before we terminate with this explicitly Lullian part of the *Opera*, a final passage must be quoted, which vividly underscores the degree to which Newton had appropriated the late medieval project of graduating and improving ethanol-based "burning waters" or "water of life" and turned it to his own purposes. In the *De secretis naturae* of pseudo-Lull there is a famous passage reworked from Jean de Roquetaillade that speaks to the marvelous odor of the prepared quintessence. According to pseudo-Lull, the fragrance of the quintessence is so subtle and sweet that it will draw unsuspecting humans to itself by its scent. More than this, it can even attract birds that happen to be in the vicinity.[38] Although it should be absolutely clear by now that Newton, unlike his medieval sources, did not think the quintessence to be an alcoholic compound, this passage reemerges in his *Opera* almost verbatim:

[35] Keynes 41, 3v–4r.

[36] Keynes 41, 4v.

[37] Keynes 41, 5v: "et convertetur in Quintessentiam gloriosam et odoriferam." See also the bottom line of 6v, where the phrase is repeated after the additional operations requisite to the making of the "glorious and odoriferous" quintessence are described.

[38] The Lullian passage is found in Lull, *Tractatus brevis et eruditus*, 35. The somewhat less flamboyant account given by Roquetaillade appears in Joannes de Rupescissa, *De consideratione quintae essentiae rerum omnium* (Basil: Gratarolus, 1561), 32–34.

Once a continuous circulation has been made for many days, open the mouth <of the vessel> which you stopped up with the said seal, and if a more-than-marvelous odor exhales, so that no fragrance on earth may be compared to it, insofar that the vessel placed in a corner of a house attracts all those who enter with an invisible bond, or if the vessel, when it is placed upon a tower, attracts all birds whose nostrils its odor has reached so that it causes them to remain around it, then, son, you will have our Quintessence, which is otherwise called vegetable Mercury, so that you may apply it as you will in the Magistery of the transmutation of metals.[39]

Newton found this passage on inebriated birds to be so compelling that he even went so far as to collate two variant versions of the Lullian text, concluding correctly that they were both fundamentally the same work.[40] In *Opus* five, this is followed by an attempt to determine the cause of the quintessence's attractive power. Following his usual assortment of Lullian commentators, which include George Ripley along with more obscure figures such as Christopher of Paris, "S. H.," and "Bross" (who appear in printed versions of the *Appendix* to Christopher's *Elucidarius*), Newton concludes that the marvelous fragrance stems from the gold dissolved in the quintessence.[41] Moreover, the gold does not enter into solution intact, but it is analyzed into its chymical principles. In order to determine the exact way in which this event takes place, Newton now turns to the "circulated salt" (*sal circulatum*) of the Helmontian tradition, which many chymists, including George Starkey, identified with the wonder-working solvent immortalized by Van Helmont himself, the alkahest.

At this point, Newton begins a subsection within *Opus* five that starts with the statement, "The liquor Alkahest agrees in all things with the Quintessence."[42] This claim is followed by several dense folios full of comparisons between the Helmontian alkahest and the Lullian quintessence, supported by paraphrase quotations from Van Helmont, Starkey, pseudo-Lull, and others. Newton begins by saying that the alkahest is prepared in a way similar to the quintessence, from a certain alkali made volatile and a liquor called "Ignis-Aqua" (firewater). As far as linkage with the quintessence goes, what Newton has in mind here is probably the Lullian "white foliated earth" that appeared in *Opus* four as a product of the digestion, calcination, and cohobation of rectified spirit and its

---

[39] Keynes 41, 6r: "Facta per multos dies continuatione circulationis, aperi foramen quod cum dicto clausorio obstruxisti, et si odor supramirabilis exeat, ita quod nulla mundi fragrantia <*illeg.*> ei comparari valeat, in tantum quod vas positum ad angulum domus vinculo invisibili trahit omnes intrantes, aut vaso posito supra turrim trahit omnes aves quarum ejus odor nares attigerit, ita quod circa seipsam stare faciat; tunc habebis fili nostram Quintessentiam quæ aliter dicitur Mercurius vegetabilis, ad tuum libitum ut applices in Magisterio transmutationis metallorum."

[40] Keynes 41, 6r: "Raymund de Quintess. p. 24, 25, 26, <*illeg.*> collat cum Secret p. 34, 35, 36." Newton's explicit observation that the two texts are the same occurs a few lines before the quoted passage, "Raymund. de Quintessent. p. 17, 18, 19, 20 collat cum libro Secretorum p. 30, 31. Nam hæ sunt editiones duæ ejusdem operis libri."

[41] See Christophor of Paris, *Elucidarius*, and *Appendix practica* to his work, in *Theatrum chemicum* (Strasbourg: Lazarus Zetzner, 1661), 6: 195–270 and 271–93.

[42] Keynes 41, 7r: "Liquor Alkahest per omnia congruit cum Quintessentia."

residual earth. As for the claim that the Helmontian alkahest should be made from an alkali plus Ignis-Aqua, Newton bases this on two widely separated sections of Van Helmont's *Opuscula medica inaudita* and *Ortus medicinae* (though both are found in the 1667 Lyon edition that Newton was using). The passage from the first text merely says in passing that the chymist should make use of "volatile salt of tartar" in the absence of a truly radical dissolvent (presumably the alkahest).[43] The reference to Ignis-Aqua, on the other hand, is part of a long didactic dream that Van Helmont recounts, in which he finds himself in a royal court where the king, seated on a brilliantly lit throne, is a personification of pure Being itself, in other words, God. His footstool is Nature, and the Porter of the hall is Intellect. The Porter silently gives the Belgian chymist a little book, which Van Helmont at once chews up and swallows. Having eaten the book, Van Helmont finds that his head has become transparent, signifying a newfound understanding of the natural world. A spirit from the King's dais then hands him a flask containing a substance called Ignisaqua—a mixture of fire and water. Van Helmont immediately knows the powers of all simple medicines in the world, and the dream continues page-upon-page with a laborious recitation of these medical simples.[44] In neither of these textual loci does Van Helmont advise that one should make the alkahest by combining volatile salt of tartar and Ignis-Aqua; this is a Newtonian reassembly of texts that have supposedly been dissociated on the principle of "dispersion of knowledge." Newton's justification for claiming the necessity of both ingredients stems from his desire to see a concordance here between pseudo-Lull and Van Helmont.

After giving some additional information on the modes of preparing the quintessence and the alkahest, Newton then claims that the two celebrated liquors "also agree in their actions on other bodies."[45] Not only do both liquids dissolve virtually anything put into them (though this may require extensive advance preparation), they also make the dissolved bodies volatile. Quoting Starkey's Helmontian textbook *Pyrotechny*, Newton says that the alkahest can render pebbles, gemstones, marchasite, rocks, sand, clay, earth, and even glass volatile. The Lullian quintessence can perform similar things without generating the ebullition and heat that the mineral acids would produce. In the case of gold, the metal must first be calcined "in the Lullian fashion," which means that it must first undergo amalgamation with mercury and then dissolution of the quicksilver in aqua fortis (nitric acid). If one places the resulting calx in boiling quintessence, Newton's sources tell him, most of it will dissolve within twenty-four hours.[46]

Reiterating again that the quintessence and the alkahest are the same thing, Newton finally comes to a sort of sticking point.[47] Van Helmont had made a

---

[43] Joan Baptista Van Helmont, *Opuscula medica inaudita* (separately paginated) in Van Helmont, *Ortus medicinae* (Lyon: Huguetan and Barbier, 1667), 105.

[44] Van Helmont, *Ortus medicinae*, 290–91.

[45] Keynes 41, 7r: "Congruunt etiam in actionibus suis in alia corpora."

[46] Keynes 41, 8r.

[47] Keynes 41, 8v: "Idem igitur sunt Quintessentia et liquor Alchaest <sic>. Et hoc menstruum est Liquor unicus quocum metallorum sulphura a mercurijs separantur Philal in Ripl. Epist. p. 14."

great deal of the fact that the alkahest dissolves other materials without undergoing combination with them. It acts *sine repassione* (without being acted on), and can be reseparated from the solution intact. This fact made the alkahest different from traditional accounts of the philosophers' stone or elixir, which was typically thought to combine intimately with base metals and convert them to gold or silver. Since Newton clearly imputed transmutational powers to pseudo-Lull's quintessence, the purely dissolutive and reseparable characteristics of the alkahest presented an impediment to their identification as one. Thus he quotes several passages from Philalethes in which "the American philosopher" argues that the alkahest provides nothing to the goal of transmutation and is in fact more difficult to prepare than the precursor to the philosophers' stone, the sophic mercury. Yet Newton being Newton, he finds a way to overcome this obstacle, at least in principle:

> But if the menstruum be sharpened with its own arsenical sulfur before its final digestion of sixty days, and be properly inflamed by a fermental power, so that it not only dissolve metals but can also ferment them and be acted upon by them in turn, it will no longer be the liquor alkahest, but our mercury.[48]

In other words, the alkahest (or quintessence) can itself become the sophic mercury if a fermental power is implanted in it by "its own arsenical sulfur." This kindling must occur directly before the long digestion that leads the quintessence to the "glorious and odoriferous" state that allows it to attract both people and birds. Needless to say, this "arsenical sulfur" has nothing to do with the toxic arsenic sulfides commonly called orpiment and realgar. Instead, "arsenic" is a synonym that Newton uses for the "white foliated earth" sublimed from the rectified spirit of "wine" that we discussed earlier, and this in turn is synonymous with the "white sulfur of nature."[49] Thus after some further examples, Newton repeats that the quintessence is useless for transmutation unless the white sulfur of nature provides it with an internal fire. Yet this must be added in moderation, for too much will burn the metals to be dissolved in it, particularly gold.

The need for a ferment leads Newton into a lengthy discussion of the proper proportion of white, foliated earth, which terminates in an actual recipe. One must take one part of "the salt sublimed as above," meaning the white, foliated earth, and place it in two parts of "spirit of wine" rectified as described above (not in the foresaid quintessence). They must then be digested for twenty-four hours, distilled seven times or more (presumably with cohobations), and then the menstruum is distilled multiply by itself in order to remove any further solids that it may deposit. After quoting several further passages from Ripley and Philalethes, Newton then draws this section to a close and passes to *Opus* six.

---

[48] Keynes 41, 9r: "tamen si menstruum ante digestionem suam <*illeg.*> ultimam dierum 60 cum sulphure suo arsenecali debite acuetur & vi fermentali debite accendatur ut metalla non solum dissolvere sed etiam fermentare possit & ab ipsis vicissim pati: tunc non erit amplius liquor Alkahest sed mercurius noster."

[49] See "noster sublimatus Arsenicum" on folio 4r of Keynes 41.

Opus Six: The Problem of Fermental Love

The foregoing discussion might make it sound as though Newton felt he had solved the problem of adding a fermental virtue to the quintessence or alkahest in order to turn it into a genuinely transmutative agency. It would be more correct to say that he now felt himself to be on the right track, but as is often the case in chymistry (and chemistry), the technical details remained to be worked out. The title of *Opus* six in the reworked version found in Keynes 41 is "The Dry and Wet Solution of the Vulgar Metals and Their Purgation, the Infinite Multiplication of the Sophic Mercury, and the Extraction of the Living Gold."[50] This does rather scant justice to the many themes covered in this *Opus*, however, which begins with precisely the problem on which Newton ended *Opus* five, namely, the fact that the alkahest (and hence the quintessence with which he identifies the alkahest) works *sine repassione*—without being acted on. At multiple points Newton says that this produces a "violent" dissolution, but he is not referring to the powerful bubbling and generation of heat that one often associates with the action of the mineral acids. Rather, he has in mind the fact that the alkahest can be reseparated from its solutes without undergoing any combination with them. It acts in a "violent" or adventitious way like a surgeon's scalpel dissecting a cadaver into pieces as opposed to the "natural" working of putrefaction and decomposition that would also return the corpse to different components. It was precisely this acknowledged feature of the alkahest that provided its main appeal to the mechanical philosopher Robert Boyle, who saw the marvelous dissolvent as an ideal means of analyzing bodies for the simple reason that it could be cleanly separated from the solutes without producing artifactual compounds.[51] But this was not Newton's goal; instead, he was seeking a menstruum that would not only dissolve bodies but combine with them and impart a generative principle; this is what he means in *Opus* five when he speaks of the conversion of the alkahest to the sophic mercury.

Thus *Opus* six begins what at first seems to be an attempt to impart the required ferment to the Lullian quintessence. Newton advises that gold calcined "in the Lullian manner" (amalgamated with mercury and then subjected to nitric acid) be ground with the white sulfur or oil described above. After multiple digestions and distillations with the quintessence, a gummy liquor is produced, which is the "soul of the sun" or gold (*anima solis*). Surprisingly, Newton then asserts that this process has not yet achieved the required result:

> But this menstruum acts without being acted upon (*sine reactione*) like the liquor Alkahest, and therefore, thanks to an absence of fermental love, the gold will be destroyed violently and will not be transmuted by vegetating.[52]

---

[50] Keynes 41, 11v: "Solutio sicca et humida metallorum vulgi ^ corum☿ purgatio & Multiplicatio infinita mercurij sophici & extractio ~~auri~~ auri vivi.

[51] Newman and Principe, *ATF*, 292–94.

[52] Keynes 41, 11v: "Sed menstruum hic agit sine reactione ad modum liquoris Alkahest ideo☿ ob defectum amoris fermentalis aurum hac ratione destruetur violenter & non transmutabitur vegetando."

The reason for Newton's hesitation lies in his Lullian source, which explicitly states that this process would lead to a medicine but not to an agent of transmutation. For the transmutational purposes, pseudo-Lull advises that one must add mercury to the quintessence before its digestion with the gold.[53] After providing another process from the same Lullian text (the *Ultimum testamentum*), this time with mercury, Newton determines that here too the quintessence works without being acted on and cannot therefore serve to transmute metals. Nonetheless, despite being violent and lacking in "fermental love," this operation is useful for softening and "opening" metallic bodies so that they may better receive the ferment.[54] In the version of this sentence given in the earlier draft provided by Dibner 1070A, Newton adds that the performance of this "opening" requires a preliminary cleansing by sublimation, a topic that he treats extensively in the remainder of *Opus* six.[55]

The putative "opening" of the metallic bodies by the Lullian quintessence provides Newton with a transition to the work of Snyders, which now becomes the focus of his discussion. As we discussed in chapter eleven, Snyders's chrysopoetic work involved two "fires," one of them hot and able to kindle the internal sulfur of bodies by a principle of sympathy, the other cold and penetrative, which rendered the metals "porous" and hence capable of easy entry by the hot fire. Newton now relates that the sequential use of the cold and hot fires first opens the metals, thus allowing the hot fire to enter them and separate their indwelling chymical principles, mercury, sulfur, and salt. According to Newton's nonstandard interpretation of Snyders, this "dry solution" actually involves liquid menstrua and is called "dry" because it results in the production of a "dry water" consisting of the purified chymical principles of each metal.[56] What follows in both versions of the *Opera* is a detailed discussion of Snyderian techniques for separating the three principles of each metals. First Newton describes the analysis of gold, then that of mercury, followed by the "humiliation" of Jupiter and exaltation of Saturn as presented in Snyders's *Metamorphosis*, and now seen explicitly as stages in the separation and purification of the chymical principles of these metals. One must wonder at the reasons behind the necessity for purifying each of these successive metals. The answer may possibly be found in a passage that Newton reproduces from Snyders's *Commentatio*, where the German alchemist speaks of an elixir made from all the metals with the exception of gold:

[53] Pseudo-Ramon Lull, *Testamentum ultimum*, in *Artis auriferae* (Basel: Conrad Waldkirch, 1610), 3: 9–10.

[54] Keynes 41, 11v–12r: "Hæc dissolutio violentia est sine amore fermentali sed emollit tamen et aperit corpora metallica & præparat ad recipiendum fermentum."

[55] Dibner 1070A, 6r: "Hæc igitur dissolutio violenta est sine amore fermentali. ~~nisi~~ Tamen hac ratione emolliuntur & aperiuntur corpora metallica & præparantur ad ~~solutionem~~ recipiendum fermentum. Sed prius a fæcibus purgari debent per sublimationem ut sequitur." The expressed need for the purification of the metallic bodies by sublimation helps explain the inclusion of much following material: in Keynes 41 this rationale has dropped out, making it more difficult to follow Newton's reasoning. This is one of several instances where the earlier draft has preserved Newton's original plan, the footprints of which he has covered over in successive reworkings.

[56] Keynes 41, 13r.

In the absence of gold, the Elixir can be made from the first matter (that is, the spirit of the world) and all the metals conjoined into one, namely from their sulfur and salt, which must be extracted with the sympathetic, metallic fire, for these magical elements must be dissolved in the spirit of the world.[57]

It is important to note here that Newton himself employs gold rarely if ever in the experimental records of his alchemy that he left in CU Add. 3973 and 3975. Quite possibly, then, his concern with dividing and purifying a variety of other metals in Keynes 41 was an attempt to realize Snyders's instructions that the elixir be composed from a metallic aggregate after each of the metals had been analyzed into its principles. It appears also that Newton believed this combining of different metals would induce them to activate and release their progenerative components, otherwise called their "sperm." As we will now see, Newton evidently believed that the processes of Snyders and Sendivogius could supply the very principle of "fermental love" that was lacking in the Lullian quintessence and Helmontian alkahest. Surprisingly, he enters into this discussion in the context of a much touted theme of eighteenth-century chemistry, the topic of elective affinity.

Elective Affinities and the Generation of Metallic Sperm in *Opus* Six

Newton is justifiably famous for his comprehensive treatment of elective affinity, the phenomenon that allows one metal to displace another that is already in a solution (often in a mineral acid). Although Newton himself did not employ the term "elective affinity," the topic and its ramifications form the core of the famous *Query 31* of the *Opticks* in its later editions, and Newton's treatment served to encourage a massive vogue for the subject in the eighteenth century. We will deal at length with the topic of elective affinity in chapter twenty-one of the present book. However, it cannot fail to attract our interest that Newton already discusses this subject in the *Opera* text, at least a decade and a half before it appeared in *Quaestio 23* of his Latin *Optice* of 1706, the precursor to the 1717 *Query 31*. The context for this discussion is precisely the need for the metals to acquire a ferment or "fermental love" so that they can lead to the formation of a transmutatory elixir. The topic of metallic fermentation comes up many times in the work of Snyders, and the analysis of Keynes 58 in our previous chapter described some of what Newton made of this difficult material. In his *Opera* manuscripts, Newton

---

[57] Keynes 41, 19r: "Defectu auri potest . . . Elixir ex materia prima (nempe spiritu mundi) & omnibus metallis simul in unum conjunctis fieri, nempe ex eorum Sulphure et sale quæ per sympathicum metallicū ignem extrahi oportet, Nam hæc magica sunt elementa et in spiritu mundi solvi." See also Newton's explicit claim in the *Index chemicus* that "all the metals enter into the composition of the stone" (*Metalla omnia compositionem lapidis ingrediuntur*), at Keynes 30/1, folio 61r. One should also note that Snyders's text lacks the term "spiritus mundi" here. See Snyders, *Commentatio*, 25, for the passage that Newton is glossing. Newton explains elsewhere in the text that the "spiritus mundi" that he added as a parenthetical gloss to Snyders's text is simply the mercury of Saturn, or a spirit distilled from it. For Newton's explanation of the term "spiritus mundi" as the mercury of Saturn, see Keynes 41, 12v and 18r.

repeats Snyders's comments at length, and then passes to the *Novum lumen chemicum* of Sendivogius.

It is "Tractate nine" of the *Novum lumen* that forms the immediate object of Newton's interest. There the Polish chymist discusses "the commixture of the metals, or the eliciting of the metallic seed." Sendivogius tells us that in order to understand the coition of the metals, so that they may emit and receive seed, we should think of the ordering of the geocentric cosmos. In their traditional Ptolemaic order, Saturn occupies the outermost sphere, followed by Jupiter, Mars, Sol, Venus, Mercury, and Luna. The influence of the planets, according to Sendivogius, descends toward the central earth but does not ascend. On the same principle, outer planets (that is, the metals represented by the traditional planetary *Decknamen*) can be easily transmuted into the inferior ones, but not the contrary. Thus, no doubt thinking of the famous transmutatory vitriol springs at Goslar and elsewhere in central Europe, Sendivogius says that iron is easily transmuted into copper, but not vice versa. This corresponds to the fact that in the geocentric cosmos, Venus is inferior to Mars. Similarly, Jupiter easily becomes Mercury, Sendivogius says, because Jupiter is second from the starry firmament and Mercury second from the earth. Finally, the outermost position of Saturn matches the innermost of Luna, meaning that lead is transmutable to silver, a fact that receives implicit support from the commonplace observation of argentiferous lead ores. Sendivogius then points out that Sol is found between each of these planetary pairs in the geocentric universe and adds cryptically that the alchemist who can "administer the nature of Sol to these mutations" will acquire a thing of greater value than any treasure. Sendivogius summarizes by adding that the chymist must not ignore which metal should be conjoined to which, nor overlook their mutual correspondences.[58]

Sendivogius's Tractate nine receives Newton's full attention in Keynes 41 and Dibner 1070A, where it becomes clear that he sees its justification as lying in the elective affinities existing between pairs of metals. It will be necessary to reproduce this critical passage from *Opus* six here so that we can examine its various features:

> To be sure, Jupiter and Mercury must be conjoined in order to generate a Sperm (*Novum lumen* tract. 9, pag. 46). And just as Mars and Luna in Magnesia first enter coition with Venus and Saturn in the *plumbum æris*, so afterwards Mars, Luna, and Mercury in magnesia have intercourse with Venus, Saturn, and Jove in the lead for generating sperm. Those which enter coitus for generating sperm must be separate before the coitus so that they enter into intercourse in the fermentation, and the motion of the fermentation is augmented by the coitus. Venus mates with Mars, Luna with Saturn, and Mercury with Jove, because an acid spirit <deserts>

---

[58] Michael Sendivogius, *Novum lumen chemicum* (Geneva: Joannes de Tournes, 1639), 45–47. By this time in his career, Newton was using the 1639 Geneva edition as opposed to the one edited by Nathan Albineus that he made use of in his earlier alchemy. See Harrison, number 1192, for Newton's copy of the 1639 edition, still found in the British Library.

Venus so that it may enter Mars, and deserts Luna so that it may penetrate Saturn, and deserts Mercury so that it may work on Jove. They also have intercourse in amalgamating because Mars may easily be mixed with Venus, Saturn with Luna, and Jupiter with Mercury. But the Mercury must first be purged.[59]

With these rich comments Newton interprets and expands on the Sendivogian correspondences between the metals. It is important to know first that for Newton in his maturity, "magnesia" typically means antimony. Thus in the Huntington Library's manuscript Babson 420, containing Newton's famous *Praxis* text (probably from the 1690s), magnesia is explicitly equated with "Antimony" and the symbol "♁."[60] How then do Mars and Luna combine in antimony? The answer is relatively simple, since in the Philalethan interpretation that Newton intimately knew, "Luna" can refer not only in traditional fashion to silver but also to the "reguline part" or "reguline mercury" of antimony, namely, the component of the crude mineral that we would today simply call elemental antimony.[61] This use of "Luna" as a *Deckname* for the hidden regulus within antimony ore appears several times in the full version of Newton's *Index chemicus*, which was composed only a short time after Dibner 1070A and Keynes 41.[62] From all this, then, it is clear that Newton is saying that iron and the reguline or mercurial part of the antimony combine when the iron acts on stibnite, with the result that a regulus sinks to the bottom of the vessel. Whether Newton is using "magnesia" here to refer to the crude antimony before their combination or the regulus containing iron is not entirely clear, nor is it significant for his present purposes. The point is simply that the iron and the hidden reguline "mercury" have mated to produce an offspring, which is the *regulus martis* of seventeenth-century chymistry.

In the same way that iron and antimony mate to produce regulus, so copper and lead are combined in *plumbum æris*. This "lead of copper" is the mysterious ore discussed by Johann Grasseus in his *Arca arcani*, and as I pointed out in an earlier chapter, Newton thought of this as a lead mineral containing copper. At any rate, Newton's point here is rather straightforward. The couples made up respectively of Mars (iron) and Luna (reguline "mercury" of antimony) in magnesia and the copper and lead in *plumbum æris* must first independently mate. Only then can their products be combined, apparently

---

[59] Keynes 41, 15r: "Iupiter utiᴑᴋ et Mercurius conjungi debent ad Sperma generandum (Nov. Lumen tract. 9. pag. 46.) Et quemadmodū primo coeunt Mars et Luna in Magnesia cum Venere et Saturno in plumbo æris, sic postea Mars Luna et Mercurius in magnesia coeunt cum Venere Saturno et Iove in plumbo ad sperma generandum. Quæ coeunt ad sperma generandum, ante coitum distincta esse debent ut in fermentatione coeant & ex coitu motus fermentationis augeatur. Coit Venus cum ♂, Luna cum ♄ & ☿ Mercurius cum ♃ quia spiritus acidus Venerem <deserit> ut Martem ingrediatur & Lunam deserit ut saturnum penetret & Mercurium deserit ut operetur in Iovem. ~~Sed Mercurius prius purgari debet~~ Coeunt etiam amalgamando quia ♂ cum ♀, ♄ cum ☽ & ♃ cum ☿ facillime miscentur. Sed Mercurius prius purgari debet." I have supplied the Latin word "deserit," which appears in the corresponding passage from Dibner 1070A, at 17r.

[60] Babson 420, 1r.

[61] See Philalethes, *Marrow*, part 1, book 3, page 43, stanza 34, and part 2, book 1, page 7, stanza 28.

[62] Keynes 30/1, folios 7r and 30r.

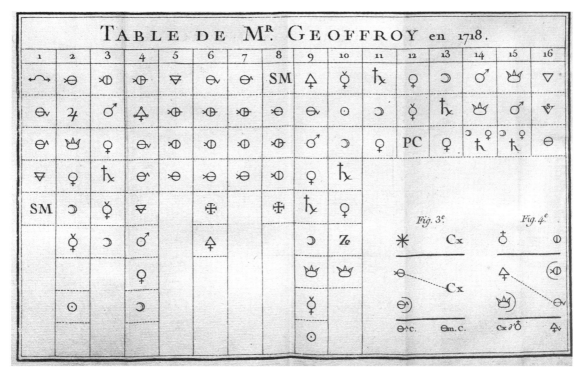

FIGURE 13.1. Affinity table or *Table des rapports* by Etienne Geoffroy, first published in 1718. The column labeled "3" has the symbol for nitric acid at the top and lists the metals in order of their descending reactivity with that "menstruum," namely, iron, copper, lead, mercury, and silver. From *Recueil de dissertations physico-chymiques, presentées à différentes academies* (Paris: Nyon l'aîné and Barrois l'aîné, 1781). Reproduced from the Cole Collection, by courtesy of the Department of Special Collections, Memorial Library, University of Wisconsin–Madison.

with the addition of quicksilver and tin, in order to provide another round of copulation and the maximum amount of fermentation. The copper in the *plumbum æris* will combine with the iron in the *regulus martis*, the reguline component of the *regulus martis* will unite with the lead in the *plumbum æris*, and the quicksilver will conjoin with the tin. How does Newton know that this matchmaking will occur as he predicts? It is at this point that he resorts to the phenomenon of elective affinity.

As he says, an acid spirit (or as we would say more simply, "an acid") dissolves iron in order to "desert" or precipitate copper. This is a simple fact of chymistry, already well known by the beginning of the seventeenth century, which we refer to today under the rubric of the reactivity series (see video clips at www.chymistry.org).[63] In 1718, Etienne Geoffroy, possibly stimulated by Newton's *Opticks*, would publish a famous table of elective affinities or *Table des rapports* that illustrates the facts of metallic solubility and precipitation in graphic form (figure 13.1). The fact that the symbol for iron is found closer to the top of the column surmounted by the symbol for "acide

[63] For at least one earlier writer on elective affinity, see William R. Newman, "Elective Affinity before Geoffroy: Daniel Sennert's Atomistic Explanation of Vinous and Acetous Fermentation," in *Matter and Form in Early Modern Science and Philosophy*, ed. Gideon Manning (Leiden: Brill, 2012), 99–124.

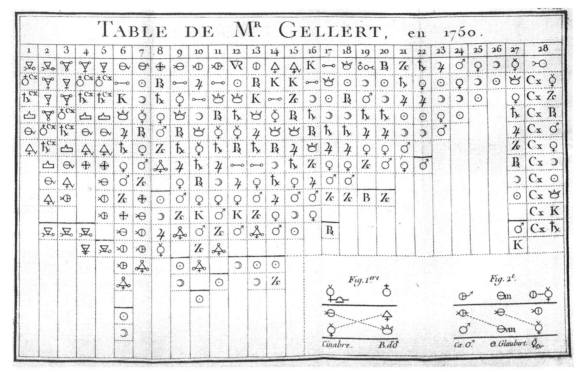

FIGURE 13.2. Table of precipitations by Christlieb Ehregott Gellert, first published in 1751. Gellert's table reverses the order of standard affinity tables, because he ranks his materials in the order of their precipitation, not their dissolution. Columns nine, ten, and eleven show that lead is more reactive than regulus of antimony with hydrochloric, nitric, and sulfuric acid. From *Recueil de dissertations physico-chymiques, presentées à différentes academies* (Paris: Nyon l'aîné and Barrois l'aîné, 1781). Reproduced from the Cole Collection, by courtesy of the Department of Special Collections, Memorial Library, University of Wisconsin–Madison.

nitreux" (nitric acid) than that of copper illustrates the iron's displacement of copper in a solution of nitric acid. Although Geoffroy's table does not record the relative solubilities of lead and antimony regulus in acids, Newton's observation that an acid spirit deserts the regulus to dissolve lead is confirmed by at least one other affinity table of slightly later date, namely, the much more extensive chart of precipitations published by C. E. Gellert in the middle of the eighteenth century (figure 13.2).[64] As for the relative solubilities of tin and quicksilver that Newton also mentions, these may be found under "acide du sel marin" (hydrochloric acid) in the table of Geoffroy.

Newton finds further evidence of an affinity between iron and copper, lead and regulus, and tin and mercury in the fact that one can "easily" make mixtures or alloys of these metals in a process that also involves amalgamation. Given the unlikely possibility of forming an amalgam of quicksilver and several of these materials, particularly iron, Newton may have been

[64] Gellert's table indicates that all three of the then-known mineral acids, hydrochloric, nitric, and sulfuric, have a greater affinity for lead than for regulus of antimony. It is important to know that Gellert's table intentionally reverses the normal order of affinities for each of the materials at the top of the columns. It must be read in reverse order, with the material having greatest affinity with the top substance found at the bottom.

making preliminary alloys and then amalgamating the alloys rather than the pure metals. The well-known artisanal practice of fire gilding armor and weapons, where iron objects were first plated with copper so that the mercury used to bond the gold leaf to the gilded object would stick, could easily have been his inspiration. Newton's younger contemporary Gellert does in fact record combinations of iron-copper, antimony-lead, and tin-mercury in his famous table.[65] It is easy to see why the ability of two metallic materials to mix intimately would supply evidence that they could "copulate," but less obvious that their respective solution and precipitation in an acid solution would provide substantiation of the same. Why would Newton have seen the fact that iron precipitates copper as evidence of an affinity between the two metals as opposed to an attraction between the metals and the acid, albeit weaker in the case of copper than of iron? The answer must be sought in the cosmic schema given by Sendivogius. In each of the three pairs of metallic materials, the superior "planet" precipitates the inferior one. Sendivogius matches the outermost, Saturn (lead), to the innermost, Luna (silver, or antimony in Newton's reading), the next from the firmament, Jupiter (tin), to the second from the earth, Mercury (quicksilver), and the third superior planet, Mars (iron), to the third inferior planet from the earth, Venus (copper). Just as the influence of the planets descends, according to Sendivogius, and does not ascend, so the superior planet-metals precipitate their inferior partners from the solution in which they were dissolved. Hence, rather than thinking in terms of a column of metals ranked in terms of their affinity to an acid at the head of the series, Newton at this point had in mind pairs of affinities corresponding to Sendivogius's cosmic scheme. There is no escaping the fact that Newton's early thoughts on the reactivity series are an attempt to justify and build on the cryptic remarks of Sendivogius.

Opus Six and Philalethes

Newton's ruminations on the Sendivogian copulation of the metals to produce sperm terminates with the advice that the quicksilver in its amalgamations must be thoroughly purged. This soon leads him into a discussion of the various methods for cleansing the sophic mercury that Philalethes gave in his *Experiments for the Preparation of the Sophick Mercury*, an extract from Starkey's notebooks that Newton owned in its printed Latin and English versions, and also in manuscript.[66] In addition, Newton recapitulates Philalethes's important gravimetric experiment for demonstrating that the sophic mercury acquires fermental virtue rather than increased weight during its

---

[65] See Gellert's table under the symbols for regulus of antimony (for its combination with lead), iron (for its combination with copper), and quicksilver (for its combination with tin).

[66] For the printed versions of Starkey's *Experiments for the Preparation of the Sophick Mercury* see Newman, *GF*, 268–69. The copies owned by Newton are found in Harrison's number 554 and 1407. A manuscript copy of the Latin text in Newton's hand has recently been purchased by the Science History Institute in Philadelphia.

"acuation" by means of repeated sublimation.[67] As Newton points out, the production of the Philalethan sophic mercury should not only purge the quicksilver of external filth but also add a metallic sulfur to it, which is the source of its fermental virtue.[68] From this Newton passes to other, more famous topoi in the corpus of Philalethes that also dealt with the "acuation" or animation of the sophic mercury, namely, its treatment with the doves of Diana and the alloy of copper and antimonial regulus that Newton, following Philalethes, typically called "the net." A close reading of this passage reveals that Newton's mature understanding of Philalethes, though quite different from the early thoughts that he expressed in Keynes 19 and other documents from the late 1660s or early 1670s, still differed strikingly from what we now know of Starkey's actual work in the laboratory. At the same time, a consideration of this passage from Keynes 41 displays remarkable parallels with some of the operations in Newton's laboratory notebooks and therefore sheds a light on the vexed problem of understanding the precise linkage between his reading and his practice at the bench:

> Otherwise the sublimation of Mercury is effected in two manners. For Mercury is united with Mars either through the ferment of the Doves, whose preparation is tedious, long, and very difficult, or through Venus in the net. For she has an affinity both for Mars and for very bright Diana, and being very eager to act, and also the sole medium by which Sol and quicksilver can be united, procures a true love between them. Mars is thus conjoined with the mercury of the regulus so that the two can never be separated until the soul of Mars be fixed into most perfect gold. And this does not occur except by the mediation of Venus, through whose association Diana separates them: although the doves of ~~Venus~~ Diana may also be used. Marrow of Alchemy, part 2, pag. 15, 16, 17, & part 1, pag. 44. So mix ♁, ♂, ♀, ♃, ♄, ♃☾ & and zinc, and you will have our Jove, and the helmet of Mercury. Amalgamate this with twice or three times as much Mercury so that the material be very thin, but not dropsical.[69]

The immediate pretext for this discussion is Philalethes's *Marrow of Alchemy*, where the "American philosopher" describes his making of the sophic mercury. As we discussed in an earlier chapter, Starkey originally thought that in order to perform the requisite sublimations of quicksilver from star regulus

[67] See Newman and Principe, *ATF*, 121–24.

[68] Keynes 41, 15r–15v: "Si mercurium his sublimationibus impraegnari velis cum sulphure metallico, sublimandus erit a proprijs corporibus per salem [nostrum] simplicem] ut rejiciat fæces & simul dissolvat corpora Postea opei <*sic*> convenit et potens est in dissolvendis speciebus metallicis."

[69] Keynes 41, 16r: "Cæterum sublimatio Mercurij duobus fit modis. Mercurius enim cum Marte unitur vel per fermentum Columbarum quarum præparatio tædiosa est & longe difficillima, vel per Venerem in reti. Hæc enim tam Marti affinis est quam ~~Mercurio~~ Dianæ nitidissimæ & amorem verum inter eos conciliat ad motum utiq; promptissima & medium unicum quo sol & argentum vivum nostrum uniantur. Mars cum mercurio reguli sic conjunctus est ut separari nequeant, donec Martis anima fixetur in aurum perfectissimum. Quod non fit nisi per mediationem Veneris per cujus associationem Diana illos separat: quamvis etiam ~~Veneris~~ ^Dianæ columbæ solæ usurpari possint. Marrow of Alk part. 2. pag. 15, 16, 17 & part. 1 pag. 44. Misce ergo ♁, ♂, ♀, ♃, ♄, ♃☾ & Zinetū, ^et habes Iovem nostrum, et <*illeg.*> galeā ☿ij. Hoc <*illeg.*> amalgama cum duplo ^vel triplo ☿ij, ita ut <*illeg.*> materia sit tenuissima sed non hydropica. ^Sublimetur ☿ a sale simplici ut fæces rejiciat & rursus amalgemetur."

of antimony, one must first alloy the regulus with twice its weight of silver, his "two doves of Diana." Between 1651 and 1654, however, Starkey had discovered that the sublimation would also work if copper was substituted for the much more expensive silver. He called his purple alloy of copper and star regulus "the net," in part because of its fine surface of crystals, which had a reticulated appearance, and in part because he associated it with the net that Vulcan used to ensnare the adulterous Mars and Venus in Ovid's *Metamorphoses*. I will deal with Newton's interpretation of the net at much greater length in the context of his laboratory notebooks, but several comments are in order here as well.

Newton's reference to "the ferment of the Doves" reveals once again that he did not accept the interpretation of the birds as simple *Decknamen* for metallic silver, even though it is highly likely that by now he had read the Latin version of Starkey's *Clavis*, the 1651 letter in which the American chymist openly revealed the secret.[70] On the other hand, Newton's experimental records reveal unequivocally that he did have a correct understanding of the net as an alloy of stellate antimonial regulus and copper.[71] He also understood that the Philalethan writings typically speak of Mars or iron as being contained within the martial regulus because iron is used to reduce the regulus from its stibnite ore. Starkey thought most of the iron (in the form of horseshoe nails added to the stibnite during its refining) was "destroyed" or dissolved into its principles in the process of reducing the regulus, though some unaltered iron would remain. Instead of thinking as we do today that the sulfur in the stibnite (antimony sulfide) was combining with the iron to produce ferrous sulfide and release the elemental antimony, Starkey thought the sulfur of the iron, which he viewed as identical to the sulfur of gold, combined with the "mercurial" component of the stibnite to create the regulus. This component of the iron was thought to be a volatile sulfur that was then carried over from the regulus into the net. But what exactly was one supposed to do with the net once the alloy of copper and regulus had been obtained? The next sentences that Newton paraphrases into Latin from the *Marrow of Alchemy* are among the most obscure that Starkey wrote. What does it mean when the *Marrow* says that Venus acts as a mediator between Mars and Diana, or that this mediation with Venus allows Diana to separate Mars from the "mercury of the regulus?"

Although the *Marrow*'s real meaning here is far from obvious even if one has read Starkey's letters and laboratory notebooks, I will hazard an interpretation. First, it is quite clear that Diana here cannot refer to the silver used in the pre-1654 sophic mercury process, since Starkey's copper process with the net was intended to supplant his earlier one with silver. We can therefore exclude the possibility that the *Marrow* is referring to an alloy of martial regulus (along with its putative volatile, ferrous component), copper, and

---

[70] Newton's copy of Starkey's *Clavis* is Keynes MS 18. For the origin of this text, see William R. Newman, "Newton's Clavis as Starkey's 'Key,'" *Isis* 78 (1987): 564–74; for a more complete version of the letter, see the edited text in Newman and Principe, *LTF*, 12–31.

[71] See CU Add. 3975, 43r: "R ♂ 9 1/4, ♀ 4 gave a substance wth a pit hemisphericall & wrought like a net wth hollow work as twere cut in."

silver. As noted a few paragraphs ago, however, the term "Luna" or "moon" appears in the *Marrow* not only as a synonym for silver but also as a *Deckname* for the regulus of antimony, and at times it could even stand for the sophic mercury itself.[72] Since Diana was universally acknowledged to be the goddess of the moon, it would not have been a great stretch to personify the celestial body by equating it with the divinity that ruled it. It is therefore not unlikely that Starkey is here engaging in a deliberate mystification by substituting Diana for the regulus. Thus when he says in the *Marrow* that Venus is "ally'd to Gold, And eke to Mars, also to *Dian* bright," he is probably referring to the net, which contains Venus (copper), "Gold" (the volatile sulfur from iron carried over into the regulus), Mars (the remnant of the iron horseshoe nails used by Starkey in reducing the regulus from stibnite), and the regulus itself in its role as Diana.[73]

Slightly less obscurity reigns in the *Marrow*'s claim that Mars is permanently joined to the mercury of the regulus "until the soul of Mars be fixed into most perfect gold." The idea is again that the sulfur or "soul" of iron was dissociated from the metal when it was broken down during the initial reduction of the antimonial regulus from its stibnite ore. Hence in the Philalethan interpretation, the soul or "volatile gold," also called "our gold," is now bonded permanently with the regulus unless it is extracted therefrom by quicksilver in the course of producing the sophic mercury by repeated sublimation from the net. As for the statement in Newton's paraphrase that "the two can never be separated until the soul of Mars be fixed into most perfect gold," this refers to a claim, particularly well developed in the Philalethan *Experiments for the Preparation of the Sopick Mercury*, that the sophic mercury can be digested by itself until it congeals, first into a *luna fixa* ("fixed silver") having the weight of gold and its resistance to aqua regia but lacking its yellow color, then into actual gold with all its normal properties, and finally, if one continues long enough, into an "Oyl as red as Blood."[74] At various points in his letters and writings, Starkey refers to this process as "extracting" the *Sol* and *Luna* from the sophic mercury, and in the Philalethan *Secrets Reveal'd* it is even said that "our Gold," meaning the hidden sulfur of iron in martial regulus, is vendible and may be sold without scruple once it has been "reduced to a Metal."[75]

A final obscurity in Philalethes's language must be put to rest before we see what Newton made of this difficult material. As Newton paraphrases the *Marrow*, the poem says that the fixation of the volatile gold or sulfur of iron "does not occur except by the mediation of Venus, through whose association Diana separates them." The information about Venus is straightforward: it is the copper in the net that allows quicksilver to amalgamate with the regulus and "extract" from it the soul or sulfur of Mars, which can then be fixed into metallic form. But what of Philalethes's claim that this association allows Diana to separate the volatile sulfur from the regulus? Obviously, Diana

---

[72] The *Marrow* refers to the sophic mercury as "our Moon" at part 2, book 2, page 23, stanza 2.

[73] The passage from the *Marrow* quoted here is found at part 2, book 1, page 15, stanza 59.

[74] Eirenaeus Philalethes, *Experiments for the Preparation of the Sophick Mercury*, 8, in Philalethes, *RR*.

[75] Philalethes, *SR*, 65.

cannot be the regulus here as it was before, since it is the regulus that is being separated into its components. Nor can Diana in her role as silver play any part, since the process employs Venus in the form of the net. Thus Philalethes presents us with a puzzle, or rather a riddle. The solution again may lie in the polysemic language that Starkey employed in his Philalethan writings. Quite possibly, "Diana" here again stands in for the moon, but in this case the moon itself more likely represents the sophic mercury than it does the regulus of antimony. If so, Philalethes is merely reiterating that the sophic mercury is able to extract the volatile sulfur of Mars from the regulus because of the amalgamation that it can undergo thanks to the help of Venus (copper). Newton believed Philalethes to be a master of polysemy, and in this case he was right.

Characteristically, Newton followed his own distinct path when confronted with these Philalethan obscurities. Immediately after paraphrasing the *Marrow*'s cryptic remarks that it is Venus's association that allows Diana to separate the volatile gold from the star regulus of antimony, Newton relates a recipe that is, I believe, original to him—"So mix ☿, ♂, ♀, ♃, ♄, ♃ & and zinc." In order to make sense of these directions, one must start at the beginning, with the word that I have translated "So" (*ergo*). The word "ergo," which could as easily be translated "consequently," "accordingly," or "therefore," can only be taken to mean that the recipe puts into practice the preceding material from the *Marrow*. But how can this set of instructions involving a mixture of stibnite, iron, copper, tin, bismuth, and zinc fulfill the allusive guidelines of Philalethes? In order to arrive at an answer, we must consider the ingredients as falling into two groups, the first made up of stibnite, iron, and copper, the second composed of the remaining constituents. Once we make this interpretive move, the rest becomes fairly simple, at least with the aid of Newton's experimental notebooks. We are quite familiar by now with the process of reducing regulus martis from stibnite by means of iron at high temperature. This accounts for the first two ingredients. The addition of copper would be required in order to turn the ordinary regulus into "the net," Newton's purple, reticulated alloy. Thus we have accounted for the presence of crude antimony, iron, and copper in the recipe. What then of the tin, bismuth, and zinc that follow?

Newton's two laboratory notebooks, CU Add. 3973 and 3975, contain a wealth of experiments involving bismuth, tin, and zinc (usually called "spelter"). Remarkably, Newton even refers multiple times to an alloy made of bismuth, tin, and bismuth ore as "Diana" in some records from 1682.[76] Although this particular product did not involve zinc (at least in its recorded form from 1682), Newton elsewhere made mixtures of regulus of Jupiter (antimony regulus reduced by tin) and zinc, copper, regulus of Venus (antimony regulus reduced by copper), and zinc, and other alloys of "spelter."[77] Given the transformation that Newton's chymical practice experienced over time, it is not at all unlikely that Newton's Diana alloy of 1682 may have undergone a modification by the time of the *Opera* text's initial drafts—composed no

---

[76] CU Add. 3973, 16v.
[77] CU Add. 3975, 75r–75v, 138v.

earlier than the second half of the 1680s—to include zinc as well. Since his first recorded experiments with zinc appear only after the end of February 1683/84, the metal may even have been unavailable to Newton in 1682.[78] Moreover, inclusion of zinc in the Diana alloy would have been an obvious move to try; like bismuth and tin, zinc is a white metal of relatively low melting point, which would have made it a natural candidate to use in an alloy involving tin and bismuth.

Thus it appears quite likely that Newton's incorporation of tin, bismuth, and zinc in his recipe was an attempt to combine a modified form of the Diana alloy found in his notebooks with the net. If so, then one can see how this recipe fulfilled the cryptic allocutions of the *Marrow*. Diana, in the form of the Diana alloy, was supposed to extract the hidden "volatile gold" or martial sulfur from the net, where it had been transported from the initial reduction of the regulus by means of iron. The product, Newton says, will be "our Jove, and the helmet of Mercury," in other words, a substance to be combined with mercury. Only at this point, in Newton's interpretation of the "American philosopher," should the alloy be amalgamated with quicksilver. My analysis of Newton's process receives further support if one looks at the following lines, where he terminates this section of *Opus* six with the following operations:

> ∧The ☿ should be sublimed from simple salt so that it cast forth its feces and should again be amalgamated. Then make a ferment in the quantity of a hazlenut of the volatile king and queen, and the water-bearer who is the father of each of them, and when all has become water, throw in a small part of the amalgam, and when it ∧has dissolved, add more and more until all the water takes on the form of the amalgam. Wash off the feces, sublime, and do this seven times adding ∧perhaps only ♂ and ♀ in the later sublimations. The living gold and living Luna should be extracted and they should copulate eleven times. Philal. in Ripl. Gates. p. 105, 106, 113, 114, 115, 116, 133, 134. Thus Mercury ∧after his purgation strikes his helmet with his caduceus and infects the Nymphs with an incantation ∧and dissolves the metallic species.[79]

Here Newton first subjects the foregoing amalgam to additional purging with common salt, a straightforward process, then makes a ferment from "the volatile king and queen" and "the water-bearer," probably with the aid of an unspecified menstruum. These *Decknamen* stem from Philalethes's *Ripley Reviv'd*, where they participate in an alchemical process that results in the drowning and dissolution of the royal couple along with the water bearer.[80]

---

[78] The first reference to "spelter" in CU Add. 3975 comes at 73v, four full folios after the date "Feb. 29 1683/4." The first reference to the material in CU Add. 3973 appears at 19r, which is dated "Apr 26ᵗ 1686".

[79] Keynes 41, 16v: ∧Sublimetur ☿ a sale simplici ut fæces rejiciat & rursus amalgemetur. Dein fac fermentum in quantitate nucis avellanæ ex rege et regina et̶ volatili & Aquario qui utrius♃ pater est, & Vbi totum aqua est injice particulam amalgamitis & ubi hæc dissoluta<illeg.> ∧est injice plus & plus donec aqua tota formam induat amalgamatis injecti. Fæces ablue sublima & hoc fac septies addendo ∧forte solum ♂ et ♀ in sublimationibus posterioribus. Extrahantur aurum vivum et Luna viva & coeant undecies. Philal. in Ripl. Port. p. 105, 106, 113, 114, 115, 116, 133, 134. Sic Mercurius ∧post purgationem suam caduceo galeam ferit, & incantando Nymphas inficit ∧dissolvit♃ species metallicas.

[80] For an analysis of this fable, see Newman, *GF*, chapter four.

In several other manuscripts from his mature period, Newton equates the king and queen with the two serpents of the caduceus, and the water bearer with the rod.[81] As we saw from our analysis of Newton's manuscript Keynes 58 in the foregoing chapter, the serpents around the caduceus were the volatile vitriol composed of salts of copper and iron sublimed with antimony compounds, while the rod, equated there with the scepter of Jove, was a complicated material involving lead, tin, and bismuth. We shall have occasion to return to these materials and processes in a subsequent analysis of Newton's *Praxis* manuscript, but for now it is sufficient to point out that his goal here is the extraction of *aurum vivum* and *luna viva*, the living gold and silver that in the Philalethan tradition refer to the volatile gold or metallic sulfur and "mercurial" component of the regulus, both of which go into the making of the sophic mercury.[82]

Newton then terminates this section by reiterating the whole process in language derived from the *Exposition of the Hieroglyphicall Figures* attributed to Nicolas Flamel. One of the images in Flamel's "Book of Abraham the Jew" showed Mercury holding "in his hand a *Caducaean* rodde, writhen about with two *Serpents*." The god then "strooke vpon a helmet which couered his head" by hitting himself with the caduceus.[83] For Newton, Mercury puts on his helmet when the quicksilver is amalgamated with the net and Diana alloy of the previous passage, after which the amalgam is then purged with salt. Once we understand that the caduceus with its two snakes and its rod are the same as the king, queen, and water bearer, then it follows that Mercury's striking of his helmet with the caduceus is equivalent to the dissolution of the king, queen, and water bearer and their amalgamation with the purged sophic mercury. The result of all this is evidently the production of an extremely powerful menstruum capable of dissolving "metallic species."

The Return of Lull and the End of *Opus* Six

The reader may well wonder where the wise man of Mallorca lies camouflaged in the dense thicket of enigmata that we have just examined. One can rest assured; Lull has not been forgotten. After explicating the conundrums of Sendivogius, Snyders, Philalethes, and Flamel, Newton returns near the end of the *Opera* text to his original topic of discussion, the quintessence of the school of pseudo-Lull. The subject is still the cleansing and purgation of the sophic mercury, as it was in the florilegium's earlier treatment of Lull, but the literary focus has changed. The immediate pretext of

---

[81] Keynes 21, 16v: "The same thing is described in the figures of Abraham ye Iew where Mercury strikes on his helmet wth his rod & saturn wth wings displayed ~~comes~~ & ^an hour glass on his head comes running & flying against him as if he would cut off his leggs. ffirst ye serpents are twisted about ye rod by fermentation; ffor these three are ye King Queen & Water bearer or ye fire ye liquor of ye vegetable Saturnia & ye bond of ^whe ☿ ~~in Philaletha~~ ^in Philaletha." See also Keynes 53, 2v, where the identification between the rod of the caduceus and the water bearer is again made.

[82] Newman, *GF*, chapter four.

[83] Nicolas Flamel, *Nicholas Flammel, His Exposition of the Hieroglyphicall Figures* (London: Thomas Walkley, 1624), 11–12.

Newton's discussion of this Lullian theme lies in his analysis of a key text that, like Dickinson's *Epistola*, was published in London during the final decades of the seventeenth century. The collection in which the tract appeared is the *Aurifontina chymica* edited in 1680 by John Frederick Houpreght, and the work itself bears the improbable title *Hydropyrographum hermeticum* (Hermetic Fire-Water Text). The *Hydropyrographum* is in some respects similar to Dickinson's later exchange with Mundanus in that it explicitly models the quest for the philosophers' stone on the practice of making alcoholic quintessences in the tradition of pseudo-Lull. Asserting that "none of all the Philosophers hath written more clearly nor better than *Raymund Lullie*," the author says that ordinary quicksilver must first be purged with a sublimation from common salt, an old and well-known technique that Newton referred to earlier in *Opus* six. The sublimate is then cast into warm water, dried, and distilled with salt of tartar. These operations will free the quicksilver from "its extraneous moisture and feculency," but they cannot liberate it from the terrestrial impurity that *"lies hid in its inmost center."* For that, one needs more drastic forms of purification, and more secretive as well.[84] This fit Newton's brief exactly, so he paraphrased the *Hydropyrographum*'s instructions for carrying out this Lullian purgation at some length:

> Nor can it be severed otherwise then by reducing it into its primum ens or materia prima by putrefaction wthout addition of any thing heteregeneal, ∧wch primum ens is a milky crystalline silvery liquor clear as the tears of the eye & if it be not thus putrefied & opened the menstruum will not be worth a fig. But w^n it is thus dissolved into its primogeneal water we may cleans its inside ∧from the extraneous water & fæculent earth by destillation, as Philosophers have described by ye rectification of spirit of wine, & cohobation upon its own earth till it come over with it, & accuation of this wine wth its own salt. And this spirit of wine thus prepared *<illeg.>* resolves new ☿ into ye primum ens or primogeneal water, whereby it is multiplied wthout end by putrefaction ∧(of 40 days) & destillation (Hydropyrogr. p. 20, 21, 22, 23, 24, 25, 26. Thesaurus p. 102, 103, 104).[85]

The author affirms that his mercury must be putrefied by itself so that no extraneous material enters into the process. In this fashion it will be able to return to its *primum ens* or first matter, a milky, yet crystalline liquor. The *Hydropyrographum* explicitly says that this process is modeled on the sharpening or acuation of "spirit of wine" (ethanol). Just as the wine is distilled to isolate its burning spirit, and the spirit is further purified from watery phlegm by cohobation with its own salt, so the mercury is acuated by processing it with materials drawn from itself. As in Mundanus and Dickinson, the model substance here is salt of tartar, made from lees of wine by calcination and lixiviation, and then used to absorb the excess water out of the desired ethyl alcohol.

---

[84] Anonymous, *Hydropyrographym hermeticum*, in Johann Frederick Houghpreght, *Aurifontina chemica* (London: William Cooper, 1680), 22–23.
[85] Dibner 1070A, 19v.

But a careful look at Newton's paraphrase reveals that he is actually discussing an additional operation beyond those necessary to acquire the *primum ens* of mercury. The passage also states that the *primum ens* may be "multiplied w^th^out end by putrefaction ^(of 40 days)^ & destillation." This information did not come from the *Hydropyrographum* itself, but from another English tract printed in the *Aurifontina*, namely, the *Thesaurus, sive medicina aurea* (Treasure, or Golden Medicine). The author of that treatise says that the mercury will putrefy into a "Milkie, Crystalline, and Silver Liquor or Water" in a period of three or four months, but after that initial dissolution has occurred, "thou mayst ever after dissolve more and more Mercury in fully fourty days."[86] In other words, the *primum ens* of mercury has the ability itself to return more purged quicksilver into the state of a "primogeneal water," and this process takes only forty days as opposed to the initial three or four months required to make the first batch of *primum ens*. Newton was fascinated by this prospect, and in fact the remainder of the *Opera* text as found in Dibner 1070A and Keynes 41 is taken up with the theme of putrefying and multiplying mercury.

Newton now assembles a set of passages from various texts that support the views of the *Thesaurus* and the *Hydropyrographum*. In particular he returns to the themes of the white, foliated earth (*terra alba foliata*) and the white sulfur of nature that he discussed in *Opus* four.[87] The fact that Newton links these products to the new narrative provided by the *Hydropyrographum* and the *Thesaurus* shows that there is more textual coherence to the *Opera* than meets the eye at first glance. Indeed, Newton now returns to Philalethes and Snyders once again, treating the multiplication of the sophic mercury as found in their work and clothing it in the quintessence-language of the Lullian school. Newton first recapitulates Snyders's analysis of the metals into the three chymical principles by means of a "dry solution." The section is extremely interesting because it shows that Newton was not merely engaging in a process of uncritical heaping up of synonyms and parallels; nor did he believe that all alchemical authors were simply saying the same thing in different language. Here he explicitly underscores Snyders's ignorance of certain alchemical themes and processes found in Philalethes, Sendivogius, and other authors:

> Snyders very little <*minime*> understood the purgation of the ☿ by means of the rod and the extraction of the living gold <*aurum vivum*> and its conjunction in the hour of its nativity.[88]

After making this surprising observation about Snyders, Newton then passes to a discussion of the German's "wet solution," which was supposed to follow the analysis of the metals into their three principles that Snyders referred to

[86] Anonymous, *Thesaurus, sive medicina aurea*, in Houghpreght, *Aurifontina chemica*, 103.

[87] Keynes 41, 17v.

[88] Keynes 41, 19v: "Nam purgationem ☿^ij^ & fermentationem per virgam & extractionem auri vivi & conjunctionem in hora nativitatis Snyders minime novit." This claim of Snyders's ignorance also occurs in Dibner 1070A at the end of 17v. There is a similar statement in Royal Society MS MM/6/5 on 6v. For "*the hour of the Stones Nativity*" see Philalethes, *Ripley Reviv'd*, 278. See also Philalethes, *Secrets Reveal'd*, 75.

as "dry." In the section of the *Commentatio* referred to here, the German chymist says that the wet solution is carried out by the mercury of Saturn, which he equates with the soul or spirit of the world.[89] Newton repeats this passage in *Opus* six, adding that "soul of the world" is identical to "the white spirit" and the quintessence.[90] That Snyders himself does not use the term "quintessence" at this point underscores the fact that Newton is still actively embroidering his Lullian tapestry with flowers from another garden. According to the final lines of the *Opera* text in Keynes 41, Snyders putrefies the metals previously "opened" by dry solution in this menstruum and thereby acquires what the simpler Lullian quintessence of *Opera* one through five could not achieve—a permanent union of the purified principles in which "the solvent remains as it were inseparably with its solute." It seems, then, that Newton thought he had found the answer to the problem posed by the absence of fermental love in the quintessence and the alkahest discussed at length in *Opus* five. Despite his having followed a different path from that of Philalethes, Snyders had supplied the quintessence with its ferment and roused the alkahest from its perennial ennui: no longer cursed by the narcissism of its one-sided action without reaction, the marvelous dissolvent of Van Helmont could now enter into a reciprocal relationship with the subject of its passion. The result, Newton thought, would be the philosophers' stone.

Conclusion

Apart from the many technical details of his chymistry that lie buried in the *Opera* florilegium, the most striking feature of this literary exercise lies in the extraordinary practical precision that Newton hoped to extract from the Lullian treatment of the quintessence. He was obviously not alone in thinking that Ramon Lull had hidden an elaborate process for making the sophic mercury beneath a deceptive discussion of alcoholic spirits; in this, Newton was preceded by Dickinson, Mundanus, and the anonymous author of the *Hydropyrographum hermeticum*, among many others. But the almost fetishistic attention to detail that Newton applied to his analysis of multiple works of the Lullian school, collating different editions of the same books to check for their variants and cross-referencing text upon text, is astonishing. And yet we must resist the temptation to ascribe this extreme concern with detail to a clinical obsession such as graphomania. Anyone who has tried to replicate old chymical processes from their description in books and manuscripts—never mind whether they involve chrysopoeia or not—will understand the problems that Newton faced. The failure to perform a single washing, precipitation, filtration, or sublimation can easily spell failure, even when one has interpreted the ingredients, apparatus, proportions, and temperatures of the process correctly. Newton's seeming infatuation with detail

[89] Snyders, *Commentatio*, 66.
[90] Keynes 41, 19v: "Et hanc fieri ait per mercurium Saturni quem vocat animam mundi, id est per spiritum album vel Quintessentiam. p. 66."

was a natural and even a correct response to the monstrously difficult task of interpretation that he faced. The remarkably fastidious analysis that he brought to bear on alchemical texts is the same precision that he brilliantly applied to experimental optics, mathematics, and the study of gravity.

The real problem was at a deeper level. For Newton, the adepts could not be wrong, even if, like Snyders, they might take a divergent path to the ultimate goal of the philosophers' stone. What this perfect comprehension implied is that even ancient or early medieval texts such as the Arabo-Latin *Turba philosophorum* concealed processes and materials that we now understand to have been unknown until much later times. Even such seemingly humble substances as distilled ethyl alcohol and the mineral acids were unfamiliar, in reality, to the author of the *Turba*. As for the pseudo-Lullian corpus with its sometimes laughable vaunting of the powers of ethanol, this exaggeration was clearly a product of the infancy in which knowledge of alcoholic distillates still lay at the time when Jean de Roquetaillade wrote his seminal work on the quintessence. For Newton it meant something else entirely, namely, that the real subject of the Lullian medico-alchemical corpus, hidden beneath a delusory story about wine, tartar, and alcoholic distillates, lay elsewhere. The fundamental problem that Newton faced was not a psychiatric disorder but an educational one; like most of his contemporaries, he had a minimal understanding of the history of science and technology. In an age when Copernicus and his early followers could think that Pythagoras upheld the heliocentric system, Newton's blindness is understandable, and for a person of his gifts, perhaps even inevitable.

At the same time, however, we must not overlook the possibility that Newton's belief in the infallibility of the adepts led him to scientific insights of his own. The discussion of elective affinities buried in Newton's *Opera* is a case in point. Would Newton have engaged in his study of affinities at all if he had not been induced to do so by his desire to understand Sendivogius and other chrysopoetic sources? An even more extensive discussion of displacements resulting from solution and precipitation occurs in Keynes 58, the same manuscript that contains Newton's ambitious attempt to put Snyders's work into practice, although Newton's comments on affinity here seem to be a précis of another author's work.[91] In reality, Newton's attempt to extract practical meaning from writers such as Sendivogius and Snyders cannot be dissociated from his own experimental enterprise, since the investigative records preserved in his laboratory notebooks are largely attempts to test and refine processes that he extracted or reworked from chymical books and manuscripts. It is time now to turn our full attention to these remarkable documents, which reveal unequivocally that Newton the chymist was in all respects the equal in experiment of Newton the physicist. Only after examining these laboratory records on their own terms will we be able fully to appreciate his integration between text and practice in the very manuscript that bears the Greco-Latin version of that word—*Praxis*.

[91] Keynes 58, 7r.

FOURTEEN

The Shadow of a Noble Experiment

NEWTON'S LABORATORY RECORDS TO 1696

The Setting of Newton's Laboratory

The reader of Newton's alchemical transcripts, synopses, indexes, single- and double-column analyses, and florilegia, in other words the material that we have been examining over the last five chapters, might easily form the opinion that these documents represent the work of a pure armchair chymist unfettered by the labors of the laboratory. Newton's decades of literary decipherment in the interest of reducing riddles to practice provide little sense of the enormous physical effort that he devoted to the aim of joining the exalted ranks of the adepts, a goal that could only be attained by the experimental acquisition of the philosophers' stone. The cause of this peculiar absence of personal experimental records in the main bulk of Newton's chymical *Nachlass* lies not in inaction on his part but in his rigorous preliminary application of textual decipherment to his sources before taking them into the laboratory. The enigmas of the sages required sustained analysis as verbal puzzles before they could receive testing at the bench. Let us now turn to the experimental side of Newton's deep engagement with alchemy. Some appreciation of his labors flows from the breathless account of his secretary in the second half of the 1680s, Humphrey Newton, who responds to queries by John Conduitt in the following fashion:

> He very rarely went to Bed, till 2 or 3 of ye clock, sometimes not till 5 or 6, lying about 4 or 5 hours, especially at spring & ffall of ye Leaf, at wch. Times he us'd to imploy about 6 weeks in his Elaboratory, the ffire scarcely going out either Night or Day, he siting up one Night, as I did another till he had finished his Chymical Experiments, in ye Performances of wch. he was ye most accurate, strict, exact: What his Aim might be, I was not able to penetrate into but his Paine, his Diligence at those sett times, made me think, he aim'd at somthing beyond ye Reach of humane Art & Industry.... On ye left end of ye Garden, was his Elaboratory, near y^e East end of y^e Chappell, where he, at these sett Times, employ'd himself ... with a great deal of satisfaction & Delight. Nothing extraordinary, as I can Remember, happen'd in making his Experiments,

FIGURE 14.1. Detail showing Newton's lodgings and garden at Trinity College, Cambridge. His laboratory may have been located in the roofed shed attached to the chapel at the right. From the 1690 *Cantabrigia illustrata* of David Loggan.

wch. if there did, He was of so sedate & even Temper, y^t I could not in ye least discern it.[1]

Beyond the picture of a sleep-deprived Newton engaged in feverish experimentation, Humphrey's account conveys the valuable information that Newton's chymical laboratory was located within his garden near the east end of the Trinity College chapel. A famous copper engraving in the 1690 *Cantabrigia illustrata* of David Loggan reveals that Newton's garden had a high wall around it, and that access to it from his room descended by means of an enclosed stair to a ground-floor loggia (figure 14.1). There is also a small, shed-like structure located within the bay of the chapel adjoining the living quarters. Traditionally, scholars have assumed this to have been the location of Newton's experimental efforts.[2] Although a recent archaeological study has thrown doubt on the claim that this small space contained Newton's laboratory, Loggan's print gives no additional contender other than the loggia itself, and no chymical apparatus appears there in his print. In a word, the precise location of the laboratory in Newton's garden remains at

---

[1] Humphrey Newton to John Conduitt, January 17, 1727/28, in Keynes 135, 2–3, from *NP*, accessed August 2, 2017.

[2] Dobbs, *FNA*, 98. Although Dobbs must be referring to the shed rather than the loggia, since she says that it was located "at the Chapel end of the garden," she incorrectly states that the structure consisted of "two stories." This error descends from J. M. Keynes, who uses the same expression in his "Newton the Man." See Dobbs, *FNA*, 98n9, where she references the relevant passage from Keynes.

present unresolved; one hopes that additional archaeological research will be forthcoming.[3]

In the absence of any visual or archaeological record of Newton's laboratory, we have little other information than that which the experimental notebooks provide. Humphrey's supplementary remarks in a letter composed in February 1727/28 are entirely pedestrian:

> About 6 weeks at Spring & 6 at the ffall the fire in the Elaboratory scarcely went out, which was well furnished with chymical Materials, as Bodyes, Receivers, heads, Crucibles &c, which was made very little use of, the Crucibles excepted, in which he fused his Metals: He would sometimes, thô very seldom,) look into an old mouldy Book, which lay in his Elaboratory, I think it was titled,—Agricola de Metallis, The transmuting of Metals, being his Chief Design.[4]

Any practicing chymist of the seventeenth century would have had "materials" on which to work; unfortunately, Humphrey says nothing about their nature. As for the bodies, heads, and receivers of his letter, they refer respectively to the three parts of a contemporary still, namely, the large flask in which material was heated, the "capital" or still head in which distilled material condensed and ran down the snout, and the vessel in which the distillate was collected. Finally, Humphrey mentions crucibles, namely, the ceramic dishes in which solid materials were heated to high temperatures. According to the amiable secretary, these were the only apparatus that Newton used on a regular basis, but this claim throws much of Humphrey's testimony into doubt. By working through Cambridge University Additional manuscripts 3973 and 3975, the principal records of Newton's alchemical experimentation, Peter Spargo has discovered references to a panoply of specialized chymical apparatus regularly employed by Newton. These include retorts, various mortars and pestles for grinding different sorts of materials, fire shovels, water baths, candles, iron plates, both open and sealed glass flasks, quills acting as small spatulas, crucibles, egg-shaped flasks with long necks, large-mouth glass vessels for catching distillates or sublimates from heating on shovels and the like, cold water baths, receivers, filtering apparatus, and a special earthenware apparatus for performing a sort of fractional sublimation.[5] Further records of Newton's apparatus appear in his purchase lists from 1669, 1687, and 1693; the lists of 1687 and 1693 reflect acquisitions

---

[3] Peter Spargo, "Investigating the Site of Newton's Laboratory in Trinity College, Cambridge," *South African Journal of Science* 101 (2005): 315–21.

[4] Humphrey Newton to John Conduitt, February 14, 1727/8, in Keynes 135, 5, from *NP*, accessed August 2, 2017. I have modified the Newton Project's transcription slightly, in accord with the reading given by Peter Spargo in his article "Newton's Chemical Experiments: An Analysis in the Light of Modern Chemistry," in *Action and Reaction: Proceedings of a Symposium to Commemorate the Tercentenary of Newton's Principia*, ed. Paul Theerman and Adele F. Seef (Newark: University of Delaware Press, 1993), 123–43, see 127. I have also changed the Newton Project's reading of the date of the letter, February 17, to February 14, the reading of David Brewster, *Memoirs of the Life, Writings, and Discoveries of Sir Isaac Newton* (Edinburgh: Thomas Constable, 1855), 2: 98.

[5] Spargo, "Newton's Chemical Experiments," 135–36.

made in London from two apothecaries, "Mr Stonetreet" and his successor in the business, "Mr Timothy Langley."[6]

Beyond the glassware and ceramic implements described in Newton's experimental notebooks and lists of purchases, the most obvious feature of his laboratory would have been the charcoal-burning furnace or furnaces located there. Like many contemporary chymists, Newton regularly performed his experiments at the high temperatures required for metallurgical operations, and the need for specialized ovens is already recognized in the Fitzwilliam manuscript, where Newton mentions that he bought "A ffurnace" and "A tin ffurnace" in 1669.[7] The furnaces pictured in the University of Chicago manuscript Schaffner Box 3 Folder 9 would also have served him well, but we now know that Newton copied these from another source rather than designing them himself, as we discussed in chapter five. Hence one cannot be certain of the design or form of his furnaces, though they may well have followed the instructions laid out by the anonymous *Treatise of Chymistry* that Newton pillaged (figures 14.2 and 14.3). It is by means of these "Vulcanian implements," along with the apparatus referred to piecemeal in his experimental notebooks, that Newton managed to produce such exotic desiderata as "liquor of antimony," "the net," and "sophic sal ammoniac." Let us now turn our attention to his remarkably precise, if guarded, instructions for making these and other material precursors to the philosophers' stone.

Introduction: Methodological Principles

To say that Newton's chymical laboratory notebooks are daunting in their complexity would be an exercise in understatement. The sheer mass of archaic procedures and operations found there is intimidating enough, but the fact that Newton often uses alchemical *Decknamen* in a highly idiosyncratic fashion adds another dimension of discouragement. In addition to using enigmatic terms such as "sophic sal ammoniac" and "the scepter of Jove," Newton even employs the contemporary language of technical chymistry in ways that are peculiar to him. The word "vitriol," for example, which normally means a sulfate (usually of iron or copper) in seventeenth-century Britain, usually means something else entirely in Newton's notebooks; for him, it is generally a complex, crystalline product made by dissolving and then evaporating a metal in a solution of aqua fortis (mostly nitric acid) that has been "sharpened" with ordinary sal ammoniac and in which stibnite, the sulfide ore of antimony, has been dissolved. The linguistic complexity of the notebooks by itself has been enough to mislead several modern researchers seriously, and this is only the beginning of the difficulties that they present.[8]

---

[6] CU Add. 3975, 174v.

[7] Fitzwilliam Museum Newton notebook, 8r.

[8] As for example Spargo, "Newton's Chemical Experiments." In his "Table 1," between pages 129 and 132, Spargo identifies Newton's "Vitriol made with ♀" as simple copper sulfate.

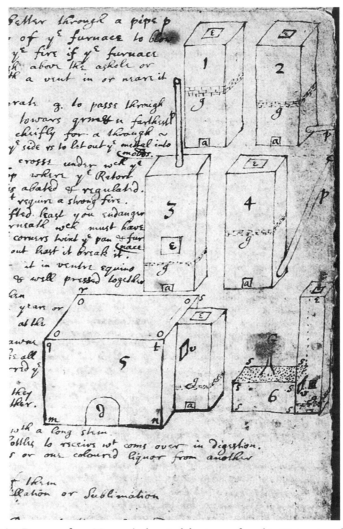

FIGURE 14.2. Furnaces from Newton's chymical dictionary found in University of Chicago, MS Schaffner Box 3 Folder 9. Their source is the anonymous *Treatise of Chymistry* found in the Sloane Collection of the British Library.

In addition to basic problems of language, one must also add that the overall purpose of the experiments is never spelled out, even when Newton adds a line or two of conclusions at the end of a given set of operations. Unlike George Starkey's notebooks, to name one example, Newton's laboratory records never give a transparent history of his past successes and failures; nor do they plot out his future course. And to make things even more difficult for historians, there is the fact that like most recorded chymical experiments and recipes, Newton's records embody layer upon layer of tacit practices and skills that he did not bother to write down. In many instances we find ourselves in a position like that of a child trying to make an exotic torte from a cookbook without knowing the proper way to crack an egg, never mind having no practical experience with such additional techniques as separating out the yolk, whipping the white, and folding in the flour.

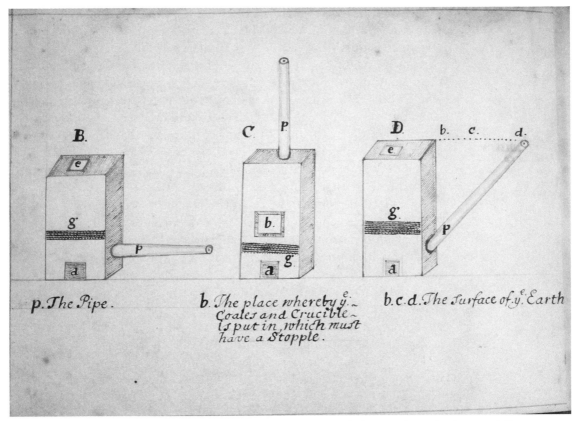

FIGURE 14.3. Three of the furnaces found in the anonymous *Treatise of Chymistry* (British Library, MS Sloane 2206) copied by Newton. The rest of Newton's furnaces are found on subsequent pages of the Sloane manuscript.

Nonetheless, historians of alchemy have learned that laboratory notebooks—when we are lucky enough to have them—usually provide the best means to understanding a given author's practice and motivations. And these are not the notebooks of just any researcher, but the laboratory records left by one of the greatest experimenters of all time. In our favor we have the extraordinary degree of precision for which Newton was, and still is, known. In addition, Newton makes it clear when he has actually performed a given set of procedures, usually using the first person past tense and sometimes even providing dates; this allows us to distinguish actual experiments from "conjectural processes" (a useful term of Starkey's) that were only planned out or copied but not necessarily put into practice. Finally, as we shall see, Newton repeats the same protocols time and time again, often making it possible to determine his motives by comparing one experiment to another.

In order to make sense of this challenging material, it will be necessary to express a caveat and then to come up with a set of ground rules. The caveat is straightforward: since Newton does not normally express his motivations outright, we must arrive at them by an indirect route. In the absence of clearly stated authorial goals, our method will often be one of reasoned inference rather than direct translation of Newton's language into our own. In some cases this means that we will interpret Newton's laboratory operations and materials in the light of their "standard use" among early modern chymists. Fortunately, writers such as Nicolas Lemery, a famous writer of chymical textbooks with which Newton was familiar, left detailed instructions explaining the panoply of operations typically employed in the laboratories of the time. And Newton's favorite authors Robert Boyle and George Starkey, whose works are heavily excerpted in Newton's notebooks, provide valuable clues. But it is not enough for us merely to read these texts and assume that we have gained a proper understanding of the material at hand. In many instances one simply cannot comprehend Newton's meaning without repeating the very processes that he describes. Hence I will employ a combined method here of interpreting Newton's text; in the fashion of a traditional historian I will decipher his words and those of his sources, but where necessary I will supplement this material with modern laboratory replication.

The path of reasoned inference supplemented by laboratory replication does not in itself provide an escape from the bewildering maze of Newton's experiments, however. The richness of this material requires us to tease it apart and to approach it from multiple, successive perspectives. Fortunately, one of Newton's collections of notes, Cambridge University Additional MS 3973, consists of a set of loose sheets arranged in strict chronological order, as shown by the occasional dates that he provides in the text. This means that we have a sequential, if partial, record of Newton's chymical experimentation, ranging from December 10, 1678, to the last recorded date of February 1696. Hence it is possible, at least in principle, to follow Newton's projects as they grow and develop over time in CU Add. 3973.

In addition there is Newton's bound chymical laboratory notebook, CU Add. 3975, which is 174 folios long and which contains the second version of Newton's famous early treatise "Of Colours" near the beginning of the manuscript, as I discussed in chapter six. This volume, which Newton acquired in bound form, is a heterogeneous collection of reading notes, records of experiments, and short essays. In certain respects, CU Add. 3975 displays the nature of a commonplace book, since much of it is organized around headwords such as "Of cold & freezing" (folio 15v) and "Of fformes & Transmutations wrought in them" (folio 32r). But there are several features that CU Add. 3975 does not share with ordinary commonplace books. In particular, CU Add. 3975 contains forty-two closely written folios describing Newton's alchemical experiments (41v–43v; 52r–80v; 136r–144r; 173r–173v), some dated, but most not. At present I will be concerned with the manuscript from folio 41v on, where Newton begins recording his own experimental notes, interspersed occasionally with passages excerpted from Robert Boyle, George Starkey, and other chymical writers. The experimental records appear for the most part to have been copied from earlier drafts (including parts of CU Add. 3973), making CU Add. 3975 a sort of master repository of Newton's efforts in the realm of laboratory chymistry. Despite belonging to a second or third generation of copying, they are entered for the most part in chronological order, as the occasional interspersed dates make clear. Among the chymical experiments in CU Add. 3975, the earliest date is May 10, 1681 (on 62r), and the last one belongs to 1693 (174v), where Newton records apparatus and chemicals bought from "Mr Timothy Langley," a London apothecary. As other scholars have argued, however, some of the experimental material found before the May 10, 1681, entry is considerably older than this explicit date; parts may even date to the late 1660s, since other, nonchymical experiments in the notebook definitely do date from that period.

There is also a final source of Newton's laboratory chymistry, and one that has been largely overlooked by previous scholars. I refer to the dissociated pamphlet made of a single folded sheet and found in the midst of the loose pages making up Boston Medical Library manuscript B MS c41 c, which, although undated, appears to stem from the earliest period of Newton's recorded alchemical experimentation. This short document will provide important clues to Newton's operational work as we move forward.

Despite these clear chronological markers, our presentation is complicated by the fact that Newton's modus operandi consisted of several distinct methods. On the one hand, he was trying to read and comprehend the extraordinarily difficult riddles provided by previous alchemical writers such as Michael Sendivogius, Eirenaeus Philalethes, and Johann de Monte-Snyders. We have already analyzed the purely literary techniques that Newton used to decipher these and other authors' *Decknamen* and enigmata in the foregoing chapters. His method included tools such as cross-referencing, substitution of one term for another, expansion of passages that he considered abbreviated, elimination of excess terms that he thought were inserted to obscure the sense of a recipe, and assembly of passages that he assumed to have been separated on the principle of the "dispersion of knowledge"; in short, the standard repertoire of decoding methods early modern alchemists used. After solving these verbal entanglements to the best of his ability, Newton would then try the results of his decipherment in the laboratory. Hence his experiments represent Newton's attempts to test the correctness of his interpretations and to modify them in accordance with laboratory experience. The integration between alchemical enigmata and laboratory experience in Newton's work can best be studied after one has a solid idea of his overall experimental methodology. I therefore provide a case study of his combined approach in chapter sixteen.

There is also another way to approach the experimental notebooks, however, and that is through the developing techniques that Newton employs, in the light of his occasional comments about their success or failure. Here we find ourselves squarely in the realm of "vulgar chymistry," the technical discipline that Newton learned from reading works such as the anonymous *Treatise of Chymistry* described in our chapter five along with the publications of Boyle, Starkey, Lemery, and many others. It is here that we encounter the protocols that emerge time and time again in Newton's laboratory notebooks, and it is the development of these repeated operations that will form the main subject of the next three chapters. As I have argued earlier, especially in chapters seven and eight, Newton's laboratory notebooks put into practice the fusion of mechanism and hylozoism that characterizes his early treatises found in Dibner 1031B—*Of Natures obvious laws & processes in vegetation* and *Humores minerales*. The fundamental status that he accorded to reactions taking place in the vapor or gaseous state is reflected in the procedures of the experimental notebooks with their lengthy chains of operations built around multiply repeated dissolution, sublimation, and distillation of minerals. Not all early modern chymists followed this path to the fabrication of the philosophers' stone; it is conspicuously different from the methods favored by Boyle and Starkey (based on the processes of the *Clavis* or similar ones), and also different from what we know about the operational methods employed by Snyders. We should see Newton's experimental notebooks as adhering to the lead of his own theory of the subterranean generation of metals, minerals, and salts, as he laid it out in Dibner 1031B.

Let us briefly consider a representative (if composite) example of these standard procedures here, in their fully mature form from the 1690s. Newton first takes a material containing one or more metals and/or metalloids, dissolves these in aqua fortis with the addition of stibnite and sal ammoniac or related compounds, and heats them until they dissolve. Normally a "calx" or heavy precipitate will remain at the bottom of the vessel during dissolution, so Newton then decants the solution, filters it, and boils it to dryness. Often before evaporating the solution he dilutes it in water in order to get another precipitate, which remains trapped in the filter paper. He then takes the salt that has been left behind after evaporation and sublimes it, usually with a sublimate that has already been prepared either explicitly from stibnite and sal ammoniac or with "sophic" sal ammoniac, a proprietary material whose exact production he never describes. He carefully notes the volatility of the newly subliming material as well as its fusibility during the sublimation, its color, texture, and very often its taste. Then Newton turns to the "calx" or precipitate that was left behind during the dissolution. He washes the calx, dries it, and then tries to sublime it, either by itself or with more of his sublimed stibnite and sal ammoniac (or with his sophic sal ammoniac), or even with the sublimate that he arrived at by heating the evaporated salt in the first step. In order to be sure that everything volatile has sublimed, Newton often employs multiple levels of heating—sometimes the gentle heat of a candle under a plate of glass, more often the higher temperature afforded by a sand bath used with a retort, and sometimes even the violence of a naked fire used to drive up a sublimandum on a fire shovel.

Often individual steps are left out and other steps intervene, but the same pattern of dissolution, evaporation, washing, precipitation, sublimation, and resublimation occurs with remarkable frequency throughout Newton's notebooks. It is clear that he is not slavishly copying and repeating the experiments of previous alchemists, since that approach would have involved him in a multitude of unrelated processes using the full scope of materials employed in early modern alchemy, which ranged from dung and urine to alcohol and May dew, and terminated with gold itself. In reality, these substances are mostly if not wholly absent from Newton's experiments. Instead of employing them, Newton has devised a standard series of tests or assays in order to arrive at the materials that he deems to be required for the alchemical magnum opus. As we shall see, this series of tests, with numerous variations, allows him to exclude particular techniques and materials with considerable confidence. The tests themselves, as well as the materials exposed to them, are subject to constant scrutiny and modification on the part of their creator. It is in fact the chronological evolution of Newton's laboratory practice taken from a largely phenomenological perspective that will form the main subject of the present chapter.

Working backward from Newton's experiments to his likely motivations requires one further operational principle worthy of mention. As we will see from our analysis of Newton's "standard reagent" liquor of antimony, he was in the habit of employing different names for the same substance. Vinegar, spirit, and salt of antimony are all terms that Newton freely exchanges with "liquor." A similar pattern often emerges from Newton's creation and use of additional names and graphic symbols. A careful reading of the laboratory notebooks reveals, for example, that a sublimate of copper created by elevating an antimonial "vitriol" of the metal with a mixture of sublimed sal ammoniac and stibnite, and then precipitating out the insoluble antimony compounds with water, is identical to "volatile Venus" ("Ven. vo." or "Ve. vo.").[9] The term "volatile Venus" is at other times replaced with the symbols ♀ and ☿, again meaning the sublimate of salt (or "vitriol") of "Venus antimoniate" (♀ ♂^{iate} or ☿).[10] The interchangeability of these terms and symbols suggests an important and powerful tool that we can employ in deciphering Newton's laboratory records, namely, Ockham's razor. As we proceed into the labyrinth of Newton's chymical nomenclature we should continually ask ourselves if the emergence of an unfamiliar term necessarily implies the existence of a new material, or if, to the contrary, the same substance is reappearing under an altered name. Applied with discretion, the principle of parsimony will sometimes be our most faithful guide.

Armed with these principles, let us therefore proceed to a consideration of Newton's laboratory records. The present chapter treats the records of Newton's earliest systematic experimentation in chymistry, from roughly 1669 up until the mid-1670s. Here I rely mainly on CU Add. 3975 and Boston Medical Library B MS c41 c, since they contain records from that period. Chapters fifteen and sixteen provide a study of three chymical projects that run throughout CU Add. 3973 and 3975 all the way up to their termination in the 1690s. All three themes deal with the preparation and purification of

---

[9]CU Add. 3975, 59r; the so-called vitriol is not simple copper sulfate, as one might expect, but a compound prepared in advance by imbibing copper with liquor of antimony, then drying or crystallizing the soluble product.

[10]For ♀ see CU Add. 3973, 16v. For ☿ see CU Add. 3973 32r. The symbol ☿ is itself built on ♭, which merely signifies the salt of Venus antimoniate, and that symbol is in turn derived from ♭, which simply means Venus antimoniate. Venus antimoniate probably refers to the unrefined product of imbibing copper with Newton's "liquor of antimony," whereas salt of Venus antimoniate is the crystalline "vitriol" extracted from it. The sublimate of Venus antimoniate refers to the so-called vitriol once it has been sublimed with an adjuvant such as Newton's "sophic sal ammoniac." In general, Newton's term "antimoniate" means a salt of a metal and antimony. This fact emerges clearly from a passage in CU Add. 3973, where Newton is discussing lead antimoniate. He describes a sublimation that began with "100gr of ♄ ♂iate (wch well dried might perhaps have weighed 95 gr.)." The fact that Newton's lead antimoniate contained 5% water points unequivocally to a hygroscopic salt, not a metallic alloy. See CU Add. 3973, 5r.

Newton's standard laboratory reagents, and these records can be examined without much recourse to Newton's explicitly chrysopoetic sources. Since the material in CU Add. 3973 is closer to Newton's original jottings made in the laboratory, and since it is more comprehensively dated, chapters fifteen and sixteen use this manuscript as the main point of departure. Finally, chapter sixteen returns to the issue of Newton's alchemical sources in order to present a case study of his integration of experiment and textual analysis. The focus of this section is Newton's work on the net, a purple alloy of martial regulus of antimony and copper that he inherited from the corpus of Philalethes and put to his own uses.

The Earliest Phases of Experimentation

The Influence of Robert Boyle and Newton's Liquor of Antimony

No better evidence for the potent influence that Robert Boyle exercised on the young Isaac Newton can be found than the first forty-one folios of CU Add. 3975. I have already argued that Boyle's work on the analysis and resynthesis of materials ranging from saltpeter to stibnite and turpentine served Newton as a powerful heuristic in his argument that white light consisted of unaltered spectral rays whose respective colors were cloaked by the illusion that their combination produced on the human retina. Not surprisingly, Newton had already assimilated Boyle's 1664 *Experiments and Considerations Touching Colours* when he composed the first draft of his own treatise "Of Colours" in his student notebook "Certain Philosophical Questions."[11] When we pass to the commonplace book–cum–laboratory record, represented by CU Add. 3975, a number of additional Boylean works are extracted or cited, among which *Essays of the Strange Subtilty, Determinate Nature, <and> Great Efficacy of Effluviums, The Usefulness of Experimental Natural Philosophy, New Experiments and Observations Touching Cold, Certain Physiological Essays,* and *The Origin of Forms and Qualities* are prominent. As we proceed more deeply into CU Add. 3975, Newton's topical headings change from entries like "Of cold & freezing" (19r), "Rarity, Density, Elasticity, <*illeg.*> Compression &c" (24r), and "Of fire, flame, ye heate & ebullition of ye heart" (26r), to explicitly chymical topics. The subject heading "Of fformes & Transmutations wrought in them" (32r) reveals that Newton was busily digesting Boyle's *Origin of Forms*; more than any other source, this text served as the immediate pretext for the transformation of CU Add. 3975 from a commonplace book to an active record of experimentation.

As B.J.T. Dobbs argued in her *Foundations of Newton's Alchemy*, however, Newton's first chymical experiment recorded in CU Add. 3975 was almost certainly inspired by another Boylean text, namely, the 1669 edition of

[11] Newton, *CPQ*, 440–42, 454–60.

Certain Physiological Essays.[12] The passage of interest to Newton appears in the midst of Boyle's discussion of "fluid bodies" and the fact that they do not always mix when contiguous. As an example of this phenomenon, Boyle dissolves quicksilver in nitric acid to arrive at a clear, seemingly homogeneous solution; the subsequent addition of lead filings produces an immediate and striking effect:

> Dissolve one Ounce of clean common Quick-silver in about two Ounces of pure Aqua fortis, so that the Solution be clear and total, then whilst it is yet warm, pour into it by degrees, lest they boyl over, half an Ounce or one Ounce of Filings of Lead, and if no Error, nor ill Accident have interven'd, the Lead will be in a trice praecipitated into a white Powder, and the Mercury reduc'd into a Mass (if I may so speak) of running Quick-silver, over which the remaining part of the Aqua fortis will swim.[13]

As Dobbs points out, this is straightforward chemistry. It is in fact a classic case of a replacement reaction, with the quicksilver coming out of the acid solution as the lead dissolves. What impressed Boyle about this reaction was the immediate separation of three distinct "bodies"—the clear acid solution on top with the quicksilver and white powder below. There is no hint of chrysopoetic intent in Boyle's description, and a few lines later he is careful to deny that the quicksilver emerging from the reaction is "the true Mercury of Lead." Nonetheless, his following admission that it is "somewhat differing from common Mercury, and fitter than it for certain Chymical uses" seems to have triggered Newton's interest. Hence we find him using Boyle's description (with the same quantities of quicksilver and aqua fortis) as the model for a grander experiment that he clearly carried out in the laboratory:

> In Aqua fortis 2℥ dissolve ☿ 1℥ or as much as it will dissolve. Then put an ounce of Lead laminated or filed into it by degrees & ye lead will bee corroded dissolving by degrees into ☿ & besides there will fall downe a white præcipitate like a limus being y^{e} ☿ præcipitated by y^{e} 🜍 of ♄. Out of an ounce of ♄ may bee got 1/3 ℥ of ☿ If the remaining liquor bee evaporated there remains a reddish matter tasting keene like sublimate. The same liquor will extract y^{e} ☿ of ♃. If ♀ bee put into it, it is presently covered wth ☿ I know not whither y^{t} ☿ come out of ye liquor or of ♀ for ye liquor dissolves ♀. Also ♀ will draw ☿ out of ye limus wch falls down in dissolving ♃ or ♄ & also out of ye liquor both during ye solution & afterward.[14]

In analyzing Newton's remarks, it will be helpful to use the word "quicksilver" for the "vulgar" material that we moderns recognize as an element, namely, the "Hg" of the periodic table, and to employ the term "mercury" for the chymical principle that went by that name. Remarkably, Newton completely disregards Boyle's rather mundane explanation of the experiment, interpreting it instead as an unequivocal separation of the lead into

[12] Dobbs, *FNA*, 139–41.
[13] Robert Boyle, *Certain Physiological Essays*, in *Works*, 2: 147; 1669, 202.
[14] CU Add. 3975, 41v.

its mercurial and sulfurous constituents. For Boyle, the white, powdery precipitate was simply the lead in another form, whereas Newton sees it as the quicksilver previously dissolved in the acid solution and now precipitated by the liberated sulfur of the lead (y^e ☿ præcipitated by y^e 🜍 of ♄). In his view, the silvery, heavy liquid that also came out of solution was not mere quicksilver, but the mercury of lead, which had been freed from its accompanying sulfur. Buoyed by this success, Newton then tried the same experiment on tin and copper. Although he seems to be completely satisfied with his isolation of tin's mercury, questions emerge when he arrives at copper; the problem for us is determining the precise nature of his doubts. According to Dobbs's reading of the passage, Newton probably deduced from the changing color of the solution (it would turn blue or green as the copper dissolved) that something other than a mere separation of the copper into its constituents was occurring. As she sees it, the blue or green color would have tipped him off to the fact that copper was simply being dissolved intact by the acid. But in fact Newton does not refer to the blue or green color at all, and it is by no means sure that its presence would have dissuaded him from the view that the mercury of the copper was being separated from the metal. After all, who is to say that the internal mercury or sulfur of copper could not be colored? So what does Newton mean then when he says, "I know not whither y^t ☿ come out of ye liquor or of ♀ for ye liquor dissolves ♀"?

Although he is silent on the subject of the changing color of the solution, Newton does refer explicitly to another phenomenon. In fact, the primary observation on which he comments is of a distinctly different nature than color change. He is openly impressed by the fact that copper is not only quickly covered by mercury or quicksilver during the metal's own dissolution but that the copper can even extract quicksilver from the limus (powdery or muddy precipitate) left respectively by the lead and tin. In order to understand the significance of this, we need to return to the beginning of the experiment. At this early stage of his alchemical career, Newton clearly thinks that the acid-quicksilver solution really has freed up the mercury of lead. But of course this means that the lead's other internal principle, its sulfur, was also liberated in the process of separation. The freed sulfur of lead then must have combined with the vulgar quicksilver in the solution to make the white precipitate or "limus," as Newton himself says. Once freed from its own plumbic mercury, the sulfur would naturally have combined with the vulgar quicksilver that was present in the solution. When Newton observes that copper can extract mercury from the precipitate that fell down when lead or tin are "analyzed," it follows—on the strength of his belief that that material was just vulgar quicksilver in combination with a metallic sulfur—that the mercury collecting on the copper when it is put into the solution could also be mere vulgar quicksilver rather than the actual mercury of the copper. Moreover, Newton probably wondered at the absence of a cupric precipitate or "limus." If the silvery material collecting on the surface of the copper were really the chymical principle mercury, why did the liberated sulfur of the metal not combine with the vulgar quicksilver in solution, as he observed that it had in the case of lead? Ironically, it is therefore Newton's own commitment to the mercury-sulfur theory of metallic composition that leads him to question whether the theory has received an ocular demonstration from the experiment with copper.

The next two experiments (or rather prescriptions for practice) in CU Add. 3975 were, like the one for the mercuries of the metals, probably inspired in part by Boyle, as Dobbs has also noted. The first of these consists of an attempt to arrive at the mercuries of the metals by baking corrosive sublimate ($HgCl_2$—here called "Venetian sublimate") with sal ammoniac and then gently heating the product either with the metal (powdered or filed) or with a regulus of the metal. This is the earliest mention in CU Add. 3975 of different metallic reguli, a theme that would soon occupy Newton at great length. Like numerous early modern chymists, Newton put great stock in the fact that a number of different metals could be used to reduce the metalloid antimony from its sulfide ore, stibnite.[15] In accordance with contemporary usage, Newton refers to these as reguli of the metals themselves—hence regulus of iron (*regulus martis*), regulus of lead (*regulus saturni*), regulus of copper (*regulus veneris*), and regulus of tin (*regulus jovis*). Additionally, he speaks of "Regulus of ♂" as something distinct; presumably he is thinking here of *regulus antimonii per se*, the metalloid reduced by means of partially calcined crude tartar (often called "black flux"). A few lines after explaining how to produce these reguli separately, Newton then says that the metals can be put into the crucible "successively according to their fusibility ♂. ♀. ♂. ♃. ♄." It appears therefore that Newton may already have been making compound reguli of multiple metals at this stage of his development, a feature of his practice that is well developed in later parts of the notebook. He advises that they be added in reverse order of their melting points: iron, which melts at about 1538°C, should be placed in the crucible with the stibnite first, and lead, which melts at about 327°C should come last.

References in his reading notes suggest that Newton thought of the production of reguli as purging the metal that acted as a reducing agent in the same way that assayers purified gold by means of stibnite.[16] This reasoning had in fact provided the major impetus to George Starkey's procedure for making the sophic mercury, namely, the creating of an ultrapure, subtle form of quicksilver by amalgamating it with metallic antimony and then repeatedly distilling it. That Newton would combine such a process of purgation with an attempt to isolate the mercuries of the respective metals is not surprising, since his goal consisted of arriving at the mercurial principle in pure form. Equally unsurprising is the fact that he follows this with a procedure for making corrosive sublimate and for detecting adulterated versions of it, since he was using that material again to arrive at the pure mercuries of the metals.

The addition of sal ammoniac to corrosive sublimate here has already drawn the attention of Dobbs, who again correctly notes a Boylean source.[17]

---

[15] See J. W. Mellor, *A Comprehensive Treatise on Inorganic and Thoeretical Chemistry* (London: Longmans, Green, 1929), 9: 350.

[16] I mean "reducing agent" in the modern "redox" sense here. In his commentary to Johann de Monte-Snyders's *Metamorphosis of the Planets*, for example, Newton says (addressing antimony), "Thou purgest yᵉ leapers (impure metals by making them regulus's)." See Cushing, 4r.

[17] Dobbs, *FNA*, 141–42.

In his *Origin of Forms*, Boyle includes a long succession of experiments in which he employs corrosive sublimate to "open" the bodies of the metals. After observing the changes wrought on silver and copper by common corrosive sublimate, Boyle describes a "new kind of Sublimate" that he made by subliming equal quantities of sal ammoniac and normal corrosive sublimate.[18] This new sublimate seemed to work more radically on copper than the normal sal ammoniac, so Boyle suggested that future researchers try it on gold. He ends the experimental report with a note of oracular excitement— "But of This, having now given you a Hint, I dare here say no more."[19] Such a hint could not miss the sharp eye of Newton, and though he does not explicitly reference the passage in his laboratory notebooks, Boyle's words appear in slightly altered form in one of Newton's early chymical dictionaries, Oxford Don. b. 15. Probably composed somewhat before the passages in CU Add. 3975 that we have been discussing, Don. b. 15 recapitulates Boyle thus—"p^{r}haps there may bee Sublimates made (as by subliming common sublimate & Sal Armoniack ∧$^{\text{well poudered together}}$) wch besides notable operations on other metalls, may act upon Gold too."[20]

Newton, like Boyle, initially hoped to use corrosive sublimate in combination with sal ammoniac to "unlock" or "open" the metals, the term "open" a verbal play on the "opening" of a cadaver by means of a scalpel to connote the analysis of a metal or other material. Although Boyle's suggestion of subliming corrosive sublimate with sal ammoniac was unusual, the "opening" of metals was part of the program common to many seventeenth century alchemists to arrive at a solvent or "menstruum" that would loosen up the structure of metals and possibly serve as a means of transmuting them. Yet it seems that Newton soon grew disappointed with corrosive sublimate as a tool for attaining this end. Apparently unsatisfied with the results of these experiments, Newton abandons them abruptly on folio 53r of CU Add. 3975 and passes to a discussion of other menstrua, such as oil of vitriol or sulfuric acid. His attention would soon shift from this well-known menstruum to a far more obscure one, namely, the substance—for I will argue it is basically one substance—hiding under the terms "liquor of antimony," "spirit of antimony," "vinegar of antimony," and "salt of antimony."

After pointing out on 53r that oil of vitriol grows hot when water is mixed with it, Newton introduces the material that he here calls "spirit of antimony," and points out that adding it to oil of vitriol also produces considerable heat. He then expands on this observation, commenting:

> The sp^t of ♂ once destilled grew warm also by mixing it wth water, & much more would ~~it after it is desti~~ it after a full separation from ye flegm by ye next preparation.[21]

18 Boyle, *Origin of Forms,* in *Works,* 5: 403; 1666, 300.
19 Boyle, *Origin of Forms,* in *Works,* 5: 404; 1666, 302.
20 Don. b. 15, 7r.
21 CU Add. 3975, 53v.

Unfortunately, Newton reveals nothing here about the method of producing spirit of antimony, but the reference to spontaneous warming on addition of water suggests that it, like oil of vitriol, contains a high percentage of one of the mineral acids, sulfuric, hydrochloric, or nitric. This suspicion is increased by the next few lines, where Newton advises that spirit of antimony can be used to "draw" or extract the salts of various metals:

> This sp^t onc<e> destilled draws y^e Salts of some metals (of ♂, ♀, Wismuth, Cobalt, ♃) but not of ♄ $^{∧yet \ wth \ heat \ it \ works \ on}$ ♄.

It is clear from this passage that the spirit of antimony is a menstruum or dissolvent, and that Newton is using it to produce, or as he thinks of it, to "extract," salts of iron, copper, bismuth, cobalt, and tin. Lead is more problematic for him, but as he says, the spirit of antimony will even work on that metal if heated.

Unlike Newton's use of corrosive sublimate, which he abandoned after a few folios, this employment of spirit of antimony continues throughout the length of CU Add. 3975 after its initial introduction on folio 53r. The material became a staple reagent in Newton's armament of menstrua. Newton was still using spirit of antimony in the mid-1690s, for he informs us in a passage of CU Add. 3973 dated February 1696 that he was working with lead ore "impregnated ten years ago wth distilled spirit of ♂ in ye proportion of 9 to 4."[22] Arriving at the concrete referent lurking behind "spirit of antimony" is a deeply worthwhile goal if we are to understand Newton's alchemy, since he employed this material prominently in his experimentation for at least two decades, and probably more. As we shall see, this elusive substance also reflected the influence of Boyle on the young Newton, though in order to demonstrate that point we must first establish its identity. This is not an easy task, first because Newton's terminology for this material varies from place to place in his notebooks, and second because the terms used for it could, at least prima facie, refer to a number of different chymical products.

First we must establish the fact that spirit, liquor, vinegar, and salt of antimony all refer to the same material. It is not at all unreasonable to suppose that these terms would serve for the same substance. In seventeenth-century chymistry, a spirit is typically a distilled fluid, or at least a material that is volatile enough to undergo distillation. The word "liquor" simply means a liquid, as in modern English. As for "vinegar," this term was already used in the Greek alchemical corpus to describe any sort of acidic material, which would include our modern vinegar as well. Finally, for our early modern chymists, the term "salt" can refer to the active saline ingredient in a liquid; it does not have to stand for a dry, crystalline substance. Hence Newton often speaks of "fluid salts" and describes them frequently as acidic. If one puts all of this together, it appears that we are looking for an acidic liquid that is either capable of being distilled or is already a distillate. But none of this provides evidence that Newton himself was using these terms as synonyms. For that we have to return to the experimental notebooks.

---

[22] CU Add. 3973, 43v.

Careful reading combined with digital tools such as latent semantic analysis reveal numerous instances in CU Add 3975 and 3973 where Newton seems unconsciously to pass from one term to the other while repeating the same protocols, which suggests that the terms mean the same thing; let us examine one such case here.[23] In a late passage from CU Add. 3975, Newton describes an experiment with regulus of tin fused with copper:

> Reg ♃ 2, Copper 4 melted together & poudered & ground fine & imbibed wᵗʰ ♁ twice, drank up yᵉ liquor slowly & with difficulty in the heat of blood, tasting acid till it was dry.[24]

The symbol "♁" combines the normal signs for salt, "⊖," and antimony, "♁"; clearly, it represents "salt of antimony." Since this material is imbibed by the alloy, it must be a liquid. Moreover, Newton explicitly says that the alloy "drank up yᵉ liquor," shorthand here for the "liquor of antimony," referring to the salt of antimony in a dissolved state. Finally, Newton says that the liquor tasted acidic, strengthening my suggestion that the liquor is an acid and hence capable of being termed "vinegar." So we have a fairly firm identity between "salt of antimony" and "liquor of antimony," and a hint that this material could also be termed "vinegar of antimony." The hint becomes a certainty if we look at two parallel passages. The first, from CU Add. 3973 (11r), describes a yellow, deliquescent salt made by adding "distild Vinegre of ♁" and several other materials to an antimonial sublimate. The parallel passage from precisely the same experiment in CU Add. 3975 (59r) is almost verbatim identical except that the distilled vinegar of antimony has become "distilled liquor of ♁." Clearly in copying the manuscript Newton replaced "vinegar" with "liquor," an unproblematic substitution if the two terms represent the same substance. The same substitution occurs at other points in the two manuscripts as well, further cementing their identity.[25]

One could continue with this line of argument, since there is plenty of additional evidence supporting the fact that Newton's spirit, liquor, vinegar, and salt of antimony are one material, but instead let us pass to a more difficult problem, namely, the concrete referent hiding behind these terms. Henceforth I will use these four terms as synonyms, unless otherwise stated, and will also refer to all four simply as "liquor of antimony" for the sake of simplicity. But finding their material identity is not as simple as it might sound, first because Newton never explicitly tells us how he made this substance, and second because terms like "vinegar of antimony" and "spirit of antimony" were used to mean different things by Newton's sources. At least one candidate for this material is found in the works of Basilius Valentinus, for example, whose pseudonymous writings were among Newton's earliest alchemical sources. Basilius's *Triumphal Chariot of Antimony*, which

[23] My discovery of parallel passages in the Newton corpus has been facilitated by using the latent semantic analysis tool developed by Wallace Hooper for *CIN*.

[24] CU Add. 3975, 138r.

[25] Liquor of antimony and vinegar of antimony are again equated in the parallel passages found at CU Add. 3973, 9r, and CU Add. 3975, 54v.

Newton owned in the English translation printed in 1678, contains a recipe for the vinegar of antimony that treats stibnite as though it were a vegetal material that could undergo fermentation by the simple addition of water. The stibnite is melted, ground, and fermented in distilled water until it releases a froth. The water is distilled, the residue sublimed, and refluxed with the water repeatedly, the residue being sublimed after each cohobation. Each time the process is repeated, Basilius assures us, the vinegar grows more acid. Is this how Newton made his antimonial vinegar? Probably not. Nowhere in the Newtonian alchemical *Nachlass* have I found Newton commenting on this process, nor is there anything similar to this procedure in his laboratory notebooks. Additionally, modern attempts to replicate the process of Basilius have so far been inconclusive.[26]

In reality, Newton's liquor of antimony was a material very different from the Basilian vinegar, but in order to demonstrate this fact, it will be necessary to present some documentary evidence that has hitherto been ignored by Newton scholars, namely, the fragment of a laboratory notebook found in the Boston Medical Library, namely, B MS c41 c. There are several strong hints that this manuscript is an early record, for example the rather basic nature of the experiments described in the text and also the fact that Newton employs the avoirdupois system of weighing there as opposed to the apothecary weights characteristic of his mature notebook entries. Boston Medical Library B MS c41 c also contains a record of what may have been Newton's earliest experiments with the material that he would come to call "liquor of antimony."[27] In the following account, he describes an experiment that provides important clues to its nature and composition. It is found on folio 1r of the manuscript, thus:

> ☿♎ ℥ 4, ☾ ℥ 4 gives liquor ℥ 3/16 $^{(= 90\ grains)}$—3 or 4 grains $^{\wedge (= 90\ grains)}$ of congealed salt wch drop<t> into y^e receiver, <*illeg.*> and in y^e neck of the Retort was 880 grains of salt besi<de> <*illeg*: what?> $^{\wedge\ was\ lost}$ y^t stuck in y^e neck of y^e retort wch might be about 30 gr <*illeg.*>.

This is a straightforward recipe for butter of antimony (or antimony trichloride in modern terminology), though Newton does not say so. When corrosive sublimate or mercuric chloride is sublimed with antimony trisulfide, the mercury combines with the sulfur and the antimony combines with the chloride to form solid butter of antimony in the neck of the retort, as Newton says. Interestingly, Newton does not use the term "butter of antimony" here for the solid antimony trichloride that collects in the neck of the retort. He may or may not have realized that he was making that material, since this is an early manuscript, but what concerns us is the fact that he calls this product a "salt" and indicates that the process also produced a "liquor."

[26] Lawrence Principe, "Preparing the Vinegar of Antimony," *Quintessentia* 2 (1981), http://homepages .ihug.com.au/~panopus/essentia/essentiaii4.htm, accessed December 21, 2015.

[27] On 1v of Boston Medical B MS c41 c, Newton describes experiments with "sal ☿ distilatus aridus," which is probably mainly butter of antimony. Newton would later use the term "salt of antimony" as a synonym for the product arrived at by dissolving stibnite in ammoniacal aqua regia, as I argue in the remainder of this chapter.

Could these be the salt of antimony and liquor of antimony to which he refers many times in CU Add 3973 and 3975?

Acting on this clue, I worked through a number of experiments in CU Add. 3973 and 3975 that involved stibnite, particularly those in which Newton refers to the formation of butter of antimony during the process. Performing these experiments revealed considerable evidence to support the claim that Newton's liquor of antimony consisted primarily of antimony trichloride dissolved in one or another acid solution. The presence of an acid is important, since butter of antimony, despite being quite hygroscopic, reacts with water to form an insoluble white precipitate of antimony oxychloride, or *mercurius vitae* as chymists called it, if its acidic solution becomes too weak. Since Newton wanted to use this material as a menstruum, he needed to keep the solution quite acid.

According to CU Add. 3973, Newton began a series of sublimations in January 1680 that started with a material that he calls "♁ once acted on by ⊕." The second of these experiments makes it clear that the choice of stibnite that had already been treated with aqua fortis was an accident. One of Newton's perennial goals was that of making the sublimate of stibnite less dark, or as he puts it, "dirty." Thus he says the following:

> The sublimate I used in these exp^{ts} being old, I made some new wth crude ♁. This was fouler then ye former & had much ~~imp~~<*illeg.*> red dusty sulphur adhering to ye top of ye glass, wch made me suspect ye former was not made by crude ♁ but by ♁ once acted on by ⊕.[28]

Newton clearly preferred the stibnite that had been acted on previously by aqua fortis to the untreated crude antimony. This was evidently a case of the fortunate accident in the laboratory leading to positive results. Thus his experiments over the next few folios specify that he began with the previously treated antimony rather than with stibnite fresh from the mine.

So how should we proceed to replicate these experiments? When crushed, stibnite is normally black, but nitric acid, the usual referent of the term aqua fortis, oxidizes stibnite rapidly to a white or yellowish powder with a vigorous reaction. Moreover, Newton's subsequent experiments make it clear that he usually mixed his aqua fortis with sal ammoniac in order to arrive at a sort of aqua regia. What we would normally make today by mixing nitric and hydrochloric acid, Newton made by providing chloride to the nitric acid in the form of sal ammoniac—ammonium chloride. Was Newton following this practice in the above experiment even though he neglects to mention the sal ammoniac? The fact that he mentions only aqua fortis would engender no surprise since he often leaves out steps and ingredients when he is following his own standard protocols.[29] Since we do not have a record of Newton's

---

[28] CU Add. 3973, 5v.

[29] Whether Newton's preliminary "♁ once acted on by ⊕" referred to the product of simple nitric acid or rather aqua regia is of little import in this particular instance, however, since the January 1680 experiments clearly do employ sal ammoniac at a later stage. In either case, the final results would have been largely the same. I have tried this experiment both using nitric acid and substituting Newton's ammoniac aqua regia for it while making the "antimony once acted on by aqua fortis." The nitric acid produces a white residue, while

PLATE 1. Stibnite from northern Romania. William R. Newman's sample.

PLATE 2. A mine in Cornwall where blue vitriol (copper sulfate) has permeated the shaft. This highly soluble material can accumulate and form stalactites when it drips down from the upper walls; dissolved in runoff, it forms the vitriol pools whose transmutative powers Newton wanted his friend Francis Aston to investigate in Europe. Photo courtesy of Simon Bone Photography.

PLATE 3. Native Wire Silver from Himmelsfurst Mine, Freiberg. Courtesy of Kevin Ward.

PLATE 4. Distilled liquor of antimony dropped into a bottle of deionized water. The clear, colorless solution immediately precipitates an insoluble white cloud of *mercurius vitae* (antimony oxychloride) on meeting the water. Prepared by William R. Newman in the laboratory of Dr. Cathrine Reck in the Indiana University Chemistry Department.

PLATE 5. Newton's liquor of antimony being produced from aqua regia (nitric acid "sharpened" with sal ammoniac) and stibnite (antimony trisulfide). A vigorous reaction takes place, leaving behind a yellowish-white precipitate. Prepared by the author in the laboratory of Dr. Cathrine Reck in the Indiana University Chemistry Department.

PLATE 6. Crystals deposited by slow evaporation of liquor of antimony. Prepared by the author in the laboratory of Dr. Cathrine Reck in the Indiana University Chemistry Department.

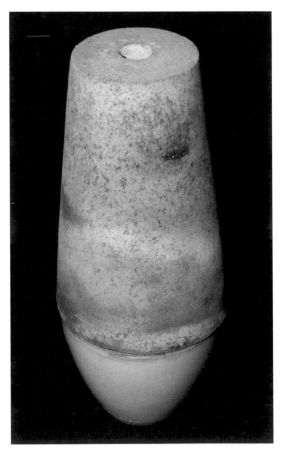

PLATE 7. Modern subliming apparatus used to replicate Newton's experiments, consisting of an inverted, drilled-out fire-assay crucible set atop a porcelain, Coors crucible (in practice, the joint would be taped). The sublimate has permeated the fire-assay crucible, as shown by the variously colored horizontal bands at different heights. Prepared by the author in the laboratory of Dr. Cathrine Reck in the Indiana University Chemistry Department.

PLATE 8. The "net," or "Vulcan's net," is a purple alloy of copper and metallic antimony. The alloy seems to have been invented by the American alchemist George Starkey (who authored most of the texts attributed to Eirenaeus Philalethes). Starkey was impressed by its fine, crystalline surface, which reminded him of network. He based the name "net" on the mythical net of bronze that Vulcan used to ensnare Mars and Venus in bed together. The alloy played a major role in the preparation of the Philalethan sophic mercury, and Newton made great use of it in his own experimentation. Prepared by the author in the laboratory of Dr. Cathrine Reck in the Indiana University Chemistry Department.

PLATE 9. Crystal of "verdigris vitriol" about one-quarter inch across, made by reacting malachite with Newton's liquor of antimony, then filtering and crystallizing the solution. Prepared by the author in the laboratory of Dr. Cathrine Reck in the Indiana University Chemistry Department.

PLATE 10. Cracked crucible used in subliming ceruse from Newton's *succedaneum* to "volatile Venus" as described in Fatio de Duillier's letter of August 1, 1693. The multiple colors on the bottom and sides of the crucible reveal the presence of lead and copper compounds as well as the partially reduced metals themselves. Prepared by the author in the laboratory of Dr. Cathrine Reck in the Indiana University Chemistry Department

process for making antimony once acted on by aqua fortis, I worked on the assumption that Newton was actually using a homemade aqua regia, and followed the proportions that he gives a folio later:

> Ian 22 I dissolved 280 gr of ☿ once acted on by ℞
> in ✳ 480g & ℞ 480gr & water 960gr. . . . [30]

Before proceeding further, it is necessary to know the concentration of Newton's aqua fortis. As it happens, Newton's great exactitude as an experimenter allows one to calculate the concentration of his solution in at least one instance with some degree of precision.[31] In the Boston Medical Library manuscript he tells the exact weights of aqua fortis and salt of tartar that were required to neutralize each other and produce eleven grains of saltpeter. One can estimate that Newton's aqua fortis, at least in that instance, had a concentration of about 32.7% nitric acid. This corresponds very closely to the concentration of aqua fortis produced in modern replications by distilling vitriol and saltpeter, so I have employed a nitric acid of this concentration in my own work.

After producing an antimony once acted on by aqua fortis, I then carried out some of Newton's further protocols, particularly that of filtering the solution and distilling it in a retort. In one of several cases where Newton does this, he says the following:

> there came after ye acute flegm ^first a liquor wch in ye air smoaked much & imbibed ye moisture of ye air, then a salt mixed wth this liquor. Which salt was as volatile as ✳ & very fusible. Most of it stuck in ye neck of ye retort like butter of ☿ & this weighed 270 gr.[32]

It is very probable that Newton's salt here was not just *like* butter of antimony but actually *was* butter of antimony, which can fume in a humid atmosphere (and it might have been accompanied by some fuming antimony pentachloride as well). Stimulated by this hint, I tried Newton's experiment. The experiment resulted in some antimony trichloride being distilled over,

---

the aqua regia yields a much yellower one. Newton's next step would either have been to sublime the residue directly with sal ammoniac, as he does with crude stibnite in the experiments that open CU Add. 3973, or to react the residue with his ammoniacal aqua regia, distill off the liquid, and then sublime the remaining solids, as he does on 6r–6v of CU Add. 3973. Hence the chloride necessary for making butter of antimony and possibly other salts would have entered into the process regardless of the starting ingredients.

[30] CU Add. 3973, 6r.

[31] Interestingly, Friedrich Dobler's experiments based on the work of Conrad Gesner produced an Aqua Fortis of 34.2% or 6.568 M. See Dobler, *Conrad Gesner als Pharmazeut* (Zurich, 1955), 94. In the Boston Medical Library MS, Newton performs a sort of titration. He notes that equal parts of oil of tartar and AF "satiate," that is neutralize, each other. Then he says that 42 grains of the two—presumably 21 grains (1.361 g) of each—yield 11 grains (0.7128 g) of saltpeter. The reaction in modern terms is $2HNO_3 + K_2CO_3$ → $2KNO_3 + CO_2 + H_2O$. Determining grams per mole and using the NO_3 as a limiting factor, the NO_3 is 61.38% of the saltpeter by weight, or 0.4375 grams. And in nitric acid, NO_3 makes up 98.41% of the compound by weight. So the weight of the HNO_3 that went into the reaction was 0.4446 grams. Now since Newton tells us that he started with 21 grains (1.361 grams) of AF, this 0.4446 grams of HNO_3 is an unknown percentage of the 1.361 grams of dilute acid. 0.4446/1.361 = 0.3267, so the AF was actually 32.67% HNO_3. This comes to 6.211 M according to the online Sigma-Aldrich molarity calculator, based on a density of 1.1968 at 20°C.

[32] CU Add. 3973, 6v.

FIGURE 14.4. Distilled liquor of antimony dropped into a bottle of deionized water. The clear, colorless solution immediately precipitates an insoluble white cloud of *mercurius vitae* (antimony oxychloride) on meeting the water. Prepared by William R. Newman in the laboratory of Dr. Cathrine Reck in the Indiana University Chemistry Department. See color plate 4.

as both qualitative analysis by means of antimony test paper and the strikingly hygroscopic character of the compound revealed. Additionally, the solution, when dripped into distilled water, produced an immediate and obvious white precipitate, which is no doubt mercurius vitae, or in modern terminology antimony oxychloride (figure 14.4).

From all of this, it is reasonably well established that Newton's liquor, vinegar, spirit, and salt of antimony were all primarily antimony trichloride, though in most cases in a solution of aqua regia and no doubt containing other materials as well.[33] But this leads to an additional question. What was Newton's inspiration for these experiments, and why was he so impressed with butter of antimony? The answer stems again from Newton's reading of Boyle, who influenced him profoundly in many ways. In *The Origin of Forms and Qualities*, Boyle describes making a *menstruum peracutum* (very strong menstruum) by pouring strong spirit of niter, that is, nitric acid, on rectified butter of antimony.[34] This was a standard way of making a chymical medicament known as *bezoarticum minerale*, which precipitated when the nitric acid was poured on. Boyle refluxed the acid with the *bezoarticum minerale* and then distilled off the liquor; he claimed that he managed to transmute gold into silver by means of this menstruum.[35] On the strength of the old dictum *Facilius est aurum construere, quam destruere* (It is easier to make gold than to destroy it), this reverse transmutation served as indirect evidence for the possibility of chrysopoeia. Newton carefully excerpted this section of Boyle's *Origin of Forms and Qualities*, as one can see from the following passage in CU Add. 3975:

> <illeg.> On yᵉ rectifyd oyle of yᵉ Butter of Antimony poure as much strong spirit of nitre as will precipitate out of it all yᵉ Bezoarticum Minerale, & wᵗʰ a good smart fire distill of all the liquor yᵗ will come over & (if neede bee) cohobate it upon yᵉ Antimoniall pouder This liquor is Mr Boyls Menstruum Peracutum.[36]

Boyle's process for making *menstruum peracutum* was obviously quite different from the one that Newton employed for his liquor of antimony. Nor does Newton ever use the term *menstruum peracutum* for his own product, despite the fact that it, like Boyle's menstruum, employed butter of antimony and a strongly acid solution. Did Newton realize this fact? The answer is surely affirmative, since in another experiment after reacting stibnite with aqua fortis and sal ammoniac, Newton says:

> To this I poured water till all yᵉ ☿ vitæ was precipitated. It took 8 or 12 times its quantity of water to cleare it well.[37]

---

[33] See Henry Watts, *A Dictionary of Chemistry* (London: Longman, Green, Longman, Roberts, and Green, 1863), 1: 332. Watts points out that sulfuric acid can form when stibnite is dissolved in aqua regia, thus opening up a range of other subsidiary reactions.

[34] For Boyle's *menstruum peracutum*, see Lawrence M. Principe, "The Gold Process: Directions in the Study of Robert Boyle's Alchemy," in *Alchemy Revisited*, ed. Z.R.W.M. von Martels (Leiden: Brill, 1990), 200–205.

[35] See Boyle, *Origin of Forms*, in *Works*, 5: 418; 1666, 351–52: "The Menstruum then I chose to try whether I could not dissolve Gold with, is made by pouring on the rectifi'd oyl of the Butter of Antimony as much strong spirit of Nitre, as would serve to praecipitate out of it all the *Bezoarticum Minerale*, and then with a good smart Fire distilling off all the Liquor, that would come over, and (if need be) Cohobating it upon the Antimonial powder. For though divers Chymists, that make this Liquor, throw it away, upon Presumption, that, because of the Ebullition, that is made by the Affusion of the spirit to the Oyl, and the consequent precipitation of a copious Powder, the Liquors have mutually destroy'd or disarm'd each other; yet my Notions and Experience of the Nature of some such Mixtures invites me to prize this, and give it the name of *Menstruum peracutum*."

[36] CU Add. 3975, 40v.

[37] CU Add. 3975, 72v.

Mercurius vitae was antimony oxychloride, which is insoluble in water. Every practicing chymist in the seventeenth century knew that the starting point for *mercurius vitae* was butter of antimony. Newton too must have realized this fact, and it is likely that he understood himself, at least early on, to be following Boyle's lead. As in the case of his early experiments with the mercuries of the metals and the sublimation of corrosive sublimate with sal ammoniac, Newton's method of producing liquor of antimony was an attempt to capitalize on the chymical hints strewn throughout the works of Boyle.[38]

The combination of laborious textual analysis, sometimes aided by computational techniques, and the physical replication of Newton's laboratory products has allowed us to solve one of the preliminary mysteries presented by his experimental notebooks, namely, the identity of the ubiquitous "liquor of antimony." Loosely inspired by Robert Boyle's *menstruum peracutum*, this dissolvent would serve Newton as a sort of standard laboratory reagent over several decades, entering into a remarkably wide variety of experiments that were intended to yield the successive ingredients required to produce the philosophers' stone. Although Newton's immediate stimulus for the liquor of antimony was probably Boyle's *Origin of Forms and Qualities*, however, the younger scientist clearly believed that the material was veiled behind the extravagant puzzles of the adepts as well. Basing himself on Starkey's *Marrow of Alchemy*, Newton's experimental notebooks openly identify the dragon or serpent that killed the comrades of the mythological founder of Thebes, Cadmus, with his liquor of antimony.[39] From Newton's perspective, Boyle had merely given a plain description of material that was already present in more hidden form beneath the cryptic riddles of the Philalethes corpus. This reinforces a valuable lesson that we have already discussed in other contexts. For Newton, there was no rigid distinction between "vulgar chymistry" and the wisdom of the adepts; the latter was simply a more elevated version of the former, elaborated into the form of complicated verbal conceits for the purpose of deluding the unworthy. As we shall see in the next chapter, this unity of method led Newton to mine the ordinary literature of seventeenth-century technical chymistry, some of it published in venues as commonplace and familiar as the *Philosophical Transactions* of the Royal Society, in the goal of acquiring the exotic secrets of the alchemical initiates.

---

[38] For Boyle's practice of hinting at chrysopoetic secrets, see Principe, *AA* (Princeton, NJ: Princeton University Press, 1998).

[39] See my chapter sixteen herein.

FIFTEEN

The Quest for Sophic Sal Ammoniac

Volatile Salt of Tartar and David von der Becke:
A Lesson in Seventeenth-Century Chymical Affinities

As we have seen, Boyle's influence is writ large in the earliest chymical experiments recorded in CU Add. 3975. Newton found inspiration in the English "naturalist's" attempts to extract mercuries from the different metals, but this was not the only practical application that he made of Boyle's comments. His first explicit use of sal ammoniac as a subliming agent probably stems from Boyle's attempts to produce "a new kind of Sublimate" by volatilizing corrosive sublimate with sal ammoniac.[1] Yet the heavy role that the English "naturalist" played in Newton's development should not blind us to the important and hitherto unnoticed use that Newton made of other chymists who were also published in contemporary scientific journals. One overlooked figure of particular significance emerges from the material immediately following the earliest chymical experiments in CU Add. 3975, which Newton probably entered in or soon after 1674.[2] Six paragraphs here describe the properties and action of sal ammoniac, especially its wet reaction with salt of tartar. The reaction, which was usually carried out with "oil of tartar per deliquium," that is, salt of tartar that had been allowed to absorb water from the atmosphere, proceeds according to the following path in simplified terms: $2\,NH_4Cl + K_2CO_3 \rightarrow 2KCl + CO_2 + H_2O + 2NH_3$. As Newton says:

> If Sal Armoniack ^be put into ~~Aqu~~ Oyle of Tartar p deliquium, its acid salt will let go the urinous & work upon the Alcaly. And the ur<in>ous thus let loos becomes very volatile so as to strike y^e nose wth a strong scent & fly all away if it be not soon inclosed in a vessel.[3]

No doubt Newton had experienced this phenomenon himself, but most of the material in these paragraphs stems from a work that had been published

---

[1] Boyle's comments appear in *Origin of Forms*, in *Works*, 5: 403–4; 1666, 299–302, as noted by Dobbs, *FNA*, 139–43. See CU Add. 3975, 41v and 43v, where corrosive sublimate and sal ammoniac are combined.

[2] There is an obvious change in ink color after line 14 of CU Add. 3975, 43r. Newton also begins putting crossbars on his Saturn symbols (as at 43v: ♄), whereas before they were unbarred. It is likely that a significant period of time intervened between the earlier portion of 43r and the immediately following material. A likely *terminus post quem* for the material taken from von der Becke is given by the publication of his epitome in the *Philosophical Transactions* of 1673/4.

[3] CU Add. 3975, 43r.

in 1672 and then epitomized in the *Philosophical Transactions* of the Royal Society in 1673/4. I refer to the *Epistola ad Langellotum* of David von der Becke, a physician of Minden, Germany. About von der Becke very little is known, but his *Epistola* belongs to a cluster of responses by various authors to Joel Langelot, physician to the Duke of Holstein-Gottorp, whose own *Epistola ad praecellentissimos Naturae Curiosos de quibusdam in chymia praetermissis* had been published in 1672. Langelot's letter had itself been epitomized in the January 1672/3 issue of the *Philosophical Transactions*, which no doubt served as a pretext for the subsequent appearance of von der Becke's abstract.[4] The reason for the excitement about Langelot's *Epistola* lay in its description of "volatile salt of tartar," one of the highly sought-for *arcana majora* described by Joan Baptista Van Helmont in his immensely influential works on chymistry. Since ordinary salt of tartar or potassium carbonate is a strikingly refractory material that melts at 891°C and decomposes at still higher temperatures, it is not the sort of thing that one would expect to sublime. But Langelot described a method by which he calcined crude tartar to blackness, immersed it in water, and then added further uncalcined tartar. According to the physician, this resulted in a bubbling fermentation; once this was over, Langelot put the material in an iron bolthead and distilled it over heat. He found that the "gross and feculent" tartar was entirely volatilized, leaving no "fixt Salt" in the bottom of the vessel. It therefore appeared, at least to some, that Van Helmont's obscure prescriptions for finding the volatile salt of tartar had been realized.[5]

Von der Becke's *Epistola ad Langellotum* is largely a commentary on Langelot's process with an attempt to explain it in terms of Helmontian chymical theory by using sal ammoniac as a model substance for understanding tartar. Since Newton extracts extensive notes on this important topic in CU Add. 3975, it would be necessary to examine his comments on general principle alone. More than this, von der Becke's account gives us an exemplary window into the fundamental understanding that dominates Newton's approach to chymistry. Following von der Becke quite closely, Newton explains the process for making volatile salt of tartar at some length. Newton's (and von der Becke's) explanation is based on the correct observation that sal ammoniac is composed of "an acid & urinous salt both wch are severally volatile enough but together they fix one another," or as we would say, hydrogen chloride gas and ammonia that combine to form solid ammonium chloride.[6] When the highly alkaline salt of tartar solution (our potassium carbonate) is added to the sal ammoniac, the "acid salt" is attracted to the alkaline material and the "urinous salt" or ammonia passes off. This spirit is what "strikes the nose" in Newton's earlier paragraph. In full accord with

[4] See Anna Marie Roos, *The Correspondence of Dr. Martin Lister (1639–1712)* (Leiden: Brill, 2015), 1: 471.

[5] Joel Langelot, "An Extract of a Latin Epistle of Dr. Joel Langelot, Chief Physitian to the Duke of Holstein Now Regent: Wherein Is Represented, That by These Three Chymical Operations, Digestion, Fermentation, and Triture, or Grinding (Hitherto, in the Authors Opinion, Not Sufficiently Regarded) Many Things of Admirable Use May be Performed, English'd by the Publisher," *Philosophical Transactions of the Royal Society* 7 (1672/3): 5052–59.

[6] CU Add. 3975, 43v.

the Helmontian tradition, von der Becke sees this displacement reaction in terms of a corpuscular schema governed by elective affinities. The particles of "acid salt" in the sal ammoniac have a greater affinity for salt of tartar than they do for the "urinous salt" in the sal ammoniac; hence, they dissociate from their urinous salt when placed in a solution of tartareous salt and form a bond with the latter. The same type of reasoning opens up a way of explaining Langelot's process for the volatile salt of tartar, reprised by Newton in the following terms:

> So if to a solution of crude Tartar in water be put ^by degrees Salt of Tartar, or Tartar calcined suppose to black, the acid spirit of ye Tartar will forsake ye Alcalisate <*illeg.*> (or urinous) to work upon ye fixt Salt of Tartar. And ye Alcalisate (or urinous) salt thus let loos becomes very volatile so as to fly sudde<n>ly away. And in ye remaining Solution will be a salt compounded of ye acid sp^t of Tartar & sulphureous or volatile part of ye Alcaly, wch salt is volatile but not more volatile then Sal-armoniack or its flowers. But by ye addition of new Salt of Tartar (perhaps after it hath been sublimed) in wch ye acid may work the urinous will be let loos & become exceeding volatile as before [7] & in the action the earthy parts of the fixt salt will be præcipitated.

Although Newton has elided some important features of von der Becke's explanation, the original epitome in the *Philosophical Transactions* is easily intelligible. Crude tartar or argol is a product that deposits over time on the interior of wine casks. As the tartar comes out of the barrel it will itself be acidic, a fact that is obvious to sense, and due to the presence of an "acid salt." Yet according to Helmontian theory, the crude tartar will also contain a more volatile, alkaline or urinous salt; as in the case of sal ammoniac, it is the mutually restraining action of these two opposed salts that provides even crude tartar with a moderate degree of fixity, preventing it from subliming away at room temperature. In order to explain the much greater fixity of the salt of tartar (potassium carbonate) made by calcining the crude tartar, the Helmontians argued that the intense heat of calcination fused and "con-coagulated" the previously volatile salts in the tartar with an additional ingredient, namely, "Earthy parts" or particles of earth found intermixed in the unrefined tartar.[8] Once the previously volatile salts have been concoagulated with the earthy particles, "they can no more rise and fly away than birds fastn'd to a rock," as von der Becke's epitome puts it.

The secret of volatilizing salt of tartar should therefore reside mainly in the art of separating the volatile acid and alkaline salts in it from the restraining earth particles that account for its high degree of fixity. According to von der Becke, this separation can be achieved most effectively by following the methods of nature herself, particularly the method of fermentation. The

---

[7]CU Add. 3975, 43r. For von der Becke, see "An Extract of a Letter, Written by David Von Der Becke, a German Philosopher and Physitian at Minden, to Doctor Langelott, Chief Physitian to His Highness the Duke of Holstein Now Regent, Concerning the Principles and Causes of the Volatilisation of Salt of Tartar and Other Fixed Salts: Printed at Hamburg, 1672," *Philosophical Transactions (1665–1678)* 8 (1673/74): 5185–93.

[8]Von der Becke, "Extract of a Letter," 5187.

FIGURE 15.1. David von der Becke's theory of the production of volatile salt of tartar. In the initial stage, crude tartar (argol) is added to salt of tartar (or partially calcined tartar) in solution. Modeling his theory on the fact that salt of tartar (potassium carbonate) decomposes sal ammoniac by combining with its "acid salt" (HCl) and causes it to release a "urinous spirit" (ammonia), von der Becke thinks that something similar will happen when the salt of tartar is mixed with crude tartar. The crude tartar will be decomposed, losing its volatile, urinous component as a vapor, while its acid salt will combine with the alkaline salt of tartar to form a new salt. At the same time, the salt of tartar will lose its earthy component, becoming a new salt with a volatility similar to that of sal ammoniac (which sublimes at around 338°C). In order to arrive at genuine "volatile salt of tartar," a further set of operations is required.

fermentation will be effected by adding a leaven in the form of crude tartar to partially or completely calcined tartar, the very process that Langelot had described in his *Epistola*. If one calcines the crude tartar to blackness, some of it should be converted to salt of tartar, a phenomenon that was well known from the artisanal production of the salt. As we have seen, von der Becke explained that crude tartar, like sal ammoniac, consisted of both an acid and an alkaline part in combination. Following the pattern of sal ammoniac's dissociation into its acid and urinous, alkaline components when placed in a solution of salt of tartar, if crude tartar is added to a solution of calcined black tartar or salt of tartar, it should dissociate and lose some of its acid salt to the more powerfully alkaline one in the calcined tartar. Again as in the case of sal ammoniac, the "Alcalisate (or urinous) salt" in the crude tartar ought to become volatile and pass off once it is freed from the inherent acid salt (figure 15.1). But as von der Becke says in the *Philosophical Transactions* epitome, this volatile product is properly not the Helmontian volatile salt of tartar, but rather the volatile salt of crude tartar.[9] And even if it were "a real volatile Salt of Tartar," there still remains another problem: if the process is carried out all at once, the bubbling fermentation of the crude tartar will release its volatile salt in such quantity and violence that it will burst the containing vessel; on the other hand, if the ingredients are added incrementally, the volatile material will escape the vessel during its successive openings. Hence one cannot hope to extract the volatile salt of tartar from crude tartar, but must attempt to derive it from the fixed salt of tartar.

Von der Becke thinks this should be accomplished by returning to the endpoint of the unsuccessful attempt to arrive at volatile salt of tartar that he just described. As the volatile, alkaline spirit of the crude tartar passes off and is lost in the process, the solution of calcined and uncalcined tartar will

[9]Von der Becke, "Extract of a Letter," 5189. The Latin version of von der Becke's letter is clearer on this point. See David von der Becke, *Epistola ad Praecellentissimum Virum Joelem Langelottum* (Hamburg: Gothofredus Schultzen, 1672), 20.

now contain a salt made up of the acid spirit of the crude tartar in combination with the "sulphureous or volatile part of ye Alcaly" from the salt of tartar or calcined tartar. It is this new salt that will now form the focus of von der Becke's strategy. During the formation of the new salt, von der Becke stresses, the salt of tartar will be freed from some or all of the earthy particles with which it was concoagulated; these earthy parts will now sink to the bottom of the solution as a precipitate. The new salt will therefore not be as fixed as salt of tartar, but neither will it be the truly volatile form of the salt. Again employing the analogy with sal ammoniac and its two components, von der Becke says that the fixed alkali of the salt of tartar is indeed freed from its restraining earthy particles by the "ferment" of crude tartar, but the acid salt of the crude tartar then refixes it in the same way that the acid and urinous spirits in sal ammoniac provide it with a moderate fixity. Thus Newton's notes on von der Becke remark that the new "salt is volatile but not more volatile then Sal-armoniack or its flowers."

In order to proceed from this new sal-ammoniac-like salt to the actual volatile salt of tartar, one must pass to the final step of the process. Von der Becke says that the new sal-ammoniac-like salt, whose alkaline component has been liberated from its earthy particles by fermentation but which still has the moderate fixity associated with sal ammoniac, must be forced in turn to dissociate. In order to achieve this, he suggests the addition of fresh, fixed salt of tartar, apparently on the assumption that the fresh salt of tartar will combine with the acid part of the sal-ammoniac-like salt and liberate its purified, earth-free alkaline constituent. If this final separation can be effected, the result will be a genuine volatile salt of tartar, unlike the volatile urinous spirit that von der Becke earlier claimed could be released simply by reacting crude tartar with calcined tartar (figure 15.2). Moreover, von der Becke is convinced that this volatile salt of tartar will combine with the water in the reacting vessel rather than escaping with such force that it would break the glass.

Von der Becke's explanation is an ingenious attempt to employ contemporary knowledge of acid-base reactions, particularly as they exhibit themselves in his model substance sal ammoniac, in the service of finding the volatile salt of tartar. The greater affinity that the acid component of sal ammoniac has for the highly alkaline salt of tartar than for its own urinous component, causing the ammoniac salt to dissociate (and release ammonia), provides a schema reminiscent of the affinity tables that would acquire immense popularity in the eighteenth century. Yet von der Becke's explanation also contains some obvious shortcomings. One must wonder, in particular, why adding new salt of tartar at the end of the process would liberate the purified alkaline component of the sal-ammoniac-like salt. After all, the alkaline component in the artificial salt had been acquired from salt of tartar, which would presumably exercise the same degree of attraction on its acid component as would new salt of tartar. Newton may have spotted this weakness in von der Becke's explanation, for when he arrives at the final part of von der Becke's account, where the German suggests "y^e addition of new Salt of Tartar," Newton adds the parenthetical qualification "(perhaps after it hath been sublimed)." In other words, Newton was suggesting that the

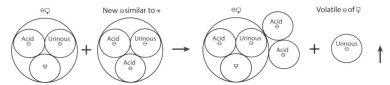

FIGURE 15.2. The second stage of David von der Becke's scheme for capturing volatile salt of tartar begins by taking the "new salt" with a similar volatility to that of sal ammoniac that he produced in stage one. Von der Becke thinks that if this "new salt" is in turn combined with fresh salt of tartar, its internal acid salt will be attracted to the more alkaline salt of tartar, causing its components to dissociate. Having previously been freed of its earthy particles, the remainder of the "new salt" will now consist of volatile salt of tartar, the material that von der Becke was seeking in the first place.

final addition of salt of tartar required that the material receive some additional treatment in order to allow it to attract the acid particles away from their alkaline partners.

A further aspect of von der Becke's letter that caught Newton's eye was the part that fermentation played in the process. Although von der Becke's interpretation of the process relied on straightforward affinities between particles of different substances, the initial choice of crude tartar as an agent for isolating the volatile alkali in salt of tartar stemmed from ideas drawn from his source, Langelot's *Epistola*, reflecting experience with "ferments" or leavens. Just as the making of sourdough bread from ordinary flour and water requires a "starter" of previously fermented dough, so the volatile salt of calcined tartar needed "the added Ferment, which is the *crude Tartar*."[10] In the view of von der Becke, it was the action of the ferment on the otherwise inactive salt of tartar that led to the salt of tartar's subtilization and "volatizing," a process that revealed itself in the form of bubbling and swelling. That Newton was fascinated by the role of fermentation in this account appears unequivocally if we examine again the single dissociated sheet that composes Boston Medical Library B MS c41 c. In this important and hitherto neglected manuscript, Newton introduces several paragraphs on fermentation with the following comments in Latin:

> Sal ammoniac promotes the fermentation of salt of tartar, for example one part of sal ammoniac, two parts of salt of tartar *per deliquium*, and three or four parts of tartar.[11]

Here we see Newton trying to carry out von der Becke's fermentation of salt of tartar by means of crude tartar, which he has accelerated by adding sal ammoniac. This "promoting" of the fermentation by means of sal ammoniac is an addition to the process by Newton. It may be that he believed von der Becke to have hidden some of his recipe under talk of sal ammoniac as a model substance when it actually had to participate directly in the volatilization of the salt

[10] Von der Becke, "Extract of a Letter," 5189.
[11] Boston Medical Library MS B MS c41 c, 1r: "✳ promovet fermentationem salis ♀; puta ✳ pt 1, salis ♀ p deliqu. pt 2 ♀ p 3 vel 4."

of tartar, or it may be that Newton believed himself to have advanced the process further than his source. At any rate, the addition of sal ammoniac would at least have led to a chemical reaction in the release of ammonia and carbon dioxide, so Newton's instructions were not vain from an empirical perspective.

Newton's modification of von der Becke's process is followed by a discussion of additional "fermentations," all of which employ salt of tartar. Capitalizing further on the reaction between salt of tartar and sal ammoniac, Newton first uses these two materials to ferment vitriol. He also tries sea salt and niter on the mixture of vitriol and sal ammoniac but finds, unsurprisingly, that the activity engendered by the combination of sal ammoniac, vitriol, and salt of tartar is more effective at producing a bubbling fermentation. He then "satiates" a mixture of oil of tartar *per deliquium*, niter, and sea salt with vitriol, probably observing the reaction that would have occurred between the potassium carbonate and the sulfate salt represented by the vitriol (whether iron sulfate or copper sulfate is unsure—either would be a "vitriol"). The choice of common salt or sea salt, niter, and vitriol alongside the previously discussed pair of sal ammoniac and salt of tartar is by no means random. The three former substances, sea salt, niter, and vitriol, were the usual sources for making the mineral acids—spirit of salt or hydrochloric acid from common salt, aqua fortis or nitric acid from niter, and oil of vitriol or sulfuric acid from vitriol. As we have already seen in the making of Newton's liquor of antimony, adding sal ammoniac to aqua fortis was a standard way of producing aqua regia, the impressive dissolvent of gold, and salt of tartar was an alkali capable of neutralizing or "satiating" all of these acids. Newton had fallen on a rich area of reactivity among different materials whose operations he was eager to discover and exploit. Hence he passes next to an exploration of aqua fortis and its action on common salt, niter, vitriol, and sal ammoniac, noting the absence of ebullition—which would otherwise reveal the presence of fermentation—when they are mixed. He finds the same lack of ebullition with spirit of salt added to these materials, but on mixing the spirit of salt with niter and adding them to "yᵉ mixed fermented salt" (probably a mixture of sea salt, niter, vitriol, sal ammoniac, and salt of tartar), Newton finds that they precipitate an "unctuous blacknes," revealing some level of activity. To round out these experiments with the mineral acids, Newton then dissolves sea salt, sal ammoniac, and niter separately in oil of vitriol, noting a reaction between sea salt and sal ammoniac. Finally, he returns to his original inspiration for this section, namely, the quest for volatile salt of tartar, with the following experiment:

> Fermented but not digested salt of ♆ not quite dry, & mixed wᵗʰ as much brick dust, yeilded ~~its~~ 1/5 of its weight (i.e. 1/10 of yᵉ whole mixture) in spᵗ & oyle & no more, although yᵉ fire was urged till yᵉ glass melted.

Here after fermenting his salt of tartar—probably with the addition of sal ammoniac, and possibly additional materials—Newton adds brick dust, a common technique of the time. The addition of powdered, difficult-to-fuse materials like brick dust, fuller's earth, and bole armeniac was a standard practice employed in the making of mineral acids by high-temperature distillation. The goal was to keep the material that one wanted to decompose in a state where it could not coalesce into a molten mass that would resist the action of heat and air. Although Newton's experiment resulted in disappointment, it reveals unequivocally that his goal, once again, was the extraction of a volatile component from the refractory salt of tartar that he tried to loosen up or "open" with an initial fermentation and then sublime or distill at very high temperature.

Newton's early experiments at volatilizing salt of tartar under the guidance of von der Becke's letter to Langelot were far from a dead end, although he failed to produce that marvelous desideratum. The experiments provide the first datable evidence in Newton's experimental notebooks of an interest in "fermentation," a process that would acquire increasing importance as his chymical work progressed. He had already discovered the allure of fermentation in *Of Natures obvious laws*, and that text even contains puzzling references to the fermentation and putrefaction of salts that may be related to the experiments described in the Boston Medical manuscript. It is certainly suggestive that *Of Natures obvious laws* claims the following:

> that salts may ^putrefy & by putrefaction ^will generate another sort of black^ish rotten substance <or> fat substance the ~~cheife~~ most fertile part of this upper crust & ye nearest matter out of wch y^e <illeg.> vegetables are extracted & into wch after death they returne. And this confirmed in that nothing promotes fermentation & putrefaction more y^n salts (~~though they hinder it as much~~ where they are incited to it.[12]

What was the basis for Newton's rather obscure claim here that salts can be made to putrefy or ferment? Was he thinking of experiments like those with the "mixed fermented salt" described in the Boston Medical manuscript? And can his claim that putrefied salts yield a "black^ish rotten substance <or> fat substance" be linked to the "unctuous blacknes" that he precipitated from the fermented salt in the Boston Medical manuscript's experiments? Finally, is his claim that "nothing promotes fermentation & putrefaction more y^n salts" in *Of Natures obvious laws* linked to the experiment of "promoting" the fermentation of vitriol with salt of tartar, sal ammoniac, niter, and sea salt? Without further evidence one would not wish to push these points, but we can at least say that both texts reveal a newfound emphasis on fermentation that is largely absent from Newton's early attempts to follow the lead of Robert Boyle. We will return to the important theme of fermentation in Newton's laboratory notebooks in due course, but first it will be necessary to discuss several materials that, like his liquor of antimony, occupied him over a period of decades.

Cambridge University Additional MS 3973

We have now made a survey of the very early material in Newton's "master notebook," CU Add. 3975 along with Boston Medical Library B MS c41 c. We are therefore in a position to engage in a systematic examination of

[12] 1031B, 2v.

CU Add. 3973, Newton's set of chronologically ordered notes for the period from 1678 to at least 1696. Yet one must acknowledge that this set of notes is manifestly incomplete, as a comparison with Newton's other main laboratory notebook, CU Add. 3975 makes clear. CU Add. 3975 was a bound composition book when Newton bought it, and it seems to have served him as a sort of master text into which he transferred earlier material. But comparing parallel passages in the two texts reveals that Newton has often reworked material as he entered it in CU Add. 3975. And because the later notebook contains dated material not found in CU Add. 3973, it is clear that he must have kept additional preliminary notes that are no longer extant. All of this makes it most convenient to use CU Add. 3973 as the basic point of analysis in the present section, supplementing it where necessary with material taken from the bound notebook.

CU Add. 3973 consists of seventeen individual sheets and booklets made by folding a single sheet and slitting it along the top edge (though some of the smaller sheets are simply folded once at the "spine" and one sheet inserted in another to make a booklet). It is important to note the dates at the beginning (and where they are present, at the end) of each booklet or sheet, since they were often composed at intervals of a year or more and represent widely differing research projects. Here I will give a brief tally of the individual pamphlets or sheets as they are ordered by Cambridge University Library along with their dating so that we can examine them in succession:[13]

Sheet 1 (1r–4v):	December 1, 1678–January 15, 1678/9
Sheet 2 (5r–8v):	January 1679/80
Sheet 3 (9r–12v):	February 1679/80
Sheet 4 (15r–18v):	August 1682
Sheet 5 (13r–14v):	July 10 (1681)
Sheet 6 (19r–20v):	April 26, 1686–May 16, 1686
Sheet 7 (21r–24v):	March 5, 1690/91
Sheet 8a and 8b (25r–28v):	December 1692—January, 1692/93
Sheet 9a (29r–29v):	April 1695
Sheet 9b (30r–31v):	February 1695/96
Sheet 10 (32r–35v):	Undated
Sheet 11a and 11b (36r–39v):	Undated
Sheet 12 (40r–43v):	Undated
Sheet 13 (44r–44v):	Undated
Sheet 14 (45r–47v):	Undated
Sheet 15 (48r–48v):	Undated
Sheet 16 (49r–50v):	Undated
Sheet 17 (51r–52v):	Undated

[13] Sheets 4 and 5 have been ordered incorrectly by the keepers of the Portsmouth Collection at Cambridge University Library. The editors of the Chymistry of Isaac Newton project have kept the library's numbering of the sheets, since this reflects their physical position in the repository; the foliation of the edited text has been changed to reflect its correct ordering. For evidence supporting this new ordering, see "Manuscript Information" for CU Add. 3973 at *CIN*.

As we will see, these records divide rather neatly into three distinct experimental programs. The first, which I will call "the quest for sophic sal ammoniac," runs from the beginning of December 1678 up until a sort of eureka moment in midsummer 1680, when Newton actually discovers the material that he terms "✱ Philosophicum"—philosophical or sophic sal ammoniac, which he also calls "oʳ ✱" (our sal ammoniac). The second program of research begins in August 1682 and consists mostly of a series of tests to determine whether stibnite refined by fusion or the ore as it comes directly out of the mine should be used in preparing Newton's antimonial sublimates. At the same time, however, Newton begins using other materials, such as the antimonial sublimate of copper vitriol to which we alluded earlier. This stibnite-testing project runs for two complete folios and has no terminal date. A third, more ambitious project is fully developed by the date that marks the start of Sheet 6, April 26, 1686, but this line of research probably began much earlier and is already present *in nuce* in the documents from 1679/80. The project, which I will describe at length in chapter sixteen, involves Newton's attempt to purify his antimonial sublimate of copper vitriol from excess antimony and sal ammoniac, and to arrive at a material that he calls "our Venus" ("oʳ ♀"). We can track the project of "extracting our Venus" from its origins up until February 1695/6, the last date in CU Add. 3973. Needless to say, a discussion of these three research projects will not exhaust the many parallel sets of experimentation in Newton's notebooks. But these three successive projects will allow us a way to follow the course of Newton's research and to show how it formed a continuous line of development rather than resolving into a desultory mass of aimless trials.

"Improving" Sal Ammoniac

As we saw earlier from our examination of Newton's alchemical theory presented by him in Dibner 1031B, he decided very early that the secrets of the art lay mainly in the area of vapors and "airs" or gases. Already by the early 1670s Newton was mining Boyle's work for ways to sublime the various metals and metalloids known to him in the hope that this would lead to intimate reactions in the vaporous or gaseous state.[14] Not surprisingly, Newton soon fell on sal ammoniac, the very material that had assumed a central role in David von der Becke's attempt to make volatile salt of tartar. Sal ammoniac has the interesting property of subliming at the easily attainable temperature of 338°C. Moreover, as I mentioned before, the substance dissociates into hydrochloric acid and ammonia gas when it sublimes, making it possible for it not only to carry up other materials but also to react with them before cooling and returning to the solid state as ammonium chloride. Sal ammoniac's ability to carry up or "volatilize" metals and metalloids already made it the object of considerable interest among the Arabic alchemists, whose

---

[14] It is worth pointing out that Newton himself distinguished between vapors and gases and even occasionally uses the Helmontian term "gas" for the latter. See his *Hypothesis of Light*, in Newton, *Corr.*, 1: 368.

works were known to Newton through Latin translations and through the works of pseudepigraphers such as pseudo-Rhazes.[15] But Newton modified the practice of earlier alchemists in one crucial respect. Instead of employing sal ammoniac alone as an adjuvant to sublimation, Newton learned that he could increase its volatilizing power by mixing it with stibnite, the sulfide ore of antimony (which Newton referred to simply as "antimony"). This is not surprising if one considers that antimony, sulfur, ammonia, and hydrogen chloride gas could all react with one another during the sublimation of stibnite with sal ammoniac to produce a variety of volatile compounds. Newton even went so far as to analyze his sublimate of antimony and sal ammoniac, finding that the ammonium chloride lost about 1/8 of its weight during the process of sublimation, a fact that he attributed to loss of its ammoniacal "spirit" (we would say its ammonia).[16] As he would say in another context, the dissociated "spirit of salt" (our hydrochloric acid) could also combine with the sublimandum.[17] Along with the ubiquitous liquor of antimony, the sublimate of crude antimony and sal ammoniac would make up one of Newton's favorite chymical tools.

As if to confirm this fact, the very first experiment recorded in CU Add. 3973 involves the sublimation of stibnite and sal ammoniac. On December 10, 1678, Newton recorded that he sublimed 240 grains of stibnite with an equal amount of sal ammoniac, leaving a residue of 130 grains below. On December 11, he used 180 grains of this sublimate to elevate iron ore, and over the next few days he played with the proportions until he managed to sublime an acceptable amount of iron ore with the antimonial sublimate. But already a problem was emerging with this method of sublimation. As Newton indicates on December 10, the sublimate looked "very red." What Newton is referring to here is Kermes mineral, a red or orange powder made up mainly of antimony sulfide and oxides formed, in this case, during the sublimation with sal ammoniac. Although Newton makes no value judgment in this passage, he complains a few folios later of the "red dusty Sulphur" that emerged during the sublimation of crude antimony with sal ammoniac.[18] This filthiness of the antimonial sublimate even becomes a point of comparison when Newton is subliming other materials, such as regulus of copper, with antimony. As he puts it, "This sublimed with ✳ gave a sublimate as foule & dirty as the sublimate of ♁ alone is," and he repeats this reproach nine lines later about another material, again using his antimonial sublimate as a point of negative comparison.[19] These complaints of dirtiness

---

[15] I refer to the author of the famous Arabo-Latin text *De aluminibus et salibus*. See Robert Steele, "Practical Chemistry in the Twelfth Century," *Isis* 12 (1929): 10–46, and Julius Ruska, *Das Buch der Alaune und Salze* (Berlin: Verlag Chemie, 1935).

[16] CU Add. 3973, 38r: "In sublimate of ♁ 6 parts of ✳ carries up 3 parts of ♁ & by letting go a good quantity of spirit of ✳ loses 1/8 of its weight so that in the sublimate of ♁ there is but 5 of ✳ to 3 of ♁. 6 parts of ✳ gives 6 1/2 of sublimate besides 1 1/2 of yellow flowers."

[17] CU Add. 3973, 34v–35r: "Mons ♀is et ☿ii works not on iron ore but sublimes in white fumes & leaves the ore tastles. Quaere whether ♄ work on it. ✳ works on it very easily & therefore the spirit of salt will stay behind."

[18] CU Add. 3973, 5v.

[19] CU Add. 3973, 10r.

occur repeatedly throughout Newton's laboratory notebooks, usually in reference to the sublimate of crude antimony made with sal ammoniac.

When Newton returned to his experimentation almost exactly a year later, it did not take him long to attempt a solution to the problem posed by the filthiness of crude antimony. In January of 1679/80, he began a series of sublimations where the antimonial sublimate was used in turn to sublime "Saturn antimoniate" (♄ ☿^{iate}), the product of imbibing lead or lead ore with liquor of antimony. Although Newton at first noted that the new sublimate was dirty, he soon found that the initial antimonial sublimate could be sublimed from salt of Mars (sal ♂^{tis}) to get a white product. Encouraged by the improvement in color, Newton then made up a fresh batch of antimonial sublimate for further use as an adjuvant.[20] As often happens in the history of experimental science, however, a lucky accident had occurred. As he puts it:

> The sublimate I used in these exp^{ts} being old, I made some new wth crude ☿. This was fouler then ye former & had much ~~imp~~<*illeg.*> red dusty sulphur adhering to ye top of ye glass, wch made me suspect ye former was not made by crude ☿ but by ☿ once acted on by Æ.

In other words, making up new antimonial sublimate and observing its increased filthiness led Newton to realize that the antimonial sublimate used in his previous experiments must have undergone a preliminary treatment with aqua fortis (Æ), namely, nitric acid (possibly "sharpened" with sal ammoniac), since it was less "foul" than the new product. So Newton now jumps into a new series of experiments in which he makes additional antimonial sublimate using "antimony once acted on by aqua fortis" rather than crude antimony. He then employs the new sublimate to volatilize "Saturn antimoniate," going through four different runs with varying proportions of sublimate and sublimandum. Subliming sixty grains of the new antimonial sublimate from forty grains of Saturn antimoniate gives Newton a caput mortuum or residue that can in turn be sublimed at a higher temperature while retaining the degree of fusibility that he desires. Placing six grains of the caput mortuum on a glass and heating to a red heat, Newton notes that in seven and a half minutes, all but a thin skin weighing less than a quarter of a grain has fumed away, leaving the glass transparent. Clearly impressed by these results, he concludes this set of operations by saying, "Whence I knew it to be ye shadow of a noble exp^t."

In CU Add. 3973, these encouraging comments are followed by a succession of experiments dated "Ian 22," or January 22, 1679/80, which all aim to capitalize on the discovery that antimony treated previously with aqua fortis could provide a cleaner sublimate than crude antimony. The first five paragraphs all begin with the same set of protocols but using different proportions of the ingredients. Newton dissolves "antimony once acted on" by aqua fortis (again, perhaps with the addition of sal ammoniac) in a solution consisting of further aqua fortis and sal ammoniac with the addition of water to dilute the acid. He then boils away the solution and sublimes a salt. Once the

---

[20] CU Add. 3973, 5r.

sublimate has been collected, he sublimes it again at various temperatures and subjects it to further tests, such as effervescence with dissolved salt of tartar. Additionally, Newton notes whether the sublimate drops a precipitate when dissolved in water; comparing the weights of the dry sublimate to the precipitate, he indicates that around seven or eight grains of the "antimonial salt" were left in solution and did not precipitate.

The quantitative character of these tests is striking and serves as an excellent example of Newton's methodology. His estimate that "about 7 or 8 grains" of antimonial salt remained in solution is not self-evident, as it is a figure derived from the use of the principle of mass balance.[21] His experiment begins with measured weights of all the ingredients: 280 grains of antimony acted on by aqua fortis, 480 grains of sal ammoniac, 480 grains of aqua fortis, and 960 grains of water. The choice of 480 grains probably stemmed from the fact that an apothecaries' ounce weighed that amount. Newton then weighs the initial sublimate produced by boiling off the acid solution—he gets 400 grains. The caput mortuum or residue at the bottom of the boiling vessel weighs 132 grains, and this diminishes to 120 grains on heating on a fire shovel over an open fire. Newton then draws a preliminary conclusion—"So that there was about 160^{gr} carried up." He must have arrived at this amount by subtracting the 120 grains of fixed caput mortuum from the original 280 grains of antimony: the 160 grains refer to the amount of antimony that sublimed. Newton then takes 60 grains of the sublimate and dissolves it in water; once he has washed and dried the precipitate, he finds that it weighs 15 grains.

So how does he arrive at the conclusion that 7 or 8 grains of antimonial salt remain in the solution? In order to answer this, one must return to the total quantity of sublimate produced by the two sublimations—412 grains (the 400 grains initially produced during and after boiling and the 12 grains given up by the test with the fire shovel). Subtracting from this the 160 grains of antimony that sublimed gives 252 grains of sublimate that must be sal ammoniac, according to Newton's logic. Comparing 160 grains to 252 grains gives a ratio of 40:63 after reducing. If we now return to the 60 grains of sublimate that were dissolved to produce 15 grains of precipitate, we can apply the 40:63 ratio to determine that the 60 grains consist of approximately 22 grains of antimony and about 38 grains of sal ammoniac. But since the sal ammoniac is entirely soluble, and there are only 15 grains of precipitate, it follows that there must still be about 7 grains of antimony unaccounted for; these must therefore be present in the solution in the form of a soluble salt.

The other experiments in this group show that Newton was producing considerable amounts of butter of antimony by means of his dissolutions of stibnite in homemade aqua regia. At one point, his butter of antimony was accompanied by a fuming, "exceedingly volatile," hygroscopic liquor. Quite possibly this contained some antimony pentachloride, which fumes at room temperature and can be produced by the action of aqua regia on the less

[21] CU Add. 3973, 6v: Newton has written "7 ^or 8 ^{gr}" and then struck through the "7 ^or."

volatile antimony trichloride.[22] Whatever the precise identity of this material, Newton was quite impressed by it, so he employed the volatile salt in a further battery of tests. Dissolving quicksilver in the salt solution, Newton notes that the salt then flowed like tallow and worked on the mercury without producing any effervescence, making an ash-colored paste. Removing some of the paste with a quill, he put this on a piece of glass over a fire to test it for evaporation. The results must have been disappointing—"The salt flew away quickly & left y^e ☿ coagulated in a hard rugged lump." Successive attempts produced similar results: the mercury stubbornly refused to be entirely sublimed by means of the volatile salt.

The tantalizing but inconclusive results of Newton's experiments with stibnite previously acted on by aqua fortis led him to return again to a sublimate made with crude antimony. Over the next few folios, one finds him doing comparative tests subliming materials like "Vitriol of ♌. Vir"—vitriol of the Green Lion—with sublimate of crude antimony (made with sal ammoniac, of course) and with "sublimate of ♂ dissolved in ♅ & precipitated wth water."[23] The latter sublimate was made first by dissolving the stibnite in Newton's ammoniac aqua regia and then diluting the solution until any materials that were insoluble in water would precipitate. The white precipitate was then filtered and dried so that it could be used as an adjuvant to subliming further substances. One can again detect Newton's goal of "cleansing" the antimony from sulfurous "filthiness." But a serious problem remained. Three parts of crude antimony–sal ammoniac sublimate sublimed from one part of green lion vitriol left one-half part below, giving Newton a sublimate yield of 50% when comparing the amount of sublimed green lion vitriol (without adjuvant) to the caput mortuum (residue). In a second test, 60 grains of crude antimony sublimate sublimed from 12 grains of green lion vitriol left a caput mortuum of 6 1/4 grains: a yield of 48%. But in contrast to these closely similar yields, 12 grains of the sublimate made from precipitate of antimony sublimed from 3 grains of the same vitriol of green lion left 2 grains of caput mortuum. In other words, the "dirty" sublimate made from crude antimony gave a considerably better yield of between 48 and 50% compared to the 33% of product carried up by the "clean" precipitate.

Newton summarized these findings with additional measurements in a new set of experiments beginning in February 1679/80. As he baldly puts it, "Sublimate of crude ♂ volatises sensibly more then sublimate of white precipitate."[24] The same result is expressed in even more forceful language in a parallel passage from Newton's other main laboratory notebook, CU Add. 3975, where "sensibly more" is replaced by "much more." What is still more interesting is the fact that Newton describes some observations in CU Add. 3975 where he tries to determine the cause of the "volatizing virtue" of crude antimony sublimate. These comments seem to reflect generalized laboratory

[22] J. W. Mellor, *A Comprehensive Treatise on Inorganic and Theoretical Chemistry* (London: Longmans, Green, 1929), 9: 476, 486.

[23] CU Add. 3973, 8v.

[24] CU Add. 3973, 9v.

experience rather than a specific experiment. First Newton precipitates some sublimate of crude antimony in water and dries the product; then he pours oil of vitriol (sulfuric acid) on the precipitate and digests it with heat "to make ye brimstone sulphur sublime."[25] Clearly Newton had observed that the acid causes some of what we would call the elemental sulfur to dissociate from the precipitate. He then washes the product to free it from oil of vitriol and sublimes it again, concluding that "this sublimate will have no volatizing virtue." Then, in order to see whether this loss of volatizing virtue is due to the absence of the lost sulfur, he takes additional sublimate of crude antimony and adds "brimstone" to it. By parity of reasoning, this should increase the volatilizing virtue of the sublimate if sulfur is indeed the cause. But Newton finds that the opposite is the case—"the volatizing virtue is thereby diminished." Moreover, if the initial addition of oil of vitriol is not followed by heating and digestion to remove sulfur, there is still a diminishment of the "volatizing virtue" despite the fact that no sulfur has been eliminated from the precipitate. From all this, Newton concludes, "tis not ye loss of ye sulphur but ye action of ye oyle of ⊕ on y^e ♂ wch destroys the volatising virtue."[26]

Despite his rather discouraging results with the product of antimony that had been treated with aqua regia and then precipitated with water, these results contained some good news, for they meant that it should be possible to remove the sulfur from Newton's antimonial sublimate without ruining its ability to volatilize other materials. Hence in the remaining experiments from 1679/80 we find him turning to sublimations made with glass of antimony and sal ammoniac as well as several reguli made with copper. Finally, he even sublimes the scoria left when martial regulus is made from crude antimony. In all cases, however, either the yield of sublimate was disappointing or the product was "redd & foule." Newton had clearly reached a roadblock, as some revealing comments from a parallel section of CU Add. 3975 make evident:

> I do not yet find any way of cleansing ye sublimate of ♂ from it's impure ♀ without destroying its volatizing virtue. If ♂ be melted w^th <*illeg.*> 1/2 1/3 or 1/4th part of ☉, the nitre does not hold down ye impure ♀ of y^e ♂ at all. But let ye whole body of y^e ♂ rise & remains it self in ye bottom without much addition of ♂.[27]

In addition to revealing Newton's puzzlement about the source of his sublimate's "volatizing virtue," this passage is probably related to the experiments in CU Add. 3973 where he tried subliming sal ammoniac from the scoria made during the production of martial regulus. It seems that Newton had hoped that the saltpeter employed as a flux for the stibnite would not only help to fuse the material but would also combine with the sulfur in the ore, resulting in a fixed or nonvolatile slag. Unfortunately, the result turned out otherwise, and the antimony left in the slag sublimed undivided, with both

---

[25] CU Add. 3975, 56v.
[26] CU Add. 3975, 56v.
[27] CU Add. 3975, 58v.

its "reguline" part and its sulfur. Here and elsewhere we see Newton using an experimentally acquired knowledge of affinities in the attempt to get chemicals to dissociate from one another by combining with a third party. For the moment, however, his attempts were blocked by the fact that cleansing his sublimate of crude antimony and sal ammoniac meant the loss of its volatilizing virtue. This situation would change when Newton returned to the bench some five months later, in the summer of 1680.

Newton's notes from February 1679/80 terminate with a succession of attempts, in Latin, to decode the *Decknamen* of his alchemical sources. I will return to these extraordinarily difficult puzzles in due course, but for the moment they are outside our scope of inquiry. Let us turn to the next collection of notes, which is dated "Iuly 10" and must refer to July 10, 1680.[28] Here too Newton writes in Latin, giving a list of five numbered headings. The excited tone of the comments, as well as their studied vagueness in describing the starting point of the processes, convey the sense of great importance that he attributed to his discovery of the sophic sal ammoniac. Newton is clearly in the grip of a "eureka moment." He begins by saying, "1 July 10. I saw the philosophical sal ammoniac. This is not precipitated by salt of tartar." Salt of tartar or potassium carbonate was a standard reagent used by chymists to precipitate metallic salts out of acid solutions. Newton employed it in his experiments of January 22, 1679/80, when he was testing the properties of the salt made by boiling a solution of antimony once acted on by aqua fortis and sal ammoniac. There he noted that the sublimate made very little ebullition when added to a solution of salt of tartar, which he attributed either to the previous loss of its "aqueous spirit" or to the possibility that this component had a stronger affinity for its "other spirit" than for the menstruum.[29] Elsewhere he explicitly uses salt of tartar to precipitate metals out of solutions in acid menstrua.[30] In this particular case, Newton's point could be that *unlike* ordinary sal ammoniac, the sophic variety can keep a heavier material sublimed with it in solution even when added to the salt of tartar solution. On the other hand, he sometimes used the salt of tartar test after already precipitating his insoluble sublimates in plain water.[31] This raises the possibility that the sophic sal ammoniac itself may have already undergone such testing with water in the present case, and that Newton had already precipitated the insoluble components of the sublimate. If that is the case, then the fact that the sophic sal ammoniac did not drop a precipitate in salt of tartar meant that it was *similar* to normal sal ammoniac, which also dissociates without leaving a precipitate. The passage is therefore ambiguous to the modern reader (though not to Newton), and hence less helpful than one might hope for determining the character of his sophic sal ammoniac.

---

[28] For the proper dating of this section (Sheet 5), see the introductory "Manuscript Information" section of the online edition of CU Add. 3973, at *CIN*. The numbered headings are found on 13r in the ordering of the edited text.

[29] CU Add. 3973, 6r. Newton was of course aware that sal ammoniac consisted of an acid spirit (our HCl) and an alkalizate one (NH_3).

[30] As at CU Add. 3973, 15v.

[31] As at CU Add. 3975, 64r.

Only with the third heading do we begin to get a sense of the nature of Newton's sophic sal ammoniac and its place within the development of his experimental program. There Newton tells us, "The white calx distilled per se emitted 20 grains of ✳ from 400 grains of calx." In other words, Newton distilled 20 grains of "✳" (sal ammoniac, or in this case, sophic sal ammoniac) out of 400 grains of "white calx" (*calx albus*). He adds that the distillation was made from the white calx per se, meaning that no ingredients were added to it. Despite the vagueness of the term "white calx," this description conveys important information. First, since the distillation was carried out per se, the white calx must have already contained the sophic sal ammoniac within itself—Newton did not intentionally add ordinary, "vulgar" sal ammoniac as a subliming adjuvant in this instance. Second, there is the term "white calx" itself. What is the sense of this vague signifier? At this point, we must employ Ockham's razor on the assumption that Newton is not making an abrupt saltation to some hitherto untried material, but is developing his practice out of preexisting protocols carried out on familiar substances. Given his dogged repetition of similar experiments with varying of proportions and the occasional isolation of variables by adding or subtracting an ingredient, development out of preexisting methods and materials seems distinctly the more probable course.

So what does Newton typically mean by "white calx"? The term "calx," though often used to mean a product of high-temperature calcination in the presence of air, had taken on an extended sense by the early modern period. Newton's early chymical dictionary, Bodleian MS Don. b. 15, is quite revealing on this point. In addition to meaning a metal powdered by the action of heat (or as we would say, an oxide), Newton uses the term to mean any solid residue that has come out from a state of dissolution in a liquid. As he puts it there:

> Præcipitation of a body out of yᵉ dissolvent into a Calx either done by abstracting <*illeg.*>(i.e. Evaporating or distilling off) yᵉ Solvent water, & thus ∧solutions of Mettalls ∧yeild Vitriolls, & saline liquors their salts ~~are turned to Vitriolls &c~~, or by putting in some body for yᵉ dissolved matter to sattle upon (as ♀ plates into a solution of ☽ in aqua fortis ∧weakened by addition of much water , or ☿ into a solution of ☉ in Aqua Regis) Or by pouring in some other liquor of a contrary nature.[32]

This dictionary entry tells us that a calx is formed either when solid material is left in the vessel after boiling off a solvent or when a finely dissolved material precipitates onto another body with which it has an affinity, or even when a precipitate is released from solution by pouring in a liquor of opposed nature (for example, salt of tartar added to an acid menstruum). In other words, the term "calx" can mean any powdery or particulate material that emerges from a solution, in addition to its more obvious sense as the product of calcination by heat.

[32] Don. b. 15, 4v.

Since Newton generally avoids high-temperature calcinations in his experimental notebooks, we should therefore expect him to use "calx" in this extended sense. So let us consider examples where "white calx" or *calx albus* occur in the early sections of CU Add. 3973 or its companion notebook CU Add. 3975. In performing this search, we should also keep in mind that "calx," in accordance with Newton's extended use, can be synonymous with "precipitate."[33] An examination of CU Add. 3973 reveals only seven possible candidates for Newton's white calx before the crucial date of July 10, 1680.[34] All of these possibilities are found between 6r and 10r, and hence stem from the winter of 1679/80, the period directly before Newton's eureka moment. The first (6r) describes a "a white fat *<illeg.>* clamy slime" that Newton had made in the previous summer (1679) by pouring multiple affusions of aqua fortis sharpened with vulgar sal ammoniac on crude antimony and heating for an extended period. He dried this slime, sublimed it, and precipitated it with water according to his usual protocols. We have already considered the second passage (6r), where on January 22, 1679/80, Newton boiled a solution of crude antimony "once acted on by ♐," sal ammoniac, and aqua fortis to dryness and sublimed the volatile salt. This would have involved a white or yellow caput mortuum or calx, as I have found by experimental replication. On 6v Newton continues this experiment by washing the sublimate of this caput mortuum to arrive at a white precipitate; as we have seen, this could also be called a white calx. Two folios later (8r), we find Newton producing a "light white calx" by subliming (with sal ammoniac) a precipitate made from antimony dissolved in aqua fortis with sal ammoniac and diluted with water. He then resublimes the light white calx with additional sal ammoniac to get another white calx. On 10v we find Newton washing a sublimate made from scoria or regulus martis (and sal ammoniac, of course) in order to get a white precipitate. This is followed, finally, by a resublimation of sublimate made from iron ore and crude antimony; the washing of the sublimate again yields a white precipitate.

All seven of these candidates for Newton's *calx albus* share a common feature—they are all produced either by dissolving stibnite in aqua fortis or aqua regia made with sal ammoniac or by subliming the same crude antimony, again with sal ammoniac, and then washing with water to produce a precipitate. The early portions of CU Add. 3975 add only one further candidate, the washing of butter of antimony to arrive at a white precipitate (primarily antimony oxychloride or "mercurius vitae").[35] Hence stibnite and

---

[33] I restrict this search to folios 1r–12v of CU Add. 3973, namely, the folios before Newton's eureka moment of July 10, 1680, and folios 1r–62r of CU Add. 3975, before the first instance of the term "sophic sal ammoniac" in the MS. But I am also excluding passages in CU Add. 3975 after 58v, for the section between 58v and 62r does not correspond to anything in CU Add. 3973 and appears to have been composed after the summer of 1680. This section already describes Newton's project for making volatile Venus, which postdates the discovery of sophic sal ammoniac.

[34] The references to a white calx or precipitate in the pre-summer 1680 part of CU Add. 3973 occur on 6r (two instances), 6v (one instance), 8r (two instances), and 10v (two instances).

[35] CU Add. 3975, 52r. One might also add the precipitate made during the preparation of Boyle's *menstruum peracutum*, which Newton discusses on 40v of CU Add. 3975. This would be a white or yellowish

sal ammoniac feature prominently in seven of the eight cases where a white calx is mentioned; the last example drops the sal ammoniac but keeps the antimony ore.[36] At this point it will again be helpful to resort to modern laboratory replication. When one carries out the first operation in the laboratory, namely, Newton's dissolution of stibnite in homemade aqua regia, a violent, bubbling reaction occurs with the release of red nitrogen dioxide gas and heat. The reaction leaves a yellow or whitish yellow precipitate, and a considerable amount of antimony goes into solution with the chloride from the sal ammoniac (figures 15.3 and 15.4). Thus the solution contains significant quantities of antimony trichloride dissolved in the aqua regia (probably along with sulfuric acid produced in the reaction), and the calx (when washed) consists of antimony compounds along with elemental sulfur. This process was later described by Newton's younger contemporary Herman Boerhaave as a method for removing the sulfur from stibnite. As Boerhaave pointed out, the metallic antimony is gradually dissolved by the aqua regia, leaving a material that he identifies as sulfur behind.[37] This certainly helps to explain the multiple uses that Newton makes of this dissolution, since the process must have opened up an opportunity of fulfilling his often expressed desire of ridding crude antimony of its sulfurous, dirty foulness. From a modern perspective the process is a relatively easy way to make a solution of antimony trichloride or butter of antimony, which may well be the main constituent of Newton's "white fat clamy slime." After subliming this product, though, Newton washed it with water and acquired a white precipitate: if the initial product was indeed butter of antimony, the precipitate would be primarily antimony oxychloride, an insoluble material formed when antimony trichloride is decomposed by water. Early modern chymists typically referred to antimony oxychloride as *mercurius vitae* for its marvelous powers, and Newton himself identifies the material by this name at a later point in CU Add. 3973.[38]

All of this evidence points to the likely conclusion that Newton's *calx albus,* from which he produced the sophic sal ammoniac, was a mixture of white antimony compounds possibly containing a certain amount of ordinary ammonium chloride. The sophic sal ammoniac would then have been a sublimate of this white calx, as Newton himself indicated on 13r of CU Add. 3973, when he said, "The white calx distilled per se emitted 20 grains of ✻ from 400 grains of calx." Our preliminary conclusion in fact receives strong and direct support from a later passage in the same manuscript where Newton performs a quantitative analysis on the sophic sal ammoniac.[39] He

---

product rich in antimony oxides, usually called *bezoardicum minerale.*

[36] Newton's description of making butter of antimony on 52r of CU Add. 3975 employs the traditional method of subliming stibnite with corrosive sublimate (mercuric chloride). But of course butter of antimony is also produced by dissolving stibnite in aqua regia, and this technique is ubiquitous in Newton's notebooks.

[37] Herman Boerhaaave, *Elementa chemiae* (Leiden: Isaac Severinus, 1732), 2: 504-6, "Processus CCVIII–CCIX," For a similar observation expressed in Lavoisian language, see also M. Fourcroy, *Elements of Natural History and Chemistry* (London: C. Elliot and T. Kay, 1790), 2: 259.

[38] CU Add. 3973, 38v.

[39] CU Add. 3973, 40v.

Figure 15.3. Newton's liquor of antimony being produced from aqua regia (nitric acid "sharpened" with sal ammoniac) and stibnite (antimony trisulfide). A vigorous reaction takes place, leaving behind a yellowish-white precipitate. Prepared by the author in the laboratory of Dr. Cathrine Reck in the Indiana University Chemistry Department. See color plate 5.

FIGURE 15.4. Crystals deposited by slow evaporation of liquor of antimony. Prepared by the author in the laboratory of Dr. Cathrine Reck in the Indiana University Chemistry Department. See color plate 6.

starts this experiment with 6 1/4 ounces of "Our ✳ freed from ♁."[40] "Our sal ammoniac" is simply another name for the sophic sal ammoniac, and the fact that it had to be "freed" from an excess of antimony, no doubt by washing and precipitation, reveals that the sophic sal ammoniac already contained that material, at least as a precursor.

In order to carry out his analysis, Newton dissolved the 6 1/4 ounces of sophic sal ammoniac in a weighed quantity of diluted aqua fortis along with 4 ounces of regulus of antimony, adding more until the menstruum was "satiated"—that is, until it would dissolve no more regulus. The idea was to test whether the regulus, in combination with the aqua fortis, would "destroy" the sophic sal ammoniac as it would have done in the case of vulgar sal ammoniac

[40] The initial quantity, 6 1/4 ounces, is confusingly written in the MS as "Q^ter of ℥vi." I have established that this must mean 6 1/4 ounces from the fact that at the conclusion of his second analysis on CU Add. 3973, 41r, Newton says "So then 40 gr of o^r ✳ was reduced to lesse then 4 by these two dissolutions." In other words, he managed to destroy all of the sophic sal ammoniac except for 4 grains that were left out of 40 grains that he started with. The first analysis left him with 2 1/2 ounces of product as described on 40v, from which on 41r he takes a sample of 16 grains: it is the 16-grain sample that undergoes analysis to be diminished to 4 grains. So the ratio 40:16:4 represents the amount of initial sophic sal ammoniac to be analyzed, the product of this first analysis, and the product of the second analysis. Since 40:16 is equivalent to 5/2, the initial amount must have weighed 5/2 of 2 1/2 apothecaries' ounces, or 6 1/4 ounces.

(as we would say in the language of post-Lavoisian chemistry, the metallic antimony would combine with the chloride of the sal ammoniac that was dissolved in the aqua fortis). When the acid solution was "satiated," this meant that the process had reached its completion; this was important since the point of the analysis was to "destroy" as much of the sophic sal ammoniac as possible. Newton then evaporated the solution to dryness to get 2 1/2 ounces of dry residue, from which he removed 12 grains. He sublimed this from 72 grains of lead ore, in order to test the results of the foregoing analysis. His choice of lead ore may have stemmed from earlier experiments where he had learned that the ore of lead would further "destroy" the sophic sal ammoniac to form a sweet material (possibly lead chloride or a related salt).[41]

The new sublimation of 12 grains of residue from 72 grains of lead ore led to the following results: an unspecified quantity of material was carried up, but washing the product showed that only 6 grains of it were soluble in water. He then resublimed the 6 grains that were soluble, and found that 4 grains rose in the second sublimation, "wch tasted keen like ✳." The sharp taste of this sublimate led Newton to conclude that these 4 grains were ordinary, vulgar sal ammoniac. Hence his conclusion states that "4^{gr} of y^e 12 were vulgar ✳." What then were the other 8 grains? This too receives an answer in Newton's final tallying of the analysis, where he scales up the conclusions drawn from the 12 grains to the entire 2 1/2 ounces left in the initial evaporation: "$^{∧one third part of}$ y^e ℥ij 1/2 wch remained after y^e evaporation, was vulgar ✳ not destroyed; one half was o^r ✳ & 1/7 part fæx." Since 4 grains or one-third of the 12 grains that sublimed from the lead ore was vulgar sal ammoniac, the 8 grains that were left must have consisted of 6 grains of undivided sophic sal ammoniac and 2 grains of impurities or "fæx." Newton's figure of 1/7 part fæx instead of 1/6 is probably intended to allow for water contained in the residue. From all of this he concludes: "So then y^e ⱯƑ destroys ye sophic ✳ as well as ye vulgar though not so much."

This fascinating analysis contains both valuable information and cause for puzzlement. First, what happened to the antimonial regulus—over 4 ounces—that was added to the menstruum at the start, since it does not figure in the final tally? Newton no doubt knew that the metallic antimony (regulus) would react with the aqua regia to form butter of antimony, and he must also have known that this compound boils at a quite moderate temperature (about 223°C by modern accounts). Hence he must have assumed that all of the butter of antimony would pass off during the boiling of the solution when he evaporated it. Thus there was no need to consider the regulus in the final tally; after all, Newton's goal was to "destroy" the sophic sal ammoniac, not to account for every ingredient's transformation during the operation.

But the experiment is revealing in a second, more profound way. As we already saw, the initial sophic sal ammoniac that went into the experiment

<hr />

[41] Such an experiment with the sublimation of sophic sal ammoniac from lead ore is found at CU Add. 3973, 40r. The sweet taste of lead chloride (PbCl$_2$) is attested by various old chemistry texts. See for example William Thomas Brande, *A Manual of Chemistry* (London: John Murray, 1830), 2: 77.

already contained an antimony compound, which Newton partially removed (probably by precipitation in water). Yet the experiment tells us another fact about the sophic sal ammoniac's initial composition. The sequence of operations that began with over 10 ounces of sophic sal ammoniac and antimony regulus gave 2 1/2 ounces of product, and this product was 1/3 vulgar sal ammoniac. Thus the initial sophic sal ammoniac must also have contained ordinary, vulgar sal ammoniac, since that material was present in the 2 1/2 ounces of solid that were left after evaporation of the acidic solution. And of course it was the chloride in the sophic sal ammoniac that converted the initial aqua fortis into an aqua regia capable of dissolving antimony regulus to form butter of antimony.

This is important news indeed! And it receives further confirmation from the following experiment in CU Add. 3975, where Newton subjects the 2 1/2 ounces of material left in the former experiment to yet another analysis. We must pay close attention to the opening words of this new experiment— "Of the said ℥ij 1/2 of oͬ ✳ I dissolved 16 gr in ⨍ 32 ᵍʳ & ▽ 32ᵍʳ." Without the symbols, Newton is saying that he took the 2 1/2 ounces of sophic sal ammoniac ("our sal ammoniac") left from the previous experiment and dissolved 16 grains of it in diluted aqua fortis. He then describes dissolving regulus in this menstruum as he did in the previous experiment, until satiated. When all is done, he evaporates the solution to get 8 2/3 grains of a keen, styptic salt, from which he takes a sample of 4 grains and evaporates all that he can over a candle; the residue left after the sublimation is less than 1/10 of its weight before the first analysis. As he puts it, "So then 40ᵍʳ of oͬ ✳ was reduced to lesse then 4 by these two dissolutions." The significance of this experiment is straightforward—at the beginning, Newton explicitly identifies the entire yield of his previous analysis, the whole 2 1/2 ounces, as sophic sal ammoniac; at the end, he says that the 1/10 of the product that remains is also sophic sal ammoniac. He makes this identification despite the fact that in the previous experiment he showed the first yield to consist of 1/3 vulgar sal ammoniac, 1/2 sophic sal ammoniac, and a remainder of impure "faex."

It seems inescapable that Newton is using the term "sophic sal ammoniac" in two ways—first, as a mixture of a certain antimony compound with ordinary sal ammoniac to improve its volatility, and second, as the antimony compound itself. What was this compound? We know that it contained antimony, but since most or all of the butter of antimony formed during the dissolution of the regulus in aqua regia boiled away during the evaporation, the trichloride is to be excluded, at least for the most part. We also know from Newton's sublimation of the sophic sal ammoniac from lead ore and subsequent washing that the material was partly or wholly insoluble in water. The volatility of this mixture would vary, depending on numerous factors in its preparation, such as purity of ingredients, dilution of acid, amount of water used in the washings, and temperature at which it was sublimed.

Although it is too early to identify the chemical composition of Newton's sophic sal ammoniac, the foregoing examination leads to the conclusion that the material was simply a variation on the antimonial sublimates that he was making with stibnite and sal ammoniac as early as December 1678 (and probably earlier). As we know, his main goal in the improvement of his antimonial sublimate was that of removing its "dirtiness" while retaining its power to act as an adjuvant in subliming other metallic materials. He seems to have met that goal, at least partly, in July 1680, a fact that led him to the breathless excitement of his "eureka moment." Later accounts of the sophic sal ammoniac confirm that it was less "foul" than the simple sublimate of stibnite and vulgar sal ammoniac, even if it may not have improved on its volatilizing power. Newton had therefore met one of the principal goals of his early chymistry—the preparation of a subliming agent that could be used to volatilize metals and their compounds without polluting them by the addition of an excessively "impure ♁." Armed with this new analytical tool, he could now proceed to additional projects.

Further Tests: Fused or Mineral Antimony?

Despite Newton's initial burst of excitement on discovering the sophic sal ammoniac in the summer of 1680, subsequent records show that he continued to tinker with different but related sublimates of crude antimony and antimonial regulus combined with ordinary sal ammoniac. His ongoing attempts to improve the cleanliness and subliming power of his antimonial sublimates, or perhaps to mitigate their ease of production, appear even in the 1690s, when he advises his Swiss friend Fatio de Duillier to use a sublimate of regulus and ordinary ammonium chloride as a volatilizing adjuvant where one might expect the sophic sal ammoniac of 1680.[42] Already in August 1682 we encounter him devising a new piece of apparatus for the purpose of testing the subliming power of a stibnite-sal ammoniac adjuvant.

The pretext for this endeavor lies in an experiment that begins by dissolving a complicated mixture of metals, salts, and ores in aqua fortis in order to create a menstruum for dissolving "Diana," the alloy of bismuth, bismuth ore, and tin that we had occasion to discuss in chapter thirteen.[43] He then evaporates and sublimes according to his usual protocols.[44] Newton finds that the sublimate produced from the Diana alloy by these means has an interesting property: sublimed from lead ore, it yields a caput mortuum that fumes for an unusually long time. After performing some seemingly unrelated experiments on lead ore, Newton then returns again to the Diana alloy, dissolving it this time in aqua fortis sharpened with a salt made from antimonial vitriol of copper previously sublimed from "Subl. of melted ♁."[45] The "sublimate of melted antimony" refers to a product made out of stibnite purified from

---

[42] For Newton's advice to Fatio de Duillier, see chapter seventeen herein.

[43] The term "Diana" for this alloy appears only on 16v of CU Add. 3973. Newton evidently made a small transcription error at the top of 15r, repeating "♃" when the second symbol should have been "♃," the ore of bismuth. This is clarified by his multiple references to the alloy, with slightly varying proportions, that follow.

[44] CU Add. 3973, 15r. Newton's chymical experiments for 1681 and the preceding part of 1682 are not found in CU Add. 3973, but occupy 62v–69r of CU Add. 3975. Some of the material between 58v and 62v could also stem from 1681, though I have not been able to determine a definite date for that section of CU Add. 3975.

[45] CU Add. 3973, 16r.

its mineral gangue by fusion and then sublimed with ammonium chloride; whether it underwent additional treatments we do not know. There can be little doubt that this sublimate was either a substitute for the sophic sal ammoniac, or perhaps another version of it. But the experiment is followed by the revealing comment— "NB ~~The~~ Therefore melted ♁ makes ye matter less fluxible & less volatile then unmelted minera of ♁." In other words, the refined stibnite used in making this sublimate was less effective for melting and volatilizing the sublimandum than the stibnite in its natural state. In the course of subliming his salt from the Diana alloy, Newton was struck by the large quantity of infusible caput mortuum that was left below. His comment attributes this fixity to the use of refined stibnite in the adjuvant sublimate. But he immediately strikes this entire comment through and decides that the matter requires further testing.

Newton's cancellation of his hasty comment provides a rare example of self-doubt in his chymical experimentation and prompts an even rarer case of a carefully described apparatus. In order to determine definitely "whether unmelted minera of ♁ makes ye matter more fluxible & volatile then melted ♁," Newton designs and builds (or has built) a special sublimatory made of earthenware in three sections.[46] According to his description, the lowest part is simply an open cylinder 6 inches wide and 3 inches deep, on which another cylinder of the same width but 5 2/3 inches deep sits. The second cylinder is closed at the top except for a round hole 2 3/4 inches wide in its middle. Finally, a third pot of the same width but unspecified height sits on top of the second; a later description suggests that it too had a central hole in its top.[47] Newton first sublimes 6 ounces of vulgar sal ammoniac with 6 ounces of "minera of ♁"—the unmelted ore—in the bottom of this apparatus for two hours, until the pot is almost red hot. The design was intended to allow different fractions of sublimate to collect in the second and third cylinders, which would allow condensation at different temperatures. Newton prefers the lower fraction found beneath the hole in the second pot, some of which he then adds to antimonial vitriol of copper and sublimes again. He washes and precipitates this sublimate and then uses it to sharpen the aqua fortis in which the Diana alloy will once again be dissolved.

After trying several different proportions of "♀," the volatile copper salt obtained by subliming antimonial vitriol of copper with mineral stibnite and sal ammoniac and then freeing the salt from excess antimony by precipitating the new sublimate in water, Newton then proceeds to the "melted ♁," that is, the stibnite previously refined from its gangue by fusion. Employing his three-part sublimatory and using similar proportions to those in the case of the unmelted ore, Newton discovers that the previously refined stibnite is less suitable for his purposes than the same mineral straight from the

---

[46] CU Add. 3973, 16r.

[47] CU Add. 3973, 29v: "About 4 a clock ye yellow flowers began to rise ^through ye hole in ye middle of ye upper pot & continued rising for 1 1/2 or 2 hours & then turned almost white, at which signe I stopt ye hole." If this is the same three-pot apparatus as before, which is quite likely, the hole spoken of here can only refer to the third and highest pot, since Newton would not otherwise have been able to see the fumes rising.

ground.[48] Although the sublimate of Diana made from the melted stibnite was whiter than the version made from the unfused ore, it was less fusible, probably indicating to him that the alloy had been "opened" less than it had with the unrefined ore of antimony.

As we have just seen, Newton's desire for a careful test of his antimonial sublimates led him so far as to devise new laboratory apparatuses for carrying out that goal. Alchemists had known since the Middle Ages that different sublimates would collect in their aludels or sublimation vessels at different heights, leaving variously colored bands on the inside of the apparatus that could be collected separately (figure 15.5).[49] And early modern chymists such as Nicolas Lemery had already devised multipot sublimatories for separating the different fractions of antimonial sublimates: Lemery's *Course of Chymistry* or some similar text may well have served as Newton's inspiration.[50] But it was probably Newton's overriding preference for natural ores and minerals fresh from the mine that led him to apply this sort of apparatus to the comparison of refined and unrefined stibnite. He had observed that his unrefined antimony ore contained "spar," which could denote anything from quartz to calcite, and it would be understandable for him to have wished for a purer form of stibnite.[51] From a modern perspective, such a choice would have seemed more than obvious. But in addition to his general preference for native minerals, there may have been another reason for him to have hesitated here before opting for the purified version of the ore. Other remarks in Newton's notebooks make it likely that the spar or some other impurity was acting as a flux in his stibnite; hence it would have been a natural move for him to have wondered whether the property of greater fusibility in the native stibnite might carry over to its sublimed products.[52] His three-pot sublimatory was probably a means of deciding between the two competing goals of greater fusibility and removal of obvious gangue. Given Newton's general preference for unrefined minerals, it is no surprise that the native ore won the day.

The specialized apparatus designed to compare refined versus unrefined stibnite encapsulates a principal feature of Newton's chymical laboratory notebooks. Although he never veered from his goal of preparing the secret ingredients of the philosophers' stone, Newton's efforts involved the continual

[48] CU Add. 3973, 16v. The proportions are not identical in the two cases: Newton used 120 grains of the sublimate of antimonial vitriol in the second case, and 116 grains in the first, and 240 grains of aqua fortis in the second as opposed to 220 grains in the first. He was presumably constrained by the quantity of materials available to him. Another strange feature lies in his use of the *unicum* symbol "☿" for the sublimate of antimonial copper vitriol. I have not found this symbol elsewhere in the corpus of Newton's alchemy. It probably represents an early form of "☿" if not a slip of the pen.

[49] Geber, for example, describes such sublimation bands. See William R. Newman, *The Summa perfectionis of Pseudo-Geber* (Brill: Leiden, 1991), 691.

[50] Nicolas Lemery, *A Course of Chymistry* (London: Walter Kettilby, 1677), "Another Antimonium Diaphoreticum," 121.

[51] CU Add. 3975, 54v.

[52] CU Add. 3973, 27r: here Newton attempts to fuse "artificial ♂" (mineral stibnite that he had purified) with iron ore in order to induce a "fermentation." He notes that the "artificial" version of the mineral, which would lack the impurities imparted by the gangue of the mineral stibnite, is much harder to fuse with the iron ore.

FIGURE 15.5. Modern subliming apparatus used to replicate Newton's experiments, consisting of an inverted, drilled-out fire-assay crucible set atop a porcelain, Coors crucible (in practice, the joint would be taped). The sublimate has permeated the fire-assay crucible, as shown by the variously colored horizontal bands at different heights. Prepared by the author in the laboratory of Dr. Cathrine Reck in the Indiana University Chemistry Department. See color plate 7.

interaction of textual interpretation, reverse-engineering of products whose identity he thought he had extracted, and testing of the results. This three-fold process remains constant throughout the thirty or more years of his alchemical experimentation, even though the notebooks display new ideas and fresh discoveries as one pages through them. For the modern reader, the main challenge to understanding Newton's laboratory records lies in the fact that he often neglects to specify the overall significance of the tests or to place them in the context of his evolving strategy. Yet in the case of Newton's ideas about sal ammoniac, one can make out a definite progression of ideas and practices. Even if the seeds of Newton's interest in this substance were already planted by Boyle, the generation-long fascination that sal ammoniac exercised on the younger scientist was heavily fertilized by his exposure to the *Epistola ad Langellotum* of David von der Becke. It was von der Becke's *Epistola* that provided Newton with an understanding of the composition of sal ammoniac from an acid and a "urinous" salt, and the German chymist's work inspired him to explore the ability of the substance to "promote the fermentation" of salt of tartar. Moreover, von der Becke viewed this "fermentation" as the action of microlevel corpuscles that associated and dissociated from one another as a result of their relative affinities; depending on the particular chymical combination of particles, one produced a compound that was either fixed or volatile.

Newton's very early decision to subject antimony to the subliming action of sal ammoniac descends neither from Boyle nor from von der Becke, however. In CU Add. 3975, it follows extensive experiments involving various reguli of antimony whose ultimate source lies in the work of Eirenaeus Philalethes.[53] It is likely that Newton's initial choice of crude antimony as the best material to combine with sal ammoniac in the quest for a subliming agent stemmed from his understanding of Philalethes as well. But Newton quickly encountered the roadblock that crude antimony and sal ammoniac produced a "dirty" sublimate, while his multiple attempts to cleanse the product of its "filth" led to a decrease in its ability to carry up other sublimanda. His years spent in attempting to solve this problem led to the discovery in 1680 of the sophic sal ammoniac. But after an initial euphoria induced by his success, Newton's enthusiasm for "our" sal ammoniac seems to have undergone a degree of cooling. Thus we find him continuing to employ antimony regulus and even stibnite in combination with sal ammoniac as subliming adjuvants throughout his career as a chymist. Sal ammoniac and antimony in both crude and refined forms continued to be fundamental components of Newton's chymical armory, and as we will see in the next chapter, both materials played a critical role in another decades-long project, namely, the attempt to make and use the substance that Newton refers to as "volatile Venus."

[53] See CU Add. 3975, 43r, where Newton is already using the Philalethan term "net" for an alloy of antimony regulus and copper.

SIXTEEN

Extracting Our Venus

Sophic sal ammoniac, for all the excitement that its 1680 discovery aroused in Newton, was not an end in itself. Like the mixture of stibnite and sal ammoniac that he had already described in his experiments from the first half of December 1678, it was a preliminary tool for subliming other metals and minerals. Indeed, Newton's sophic sal ammoniac appears to be a refinement of his previous practice with crude antimony, though in the case of sophic sal ammoniac Newton began with a white antimonial calx rather than with stibnite per se. Similarly, his experiments with the three-part clay sublimatory had the goal of settling which was better—a subliming adjuvant made with crude antimony direct from the mine, or one composed of stibnite that had been separated from its gangue by fusion. We will now pass to a different but equally important subject, namely, the volatile salt or salts that Newton made from copper. Both Philalethes and Snyders had stressed the role of copper in producing the philosophers' stone, and Newton's experimentation would confirm that the red metal yielded highly intriguing results in the laboratory.

It is possible that Newton's goal was to arrive not merely at a salt of copper in the modern sense, where the metal combines with other elements to yield a compound, but rather in the traditional alchemical sense whereby the copper is decomposed and its internal, constituent salt is released. Unfortunately, a serious interpretive problem lies in wait for us, namely, the fact that Newton, like his Helmontian contemporaries (particularly Starkey), used the term "salt" in both senses. Nor does the fact that Newton speaks of "extracting" his salt of copper solve the problem. "Extraction" did not necessarily imply the simple isolation of a preexisting material, as one might reasonably expect from its ordinary modern meaning. Newton frequently uses the term in an operational sense where it is coterminous with digestion or imbibition in a solvent, sometimes followed by evaporation. As Newton makes clear in one of his chymical dictionaries, an "extract" could therefore be synonymous with an "infusion," or even a "balsome."[1] The resulting ambiguity is thrown into high relief by passages

---

[1] Don. b. 15, 3r.

in the laboratory notebooks like the following, where Newton says that a solvent acting on an ore "extracts a salt, turning almost all ye ore to salt."[2] There is no easy way out of this inherently ambiguous situation. One thing, however, is clear. Following the pathways of Newton's remarkably careful analytic procedure reveals that he was intent on "extracting our Venus," a volatile material that had to be denuded of superficial, saline accretions. Whether "our Venus" meant an internal constituent of the metal copper, or rather a compound of the metal, Newton spent years of effort trying to arrive at the substance in pure form. At the same time, however, Newton viewed his volatile salt of copper as a means to an end. Like the sophic sal ammoniac, the copper-based material that Newton would come to call "volatile Venus" was supposed to act on other metals, minerals, reguli, and salts in a multitude of complex ways. It was in all likelihood identical to the "volatile vitriol" that Newton thought he found in the work of Snyders, as we discussed in chapter eleven. For now it is enough to describe the years of experimentation that Newton devoted to preparing and purifying this marvelous substance.

The story seems to begin around December 19, 1678, when Newton separately sublimes copper and iron with his customary adjuvant composed of stibnite and sal ammoniac.[3] Although there is no sign that he was immediately impressed with the results, things begin to change when, a little over a year later, Newton starts subliming the same sal ammoniac–stibnite mixture from "vitriols" of these and other metals. As is usually the case in his notebooks, when Newton uses the term "vitriol," he is thinking of metals imbibed with and dissolved in his liquor of antimony and then allowed to evaporate and crystallize.[4] Thus in February 1679/80, he records that he sublimed vitriols of lead, iron, copper, and green lion with his usual mixture of stibnite and sal ammoniac in a proportion of two parts adjuvant to one part vitriol. The sublimate of impregnated copper (that is, of antimonial vitriol of copper) receives by far the longest description. Newton says that it is fusible and almost entirely volatile, meaning that it can in turn be sublimed with very little residue left behind. When made with a lesser proportion of the sal ammoniac–stibnite adjuvant (3 parts of this to 2 of impregnated copper), Newton finds that the sublimandum boils and bubbles in the process of volatilizing. The caput mortuum is almost insoluble, and the part that does dissolve "does not look blew," which differentiates it from many copper compounds. Additionally, in the case of the 2:1 proportion, the copper sublimate is only partially soluble in water; with his usual precision, Newton indicates that 1/8 of the sublimate by weight does not dissolve in an aqueous solution. Finally, Newton adds that the copper sublimate—apparently made in either proportion—is white.[5]

[2] CU Add. 3975, 54r.
[3] CU Add. 3973, 1v. This was before Newton's 1680 discovery of sophic sal ammoniac, so the starting ingredients are unambiguous.
[4] See CU Add. 3973, 9r.
[5] CU Add. 3973, 9v.

At around the same time, we begin to see a growing excitement in Newton's description of this sublimate made from antimonial vitriol of copper. A passage from CU Add. 3975 records the following information:

> salt of ♀ rises in subliming wᵗʰ a rushing wind, so as to require a retort to distill it in, & is of all salts most freely volatized.[6]

This important passage reveals the main cause of Newton's enthusiasm for the volatile "salt of ♀" (by which he still means the sublimate of antimonial vitriol of copper). He found that the material sublimed more easily than any of the metallic salts with which he was familiar and could therefore be expected to serve as an effective volatilizing adjuvant for other materials. In fact, the volatile salt of copper becomes Newton's standard of comparison when referring to other sublimable salts.[7] Thus we find him immediately using the substance to aid in the sublimation of "clay of Lead mines," spelter (zinc), and lead ore.[8] In this same period, which we can definitely place between early 1680 and the spring of 1681, Newton adopts the term "volatile Venus" ("ve. vo." or "ven. vo": abbreviated forms of "venus volatilis" or "venus volans") for his sublimate of antimoniate copper vitriol. But at the same time as his interest in this material grew, problems also emerged. Just as Newton's early stibnite–sal ammoniac sublimate had been plagued by the foul dirtiness of sulfur, so the volatile Venus had its own persistent impurity, namely, the very sal ammoniac that played an essential role in its production. And just as Newton developed sophisticated analytical techniques to rid his sophic sal ammoniac of sulfur, so he worked over a succession of years to cleanse the volatile Venus of its unwanted adulteration. We already encounter Newton's first attempt to purify the volatile Venus only a few lines after his experiment in which he used it to "raise" the clay of lead mines. This is one of the precious examples in his chymical notebooks where Newton gives a plenary explanation of his motives for carrying out an experiment, so I will quote it in full:

> Sublimate of Venus made wᵗʰ Subl. of ♂, dissolved & philtred to separate yᵉ ♂ & dried & mixed ^either with iron filings or with spar would not rise in a second sublimation but stayd behind wᵗʰ yᵉ iron or spar & made yᵉ spar of a keen tast. The design was to separate yᵉ ✳ from yᵉ salt of ♀ but yᵉ ✳ did not fasten of yᵉ spar nor much on yᵉ iron, but rose alone wᵗʰ out yᵉ ♀. And if Spar & ✳ were taken alone, yᵉ ✳ rose from yᵉ spar wᵗʰout being destroyed by it.[9]

This experiment was carefully constructed as a means of separating the excess sal ammoniac from Newton's volatile copper salt. As we have already seen, Newton knew that sal ammoniac consisted of two components, an

[6]CU Add. 3975, 58r. This passage finds its parallel in CU Add. 3973, 9v, but without the information about the "rushing wind."

[7]As at CU 3973, 19r, where Newton says that the volatile salt of copper is at least as volatile as a salt of spelter that he has just prepared, or perhaps more volatile. For another example, see CU Add. 3975, 75v.

[8]CU Add. 3975, 58v, 60r.

[9]CU Add. 3975, 59r.

acid spirit and a urinous spirit in combination (hydrogen chloride and ammonia). His hope was that either iron filings or spar would selectively "fasten" on to one of these spirits and cause the sal ammoniac to disintegrate, thereby freeing the volatile copper salt from it. He begins by washing and filtering the copper sublimate (again, the sublimate of antimonial vitriol of copper) to remove what he elsewhere calls its "gross antimony," meaning both unreacted stibnite and *mercurius vitae*.[10] Having carried out this preparatory step, he tries subliming the dried volatile copper first with iron filings and then with "spar." The imprecise term "spar" could mean any sort of clear, faceted mineral in the seventeenth century; it could refer to either quartz, calcite, barite, or even one of the minerals that nowadays terminate with the term "spar" (for example, "brown spar" or crystalline siderite).[11] The use to which Newton tried to put his spar, along with iron, strongly suggests that the material was calcite (calcium carbonate), however. Both iron and calcite react vigorously with hydrochloric acid, the spirit of salt that Newton knew was combined with ammonia (his urinous spirit) in sal ammoniac. In fact, both of these materials react with the acid far more readily than copper does.[12] Thus Newton was employing an implicit knowledge of elective affinities in the attempt to destroy the sal ammoniac conjoined with his salt of copper by forcing its spirit of salt to combine either with the calcite spar or with the iron. As Newton says, the volatile copper did not sublime a second time but remained fixed with the iron or spar, thus defeating his hopes. The copper salt had apparently reacted with the spar or iron to become nonvolatile while the sal ammoniac simply sublimed by itself, intact. Then, in order to confirm the correctness of his observation, Newton tries subliming the spar with sal ammoniac alone, and again notes that the salt is not "destroyed" by it, but simply rises undamaged; again no reaction occurred. Hence his goal of removing the sal ammoniac from the volatile salt of copper by means of spar (or iron) was a failure.

Here Newton spells out a modus operandi that permeates many of his experiments, though he is seldom this explicit. He is trying to exploit his knowledge of elective affinities in the goal of segregating one substance from another that is associated with it. By adding a material that he suspects of having a greater affinity with the adulterant than with the desired substance, he hopes to separate and dispose of the adulterant. We have already seen this approach in the letter of David von der Becke that Newton carefully excerpted, where the decomposition of sal ammoniac by potassium carbonate was used as a model for arriving at a volatile salt of tartar. This is the same approach that would come to its full fruition in *Query 31* of the 1717 *Opticks*, which served as a clarion call to the compilers of elective affinity tables in the eighteenth century. Newton's exploitation of elective affinities is already illustrated by his early attempts to separate the "foul" sulfur from

---

[10] As at CU Add. 3975, 144r, where Newton speaks of "☿ (freed from yᵉ gross ♁."

[11] For stibnite gangues, see Charles H. Richardson, *Economic Geology* (New York: McGraw-Hill, 1913), 171–73. For brown spar in conjunction with stibnite, see Chung Yu Wang, *Antimony* (London: Charles Griffin, 1909), 48.

[12] Pure copper does not react with hydrochloric acid, though copper oxide will do so.

stibnite by reacting the antimony ore with niter.[13] As with sal ammoniac, however, the sulfur proved difficult to separate, and Newton found that the crude antimony remained intact despite his efforts. As we proceed through Newton's laboratory practice, we will see him employing more substances and more complicated procedures in his attempt to free his volatile copper sublimate from its inborn sal ammoniac.

Immediately after his failed attempt with iron filings and spar, Newton describes another attempt to separate the volatile salt of copper from its sal ammoniac. Again he tries to capitalize directly on his knowledge of sal ammoniac's composite nature. From reading earlier chymists such as Boyle and von der Becke, Newton had learned that sal ammoniac could be "destroyed" by salt of tartar to release its urinous spirit. As we would say, the alkaline carbonate reacts with the ammonium chloride, leading to an emission of ammonia gas and formation of potassium chloride and carbon dioxide with water. The present experiment makes use of this reaction in the hope that the copper salt will be liberated of its sal ammoniac. But unfortunately for Newton, this is not the way things turn out:

> Salt of tartar, as it destroys yᵉ ✲, so it ~~holds down the <illeg.>~~ precipitates yᵉ ☿ Venus in a blew form & ~~therefore~~ holds it down & therefore is no fit medium to separate yᵉ salt of ♀ & ✲.

The problem in this experiment is that the salt of tartar, like the spar, combines with and "holds down" the very salt of copper that Newton wants to liberate. What Newton needed was a material that would combine with one of the two constituents in the sal ammoniac without at the same time reacting with the salt of copper. For the time being, however, he was sufficiently impressed with the volatilizing power of his copper sublimate that he would grant it the name "volatile Venus" and use it directly as a means of "spiritualizing" other metals and minerals.[14]

Newton's experimental work using volatile Venus as a subliming adjuvant continued for several years without intermission. For now, let us consider the remaining paths that Newton employed in his analytical chymistry to isolate the volatile salt of copper. Soon after an entry dated February 29, 1683/4, Newton reveals that he is still tinkering with the proportions of the ingredients in his volatile Venus. He carries out three runs of sublimations with 12 parts antimonial vitriol of copper "well dryed" and different amounts of sophic sal ammoniac. In order to raise all of the "⊕olick spirit" (vitriolic spirit), Newton finds that a larger proportion of sophic sal ammoniac is better, and he settles for a 3:2 proportion. At this point, however, Newton does something unexpected. Up until now, he has been focusing almost exclusively on the *sublimate* of the antimoniate copper vitriol and the sophic sal ammoniac. Now he turns to the caput mortuum left behind at the bottom of the apparatus. He notes that when placed at a red-hot heat on a

---

[13] CU Add. 3975, 58v.

[14] As at CU Add. 3975, 61r, at the end of an experiment with spar or spelter—"Whence Spʳ is not to be spiritualized ~~by~~ immediately by ~~spt of Ł~~ Ven. vol."

piece of iron, almost all of the caput mortuum fumes away. This leads him to the idea that perhaps he could sublime all of it with "^2 or 3 or 4 <*illeg.*> times its weight of fullers' earth, & this sublimate will be pure for fermentation."[15] Hence this interesting passage tells us two important facts. First, Newton's focus on the caput mortuum rather than the sal-ammoniac-rich sublimate suggests that he sees it as a way of arriving at a volatile Venus free of sal ammoniac. And second, he makes it clear that he wants to use this pure form of volatile Venus as a way of promoting fermentation, a favorite theme of Newton's that we have already encountered.

A folio after this experiment, Newton does six more runs of sublimation with the sophic sal ammoniac and antimoniate vitriol of copper. Again he chooses the 3:2 proportion, partly because its caput mortuum "melted with a less heat." He then mixes this with powdered glass, apparently following up on the earlier suggestion of using a refractory powder to encourage sublimation, and tries to sublime it without success. Having failed with the caput mortuum, he then tries to purify the sublimate. His techniques, while by now familiar, are interesting for their level of detail. Since the volatile Venus was sublimed with sophic sal ammoniac, it contains a large percentage of that material; therefore, the copper sublimate can be used for "sharpening" an aqua fortis to make an aqua regia. Newton thus puts an unspecified amount of the volatile Venus to an equal weight of aqua fortis and introduces stibnite until it is "satiated." He then adds water to the solution "till all yᵉ ☿ vitæ was precipitated. It took 8 or 12 times its quantity of water to cleare it well."[16] Newton's goals here are quite evident. He adds the stibnite to the initial solution of acid and volatile Venus in order to destroy the sal ammoniac still present. He then dilutes the acid solution with a large amount of water in order to rid the volatile Venus of "gross antimony" by precipitating the *mercurius vitae* or antimony oxychloride. His next step is to evaporate the filtered solution and to sublime the separated salt. Despite his careful efforts, however, Newton finds that the salt is still accompanied by some spirit of antimony; apparently, the sophic sal ammoniac had not entirely destroyed it. He concludes by planning a further set of operations:

> There arose with it a small quantity of spᵗ of ♁ wᶜʰ I conceive may be separated by rectification, or by boyling yᵉ salt wᵗʰ a little spar & filtring it.[17]

Newton had noticed, probably in the late 1670s, that spar (again probably calcite) would dissolve in his antimonial liquor or "vinegar" to form a fixed salt.[18] Here he is attempting to exploit the reaction in order to separate the resulting salt from the volatile Venus.

The experiments that we have examined seemed encouraging, and in fact Newton even thought sometime after May 16, 1686, that he had managed

[15] CU Add. 3975, 71r.
[16] CU Add. 3975, 72v.
[17] CU Add. 3975, 73r.
[18] CU Add. 3975, 58v: "Spar, a good part of it dissolves readily in Vinegre to a salt, almost all of it in ♒." A comparison with CU Add. 3975, 54v reveals that the "vinegar" here is "undistilled vinegar of antimony."

to isolate his volatile Venus from sal ammoniac by subjecting it to stibnite and then subliming it from white lead (lead carbonate made by exposing a leaf of the metal to vapors of vinegar).[19] Nonetheless, he was still not entirely satisfied, possibly because of his earlier failure to sublime the entire caput mortuum left from the manufacture of the volatile Venus. Unfortunately, our record of his experimentation is incomplete, so we may never know exactly what led to his final bout of efforts to arrive at a pure form of volatile Venus, free from sal ammoniac (sophic or otherwise). At any rate, almost a full decade after the foregoing experiment, we find Newton, around February 1695/6, performing an analysis of his volatile Venus, which he now refers to as "✳🜔" ("sublimate of antimonial salt of copper" in his graphic shorthand). He does not reveal the means of the analysis, but the end results are interesting—"✳🜔 conteins in it ~~8/25 of~~ between 1/3 & 4/15 or about 3/10 of ♁ & *<illeg.>*~~7/~~*<illeg.>* 7/20 of ✳ & 7/20 of 🜔."[20] From this we see that the ✳🜔 and 🜔 were distinct but related materials. The star superimposed over the antimonial cross symbol indicated that sophic sal ammoniac had been added as an adjuvant and then the mixture volatilized to produce a sublimate. Newton specifies that 7/20 of the sublimate by weight consisted of unreacted sal ammoniac (presumably vulgar). The remainder was split between 3/10 "antimony" (it is not clear which compound or compounds of elemental antimony are meant here) and 7/20 🜔, meaning the antimonial salt of copper itself.[21]

This analysis of volatile Venus signals the beginning of a renewed attempt by Newton to isolate the volatile copper salt from the residual ingredients used in its preparation. Typically for him, Newton tries to employ his chymical knowledge to arrive at the most fundamental level of the problem. Thus, after making some initial comments about the relative solubility of sophic sal ammoniac and volatile Venus in aqua fortis, Newton launches into a project for isolating the volatile salts produced when sophic sal ammoniac is dissolved. By arriving at a more profound knowledge of these salts, he no doubt thought that he would acquire useful knowledge about the volatile Venus itself. First he dissolves measured amounts of sophic sal ammoniac in aqua fortis and water, then adds stibnite, as usual, in order to "destroy" the sal ammoniac. He filters the antimony calx from a solution that evaporates to a fat, clammy salt, and notes that the calx has now gained 20 grains when compared to the initial stibnite. He then sublimes two successive salts from the fat, clammy material; the second one sounds to the modern ear like butter of antimony; it congeals in the neck of the retort and soon reveals its deliquescence. Newton now reveals that his real interest here is still volatile Venus, as he reports that the heavy, fusible salt is "something more fusible than 🜔." Newton then tries several other tests. Subliming the fusible salt from lead ore produces no further salt, nor any increase in weight or fusibility in the

[19] CU Add. 3975, 79v.

[20] CU Add. 3973, 32r.

[21] This could perhaps be the same thing as "fixed 🜔," which Newton refers to at CU Add. 3975, 80v: "🜔 fixed conteins 6 of ♀ & 19 of ♁, in all 25, so that ♀ is 1/4 of ye whole." But this is problematic, since the material does not sound like a salt, which the symbol "🜔" is meant to imply.

ore; clearly, the salt had not "wrought" on the mineral. Similarly, subliming the fusible salt from calx of the sublimed net (Newton's alloy of martial regulus and copper) yields only a heavy sublimate that "tasted stiptic & vitriolic☿." This meant that the sublimate contained unreacted vitriol, an undesirable contaminant. All of these tests lead Newton to a fusillade of negative conclusions:

> Whence this salt is not to be used for volatising metalls nor ought to be mixed wth ☿. Whence also ye mixture of this salt wth y^e ☿ made it less potent for volatising as I found in some former exp^{ts}. Whence therefore crude ♂ is not to be used for destroying the ✳ in y^e ☿.[22]

Newton's rejection of the heavy, fusible salt, presumably butter of antimony, is based mainly on the fact that it reduces the "volatizing power" of certain other materials. But as we have seen already, he was also using the very techniques employed here to destroy the sal ammoniac in his volatile Venus, for he knew that the "reguline part" of antimony would combine with the "acid spirit" in the sal ammoniac (sophic or otherwise) and thereby destroy that salt. His sudden realization that the heavy salt produced by these operations reduced the volatile Venus's usefulness as a subliming adjuvant must have been a rude shock. Once again, Newton had hit a roadblock.

Around the same time that Newton discovered the problem with using crude antimony to destroy the sal ammoniac in volatile Venus, he also fell on another cause for dismay. Perhaps as a result of his new and disturbing discovery, Newton tried dissolving the volatile Venus in a larger amount of water than he was accustomed to using. Probably suspecting that butter of antimony was implicated, he decided to test for the *mercurius vitae* that would accompany an increase in wash water:

> ☿ dissolved in a little water let fall 166^{gr} of ^♂ial precip & by ye addition of ^much more water it let fall 33^{gr} more in all 199 or 200 grains & ye remaining salt when dried weighed 415^{gr}. So y^t ye first feces ye 2d feces ye whole feces & the salt were as 5. 1. 6. 12 1/2.[23]

Shortly after this realization that ☿ behaved alarmingly like butter of antimony when subjected to repeated washing, CU Add. 3973 introduces a series of further experiments with volatile Venus. Again, Newton is intent on isolating the volatile copper salt by destroying the sal ammoniac, but now he is compelled to use some material other than crude antimony to carry out the task. First Newton takes the washed volatile Venus, dissolves it in diluted aqua fortis, and adds bismuth (♃) instead of stibnite. After his usual protocols he gets a sublimate, but when washed, it produces a precipitate that in turn releases a styptic salt. Newton decides that the material is impure, and therefore comes to the following conclusion: "So then the exp^t succeeds not wth ♃ or any white metal but must be done wth Reg ♂ or wth ♀."[24] Not only

[22] CU Add. 3973, 33r.
[23] CU Add. 3973, 33v.
[24] CU Add. 3973, 34v.

bismuth but the other "white metals" as well have been ruled out, leaving the regulus of antimony and the red metal copper as alternatives.

Predictably enough, Newton then tries the same experiment, but substituting copper for the bismuth used before. This time he receives an inordinately small amount of volatile salt and a correspondingly large caput mortuum. Thus he concludes, "So then ye exp^t succeeds not wth ♀ but must be done wth Reg ♂."[25] Interestingly, Newton does not pass directly to regulus of antimony as one might expect, however, but interjects several experiments with volatile Venus and iron ore. He notes that sophic sal ammoniac works on iron ore very easily, leaving its "spirit of salt" behind.[26] Since this opens an obvious path to the destruction of sal ammoniac, Newton decides to see if volatile Venus will work on the ore. Performing a simple sublimation of his volatile Venus directly from iron ore, he finds that the product is insufficiently fusible and leaves "a little saline Stiptic fex," indicating vitriolic impurities. In order to confirm these disappointing results, Newton then substitutes sophic sal ammoniac for the volatile Venus and finds that an inadequate amount of sublimate is carried up from the iron ore. All of these results point to an inescapable conclusion: "So y^n iron ore is not to be used for separating o^r ✳."[27]

Around the same time, Newton had also begun experimenting with the regulus of antimony that he had proposed as an alternative to the bismuth, copper, and iron ore, which were all now ruled out as means of destroying the sal ammoniac in volatile Venus. He describes an experiment that follows the usual protocols of employing an aqua fortis sharpened with volatile Venus in order to dissolve the material being tested (here regulus of antimony), followed by washing, precipitation, drying, and sublimation. When he then sublimed the resulting "compact" salt from the calx of the net, the results seem to have been no cause for celebration—a white salt rose in flowers and left about 20% of the sublimandum as a red calx.[28] This experiment is followed by another in which Newton compares the relative ability of reguli of antimony and bismuth to destroy sal ammoniac, so he was clearly still not satisfied.[29] Probably in an attempt to refine his process with regulus of antimony, Newton then tried a similar experiment substituting regulus of tin (mostly metallic antimony alloyed with a little tin remaining from reducing stibnite with that metal). This too led to failure; the salt that Newton sublimed after his usual operations was hygroscopic, revealing that some regulus (or rather antimony trichloride) managed to remain with it.[30]

Having grown somewhat disenchanted with the regulus of antimony, Newton now turned to the "white calx of ♂ wch remained after solution in ♈," the insoluble residue, rich in antimony oxides, that is left when stibnite

[25] CU Add. 3973, 34v.
[26] CU Add. 3973, 35r.
[27] CU Add. 3973, 35v.
[28] CU Add. 3973, 35r.
[29] CU Add. 3973, 35v.
[30] CU Add. 3973, 36r.

is reacted with aqua fortis without the addition of sal ammoniac. Here again it seems that he was hoping to use his knowledge of elective affinities to separate "our Venus" from the sublimate of our vitriol. Thus Newton first dissolves the ☿ (freed from gross antimony) in aqua fortis to make a menstruum. Then he adds the white calx of antimony, perhaps hoping that during its dissolution, something like a modern double displacement reaction will occur; thus, the "reguline part" of the antimonial calx would combine with the acid part of the sal ammoniac, and the aqua fortis, now able to act on the freed alkaline component in the sal ammoniac, would let go of "our Venus." What actually happens is described in the following words:

> If in ye same 🜆 the white calx or fex of y^e^ ♁ *<illeg.>* wch remains after y^e^ ♁ is dissolved in 🜅 be put in by degrees: the liquor takes up much of this fex & extracts a fluid distillable ♁ & the fex drinks up ~~the almost all~~ ^much of^ y^e^ ☿ & coagulates it so that you cannot separate them by philtring; & ~~th~~*<illeg.>* if the 🜖 ~~rest~~ wch together with the fluid 🜔 of ♁ is separated by philtring be distilled the 🜔 of ♁ distills over in a fluid form & leaves ^all^ the ~~rest~~ 🜖 coagulated in a white ~~form~~ colour & form like ye caput mort of 🜅 wch being afterwards urged in a great heat so as to be almost red hot would not ascend into ye neck of ye retort. Nor did it upon a red hot iron emit any fume. So then the white calx of ♁ made *<illeg.>* dissolving ♁ in 🜅 is not to be used for extracting o^r^ ♀.[31]

In short, Newton explains his failure as follows. The aqua regia dissolved part of the white calx or "fex" of antimony and "extracted" and combined with the antimonial salt that was released by it (or rather formed a salt with the "reguline part"), while the remaining calx or "fex" in turn began to combine with the volatile Venus. It seems that the volatile Venus and calx had at least as much affinity for each other as either had for the aqua regia. Yet not all of the volatile Venus had combined with the white calx. When the solution, which now contained aqua regia, fluid antimonial salt, and volatile Venus, was filtered and evaporated, the volatile Venus was left at the bottom in the form of a fixed white material like the caput mortuum left behind when vitriol is distilled to make one of the mineral acids. Apparently, the process had somehow acted on the volatile Venus to render it nonvolatile.

At this point we must leave Newton's decades-long attempt to "extract" the volatile salt of Venus, for the available records break off at this point. What was he actually trying to do in terms of post-Lavoisian chemistry? Attempts to replicate Newton's experiments show that one can indeed make a "vitriol" by imbibing copper powder with the solution that Newton referred to as "undistilled" liquor of antimony. If the crude, green product of the reaction is dissolved in water and then filtered and evaporated over heat, a dark reddish brown solid emerges, which quickly turns green in the presence of

[31] CU Add. 3975, 140r–140v. There is a shorter version of this experiment, without as much explanation, at CU Add. 3973, 39v. Although less illuminating in this instance, the version in CU Add. 3973 is important for showing the chronological development of Newton's experiments. He has reordered them in CU Add. 3975 so that the material on CU Add. 3973's 32v–33r comes *after* the experiment on 39v.

the air, indicating the formation of a hydrate.[32] Subliming this "vitriol" with a sublimate of stibnite and sal ammoniac produced several distinct bands of new sublimate in the inverted glass funnel that served as a sublimatory. X-ray diffraction analysis of the lower band indicated the presence of about 50% ammonium chloride. The remaining compounds consisted mainly of metallic double salts including ammonium tetrachlorocuprate and ammonium heptachlorooxodiantimonate; additionally, triammonium hydrogen disulfate (letovicite) and a small amount of unidentified material were detected.[33] Exotic chemicals indeed to be synthesizing and refining in the seventeenth century! Further replication and analysis are required before we can be certain of the precise compound that Newton identified as "volatile Venus." But we can be sure of one thing. His years of labor at extracting "our Venus," directed at producing this material and isolating it from its accompanying sal ammoniac and other impurities, were producing interesting results.

The Net: A Case Study of Newton's Laboratory Practice
and Textual Decipherment

So far this chapter has reconstructed Newton's chymical laboratory work from a phenomenological, "bottom up" perspective, intentionally steering away from his understanding and use of chrysopoetic literature. This approach has uncovered a wealth of experimental projects, in particular his attempt to make and purify volatile Venus. But we are still left with at least one serious question. In many if not most of the chymical experiments that Newton performs with these derived materials, he employs them on a considerable variety of metals, minerals, acids, and salts, either as subliming adjuvants or as means of sharpening menstrua. What directs his choice of one particular metal, regulus, "vitriol," or volatile metallic salt as opposed to another one? In some instances, as we have seen, Newton chooses his materials on the basis of their relative affinities for one another. But this in itself is far from explaining the seeming chaos of bewildering choices that opens up once we pass beyond the similarity of the often repeated protocols—the dissolutions, precipitations, washings, sublimations, and other techniques that populate the notebooks. It is at this point that we must diverge from our investigation of the notebooks purely on their own terms and begin to interrogate Newton's literary sources for additional clues. This examination will certify a claim that I have already made several times: Newton's choice of materials and techniques was often a result of his attempt to reverse-engineer

---

[32] The experimental replication of "volatile Venus" was first carried out by me with Joel Klein at the "Making and Knowing" laboratory at Columbia University (we thank Pamela Smith for allowing us to use the laboratory facilities). A fuller description of the replication will be published in the near future.

[33] This analysis was performed by Professor David Bish of the Indiana University Department of Geological Sciences. Professor Bish and I are engaged in a longer project to produce and analyze the materials described in Nicolas Fatio de Duillier's remarkable letter to Newton of August 1, 1693, and to see how closely these materials map onto Newton's experiments recorded in his laboratory notebooks. Fatio's letter to Newton is described in chapter seventeen of the present book.

the creations of earlier chymists, based on the allusive descriptions of their appearance and properties found in the literature of chrysopoeia.

An excellent example of Newton's efforts to recreate the products of the adepts may be seen in the case of the "net" of Vulcan, or *Rete* in Latin, which happens to be one of the most straightforward *Decknamen* in his notebooks. Some of Newton's earliest surviving chymical laboratory notes, probably dating from the early to mid-1670s, describe experiments for making an alloy of regulus martis and copper that "gave a substance wth a pit hemisphericall and wrought like a net wth hollow work as twere cut in."[34] After four trials with different amounts of regulus martis and copper, Newton decides that the proportion of 4 parts copper to 8 1/2 or 9 of regulus gives the best results. The choice of regulus martis and copper, as well as the very term "net," both stem from the works that George Starkey wrote under the name of Eirenaeus Philalethes. In his *Marrow of Alchemy*, Philalethes had interpreted Book IV of Ovid's *Metamorphoses* as the vehicle of an encoded recipe for the philosopher's mercury. In his poetic study of universal transformation, Ovid related the story that the blacksmith god Vulcan was cuckolded by Venus when she took Mars as a lover. When the embittered Vulcan discovered this shameful alliance, he fabricated a wondrous net of bronze in which the gods of war and love were ensnared and displayed for the edification of the other residents of Olympus.[35] As Philalethes put it in more colorful language in stanza 60 of this section of the *Marrow*:

> *Vulcan* will jealous wax, and over-spread
> His Net to catch his Spouse with *Mars* in act,
> The limping Cuckold greev'd to feel his head
> With Horns adorn'd, and hoping this compact
>   To dash, doth show the Lovers both intrapt
>   Within his Net, in which they both are wrapt.[36]

Playing on the traditional alchemical referents of Mars, Venus, and Vulcan—iron, copper, and fire—Philalethes decoded the myth as a recipe for an alloy consisting of regulus martis (made by reducing stibnite with iron) and copper fused at a high temperature. As he pointed out elsewhere in the *Marrow of Alchemy*, the product was "Infolded just as in a Net," that is, it had a crystalline surface that looked, as Newton would later say, "like a net wth hollow work as twere cut in."[37] The attractive, purple alloy has been reproduced in a modern laboratory in the proportions recommended by Newton (figure 16.1).

---

[34] CU Add. 3975, 43r. Folios 41v–43r employ the unbarred Saturn symbol (♄) characteristic of Newton's early notes. The ink is also darker than the study of sal ammoniac beginning on 43r and continuing for several folios that relies heavily on David von der Becke's *Epistola ad Ioelem Langelottum*, which was published in Latin in 1672 and synopsized in the *Philosophical Transactions* of 1673/74, as we discussed in the previous chapter.

[35] Ovid, *Metamorphoses*, book 4, lines 167–89 in Ehwald's edition. See R. Ehwald, *Die Metamorphosen des P. Ovidius Naso* (Berlin: Weidmannsche Buchhandlung, 1903), 1: 164–65.

[36] Philalethes, *Marrow*, part 2, book 1, page 15, stanza 60.

[37] Philalethes, *Marrow*, Part 2, Book 1, p. 16.

FIGURE 16.1. The "net," or "Vulcan's net," is a purple alloy of copper and metallic antimony. The alloy seems to have been invented by the American alchemist George Starkey (who authored most of the texts attributed to Eirenaeus Philalethes). Starkey was impressed by its fine, crystalline surface, which reminded him of network. He based the name "net" on the mythical net of bronze that Vulcan used to ensnare Mars and Venus in bed together. The alloy played a major role in the preparation of the Philalethan sophic mercury, and Newton made great use of it in his own experimentation. Prepared by the author in the laboratory of Dr. Cathrine Reck in the Indiana University Chemistry Department. See color plate 8.

But Newton was not satisfied with the mere making of the net; this material was a means to a much greater end. In the alchemy of Philalethes, the net is a medium by which the star regulus of antimony is combined with quicksilver; the purging of the quicksilver by the antimony is supposed to produce an ultrapure sophic mercury. Of course this process requires that one first obtain the star regulus of antimony, and Philalethes is careful to describe the production of this substance by employing another passage from Ovid's *Metamorphoses*. In Book III, the Latin poet had given a detailed description of how the ancient city of Thebes was founded by the mythical hero Cadmus, who had been banished from his native Tyre. After a consultation with the Delphic oracle in which he is advised to build on a spot where he sees a wandering, ownerless cow lie down, Cadmus and his companions bless their good fortune when they encounter this providential bovine. They soon have cause to regret their fate, however, for the companions of Cadmus meet a poisonous dragon in a cave where they have gone to collect water, and they are summarily slaughtered by the monster. Learning of this sad outcome, Cadmus manages in turn to dispatch the dragon after a pitched duel that ends when the dragon is impaled against an oak tree by the hero's spear. Athena then appears and advises Cadmus to sow the dragon's teeth in the surrounding soil; when he does this, an army of men emerge from the ground and begin at once to fight among themselves. All but five of these strange soldiers kill one another off; the remaining ones go on to cofound Thebes with Cadmus.[38]

In *The Marrow of Alchemy*, Philalethes makes it quite clear that for him, Cadmus refers to iron and the dragon to stibnite; as for the oak on which the dragon is pinned, at least one of Philalethes's followers interpreted it as an alchemical furnace.[39] In short, the myth is an encoded telling of the process for reducing crude antimony to its metallic form. The eating of the companions of Cadmus (and of Cadmus himself in Philalethes's retelling of the myth) again refers to the incorporation of iron nails in the molten regulus in order to reduce the metallic regulus; if this is allowed to cool slowly under a thick slag, it can crystallize into the famous star regulus of antimony. All of this material is relayed in relatively straightforward language in *The Marrow of Alchemy*; Philalethes even describes the antimony-dragon as "Sable-coloured with Argent veines," an unambiguous description of stibnite. Mars must embrace this material in order to produce a shiny, "metalline" material that will be inscribed with a "stellate" seal, obviously the star regulus.[40] Clear as this language may have been, its obvious sense was entirely too simple for Newton. Although he certainly understood the allusions to stibnite and regulus martis, since he refers to these materials in his own recipe for the net, Newton thought these passages from *The Marrow of Alchemy* contained a deeper meaning as well. If we return to the passage where Philalethes describes Vulcan's making of the net, this is followed immediately by another allusion to Cadmus in the next stanza. It is again necessary to quote Philalethes here (stanza 61):

> Nor may this seem a Fable; first observe
> How *Cadmus* is by our fierce Beast devour'd,
> Whom after piercing stoutly doth deserve
> A Champions name, for (by might overpower'd,)
>> This serpent ('gainst an Oke) with deadly spear
>> Transfixeth, whom erst every one did fear.[41]

Since this passage came immediately after Philalethes's description of Vulcan's net, and for that matter after the alchemist's straightforward description of mineral antimony and its reduction by iron, Newton evidently felt that it must contain additional information about the process that would eventually lead to the philosophers' stone rather than serving as mere idle repetition. Thus for Newton the focus of the story shifted from the killing of the dragon or serpent by Cadmus to the fixing of the serpent on the oak. Instead of referring simply to the production of the star regulus by means of iron, as in Philalethes's account, for Newton the passage alluded to a process that should be performed on the oak itself. Misleadingly

---

[38] Ovid *Metamorphoses*, book 3, lines 1–137 in Ehwald's edition. See Ehwald, *Die Metamorphosen*, 1: 117–25.

[39] See Newman, *GF*, figure 3E.

[40] Philalethes, *Marrow*, Part 2, Book 1, p. 4.

[41] Philalethes, *Marrow*, part 2, book 1, page 15, stanza 61.

using the same language that he had employed earlier for the making of the star regulus, Newton's Philalethes was now advising the alchemist to transform the oak of Ovid's story by means of a serpent. Newton's new interpretation of the Cadmus myth appears both in his notes from early 1678/9 as found in CU Add. 3973 and in more fully fleshed out form in CU Add. 3975. It will be useful to quote the earlier passage here and then to look at the more complete version of the experiment, which contains some interpretation:

> On ye net poudered I poured undistilled ~ liquor of ♁, it soaked almost all into it w^{th}out extracting any considerable quantity of salt & ye salt wch it (wch was inconsiderably little) did not look blew) I poured on ye same, distilled liquor of ♁ so much as filled ye pores & it drunk up also all that. more quickly then ye former. Whence I understood ye oak must be first prepared in a metallick form & then ye serpent fixed to it, but w^{th}out adding any heterogeneous salt.[42]

Suddenly we find ourselves back in the now-familiar setting of Newton's laboratory, where he is employing his favorite reagent, the liquor of antimony (or vinegar of antimony as he calls it in the parallel passage from CU Add. 3975). His process consists of imbibing the net with this menstruum in the hope of "extracting" a salt. And one can see that Newton here unequivocally identifies the net with the oak on which Cadmus transfixed the dragon or serpent. Here, however, the serpent no longer refers to stibnite (as it did to Philalethes), but to the liquor of antimony itself. The goal, clearly, is that the liquor of antimony extract (or produce) a salt in combination with the net or oak. Newton concludes here that the more rapid absorption of the distilled liquor of antimony as opposed to the undistilled version means that no "heterogeneous salt" should be added (as would be present in the solution before its purification by distillation).

If we turn to the parallel passage in CU Add. 3975, additional details emerge. It is clear that Newton repeated the experiment and ruminated on its significance for his future practice. As in CU Add. 3973, he recounts in the later version that the extracted salt was not blue, but he adds that this means there was "no extraction of copper." The salt that did remain in the bottom of the vessel after the solution was distilled probably consisted of "the spar in the Vinegre," in short, antimony gangue that traveled into the undistilled liquor of antimony (the "Vinegre" here) when the initial stibnite was dissolved in aqua regia. Newton then repeats that a subsequent addition of previously distilled liquor of antimony led to rapid absorption of the liquid by the net. In the expanded version of the experiment he adds the interesting fact that this was only the case when the net had first been subjected to *undistilled* liquor of antimony; when the *distilled* version was added initially, it was not absorbed at all. His conclusion is strikingly different from the one in the earlier telling of the experiment:

---

[42] CU Add. 3973, 9r. The parallel passage is found in CU Add. 3975, 54v.

Whence I understood ye oak must be first prepared in a metallic form, &
then the serpent undistilled fixed to it & if need be, more serpent either
distilled or undistilled added. & then all melted together.[43]

From this we can see that Newton's understanding of the Philalethan passage
had changed. While the oak still referred to the net alloy, and the serpent to
liquor of antimony, Newton's experimental results led him to the conclusion
that the undistilled version of the liquor had to be employed before the distilled. He adds, finally, that the net and liquor, after drying, should be fused
together, a conclusion that is lacking in the earlier form of the experiment.

The net or oak and the serpent "transfixed" on it therefore provide a rare
and precious case where Newton unambiguously identifies the material referents into which he translated the *Decknamen* of previous authors. If we
proceed further into his notebooks, we can build on these solid data to expand our knowledge. Doing so reveals that the oak played a central part in
Newton's alchemical practice. This is particularly evident in the case of two
experiments from July 18 and 19, 1682, which are recorded in CU Add.
3975. In the first of these, Newton begins with the oak, which he says is
equivalent to "Reg ♂ ♀ ☿." The explicit presence of crude antimony alongside regulus, iron, and copper suggests that Newton may have tinkered with
his earlier recipe for the net, but the material remains basically the same. At
any rate, Newton takes the oak and imbibes it as before with liquor of antimony ("vinegre of ☿"). This time, however, he sublimes the product of this
imbibition—possibly the dried salt—with another material that he refers
to as "⊕ vol. philtr." This appears to be identical to an unspecified type of
vitriol that Newton volatilized in the previous experiment with sophic sal
ammoniac and then washed, filtered, and dried.[44] Quite possibly it is the
same "vitriol" of copper that Newton made by imbibition of the metal with
liquor of antimony in the production of volatile Venus. After subliming the
imbibed oak with this volatile vitriol, Newton sublimes the product with
additional sophic sal ammoniac. He is careful to point out that during each
sublimation the material at the bottom of the vessel "grew moist" and then
boiled. The experiment comes to an abrupt halt with the experience common to chemists of all periods—"& y^n ye glass broke."

On the next day, Newton resumed his experiment with the oak. This
experiment is particularly interesting for what it shows us about Newton's
goals. As we know, he was quite confident that he had deciphered the composition of the Philalethan net or oak. What we now observe is Newton
extending his reverse-engineering in order to determine the proper use to
which the oak should be put. In order to do this, he repeats the experiment
multiple times while removing variables and observing the results. As we
already saw, Newton was particularly interested in the boiling of the sublimandum when heated. His first variation, then, is to forego the imbibition of the oak and to leave out the volatile vitriol. Subliming the oak with

[43] CU Add. 3975, 54v.
[44] CU Add. 3975, 67v–68r.

sophic sal ammoniac alone, Newton records that the sublimandum does not boil. He then imbibes the oak with liquor of antimony and sublimes again, with the same amount of sophic sal ammoniac as before. This time he gets one-third more sublimate than he did without imbibition, and this yield is accompanied by easy fusion and considerable boiling. Nonetheless, an additional sublimation of the caput mortuum reveals less boiling and less sublimate than Newton acquired the day before when using volatile vitriol with the imbibed oak. Thus he concludes that "it seems yᵉ ⊕ promotes both yᵉ volatility & fusibility." As we progress further into this experiment, it appears increasingly probable that Newton was not merely making a neutral observation here but commenting on the motives of an unnamed alchemical source. Having determined Philalethes to be discussing vitriol, he is now intent on discovering the particular role that the vitriol plays in the processes at hand. Before coming to his final conclusion, however, Newton returns to the experiment from 1678/9 that we discussed before, where he first made his identification of the oak as the net.

Building on the 1678/9 experiment, Newton now tries imbibing the oak a second time with both undistilled and distilled liquor of antimony. This time, however, he notes from its taste that the salt extracted by this imbibition is itself a vitriol and can be extracted by washing with water. Next Newton follows up his earlier suggestion that after the serpent is fixed to the oak, all should be "melted together." In other words, he imbibes the oak once with the liquor of antimony and then fuses the product, getting a white alloy that is "grained almost as metals melted wᵗʰ ☿ used to bee." What comes as a surprise is Newton's evident satisfaction with this product, since the immediate source of his interpretation is far from clear. Yet his final remarks to the experiment reveal unequivocally that he viewed it as a success:

> And this I conceive to be yᵉ right preparation of yᵉ Oak. But I do not think it is to be volatized wᵗʰ Venus because ♀ yᵉ addition of more reg. of ♂ will volatize it better. Tis rather designed for a clean sulphur to joyn in fermentation wᵗʰ ☿.[45]

Newton's affirmation of his way of preparing the oak is followed by the assertion that it should not be sublimed with "Venus." This suggests rather strongly that the volatile vitriol in the experiment of July 18 was actually a vitriol of copper, and that Newton's comment is intended to provide an alternative to the conclusion that the oak should be sublimed with that substance. When he adds that more regulus of antimony "will volatize better," he is presumably referring to further regulus added to the copper in the initial production of the alloy. Since the oxide of metallic antimony produced during its fusion does in fact sublime quite readily, his conclusion is understandable. But the final sentence in this experimental report is its most remarkable feature, for here we receive a glimpse of Newton's overarching goal for the oak. As he says, it is "designed," presumably by Philalethes and his school, as a clean sulfur to join with mercury in fermentation.

[45] CU Add. 3975, 68v.

There is one final element of the Cadmus myth that requires comment, for it reveals the striking literalness with which Newton approached the task of interpreting alchemical enigmata. A section from CU Add. 3973 that must have been composed during or after February 1695/96 records a succession of experiments where Newton sublimes the oak, an unspecified vitriol, and sophic sal ammoniac in varying proportions.[46] Strewn among Newton's usual observations about fusibility, volatility, color, and taste, one finds the following remarkable comment—"And in all these ye matter sublimed like Dragons teeth." In the context of the oak, this can only be an allusion to the myth of the soldiers born from the teeth of the dragon killed by Cadmus. Apparently Newton viewed the spiky appearance of his sublimation product as a clue that he was correct in his interpretation of the dicta of the masters. For him, the sublimed material, usually described as "flowers," becomes the dragon's teeth of Cadmus's companions, the mythical founders of Thebes. This provided further confirmation that Newton's interpretation of the net was the correct one and that he was indeed on the path of the adepts.

Conclusion

In the last three chapters we have seen how Newton used the rich and sophisticated techniques of early modern chymistry to produce and purify a variety of compounds that he hoped to employ in his decades-long chrysopoetic quest. Although I have at times referred to these techniques as belonging to the early modern domain of "vulgar chymistry," it would obviously be a serious mistake to erect a rigid boundary between that field and "hermetic philosophy," as some previous historians have done. After all, it was techniques such as dissolution in acids, sublimation with sal ammoniac, and a host of other operations with a long history in alchemy ranging from dry processes such as calcination and trituration to the wet operations of imbibition, distillation, cohobation, and precipitation that formed the backbone of Newton's experimental procedure. As we have just seen in Newton's experimentation with the net, these techniques provided the practical basis to his decipherment of such chrysopoetic classics as the Philalethan *Marrow of Alchemy*. The fact that Newton employed such operations and more in his quest for the philosophers' stone shows that it would be chimerical to view these and similar processes as something radically distinct from alchemy.

And yet chymistry had evolved in the early modern period to the point where many authors, Boyle and Newton included, distinguished between the elementary operations described in contemporary chymical textbooks from a more advanced knowledge that they hoped would lead to greater secrets—*arcana majora*—such as the volatile salt of tartar that captivated Newton in the 1670s. It is no accident that this marvelous substance, one of the many desiderata that Helmontian chymistry erected as research projects, appears in his notebooks. Newton's own laboratory records are marked

---

[46] CU Add. 3973, 44r–44v.

in several ways by the influence of Helmontian chymistry, even though the "noble Bruxellian" is seldom explicitly mentioned in his notes. Van Helmont and his followers, particularly George Starkey and to a lesser extent Robert Boyle, had already adopted an explicit reliance on mass balance as a means of tracking the course and progress of reactions.[47] The emphasis on precise weighing of ingredients and products that we see in Newton's experimentation is a logical development of this Helmontian emphasis, although Newton carried his quantification to a level of precision to which few other researchers in chymistry or any other scientific field chose to adhere.[48]

There is another area as well in which we can chart the influence of Helmontian chymistry on Newton. I refer to the assumption, already explicit in the response to Langelot by David von der Becke, that chymical species are made up of robust corpuscles with differing affinities to one another. Von der Becke's attempt to divide salt of tartar into its putative acid and alkaline components was based on the dissociation of sal ammoniac into its "acid spirit" and volatile "urinous salt," as we have already seen. These constituents were viewed by him as consisting of particles (*particulae*) that associated and dissociated according to their attraction for one another or for other chymical particles. The same language and ideas appear when von der Becke speaks of crude tartar and salt of tartar: once these two materials are mixed, a fermentation will ensue in which a motion of saline particles (*particulae salinae*) continues until the point of neutralization (*punctum saturationis*) is reached. At that juncture, every particle of fixed salt of tartar (*particula Salis Tartari fixi*) will be conjoined to the acid particles of crude tartar (*conjuncta cum particulis acidis Tartari crudi*).[49] Von der Becke's chymical atomism is not merely the product of reading Robert Boyle or other midcentury mechanical philosophers. It descends instead from the qualitative corpuscular theories of medieval alchemy, refined and further developed by previous chymical atomists such as Van Helmont, Daniel Sennert, and the important but understudied figure Angelus Sala.[50]

It is no exaggeration to say that the corpuscular Helmontian chymistry von der Becke exemplified is also the approach underlying most of the experiments in Newton's chymical notebooks. Rather than thinking in terms of a total metamorphosis of one material into another, Newton focuses his attempts on the separation of preexisting substances from each other and

[47] See Newman and Principe, *ATF*, 35–155, where the contribution of Van Helmont to the understanding of mass balance is described in detail. See also Georgiana D. Hedesan, *An Alchemical Quest for Universal Knowledge* (London: Routledge, 2016).

[48] See Georgiana D. Hedesan, *An Alchemical Quest for Universal Knowledge* (London: Routledge, 2016), xiv–xix, 86–104, where the current scholarly views on Van Helmont and his matter theory are summarized. See also William R. Newman, "Alchemical and Chymical Principles: Four Different Traditions," in *The Idea of Principles in Early Modern Thought: Interdisciplinary Perspectives*, ed. Peter Anstey (New York: Routledge, 2017), 77–97.

[49] David von der Becke, *Epistola ad Praecellentissimum Virum Joelem Langelottum* (Hamburg: Gothofredus Schultzen, 1672), 19–20.

[50] For the topic of chymical atomism generally, see Newman, *AA*. For Angelus Sala, see Urs Leo Gantenbein, *Der Chemiater Angelus Sala, 1576–1637: Ein Arzt in Selbstzeugnissen und Krankengeschichten* (Zurich: Juris, 1992).

their reassociation in differing combinations, as governed by their mutual affinities. This is the method that we see employed time and time again in Newton's attempts to arrive at such desiderata as the sophic sal ammoniac and "our Venus." Even though Newton was far from being an acolyte of Van Helmont himself, his main laboratory notebook, CU Add. 3975, contains at least twenty pages of extracts from the major English-speaking Helmontian spokesman of the seventeenth century, George Starkey.[51] These passages, which are taken from Starkey's iatrochemical work *Pyrotechny Asserted and Illustrated*, merge seamlessly with additional notes that Newton took from the New England alchemist's chrysopoetic texts written under the name of Philalethes.[52] This should come as no surprise, since the same theory and practice underlies both the medical and aurific works that Starkey wrote.[53] And in addition to the self-styled Helmontians Starkey and von der Becke, we must also reckon with the influence, once again, of Robert Boyle. Boyle himself drew heavily on Van Helmont, as well as other popular chymical atomists such as Sennert, incorporating important features of their matter theory into his own mechanical philosophy.[54]

In the following chapter we will see how Newton hoped to employ exotic chymical products such as the Philalethan net and "our Venus" with the staple laboratory reagents that he perfected early in his chymical career. These included above all Newton's liquor of antimony and the mixture of antimonial compounds and ammonium chloride that underwent "cleansing" from sulfur to become his sophic sal ammoniac. Although Helmontian influences can also be found in the material that we will cover in the next two chapters—particularly in the exalted role that both the followers of Van Helmont and Newton ascribed to fermentation—an understanding of this material will require that we cast a careful eye again on Newton's chrysopoetic sources. As we shall observe, the experiments found in these records consist largely of attempts to test his preliminary conclusions arrived at by reasoning out the riddles of the adepts, and to attain thereby the summum bonum of traditional alchemical practice, the philosophers' stone. Hence we must immerse ourselves full force in the enigmata of Newton's alchemical masters.

---

[51] CU Add. 3975, ff. 34r, 82v, 88v, 92r–92v, and 106r–113r. These extracts all come from Starkey's 1658 *Pyrotechny Asserrted and Illustrated* (London: Samuel Thomson, 1658).

[52] The notes from Starkey's *Pyrotechny* in CU Add. 3975 end on 113r, and a succession of chrysopoetic headings beginning with "Gross Ingredients" starts after several blank pages on 115r. Similar headings continue for some folios, and on 123r, Newton fills out the entry "Of yᵉ work wᵗʰ common ☉" with notes largely taken from Philalethes.

[53] This point is argued in Newman, *GF*, 170–88.

[54] See Newman, *AA*, 157–89, where Boyle's debt to Daniel Sennert's atomism is discussed at length.

SEVENTEEN

Nicolas Fatio de Duillier, Alchemical Collaborator

Although Newton's laboratory notebooks record his efforts over decades to produce and purify such exotic materials as sophic sal ammoniac and volatile Venus in the solitary confines of his laboratory at Trinity College, his chymical work was not always performed in isolation. The last few years of his cloistered academic life in Cambridge saw Newton entering into an alchemical collaboration of lasting significance to him, and after his departure for London in 1696 to serve as warden of the Royal Mint, the now famous "public intellectual" took on yet another partner in the furtherance of his chymical projects. There were no doubt other interactions with alchemists as well, such as the Captain Hylliard who contacted Newton after his acceptance of the position at the Mint, and an anonymous "Londoner" who visited Newton on "Munday March 2d or Tuesday March 3 1695/6" to inform him of the process of Basilius Valentinus and Jodocus van Rehe on vitriol.[1] But these shadowy encounters hold a distant second place to the collaboration Newton undertook with Nicolas Fatio de Duillier between 1689 and 1694 and another with the distiller William Yworth in the first decade of the eighteenth century. The present chapter charts out the first of these two collaborative projects, linking them to Newton's florilegia and, where possible, to his experimental notebooks.

The well-educated younger son of landed gentry in Geneva, Nicolas Fatio de Duiller came to England in quest of patronage in 1687, when he was

---

[1] For Captain Hylliard, see chapter nineteen herein. As for the anonymous "Londoner," Newton made two transcripts of their encounter, found in Keynes 26 and Schaffner Series IV, Box 3, Folder 10. Newton asked the London chymist about the process of "Dʳ Twisden" on vitriol. This was presumably Dr. John Twysden (1607–1688), a well-known London physician. Another of Newton's manuscripts, Keynes 50, contains material copied from Twysden and titled "Jodoci a Rehe Opera Chymica." This is largely a commentary and synopsis of Jodocus van Rehe's "Process" as found, for example, in Georg Andreas Dolhopff, *Lapis mineralis: Oder die höchste Artzney, Auß Denen Metallen und Mineralien Absonderlich dem Vitriolo* (Straßburg: Georg Andreas Dolhopff, 1681), 102–13. Newton's MS Keynes 50 also contains transcripts of letters attributed to "A. C. Faber" and addressed to Twysden. This is possibly the German chymist Albert Otto Faber, who resided in London in the 1670s and 1680s. For Twysden and Faber, see the online *Oxford Dictionary of National Biography*.

twenty-three. This year of course coincided with the publication of Newton's *Principia*. A talented mathematician in his own right, Fatio was able to appreciate the significance of this foundational work, and he communicated the news of its growing impact to Christiaan Huygens, with whom he had already been in communication. In the meantime, Fatio was making inroads in the English scientific scene and was admitted to the Royal Society in 1688. He established contact with Newton and was soon granted access to the scientific papers of the Cambridge savant; Fatio hoped to help him prepare a second edition of the *Principia*.[2] Mathematics and gravity were not the only subjects of discussion between the two men, however. Already in 1689 Newton and Fatio were speaking of alchemy, as revealed in a letter of October 10 from that year. There Newton provides the following guarded comments, which are rendered even more oblique by the fact that someone has cut holes in the letter at various points:

> I am extreamly glad that you —— friend & thank you most heartily for your kindness to me in designing to bring me acquainted wth him. I intend to be in London ye next week & should be very glad to be in ye same lodgings wth you. I will bring my books & your letters wth me. Mr Boyle has divers times offered to communicate & correspond wth me in these matters but I ever declined it because of his —— & conversing wth all sorts of people & being in my opinion too open & too desirous of fame.[3]

From this letter we learn several important things. First, it seems that Fatio was already introducing Newton to unknown friends. As we will see, such acquaintances of the young Genevois soon came to play an important role in Newton's alchemical research. Second, the comments about Boyle provide a window into Newton's relationship with the famous chymist and mechanical philosopher. It is striking that Newton claims to have rejected Boyle's overtures, which evidently concerned chrysopoeia, and that he attributed his chariness to Boyle's being "too open & too desirous of fame." Newton had been in personal contact with Boyle since approximately 1673, and some correspondence between the two survives from the late 1670s and 1680s.[4] But Newton may well have resisted open discussion of his alchemical secrets with Boyle, for we know from another source that he had long been disturbed by the loose talk of his older compatriot. In a famous letter of April 1676 to Henry Oldenburg, secretary of the Royal Society, Newton

---

[2] For the basic facts of Fatio de Duillier's life, see the online *Oxford Dictionary of National Biography* and Scott Mandelbrote, "The Heterodox Career of Nicolas Fatio de Duillier," in *Heterodoxy in Early Modern Science and Religion*, ed. John Brooke and Ian Maclean (Oxford: Oxford University Press, 2005), 263–96. See also Karin Figala and Ulrich Petzoldt, "Physics and Poetry: Fatio de Duillier's 'Ecloga' on Newton's 'Principia,'" *Archives internationales d'histoire des sciences* 37 (1987): 316–49.

[3] Newton to Fatio de Duillier, October 10, 1689, in Newton, *Corr.*, 3: 45.

[4] Newton to Henry Oldenburg, December 14, 1675, in Newton, *Corr.*, 1: 393, where Newton says, "Pray present my humble service to Mr Boyle wn you see him & thanks for ye favour of ye convers I had wth him at spring." But *Corr.*, vol. 1 (Oldenburg to Newton, September 14, 1673), p. 305, suggests that Newton already knew Boyle at that date: "I herewth send you Mr Boyle's new Book of Effluviums, wch he desired me to present to you in his name, wth his very affectionat service, and assurance of ye esteem he hath of your vertue and knowledge."

refs to a recent publication by Boyle in the *Philosophical Transactions* on the "incalescence" (heating up) of a special animated mercury when it is amalgamated with gold. Modern scholarship has shown that Boyle's animated quicksilver was a version of the sophic mercury of George Starkey, and that the English virtuoso probably published his findings in the hope of attracting adepts to help him probe its secrets further.[5] Although Newton provides a detailed explanation to Oldenburg of his reasons for doubting the effectiveness of Boyle's animated mercury, which we will examine later, he is nonetheless clearly worried by Boyle's eagerness to publish. Thus he concludes his letter with a thinly veiled admonition, saying, "I question not but that ye great wisdom of ye noble Authour will sway him to high silence."[6]

Newton's letter to Fatio in October 1689 forms the first evidence of what must have been an extensive chain of correspondence between Newton and his young friend. Unfortunately, the record is now quite incomplete, but several important letters and another document of great significance have recently emerged. We will deal with those in due course, but first it is necessary to recount the events leading up to the most serious period of alchemical collaboration between the two would-be adepts, which occurred in 1693. It is clear that Newton and Fatio discussed many topics beyond chymistry, physics, and mathematics, including issues of patronage and medicine. It was not only Fatio but also Newton who was seeking preferment as the 1680s drew to a close. A letter from Fatio dated February 24, 1689/90, informs Newton that he had spoken to John Locke about finding Newton a post as tutor to the Earl of Monmouth's son.[7] Soon after this, Fatio must have returned to the European continent, for Newton reports to Locke in a letter dated October 28, 1690, that the Genevois has gone to Holland and that he has had no communication with Fatio for six months.[8] By early September of the following year, Fatio had returned to London, whereupon he resumed his association with Newton.[9]

Our next surviving exchange from Fatio to Newton consists of a dramatic letter dated September 17, 1692, in which Fatio reports that he has contracted "a grievous cold" and fears for his life. The letter is of interest for the detailed medical information that it contains, presented in an almost clinical style. Fatio says that he experienced a sensation like the "breaking of an ulcer, or vomica, in the undermost part of the left lobe" of his lungs. He speaks of lurching forward involuntarily on "a momentaneous sense of something bigger than my fist moving" when the abscess seemed to break. He follows this with the information that his pulse is now good, but that he is feverish. A medication called "Imperial Powders" has apparently proved ineffectual, though in a subsequent exchange Fatio would change his mind about this.[10]

[5] Principe, *AA*, 159–79.
[6] Newton to Henry Oldenburg, April 26, 1676, in Newton, *Corr.*, 2: 2.
[7] Fatio to Newton, February 24, 1689/90, in Newton, *Corr.*, 3: 390. See also 79.
[8] Newton to Locke, October 28, 1690, in Newton, *Corr.*, 3: 79.
[9] See William Andrews Clark Memorial Library, MS F253L 1691, a letter from Fatio to Newton, dated September 3, 1691. See also Fatio to Huygens, September 8, 1691, in Newton, *Corr.*, 3: 168.
[10] Fatio to Newton, September 17, 1692, in Newton, *Corr.*, 3: 230.

On receiving Fatio's alarming report, Newton at once responded in kind. His very concerned letter of September 21 begs Fatio to see a doctor "before it be too late" and offers money if necessary. Within a day of receiving Newton's letter, however, Fatio's health had improved, a fact that he now attributed to his self-medication. The Imperial Powders had proven to be an effective sudorific, and the only other sign of their activity had been that they also raised pimples on Fatio's lips. Fatio seems to have interpreted the pimples as a good sign; perhaps their formation meant that the peccant matter of the disease was being drawn forth. Meanwhile, his fear of a pulmonary abscess had turned out to be unfounded, although a "fomes of an exceeding sharp and troublesome matter," presumably meaning a discharge of phlegm, had developed, accompanied by a severe headache.[11] It is perhaps significant that the Imperial Powders, which may have been a chymical medicament, were among some items that Newton would purchase from Fatio in the following year.[12]

Soon after his medical crisis, Fatio would receive an invitation from Newton to come and live in Cambridge.[13] The country air would be better for his health, and Newton would help him find lodging. Although Newton's concern for Fatio has been seen as evidence of an infatuation on his part, the strong feeling that emerges from his letters may stem as much from the two men's mutual attraction for the "the noble virgin alchymia" as for each other.[14] The aurific art reemerges in a letter of February 14, 1692/93, in which Newton offers to recompense Fatio for twelve doses of his Imperial Powder and "two Chymical books" that the Genevois had left with Newton from a previous visit. A subsequent letter reveals that the books consisted of the "French Ch: Biblioth," that is, the two-volume *Bibliothèque des philosophes chymiques* published in 1672–78 and attributed to "Le Sieur S. D. E. M."[15] It has been argued persuasively that Fatio was helping Newton to improve his command of French during this period for the purpose of reading alchemical texts in that language, and the two may even have studied the *Bibliothèque des philosophes chymiques* together.[16] As we will shortly see, Fatio's command of his native language would prove valuable not only for

[11] Newton to Fatio, September 21, 1692; and Fatio to Newton, September 22, 1692, in Newton, *Corr.*, 3: 231–32.

[12] There were many different recipes for "Imperial Powders" in the seventeenth century, but two of them are found in a chymical text that Newton knew well, the *Thesaurus et armamentarium* of Hadrianus à Mynsicht. See the English translation under the title *Thesaurus and armamentarium medico-chymicum; or, A treasury of physick with the most secret way of preparing remedies against all diseases . . .* (London: Awnsham Churchill, 1688), 88–89, and 104–5.

[13] Newton to Fatio, January 24, 1692/93, in Newton, *Corr.*, 3: 241.

[14] Alchemy was often referred to as a "virgin" because of the difficulty of attaining success in the art. A title playing on this theme was *Die edelgeborne Jungfer Alchymia* published in Tübingen in 1730.

[15] Newton to Fatio, February 14, 1692/93, and Fatio to Newton, March 8, 1692/93, in Newton, *Corr.*, 3: 245 and 261. The first volume of Newton's copy of the *Bibliothèque des philosophes chymiques* survives in the Wren Library at Trinity College (shelf-mark Tr/NQ.16.94). This is Harrison no. 221.

[16] Karin Figala and Ulrich Petzold, "Alchemy in the Newtonian Circle," in *Renaissance and Revolution: Humanists, Scholars, Craftsmen, and Natural Philosophers in Early Modern Europe*, ed. Judith V. Field and Frank A.J.L. James (Cambridge: Cambridge University Press, 1993), 173–91, see 174–79. See also Dobbs, *JFG*, 170–85.

such literary endeavors but also in negotiating the cosmopolitan alchemical scene of London under William and Mary.

The chymical collaboration between Newton and Fatio took on a new momentum in the spring of 1693, as reflected in several well-known letters as well as some that have received no scholarly attention until the present. They tell a fascinating story, and one that had serious implications for Newton's research, as reflected in his experimental records and in his late alchemical florilegium, *Praxis*. The first evidence appears in an often quoted epistle of May 4, in which Fatio informs Newton of alchemical marvels produced by a new acquaintance of his in London. Fatio's chymical friend has succeeded in producing an animated mercury that leads to the growth of metallic trees in a laboratory vessel. The passage is important for Newton's subsequent work and must be quoted in full:

> If You be curious Sir of a metallick putrefaction and fermentation which lasts for a great while and turns to a vegetation producing a heap of golden trees, with their leaves and fruits I can acquaint you with it having seen it and having been told by the owner how he made it. You remember Sir how in an experiment I proposed heretofore to You I purified ☿ with some f. of L. My friend, who is a new acquaintance of mine, and a good and upright man, taketh the natural oar it self of that mineral and powders it and grinds it with common ☿ vivified out of cinnabar that it may be more genuin. The vessel he grinds it in is a wooden mortar that hath a pestill allmost so big as to fill exactly the mortar; there remaining only a thickness of a crown between the pestill and the mortar. I believe he uses an engine to grind his matters together with more force But the chief point is that they grow hot by the action, tho perhaps not sensibly. Then he separates by a Shamey Skin from the ☿ a black powder, which he acknowledgeth to contain at first some ☿ but he saith it contains very little of it afterwards. For he begins that work again sometimes, and at last instead of the mortar he useth only a bottle, as You know, and shakes the matters together.[17]

This is the first recorded instance in the Newton-Fatio correspondence of metallic trees, which bring to mind the sophic mercury of George Starkey and the dendritic "vegetation" that could be produced by mixing that material with gold and heating the amalgam. It is possible that Fatio is thinking of the Philalethan sophic mercury, but he refers to his own attempt to purify quicksilver with "some f. of L." and says that his friend used "the natural oar it self of that mineral." What could this mineral be? One can speculate that it may not have been stibnite, as in Starkey's process, but lead, and thus "f. of L." might refer to filings or even flowers of that metal.[18] At any rate, it was not the metallic trees that caught Newton's attention, but rather the specific means of purifying the quicksilver, as we shall see. As for the emphasis on grinding with a specially fabricated mortar and pestle, this brings to mind the famous 1672 *Epistola* of Joel Langelot, court physician to the Duke of

[17] Fatio to Newton, May 4, 1693, in Newton, *Corr.*, 3: 265–66.
[18] I owe the suggestion of filings to Lawrence M. Principe, private communication.

Holstein-Gottorp, which we had occasion to consider in chapter fifteen. Langelot had described a geared "philosophical mill" and another mechanical grinding device there; the latter was a crank-operated pestle attached to a metal mortar for reducing gold, antimony, and other materials to a fineness that would make them more capable of combining with other substances.[19] Langelot's *Epistola* had raised a sensation in Europe, and the *Philosophical Transactions* had even published an English digest of the work in 1672/3.[20] Since Newton was deeply interested in the responses to Langelot's *Epistola*, one must wonder whether Fatio or his source was trying to capitalize on this interest by appealing indirectly to the German's physician's philosophical mill.

But it is Fatio's remaining lines that generated a profound response on Newton's part. The purification of quicksilver by means of a ground mineral leading to the separation of a black powder from the mercury by squeezing the latter through chamois leather (Fatio's "Shamey Skin"), and the alternative method of purification by shaking "the matters" together in a bottle, captivated Newton's interest. These techniques reemerge in Newton's famous *Praxis* florilegium found in Babson MS 420, where he explicitly credits them to Fatio. "This pouder," by which Newton presumably means the initial mineral rather than the black excrescence emitted during the quicksilver's purification, "amalgams wth ☿ & purges out its feces if shaken together in a glass [Epist. N. Fatij]." In another draft of *Praxis*, Newton explicitly refers to the powder as a "black fire" and equates it with the sympathetic fire of Snyders.[21] As the reader will recall, this material was one of three "fires" or materials that Snyders claimed were essential for the making of the philosophers' stone. We will return to the topic of the three fires shortly, but for the moment let us consider Newton's major laboratory notebook, where the topic of Fatio's mercury purification occurs once again. Here it appears without acknowledgment of the young Genevois, but again one can detect the importance with which Newton invested the process. Interestingly, Newton has canceled the passage, which he may well have done after it ultimately failed to lead to the philosophers' stone:

The<*sic*> is a black pouder wth wch if common ☿ be amalgamed it fumes wth a white stinking smoke &<*illeg.*> & casts out a copious filth if shaken in a glas & separated by a cloth or leather. This being repeated severall times y^e ☿ in a weeks time<*illeg.*> will become exeeding pure, & when it is moderately pure it ceases to smoak. There is a mineral white stone almost like marble

[19] Joel Langelot, *Epistola ad praecellentissimos Naturae Curiosos de quibusdam in chymia praetermissis* (Hamburg: Gothofredus Schultzen, 1672). A plate of the philosophical mill is found between pages 16 and 17, and an illustration of a crank-operated, steel mortar and pestle on p. 27.

[20] Joel Langelot, "An Extract of a Latin Epistle of Dr. Joel Langelot, Chief Physitian to the Duke of Holstein Now Regent: Wherein Is Represented, That by These Three Chymical Operations, Digestion, Fermentation, and Triture, or Grinding (Hitherto, in the Authors Opinion, Not Sufficiently Regarded), Many Things of Admirable Use May Be Performed, English'd by the Publisher," *Philosophical Transactions of the Royal Society* 7 (1672/3): 5052–59.

[21] Babson 420, 12r and 8v.

very heavy wch of its self resolves by digesting in a due heat. This is the first matter.[22]

There are several very interesting features about these recurrences of Fatio's method of purifying mercury. The squeezing of the quicksilver through leather and the shaking in a bottle ("a glas") are the same as before, but now Newton adds that this results in the release of "a white stinking smoke" along with copious filth. Moreover, Newton now says that the process requires a week to complete, which was also absent from Fatio's letter of May 4. From all of this it appears likely that Fatio was embellishing the original story as time went by, either orally or in letters that we no longer possess. If we now pass to another passage in the same letter, the identical pattern will be seen to emerge. After supplying more information on the "heap of trees" produced in a sealed glass by fermenting the philosophical mercury with gold, Fatio adds that the mercury can even be induced to undergo a metamorphosis without the addition of the noble metal:

> This same without any addition whatsoever being put in a glass vessell with a long neck boyls and turns to a clear transparent water that doth not wet a cloth and congeals in the cold air like a heavy cristall or salt. It shuts the vessell of it self by a part of it sticking in a liquid form some where in the neck, from whence it rains perpetually upon the ☿ that is underneath; and they both turn in a few hours to those white and red powders which I did once shew to You; or at least to a matter that may be reduced to those powders, only by grinding it. Is not perhaps heat arising from friction and motion one of the Philosophers fire?[23]

All of this sounds remarkably simple. The purified mercury, if placed by itself in a long-necked flask and boiled, will turn transparent and harden in the cold, leading eventually to the production of a red and white powder. Since the philosophers' stone itself was supposed to have a red and a white form, depending on whether it was intended for chrysopoeia proper or rather for making silver, this was an obvious inlet to something important. Indeed, it appears that the mercury could cure humans as well as metals, for Fatio mentions later in the letter that his anonymous friend is going to use it to heal his ongoing pulmonary problems. But the passage on the mercury's self-sealing property, by which it sticks in the neck of the flask and closes it, thus creating a perpetual "rain" or refluxing, is particularly significant. As we will see, this portion of Fatio's process would become a key part in a fantastically complex set of related procedures in a later manuscript. The initial simplicity of Fatio's account, whereby the entire success of the process depended on the grinding of the mercury and the friction that he interpreted as the "Philosophers fire," would give way to a byzantine panoply of operations in which one can only barely make out the basic procedures outlined in the May letter. The anonymous adept, or perhaps Fatio himself, would elaborate

---

[22] CU Add. 3975, 136r.
[23] Fatio to Newton, May 4, 1693, in Newton, *Corr.*, 3: 266.

this chrysopoetic project over the next few months until it acquired a Rabelaisian girth.

The next scene in the unfolding drama of the Newton-Fatio collaboration occurs in a letter written by the young aspirant and sent from London on May 18. Fatio begins with news that Newton had already heard: the former's mother has died, and he has been trying to sort out his inheritance. Alas, as a younger son with sisters on whom dowries had to be settled, Fatio has found that his allocation is inadequate. Thus he is casting about for a way to make a living. One possibility would be to obtain a medical degree, and to employ chymical medicines as a source of income. Here the unnamed friend of Fatio reenters the picture. The friend has had wonderful successes with his alchemical medicine, giving it to more than ten thousand patients, sometimes as many as five hundred in a single day; moreover, it is cheap to prepare.[24] Hence Fatio is reconsidering the offer that the unnamed adept made to reveal the secret to him in full. In the meantime, Fatio has learned more about the processes for making the marvelous medicine, which he recounts in a plenary fashion. The details are highly pertinent to our story, as they contain information that will make it possible to penetrate further into the project that Newton and Fatio would soon jointly undertake:

> His menstruum or ☿ is an ordinary ☿ prepared wth boyling it in a wooden vessel exceeding close for several days with some rain gathered while the Sun is in ♈ or ♉). After that it is distilled with a wonderfull art and a great deal of trouble. Out of 5 ℔ of matter there comes into ye recipient 4 1/2 ℔. Yet in ye recipient You see nothing at all. It is full of vapours, wch must not have the least vent; else the whole would fly away. The recipient being carried in a cold place is there severall days before the vapours settle into the 4 1/2 ℔ of liquor. That liquor must be still prepared with an extraordnary care. The times must be kept so exact that if you loose but one moment you loose all your operation and very often all your matter. I believe it is distilled about 10 times. It is a most powerfull menstruum. It dissolves all metals and gems. It may be drawn off from allmost all sorts of bodies without loosing sensibly either its force or quantity. By its means my friend prepares his medicine wch in seven months time goes through 4 putrefactions and may still be made by new putrefactions better and better.[25]

Although the friend and the mercury appear to be the same ones as before, the process has suddenly become more complex. The quicksilver must now be boiled for days on end in rainwater collected in the spring, in a wooden container, and then subjected to a troublesome distillation. The surprising result will be that four and a half pounds of an invisible vapor pass into the recipient, where they gradually condense into a liquor. But the process is still not over—Fatio thinks that ten distillations in all are required, and at least four successive putrefactions. The result will be a marvelous menstruum that

---

[24] Fatio to Newton, May 18, 1693, in Newton, *Corr.*, 3: 269.
[25] Fatio to Newton, May 18, 1693, in Newton, *Corr.*, 3: 268.

can dissolve any metal or gem, and which may then be distilled intact without undergoing any combination with the solute. This is precisely the sort of chymical agent that Newton had been looking for in his private research, as we saw in our analysis of the Lullian *Opera* text that he composed at some point after 1686. A large portion of that florilegium was devoted to the Helmontian alkahest, which was renowned for its supposed ability to dissolve any material *sine repassione*, that is, without itself being acted upon. Newton even reported there, on the authority of Starkey's *Pyrotechny*, that the alkahest could dissolve gems, along with marchasites, pebbles, rocks, sand, clay, earth, glass, sulfur, and of course metals.[26] As we learned in chapter thirteen, a major question in Newton's mind when composing the *Opera* was whether the Helmontian alkahest was identical to the Lullian quintessence, and hence to the sophic mercury that was supposed to lead to the philosophers' stone. It must have appeared in 1693 that Fatio's friend could supply the answer.

A number of previously unexamined documents throw the next beam of light on the unfolding collaboration between Fatio and Newton, which now came to include the former's anonymous friend as an active player. The letter that we will now inspect, composed by Fatio in August 1693, is in several respects extraordinary. First, it provides the only instance known to me in which there is evidence that the tight-lipped Newton willingly revealed his chymical secrets to another party. Suddenly, we find Newton not only receiving information secondhand from Fatio's friend but also actually imparting his own hard-won knowledge through Fatio to the adept. Nor was this pedestrian material that Newton chose to release. It was in fact a close variant of the very set of processes that he had developed in Keynes 58, the manuscript that builds a chymical practice on the work of Snyders, as we discussed in chapter twelve. The reader may recall that Newton there outlined a series of processes for making such desiderata as the scepter of Jove, the two serpents entwined about the caduceus of Mercury, and even the central rod of the caduceus itself. The letter that we are about to describe belongs to the same project, and this enterprise would culminate in Newton's composition of *Praxis*, a manuscript that we will consider in due course.

Dated August 1, 1693, Fatio's latest letter, like the previous two, was sent from London.[27] The letter begins by Fatio passing on a request from his chymical friend. Fatio says that he had translated into French a previous (now lost) letter from Newton to the adept. The Francophone friend, however, was perplexed, for Newton had described the preparation of a vitriol there without indicating its overall purpose in the plan of research. In order to clarify the problem, Fatio then quotes a Latin passage from Newton's lost letter. Examining this passage will illuminate the very processes on which Newton hoped to pin his synthesis of Mercury's caduceus. I therefore translate Newton's Latin text as follows:

---

[26] Keynes 41, 7v.

[27] I thank Scott Mandelbrote for bringing the present letter, William Andrews Clark Memorial Library, MS F253L 1693, to my attention and for providing me with a photocopy of it. As he is currently preparing an edition of Fatio's correspondence, I do not transcribe the entire letter here.

After this first operation, our pulverized earth should be put in ℞. When it has begun to boil somewhat and the earth has begun to be dissolved, and the menstruum by exercising its activity has heated up in some measure, the clear menstruum should be poured into another vessel where there is pulverized *viride æris*. The menstruum can first be evaporated down to half, but that is not necessary, and on account of the poisonous fumes it is preferable to avoid that operation. Let the menstruum be digested with the *viride æris* and a vitriol will be extracted. Regulus of antimony should be sublimed with sal ammoniac (I suppose one part of regulus with two of sal ammoniac), and three parts of this sublimate should be drawn from two of ~~white lead~~ ceruse. Then a sugar is extracted from the lead *<Fatio adds in English: I suppose by this is meant the white lead or Cerusse>* with rainwater (if necessary, as much distilled vinegar is added as suffices, but it is preferable not to use vinegar). One may melt his sugar at the fire when it has dried so that the vinegar that you used evaporate; and two or three parts of our earth as well as one part of iron should be mixed with one, two, or three parts of it. Then there should be a slow boiling <of the mixture> in a sealed location at a sufficiently hot fire. Then take etc.[28]

Newton's recipe for the most avoids encoded language, but he begins with the oblique advice to take "our pulverized earth" and immerse it in aqua regia. No doubt Fatio was instructed orally to inform his friend of the mineral hidden behind "our earth." If Newton's laboratory notebooks are any guide, this must surely refer to stibnite subjected to his homemade aqua regia consisting of nitric acid fortified with sal ammoniac. As we know, this results in a reaction involving the heat and bubbling mentioned in the letter, followed by a slow dissolution of the now oxidized stibnite in the aqua regia. Newton then says to pour off the "clear menstruum." This is without doubt the standard menstruum that appears time and time again in his laboratory notebooks, referred to variously as liquor, spirit, vinegar, or even salt of antimony.

Then Newton says to extract a vitriol from *viride æris* by dissolving it in the acid solution. This is at first peculiar, since *viride æris* is usually taken to mean verdigris or copper acetate, made by subjecting copper to vinegar. Except for the occasional purpose of "extracting" sugar of lead from white lead or lead ore, Newton made little use of actual vinegar in his experimental notebooks.[29] Moreover, the term *viride æris* is not mentioned in his

---

[28] William Andrews Clark Memorial Library, MS F253L 1693, 1r: "Post hanc primam operationem Terra nostra pulverisata ponatur in ℞. Ubi paululum ebullierit et Terra coeperit dissolvi, et menstruum agendo nonnihil incaluerit offundatur menstruum clarum in aliud vas ubi est viride æris pulverisatum. Potest Menstruum prius ad medietatem evaporari, sed non opus est, et ob noxios ejus fumos præstat hanc operationem omittere. Digeratur Menstruum cum Viridi æris, et extrahetur Vitriolum. Sublimetur Reg. ♁ⁱ cum ✳ (una pars Reg. puto. cum duabus ✳) & hujus Sublimati partes tres abstrahantur a duabus plumbi albi cerussæ. Deinde a plumbo [I suppose by this is meant the white lead or Cerusse] cum aqua pluviali (addito si opus est aceto destillato q. s. sed praestat aceto non uti) extrahatur Saccharum. Hoc Saccharum ubi aridum fuerit liquescat ad ignem ut acetum quo usus fueris evanescat; et cum ejus parte una vel 2 vel 3 misceantur partes 2 vel 3 terrae nostrae et pars 1 ferri, et ad ignem satis validum in loco clauso fiat ebullitio lenta. Deinde accipe &c."

[29] For example, at CU Add. 3975, 53v, 79r, and 79v.

laboratory notebooks at all. Verdigris does appear there (in variants such as "verdegrece" and "verdigreas"), but always in passages borrowed from Robert Boyle rather than in Newton's own experimentation.[30] This is itself unsurprising given Newton's predilection for using native minerals as the starting points of his operations. Copper acetate is hardly found beneath the soil at all, since the vinegar or acetic acid required to make it is typically a product of organic rather than mineral decay. But there is another sense in which the term *viride æris* could be used in the early modern period, namely, to mean a green copper mineral. In a very telling note on Basilius Valentinus, where the putative Benedictine says to make a vitriol from "digged Verdigreece," that is, mineral verdigris, Newton substitutes "digged viride æris." Thus for him the terms verdigris and *viride æris* were synonymous, and both could refer to a mineral as well as an artificial variety.[31] The green copper mineral that Newton has in mind is hard to determine with precision, although malachite, chrysocolla, and even botallackite are all obvious candidates.[32]

There is one other highly significant locus where *viride æris* appears in Newton's manuscripts, namely, in a passage that we already examined from Keynes 58. Comparing the multiple drafts of Keynes 58 to Fatio's August letter yields rich rewards. In a later section of Keynes 58, Newton begins his directions by saying that a "double vitriol" (comprising the two serpents around the caduceus) is first "extracted," or made by subjecting the ores of copper and iron to the corrosive action of a menstruum. The dried salts are then dissolved again in the same menstruum along with "green ♀ or its salt together with Cupid."[33] The term that I have translated as "green ♀" is in the original "viride ♀," and since ♀ is the standard symbol for copper, the expression is merely another way of saying *viride æris*. Moreover, the similarity between the processes described at the beginning of Keynes 58 and those in Fatio's August letter is striking. In both cases, a vitriol (or perhaps two vitriols in Keynes 58) is made from an ore or ores by Newton's liquor of antimony, which he typically fabricated by using the technique described in Fatio's letter of reacting stibnite with aqua regia and then decanting the solution. And in both cases *viride æris* plays a part, though in Keynes 58 it is reacted with the double vitriol, while in Fatio's letter the *viride æris* is the starting material from which the vitriol is made. It is likely that Newton was presenting

[30] I find eight references to verdigris in CU Add. 3975, at 16r, 46r, 47r, 48r, 48v (twice), 49r, and 51r. All of these are taken from Boyle. I can find no references to the material in CU Add. 3973.

[31] For Newton's substitution of terms, see his Basilius Valentinus commentary in British Library Add. MS 44888, 5r. For the original text on which Newton is commenting, see Basilius Valentinus, *Last Will and Testament* (London: T. Davis, 1658), 128. This is actually part of the "Elucidation" found in the *Last Will and Testament*; it is not by the original author of the Basilian corpus, as we discussed in chapter five. For the sake of simplicity, I will henceforth refer to the various pseudonymous writers found in the *Last Will and Testament* collectively as "Basilius Valentinus" or simply "Basilius." After all, this is how Newton and his contemporaries referred to them.

[32] Botallackite, a polymorph of atacamite, is much less common than either malachite or chrysocolla, but it is in fact found in Cornwall. The very name botallackite is an eponymous formation derived from the Botallack mine in Cornwall. See https://www.mindat.org/min-732.html, accessed August 14, 2017.

[33] Keynes 58, 3r.

FIGURE 17.1. Crystal of "verdigris vitriol" about one-quarter inch across, made by reacting malachite with Newton's liquor of antimony, then filtering and crystallizing the solution. Prepared by the author in the laboratory of Dr. Cathrine Reck in the Indiana University Chemistry Department. See color plate 9.

Fatio's unnamed friend with a simplified version of his process for making the caduceus of mercury in order to get the adept's opinion on the various stages. In short, the ingredients given in the letter are *succedanea*, to use a term employed by George Starkey and other early modern writers on pharmacopoeia. A *succedaneum* was a cheaper or more easily available substance that could be substituted for its rarer counterpart.[34] Thus the ingredients of the August 1 letter were easier-to-obtain products that would display most of the properties of the materials employed in Newton's laboratory without necessarily being fully identical to them. The term *viride æris*, which could mean either ordinary verdigris made with vinegar or the "digged," mineral variety, may either have been an example of studied ambiguity on Newton's part or a recognition of the fact that either substance would suffice. I have in fact prepared "vitriols" from each of these materials (figure 17.1). But the use of *succedanea* does not exhaust Newton's caginess, a fact to which the adept's bewilderment testifies.

Fatio's friend had good reason to be puzzled. As Fatio himself points out, the production of the vitriol is followed by a set of seemingly unrelated processes. First, one part of antimony regulus is sublimed with two parts of sal

[34] George Starkey, *Pyrotechny Asserted* (London: Samuel Thomson, 1658), 150, 165–66.

ammoniac. This is a simplified version of Newton's sophic sal ammoniac, from which the external, crude sulfur of the antimony has been removed by reducing it to regulus. Then this sublimate is drawn from white lead or ceruse (mostly lead carbonate), presumably by subliming the two together. A "sugar" or sweet compound is subsequently extracted from the sublimed ceruse either with water, which is preferable, or with vinegar if necessary. Newton may very well have produced a sweet salt of lead by subliming the ceruse with sal ammoniac and antimony.[35] His half-hearted appeal to vinegar suggests that the process did not always succeed and that it was sometimes necessary to resort to "vulgar" sugar of lead, our modern lead acetate (which is in fact very sweet, though toxic). Newton's laboratory notebooks reveal that he sometimes made similar sublimations of lead ore with sophic sal ammoniac to produce a sweet salt; in cases where he was unsatisfied with the amount, he would perform an additional extraction with vinegar.[36] Fatio's letter concludes Newton's recipe with the injunction that the sweet lead compound be purified of its vinegar if that material was used, and then sealed up and heated with a mixture of "our earth"—presumably stibnite again, and iron.

Unfortunately, Newton's directions break off at this point, but Fatio has provided us with enough to see that his friend's confusion was appropriate. The vitriol produced in the first operation does not reenter at a later stage. Why did Newton describe its production at all? In order to answer this question, we may return again to Keynes 58, where Newton first describes the serpents around Mercury's caduceus and then a "dry water" made by digesting the double vitriol with lead and "its menstrue once & again." The product of this digestion eventually becomes "a black pouder," which goes on to form the basis of Jove's scepter and possibly the rod around which the snakes of the caduceus are entwined. In a marginal note found in Keynes 58, Newton equates this black powder with the product of a calcination or putrefaction described at length in Philalethes's *Ripley Reviv'd*. As the "American philosopher" says there, the regimen of calcination results in a product characterized by "the intire Blackness and Cimmerian utter Darkness of compleat Rottenness."[37] This was something that Newton had been seeking out for a long time, as it was the end result of "the first gate" and the beginning of further alchemical regimens that promised to lead to the

[35] Lead chloride, though only sparingly soluble in water, is said to taste sweet. See William Thomas Brande, *A Manual of Chemistry* (London: John Murray, 1830), 2: 77. See also Leopold Gmelin, *Hand-Book of Chemistry* (London: Cavendish Society, 1851), 5: 115, on the sweetness of lead salts in general.

[36] As at CU Add. 3975, 78v–79r: Le. o. & ✱ ana 24 pts leave 15 1/2 if sublimed in a glas or 15 1/4 or 15 1/6 if subl. in yᵉ open air. ffor in yᵉ air yᵉ Le. o. impregnated rises more easily then in a glass: & thence looks more white after yᵉ subl. is ended. Let it be in a glass retort & yᵉ heat so big as to rais the ✱ but not to make yᵉ Le. o. melt & yᵉ ✱ will leave yᵉ Le: o. first round about yᵉ bottom & sides retiring into yᵉ middle above & there growing less & less till all be sublimed: Which will be known by yᵉ ceasing of yᵉ fumes. If any remain unsublimed the ^salt core above must be cut out after yᵉ glass is broken & kept for a new Sublimation. The remaining calx in boyling water let go a salt sweetish. It required <*illeg.*> ^very copious solvent, but dissolved more easily in ~~vinegre a little~~ a much smaller quantity of vinegre."

[37] Eirenaeus Philalethes, "The Vision of Sr George Ripley, Canon of Bridlington, Unfolded," in Philalethes, *RR*, 19. For more on the first gate, or calcination, see "An Exposition upon the First Six Gates of Sir George Ripley's Compound of Alchymie" in the same volume, pp. 97–188.

philosophers' stone. It is virtually certain that Newton composed Keynes 58 before the spring of 1693, since there is no mention of Fatio in that document despite its conceptual relationship with the material in the August letter. In the closely related text of *Praxis*, on the other hand, Newton explicitly refers to Fatio as the source of the wonderful black powder that purifies mercury by being shaken with it in a bottle. In the same text, Newton also explicitly identifies Fatio's black powder with the product of "the calcination wch they call ye first gate," in other words, Philalethes's rotten black material.[38] By putting all of this together we may arrive at the hypothesis that Newton probably intended the vitriol made in the first part of Fatio's letter to be combined with the mixture composed of sugar of lead, stibnite, and iron in the second part, just as he advised to combine the two vitriolic serpents with lead and its "menstrue" in Keynes 58. He may well have been testing Fatio's friend, who should have known what to do if he was indeed an adept.

My hypothesis that Newton intended the lead compound and the vitriol to be combined is unequivocally borne out by another source. An undated and unaddressed letter in the printed *Correspondence of Isaac Newton* indisputably contains Newton's response to Fatio's question of August 1. This very short epistle, containing only nine lines in its printed form, consists mostly of Newton's rephrasing of his original recipe for Fatio's friend. Newton recapitulates the problematic text as follows, and after mentioning the sublimate of sal ammoniac and antimony regulus, he adds five key Latin words that make everything clear, namely, "first from two parts of the vitriol, then" (primum a duabus Vitrioli deinde) as follows:

> And three parts of this sublimate should be drawn first from two parts of the vitriol, then from three or four of the ceruse. Then a sugar is extracted from that ceruse with rainwater (if necessary, as much distilled vinegar is added as suffices, but it is preferable not to use vinegar).[39]

Thus Newton intended the vitriol of *viride æris* to be sublimed by means of the antimonial sal ammoniac, and the sublimate produced by this operation would then be used to sublime the ceruse. Once this had been accomplished, a salt would be extracted from the new sublimate containing vitriol, ceruse, and antimonial sal ammoniac. All of this fits closely with the protocols laid out in Keynes 58, though again, Newton was providing Fatio's friend with simplified *succedanea* instead of ingredients identical to the complicated reagents that he produced in his own laboratory. The role of the vitriol as a subliming agent in Newton's revised instructions also suggests strongly that it was acting as a *succedaneum* for the volatile Venus of his laboratory notebooks in the same way that the mixture of sal ammoniac and regulus was standing in for the notebooks' sophic sal ammoniac.

---

[38] Babson 420, 12r.

[39] Newton to [Fatio], no date, in Newton, *Corr.*, 7: 367: "The passage in ye ℞ is to be thus intended.—Et hujus sublimati partes tres abstrahantur primum a duabus Vitrioli deinde a tribus vel quatuor cerussæ. Postea de cerussa illa cum aqua pluviali (addito si opus est aceto destillato q.s. sed præstat aceto non uti.) extrahatur saccharum."

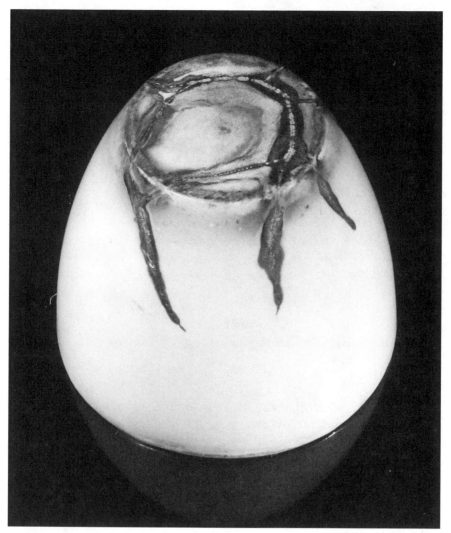

FIGURE 17.2. Cracked crucible used in subliming ceruse from Newton's *succedaneum* to "volatile Venus" as described in Fatio de Duillier's letter of August 1, 1693. The multiple colors on the bottom and sides of the crucible reveal the presence of lead and copper compounds as well as the partially reduced metals themselves. Prepared by the author in the laboratory of Dr. Cathrine Reck in the Indiana University Chemistry Department. See color plate 10.

Interestingly, modern replication has revealed that Newton's methods as decribed in Fatio's August letter are indeed an effective means of subliming certain compounds of lead (figure 17.2).[40]

What was one supposed to do with this concoction once the sublimation was complete, the sublimate had been washed, and the crystals boiled with iron and "our earth"? A subsequent passage in Fatio's letter of August 1 suggests that the final product was meant to be combined with quicksilver:

[40] See the forthcoming article by the author and Professor David Bish of the Indiana University Department of Geological Sciences on replicating and analyzing Newton's products.

> I suppose Sir that it would be easie to make ☿ to mix with this composition, provided one did use ☿ coagulated with the vapours of streams coming from lead just congealed after fusion For that ☿ might be powdered and so mixed pretty well by bare trituration with the other materialls, and the lead, that dos congele it, might be a medium to make it to unite with the saccharum ♄^{i}, and by consequence with the rest of ye mixture.[41]

Fatio suggests here that quicksilver be first amalgamated by subjecting it to the vapors of hot lead, an old alchemical desideratum, before combining it with "this composition," meaning the product that would be obtained from Newton's instructions. Fatio believed that the lead in the amalgam would make it easier for the quicksilver to unite with the sugar of lead in the "composition." The same principle of mediation underlay Philalethes's sophic mercury, which employed either silver or copper as a medium between the quicksilver used there and the regulus of antimony from which it was distilled. Fatio's gratuitous advice strongly suggests that the missing conclusion of the Latin recipe involved mixing the product with quicksilver, and this too conforms with what we know of Newton's chymical practice. An experimental record from June 1693 found in CU Add. 3973 shows that Newton was carrying out exactly this sort of process less than two months before Fatio's August letter:

> Iune 1693. The two serpents ferment well wth salt of ♄ ♃ & <*illeg.*> ♀ better wth salt of ♄ & ♀ best wth salt of ♄ alone. ☿ added ferments much more in all three cases & volatizes y^{e} mass: but better in y^{e} 2^{d} case y^{n} in y^{e} 1st & best in y^{e} last. To y^{e} 2 serpents 24^{gr} <*illeg.*> $^{\wedge \text{I added}}$ ☿ of ♄ 24^{gr} ~~added~~ by degrees & when y^{e} fermentation was over I added ☿ 16^{gr} & y^{e} matter swelled ^{much} wth a vehement fermentation $^{\text{then before}}$ & <*illeg.*> in two or three hours sublimed all to y^{e} top except 3 grains wch remained below ^{spongy} in form of a ^{dark} cinder, ~~& there was~~ & there was 9 1/4^{gr} of running ☿ besides a little that stuck in y^{e} neck of y^{e} glass wch might amount to a grain more so y^{t} y^{e} 2 matters dissolved about 1/8 of their weight of ☿ <.>[42]

Recalling that the two serpents refer to the double vitriol of copper and iron described in Keynes 58, we can see that in June 1693, Newton was testing the ability of a vitriol to "ferment" respectively with salts of lead, tin, and copper. He found that the fermentation of the vitriol was most successful when performed with salt of lead. Whether Newton was subliming the vitriol before carrying out the fermentation as in Fatio's letter is not specified by his report, though it is far from impossible. At any rate, Newton followed these experiments with another one where he fermented equal parts of the vitriolic serpents with a material that he calls "mercury of lead" (☿ of ♄). From an earlier section of CU Add. 3973 it appears that this so-called mercury of lead was made by first extracting a salt from lead ore that had been imbibed with Newton's standard liquor of antimony and then burning off the "common 🜍

[41] William Andrews Clark Memorial Library, MS F253L 1693, 1r.
[42] CU Add. 3973, 28r.

in blew fumes."[43] In other words, the mercury of lead may have simply been a lead salt that had been deprived of its sulfur, leaving a material that Newton believed to be composed primarily of the chymical principle mercury. In short, it probably differed little (if at all) from the lead salt used in the first experiments in the June 1693 laboratory report, where salts of lead, tin, and copper were fermented with the vitriolic serpents. At any rate, once the mercury of lead (or lead salt) was combined with the serpents, Newton then added 16 grains of common quicksilver. The result was promising indeed: "ye matter swelled ^{much} wth a vehement fermentation." To Newton it must have looked as though he was well on the way to producing the caduceus and then fermenting it with quicksilver. As he had advised in Keynes 58:

> Wth this rod & ye two serpents (double s^{pt}, or rather ⊕ of ♂ & ♀ extracted wth ye juice of saturnia) ferment ☿ & cleans it<.>[44]

Newton's instructions for Fatio's friend therefore consisted of simplified instructions for first making at least one of the serpents and then the rod. In his typical fashion, Newton initially left out critical material, above all the crucial fact that the vitriol and the lead-based rod had to be combined and also the method whereby that combination was to be effected. It was only after Fatio confronted Newton on behalf of his bewildered friend that the celebrated scientist coughed up the required information. And only after the materials described in Fatio's letter had been combined could one hope to obtain the black powder that would "ferment ☿ & cleans it." At this point the alchemist would be in a position to hand Mercury his caduceus, thus acquiring the means to progress to the philosophers' stone.

Who was Fatio's mysterious friend to whom Newton imparted such sensitive, even if incomplete, information? Interestingly, Fatio's letter of August 1 provides us with some clues. After finishing the chymical part of the letter, Fatio passes to an entirely different subject. The disastrous Battle of Landen had taken place on July 29 (New Style), and Fatio laments the British rout at the hands of the French as described in the *London Gazette*.[45] After passing on additional gossip about Queen Mary's distractedness at the loss, Fatio returns to the subject of his alchemical friend. His comments reveal that the friend, whom we already know to have been Francophone from the fact that Fatio translated Newton's letter into French for him, might soon have to leave the country, apparently as a result of the French and English hostilities:

> My Friend has been again interrupted, and obliged to go a second time to his Regiment. They have now orders to be in a readiness, in case they should be sent for from Flanders. So that it will be winter before I can think to begin to learn the preparation of his Remedy.[46]

---

[43] CU Add. 3973, 15v.
[44] Keynes 58, 2r.
[45] See the *London Gazette* for July 27–July 31, 1693 (Old Style).
[46] William Andrews Clark Memorial Library, MS F253L 1693, 1v.

Why would Fatio's Francophone friend, living in London, be required to report to a regiment in Flanders, where the Battle of Landen had just taken place? The obvious answer is that he was a Protestant serving in one of the Huguenot regiments in the England of William III. Fatio himself was a Protestant, so it is no surprise that his French-speaking friend would be of the same persuasion. It is this alchemical soldier's imminent departure that Fatio fears will lead to the postponement of his chymical tutorial. This important new information provides a possible key into an even more exact identification of Fatio's adept.

Another unpublished document in the hand of Fatio casts additional light on the likely identity of Fatio's friend. Possibly part of a letter from the Genevois to Newton, this text is found in a photocopy kept in the British Library.[47] It consists of detailed descriptions for making lute, the clay-like material that chymists used to seal their glassware and to reduce thermal shock. For us the important thing is not these interesting technical directions, but rather a paragraph found on the second sheet of the photocopy:

> Monsieur de Tegny, a Captain in the French Regiment of Cambon, <*the comma is corrected to a period*> who ^He <*in second hand*> marryed one Monsieur de Grancey's Sister, & he was a Gentleman of a good estate in France; his lands were in Poictou about Tegny, within three or four miles of a place where they dig out some excellent Antimony.[48]

Was this Monsieur de Tegny the source of the detailed instructions that Fatio gives for making lutes? Very likely he was, since it was common practice in the literature of secrets and recipes to follow the directions with their source. Moreover, the added fact that M. de Tegny stemmed from an estate in Poitou where crude antimony was mined strongly suggests an alchemical context for the brief biographical information Fatio supplied. As we also learn from this note, Tegny was a captain in the "French Regiment of Cambon." This is surely a reference to François Dupuy de Cambon, who under King William commanded a Huguenot regiment that was raised in 1689. Cambon died in the summer of 1693, but he was still living at the very beginning of August when Fatio indicated that his friend might have to depart for Flanders.[49] Hence Captain de Tegny could well be Fatio's anonymous friend. But this raises a further interesting point. Despite the fact that muster lists and other records of Cambon's regiment survive, there seems to be no record of a M. de Tegny among his officers. Nor for that matter have I been able to find a locality called "Tegny" in the former province of Poitou. Even accounting for variant spellings, Fatio surely cannot have meant Teigny

[47] British Library RP 2692 is a photocopy consisting of two sheets. According to a modern hand at the bottom of the first sheet, it is a "draft corrected by Newton." The correcting and adding hand does look like Newton's, but there is too little of it to be conclusive, in my opinion. The main text itself is written in the very careful, even beautiful, hand of Fatio. There is no indication of the current owner in the BL RP finding list. The BL finding list says it was "Dereserved 9 February 1991."

[48] British Library RP 2692, sheet two of photocopy.

[49] David C. A. Agnew, *Protestant Exiles from France, Chiefly in the Reign of Louis XIV* (s.l.: s.e., 1886), 2: 87–88. See also Matthew Glozier, *Huguenot Soldiers of William of Orange and the Glorious Revolution of 1688* (Brighton: Sussex Academic Press, 2008), 156–57.

in the modern-day department of Nièvre, or Treigny in the department of Yonne, since both of these principalities are found in Bourgogne-Franche-Comté, nowhere near Poitou.

All of this raises the possibility that Fatio was taken in by a trickster, or that Fatio himself created the mysterious captain as a ruse for beguiling Newton. On the other hand, the records are sufficiently incomplete to countenance the possibility that there was a Captain named Tegny among the ranks of Huguenot soldiers in Britain during the early 1690s. What we can say with certainty is that the collaboration between Fatio and Newton continued on for a considerable time after the letter of August 1, and that the alchemical project in which they were engaged took on an ever more labyrinthine aspect as time progressed. The continuation of their alchemical communications well into the fall is certified by the existence of another unpublished letter, this time a dated fragment from Fatio that Newton copied. The passage concerns "Terra Sigillata" and "Terra Lemnia," earths typically sold by pharmacists in the form of stamped medallions. Fatio had been visiting the apothecaries' shops of London, and in his typical breathless fashion he reports on the marvelous properties of these earths. From one of them it is possible to extract a type of niter that is far different from the vulgar sort. The earth seems dry at first but is actually full of an unctuous moisture; thus when it is heated, the earth releases a dangerous, white, choking fume, which condenses into a water. This earth may even contain the "first matter" or starting point for making the philosophers' stone. Newton thus concludes Fatio's report with the following comments:

> This Mr Fatio sent me in a letter Novem 14. 1693, from a Gentleman who thinks this earth to be y^e matter out of wch y^e $\overline{\text{Phers}}$ prepare their ☿, <deletion> For, saith he, there are in y^e $\overline{\text{Phers}}$ writings divers passages wch exclude ♂ & all metals. By distilling it in a Retort wth a double Receiver the spirit settles in y^e first receiver & the flegm goes on further into y^e second.[50]

Fatio's report from November 14 was thus partly based on a conversation with a gentleman who had his own ideas about the proper material on which to begin the work of making the sophic mercury. Since the philosophers had explicitly ruled out antimony and all the metals, he turned to the earth with the choking white fumes, from which he apparently extracted a spirit and an "extreamly stinking" salt. Nothing more is known about this gentleman acquaintance of Fatio's, but the collaboration between the Genevois and Newton had at least one further episode, which seems to have included both Fatio's original chymical friend (possibly M. de Tegny) and another, probably Anglophone participant. The evidence is found in yet another document, but unlike the ones that we have been examining, it is not a personal letter. Instead, it is a short but dense alchemical treatise titled "Three Mysterious Fires," today preserved in the Columbia University Library.

---

[50] University of Texas, Austin, Harry Ransom Center 182, 1r. The manuscript was part of Sotheby Lot 18.

"Three Mysterious Fires"

The Columbia University Newton manuscript, "Three Mysterious Fires," receives its modern title from the opening words of the document, "The first thing wch must be understood are the three mysterious fires." From the content it is clear that these are the three fires of Johann de Monte-Snyders's *Commentatio de pharmaco catholico*, a text that we considered earlier in this book. The title derives from the three "fires" or substances required for Snyders's process, which we can here call the fusory fire, the sympathetic fire, and the cold, metallic fire. "Three Mysterious Fires" is divided into two parts, an initial English text followed by more complicated material in Latin (for all quotations see appendix three herein, where the text is reproduced and the Latin translated). Even without taking additional evidence into account, it is obvious that Newton did not compose the English text. The assertive and cocksure tone of the author sounds completely different from Newton's own laboratory notes and expositions of difficult authors such as Snyders.[51] Consider the following passage, where this non-Newtonian feature shines forth with particular brilliance. The "He" is Snyders:

> He says that sulphur & Niter are two violent fires but y^t if one knows how to reconcile them nothing but God can hinder us from obteining *<illeg.>* health & riches & that it is the only thing wch he had reserved *<illeg.>* ~~kept secret~~ to himself & to those whom God has elected to it. He does not dissemble, for the truth is that ♄ & niter are the two contrary fires wch being united are able to penetrate any metal whatsoever.[52]

If one excludes texts that Newton was copying verbatim, this passage is utterly unlike anything that one finds in his reading notes, florilegia, or records of his experimentation. In short, we are safe to regard this English text as the product of someone other than Newton. Nor does it resemble anything written by Fatio, either during his exchanges with Newton or even in his own extensive alchemical notes found today in Geneva.[53] When we come to the six densely written Latin paragraphs that follow, however, things become far less simple. A host of purely formal reasons militate against these being the product of Newton's pen alone, as I argue in appendix three. But there is every reason to see them as a collaborative effort shared by Newton, Fatio, and the adept who may have been M. de Tegny. As we will see, the Latin paragraphs represent the culmination of the project that the three alchemical aspirants had been carrying out for some months. The work had grown into a fantastic tapestry of operations, most of which were probably never carried to fruition.

What the English section of "Three Mysterious Fires" describes is an interpretation of Snyders that we already encountered in chapter eleven. It is

---

[51] I am not the first person to notice this fact. In an unpublished paper that he has kindly allowed me to peruse, John Young comes to the same conclusion about the English part of the letter.

[52] Columbia University Library, Three Mysterious Fires MS, 1r.

[53] Fatio's alchemical manuscripts are found in the Bibliothèque de Genève. I have examined MSS Fr. 603, 605, and 609.

in fact quite similar to the accounts that we met with in "The Secret of the Author of the *Metamorphosis of the Planets*" and in Kenelm Digby's report of what Snyders told him. The author equates Snyders's fusory fire and cold, metallic fire with regulus of antimony, and the sympathetic fire with the deflagrating "yellow powder" consisting of niter, sulfur, and salt of tartar. In fact, the yellow powder forms the centerpiece of the English section of "Three Mysterious Fires." Here the powder is explicitly allowed to "fulminate" with "regulus of ♂ joyned wᵗʰ <*illeg.*> gold & all the other metals." The idea behind the interpretation here is that the intense heat and rapid deflagration of the yellow powder applied to the compound regulus "can easily teare its members" and divide the gold into the three chymical principles— sulfur, salt, and mercury. All of this is in agreement with what one might call the "standard" view of Snyders, even though this interpretation does not seem to have been shared by Newton in other, more authentic parts of his chymical corpus. His laboratory notebooks are the most authoritative part of his corpus overall, since they definitely reflect his own experimental work, and CU Add. 3975 was his "master repository" where he recopied earlier experimental material that he viewed as especially important. Yet in CU Add. 3975, Newton explicitly states that the niter and sulfur of Snyders's yellow powder are themselves *Decknamen* for the doves of Diana:

> These doves are first to be enfolded in yᵉ arms of ♀ ~~p 54~~ Secr. Rev. p 54. Snyders calls these sulphur & niter & says they are first to be united & then <*illeg.*> by their fiery spirit ^metal is to be burnt, & this he makes yᵉ key. p 65, 71.[54]

The page references for Snyders given here include the very passage from the *Commentatio* paraphrased by the English section of "Three Mysterious Fires" that I quoted above, where the anonymous commentator describes niter and sulfur as "contrary fires." Hence in Newton's laboratory notebook he is directly contradicting the literal interpretation of Snyders's deflagrating powder given in "Three Mysterious Fires."

Nonetheless, the subsequent Latin paragraphs of "Three Mysterious Fires" build on the interpretation given in the English part that precedes them. Does this mean that Newton had no part in writing them? Not at all. The Latin passages represent Newton's attempt, along with his collaborators, to explore the interpretation given by the anonymous Anglophone chymist to Snyders. They display several features that are characteristic of Newton's idiosyncratic alchemy, such as his symbols for iron ore and copper ore, ♂ and ♀, which I have encountered in no other author. The Latin section also employs the fractional proportions for ingredients that pervade Newton's experimental notebooks, as in the series where he gives the ratios of ingredients as "2, 1, 1, 4/3 or 4/5 or 4/7."[55] But there is further evidence that conclusively reveals the input of Fatio and the friend who may be M. de Tegny in "Three Mysterious Fires." At least two passages in the Latin text

[54]CU Add. 3975, 123v.
[55]Columbia University Library, Three Mysterious Fires MS, 1v.

NICOLAS FATIO DE DUILLIER · 387

contain explicit borrowings or elaborations taken from material found in Fatio's letters of May 4 and May 18. I will present this material in the following discussion as we come to grips with the processes themselves, but first we must briefly revisit the English text to see how it concludes.

After describing the analysis of an antimony-gold alloy into its principles by means of his deflagrating yellow powder, the author of the English section passes to a succession of processes based mainly on Snyders's *Metamorphosis of the Planets*. These are modeled closely on chapter fifteen of the *Metamorphosis*, where Vulcan and Mars reassemble and reanimate Venus after she and Phoebus have been burned to death by the "artificiall firework" (the yellow powder). First the Anglophone author advises to collect all the samples of slag produced by reiterate deflagration of the antimony-gold regulus with the yellow powder and then subject them to the following steps:

> Put them into ^very clear water till all be dissolved Philtrate ye whole. There will pass a very clear water. Put it ~~by its~~ apart & that is the drink of wch Mars cannot drink & into wch throwing some vinegre of white wine he saw that out of water fire did come, & y^t ye water was immediately changed & became a thick essence of a deep red. Then he said, O Venus, my lovely Venus thy beauty belongs to none other but <*illeg.*> me. There will remain some feces in the philtre wch you must well wash & even cause to boyle that there may remain none of the salts; & throw again some Vinegre till nothing more will precipitate, & the feces that remain after you have well dried & ground them, you must must <*sic*> reverberate with ~~the~~ hallf as much flowers of ♁: after wch the salt may be easily extracted even wth the spirit of vinegar. It is better to do it wth ye mercurial spirit.

Here the anonymous interpreter has taken Snyders's directions quite literally. Where Mars refuses to drink the ashes of Venus (and Phoebus) dissolved in well water, this is simply the combined slag of the deflagrations dissolved in "very clear water" and filtered. In Snyders's statement that Vulcan then added white wine and a beautiful, red substance emerged, the interpreter thinks white wine vinegar is meant, whose addition will lead to "a thick essence of a deep red." The reiterated addition of vinegar followed by filtration, boiling, and a subsequent stint in a reverberatory oven with sulfur then allows him to remove the remaining salts from the "essence." At the end of the passage, the interpreter adds that although vinegar will do the job, "It is better to do it wth ye mercurial spirit." This seems at first paradoxical, since he has nowhere mentioned any mercurial spirit in his process up to this point. It is precisely this lacuna that the Latin text then proceeds to fill.

The first paragraph of the Latin text describes an intricate set of operations that are supposed to result in the formation of the very "spirit of mercury" alluded to in the English text as a better alternative to wine vinegar. It is at this point that "Three Mysterious Fires" deviates sharply from the standard interpretation of Snyders such as we met it in the *Secret* and in Digby's work. Here and in the rest of the Latin sections, one encounters a succession of "conjectural processes," to employ George Starkey's useful expression, that border at times on the surreal in their detail and repetition. Newton and his coworkers were hard at work here, and some features of this material, particularly the elaborate descriptions of specialty glassware and other apparatus, as well as the quantity of ingredients and product, come from his collaborators and repeat features found in Fatio's letters. The instructions for making spirit of mercury begin with the instructions to take copper ore and lead ore that have previously been sublimed with corrosive sublimate and to allow them to absorb water from the atmosphere until they are wet. These ingredients may well have undergone a preliminary purification or preparation not mentioned in "Three Mysterious Fires." The first few lines of this passage probably represent a reworking and integration of Newton's Latin instructions found in Fatio's letter of August 1, since they rely on the same materials—mineral *viride æris* (copper ore), lead, though in the form of the metal most characteristic of Newton's laboratory notebooks, namely, lead ore, rather than the ceruse of the August letter, and quicksilver both in its free state and in the form of corrosive sublimate:

> From two parts of the ore of Venus ^and an equal amount <of the ore> of Saturn previously wetted per deliquium (from which mercury sublimate has been elevated in the proportion 3 to 1), let three parts of mercury be sublimed; let the sublimate be elevated from the same mineral, wetted again by deliquescence, once or twice.

One must then sublime quicksilver with this product, and the sublimate produced by that operation must be resublimed with the ores once or twice. These reiterated sublimations in turn yield a product that must itself be sublimed with a regulus made of lead, antimony, copper, iron, and gold. Additional resublimations of that product follow, and then a digestion during which putrefaction occurs. These reiterated sublimations are extremely reminiscent of Newton's experimental notebooks, even though he had largely abandoned the use of corrosive sublimate by the late 1670s. Its reemergence here is probably a contribution of his coworkers. The result of this complicated series of operations will be a red liquid, which the text assures us "is the spirit of mercury."

The rest of the page is occupied with a description of the uses to which this spirit of mercury can be put. Following up on the advice of the English part of the text, the authors say that one should use the spirit of mercury in place of the white wine vinegar mentioned there. The spirit of mercury along with highly rectified spirit of wine (ethanol) allow one to extract a salt, and after the spirit of mercury is then employed to separate out the sulfur of the anatomized gold, the sulfur and salt are reconjoined. Since the spirit of mercury provides the chymical principle "mercury" to the purified salt and sulfur, what we are witnessing is the reassembly of the gold's chymical principles, which were separated during its deflagration in a highly purified state.

The remaining Latin paragraphs, which are found on a separate folio, drop all explicit discussion of Snyders and are closer to the content of the Fatio-Newton letters. The processes described at this point are byzantine in their complexity, but a careful eye can pick out passages that clearly owe a debt to the epistolary exchange of spring and summer 1693. The numbered text begins in the following fashion:

> 1. Take ☿ 3, ♂ 1, ♀ 1.^1 part of ore of ♄ can also be added. Let them be digested for three or four days and the caput mortuum will pass into a mucilage in a cold place and let them be sublimed. The whole weight of the corrosive sublimate will ascend. Let the mucilage be mixed with mercury and a fermentation will occur, and the mercury will be incorporated with the mucilage.

Thus we are advised to sublime iron ore, copper ore, and lead ore with corrosive sublimate, and then allow the product to digest until the caput mortuum becomes a slime or mucilage. The metals involved, mercury, iron, copper, and lead, are all found in the August 1 letter, as is the advice (implicit in Fatio's comments concerning mediation of lead and mercury) to ferment the product with metallic quicksilver. But this is only the beginning of the process as described in "Three Mysterious Fires." It is followed by an array of operations and apparatuses that exceed even Newton's normal capacity for experimental invention. One of the many sublimations from the ores, for example, requires a precise arrangement of specially fabricated equipment. The fermented quicksilver must first be subjected to a preliminary sublimation and then subjected to the following steps:

> Let ^3 parts of the sublimate be put ^with one part of gold purified through antimony and reduced to powder into an alembic with a spherical head and an open tube in the top of the sphere and let it be digested in a totally sealed-up furnace in whose lid there should be three or four holes for regulating the heat ^and for allowing the tube to pass through. The heat should be so great that the matter flow and part of the matter will ascend to the sides of the glass in the form of a colored ring, and finally the ring will be separated from the remaining matter ^(which is already in a hollow form) and it will ascend up into the tube and seal it shut. The heat should be augmented no further lest the ring ascend upwards or the glass be broken. Where the ring has then clogged the tube, a sort of perpetual rain will fall down consisting ^of very small drops onto the matter below.

Here we encounter a more elaborate form of the peculiar self-sealing apparatus described in Fatio's letter of May 4, wherein a mercury compound "shuts the vessel of it self" by "sticking in a liquid form some where in the neck." And where Fatio said on May 4 that as a result of this intended blockage, the material "rains perpetually upon the ☿ that is underneath," "Three Mysterious Fires" also speaks of a "perpetual rain" falling down on the matter below. Despite the idiosyncrasy of this self-sealing glassware, one might conceivably be able to attribute this similarity in language to coincidence were it not for two other features. These additional points emerge from the second numbered section, which follows on the heels of the just quoted passage:

> 2. The sublimate that will have ascended from the two ores after a proper preparation should be put in a retort with a big receiver and with two intervening aludels. *<image of apparatus>* Let it be distilled and the retort will be filled with white fumes. When the fume has ceased and the

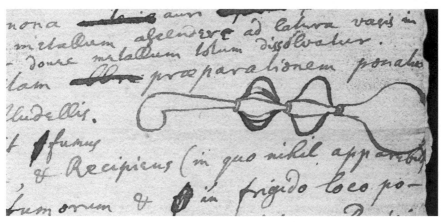

FIGURE 17.3. The receiver described in Newton's "Three Mysterious Fires" with its intermediary aludels. The apparatus is first alluded to in a letter sent by Nicolas Fatio de Duillier to Newton on May 18, 1693. From Newton's "Three Mysterious Fires" manuscript in the Columbia University Library.

retort is clear without whiteness it should be removed from the fire and the receiver (in which nothing will appear) should be skillfully removed and very quickly sealed up so that no fume escapes, and it should be put in a cold place; after three or four days the vapor will start to be condensed and run down the inside of the globe in the form of a fat water; from five pounds of the prepared sublimate you will have four and a half pounds of this water.

The two ores here are presumably the mineral of iron and copper described at the beginning of the paragraph numbered "1." These are distilled through multiple aludels, which Newton has illustrated (figure 17.3) into a large receiver or glass bottle. Oddly, their fume, which is white when it first comes out of the distillation retort, is said to be completely invisible by the time it passes through the two aludels and arrives at the bottle, or as Newton puts it, at "the receiver (in which nothing will appear)." After three or four days, if the vapor has not been allowed to escape, the invisible spirit will condense. Out of five pounds of the prepared sublimate one will receive four and a half pounds of a "fat water." Here again, features from Fatio's May 18 letter have been fleshed out in much fuller form. As Fatio said in the letter:

> Out of 5 ℔ of matter there comes into ye recipient 4 1/2 ℔. Yet in ye recipient You see nothing at all. It is full of vapours, wch must not have the least vent; else the whole would fly away. The recipient being carried in a cold place is there severall days before the vapours settle into the 4 1/2 ℔ of liquor.[56]

The identical quantities of beginning and ending ingredients, along with the fact that the May 18 letter and "Three Mysterious Fires" present the same

[56] Fatio to Newton, May 18, 1693, in Newton, *Corr.*, 3: 268.

description of the strange, invisible spirit that gradually condenses into a water, lead to the ineluctable conclusion that they both describe the same fundamental practice. The difference is that between May 18 and the unspecified date of "Three Mysterious Fires," the seemingly simple seed planted by Fatio's letter has hypertrophied and ramified into a tangled tree. Nor is the process completed yet.

What follows next is a classic separation of the "fat water" produced in this stage into its three chymical principles, followed by their recombination. The fat water must first be "digested in small glass spheres at a temperate heat until it putrefies," whereupon it first turns black, then white. It is then separated from its dregs and digested four or five more times. Following this, a spirit, the mercurial principle, is "abstracted" from the dregs and repeatedly cohobated on them. The spirit gradually turns red, and an oil, the sulfurous component, is separated from it, which is set apart for the time being. The spirit is now used to extract a salt, obviously the saline principle, from the foregoing dregs, which have in the meantime been subjected to a reverberatory fire. Once the salt has been extracted, it is imbibed with the spirit multiple times until they turn into a single, fusible material. At this point, the red oil reenters: the text says to imbibe it with the fusible salt-spirit combination "with interposed digestions of three days." If all goes well, the result will be a coagulated material that melts at a modest temperature and in which gold melts "like ice in warm water."

Incredibly, "Three Mysterious Fires" has still not finished its tortuous sequence of operations. After describing further strange apparatus, the text breaks off in midstream, as though Newton ran out of time in copying the document that lay spread out before him:

> The sublimate is prepared by putting it in a wooden globe whose upper orifice is sealed with a screw and by cooking this vessel in boiling rainwater in another vessel whose lower part is earthen and upper part is glass. Let it be cooked for about eighteen hours, and the sublimate will soften and . . . through the wood.

We encounter here a wooden globe whose upper orifice is sealed with a screw. This must be boiled in rainwater in another vessel "whose lower part is earthen and upper part is glass." All of this appears to be an elaboration of Fatio's observations from May 18 that his friend's mercury should be prepared by "boyling it in a wooden vessel exceeding close for several days with some rain." Although "Three Mysterious Fires" breaks off shortly after this, enough remains to see that the choice of wood apparently derived from the need to have either the softened sublimate or the rainwater penetrate through its pores—an unusual apparatus indeed!

The extremely fine detail of this account should not delude the reader into thinking that anyone actually performed the entire experiment, even if individual parts of it may have been carried out. Unlike Newton's laboratory notebooks, the language here is unremittingly prescriptive, and the putative results of the operations all fall into the future tense. Moreover, the evidence of rapid copying on Newton's part further (see appendix three) undermines any potential claim that "Three Mysterious Fires" was a product of his own laboratory. Instead, the document represents the collaborative effort of four participants: Newton, Fatio, the unnamed French friend who may possibly be identifiable as M. de Tegny, and the English-speaking commentator on Snyders whose analysis provided the initial impetus to the text as a whole.[57] "Three Mysterious Fires" was a blueprint for their research, but not a record of experiments successfully performed. When was this manifesto composed? Since the Latin part of "Three Mysterious Fires" incorporates elements taken from Newton's directions as recapitulated in Fatio's August 1 letter, it is unlikely to have been written before that date. Another feature supporting August 1 as a *terminus post quem* lies in the fact that Snyders does not appear in that letter (or in Fatio's previous correspondence); it was probably Newton who brought the elusive German into the discussion, given his long-standing fascination with Snyders's work. As for a final date by which "Three Mysterious Fires" might have been composed, one can suggest the date of Fatio's letter on Terra Lemnia and Terra Sigillata—November 14, 1693. By then Fatio's French soldier-adept had been supplanted by a new gentleman and his "extreamly stinking" salt. Had the Frenchman been called away to join his regiment in Flanders by that time? We have no way to know, but one suspects that by November the project had run its course.

In addition to casting a powerful beam on Newton's activities with London alchemists in late summer or fall of 1693, "Three Mysterious Fires" helps to illuminate several other features of Newton's life and work. Both the August letter and probably "Three Mysterious Fires" itself are products of a melancholy period in Newton's life, his so-called Black Year.[58] In September 1693, Newton wrote to his friends Samuel Pepys and John Locke, famously accusing them of embroiling him with women and of involving him in unsavory attempts at acquiring patronage. In the letter to Pepys, dated September 13, Newton expresses his desire to terminate their friendship; three days after this he apologizes to Locke for having wished that the amiable philosopher of Oates were dead.[59] Newton's strange behavior led to widespread reports that he was suffering from what the physicist Jean Baptiste Biot would later call a "derangement of the intellect," or nervous breakdown.[60] Indeed, in a celebrated letter to Locke of October 15, Newton describes his own symptoms:

> The last winter by sleeping too often by my fire I got an ill habit of sleeping & a distemper wch this summer has been epidemical put me further out of order, so that when I wrote to you I had not slept an hour a night for a fortnight.[61]

[57] It is possible, of course, that the Anglophone commentator was not physically part of the group behind "Three Mysterious Fires." His participation may only have been secondhand, through a transcript of his work made by Newton.

[58] Manuel, *PIN*, 213–25.

[59] Newton to Pepys, September 13, 1693, and Newton to Locke, September 16, 1693, in Newton, *Corr.*, 3: 279–80.

[60] For Biot's expression "dérangement de l'esprit," see his entry on Newton in *Biographie universelle* (Paris: L. G. Michaud, 1822), 31: 169.

[61] Newton to Locke, October 15, 1693, in Newton, *Corr.*, 3: 284.

Newton's laboratory notebooks do in fact reveal six and a half pages of densely written reports describing experiments carried out in December 1692 and January 1693, so it would be no surprise if he took to falling asleep at his chymical furnaces.[62] This habit soon developed into a generalized insomnia, accompanied by an unspecified "distemper" or indisposition that emerged as an epidemic in the summer, further contributing to his inability to sleep.[63] Although modern attempts have been made to diagnose Newton's ailment as mercury poisoning, he was well aware of the risks associated with ingesting or inhaling the metal and its compounds, and even copied out a passage from Boyle about its danger to "ye genus nervosum (i.e., compages of ye nerves)."[64] It seems odd that he would not have recognized the symptoms of mercury poisoning, such as tremors, excessive salivation, and loosening of teeth, if they had presented themselves.[65] Nonetheless, obsessive chymical research during this period may well have contributed to Newton's period of ill health, as the comment about sleeping by his fire seems to suggest. The fact that we now have good reason to think this intense activity to have continued at least throughout the summer supports some degree of linkage to the "derangement" of September, though probably rather on account of exhaustion than heavy metal poisoning.

The unpublished letters also reveal that Newton continued to interact with Fatio in alchemical matters until late fall of 1693 if not beyond, a fact that was previously unknown. Westfall speaks of an abrupt rupture between the two men probably occurring before the end of September; we now know that this was not the case, as they were still discussing alchemy on November 14.[66] Thus any attempt to link Newton's "derangement" to a precipitous break with Fatio around the time of the letters to Pepys and Locke can no longer be countenanced. In fact, one cannot avoid the suspicion that previous writers on Newton may have overdramatized both his reaction to Fatio and his strange behavior of 1693. A similar charge

[62] CU Add. 3973, 25r–28r.

[63] Newton's use of the term "epidemical" is perplexing. One might be tempted to think that he meant that the aforementioned insomnia had become chronic except that "epidemical" was not synonymous with "chronic" even in the seventeenth century; Newton seems rather to be saying that the distemper from which he suffered was widespread among the populace. A minor epidemic identified in the nineteenth century as influenza struck southern England at the beginning of October 1693, and it is possible that Newton might have been suffering from an early case. See Charles Creighton, *A History of Epidemics in Britain* (Cambridge: Cambridge University Press, 1884), 337–39, 420.

[64] CU Add. 3975, 86r.

[65] The most sustained argument that Newton was suffering from heavy metal poisoning appears in P. E. Spargo and C. A. Pounds, "Newton's 'Derangement of the Intellect': New Light on an Old Problem," *Notes and Records of the Royal Society of London* 34 (1979): 11–32. The two authors performed analyses on three locks of hair that are claimed to have come from Newton. Even if the hair is authentic, however, it may well have been collected on his death in 1727, as the authors admit on p. 23. This means that some thirty-four years would have elapsed since his experiments of 1693, making the likelihood that the hair samples reflect Newton's physical constitution at the time of his "derangement" unreliable at best. See also the cautionary material on historical hair analysis in Leonard J. Goldwater, *Mercury: A History of Quicksilver* (Baltimore: Work Press, 1972), 143–44.

[66] Westfall, *NAR*, 539. Westfall states that Fatio had probably not communicated with Newton for more than a year as of September 29, 1694, despite Fatio's claim made to Christiaan Huygens on that date that it had been "more than seven months" since he had had news from Newton.

of hyperbole can also be leveled at Westfall's treatment of *Praxis*, the late alchemical treatise in which Newton actually refers to Fatio's letter of May 4, where the Genevois spoke of the wonderful powder that could purify quicksilver merely by being shaken with it in a bottle. Let us now pass to a discussion of that text, which has been seen by some as the apotheosis of Newton's alchemical career.[67]

---

[67] Westfall, *NAR*, 529–33; Dobbs, *JFG*, 17, 71, 171–72, and 293–305.

EIGHTEEN

Praxis

Previous writers on Newton have described the manuscript *Praxis* as his most important alchemical text, possibly related to the nervous collapse of his "Black Year." This claim is particularly pronounced in the work of Westfall, who sees the treatise as marking the climactic culmination of Newton's long alchemical career. He argues that Newton composed the text in the spring and summer of 1693, "a time of great emotional stress," and that Newton's temporary loss of judgment may have led to the "extravagant claim" of chrysopoetic success found in the document. For Westfall, the ultimate failure of the processes described in the treatise led to a permanent disillusionment with alchemy on Newton's part. Accordingly, Newton's embarkation for London in 1696 for a career in the highly public life of the Royal Mint was essentially the end of his involvement with alchemy, despite the existence of a few alchemical notes written by him in that period.[1] There are, however, a number of severe problems with Westfall's conclusions. First, the evidence that he proposes for the date at which *Praxis* was composed, spring and summer of 1693, really consists only of a *terminus post quem* based on the reference to Fatio found in the text. In short, *Praxis* could have been composed at a considerably later date than Westfall thinks, perhaps even after Newton's removal to London. Second, we now know from the careful work of Karin Figala and Ulrich Petzoldt that Newton did in fact continue his alchemical research in London after becoming warden of the Mint. He even took on another collaborator, the interesting immigrant from Rotterdam, William Yworth. As we will see, *Praxis* displays characteristics that suggest both Newton's work with Fatio and his interaction with Yworth; it is therefore possible that the text was composed even as late as the time of this second collaboration. Finally, the text of *Praxis* is by no means as extravagant as Westfall makes it out to be. The statements that he takes to be claims of alchemical success are actually conjectural processes rather than affirmations of experiments that were carried out. If the reader approaches *Praxis* with a working knowledge of the other Newtonian alchemical notes

---

[1] Westfall, *NAR*, 530–31.

and florilegia that we have analyzed up to this point, the text appears no less rational than they. Indeed, a compelling way to think of *Praxis* is as a sort of continuation to the operations described in Keynes 58.

The text of *Praxis* consists of five chapters and is found in Babson 420, which contains two successive drafts of chapters four and five (folios 6v–10v make up the earlier version, 11r–15v the later). The tract receives its title from a heading that Newton supplied at the top of the first chapter. Only the second draft allocates a separate heading to the material in chapter five, and this provides an important clue to the structure of the document. Like the treatise as a whole, chapter five also bears the title "Praxis," suggesting that Newton originally thought of the entire text as being practical in nature, but then changed his mind and decided to save this designation for chapter five alone. As we will see, this decision was a good one, for in reality the first four chapters do not present a linear alchemical practice but rather identify different materials to be used in the overarching set of operations described in chapter five. The structure of the text with the rewritten version of chapters four and five is as follows:

> 3r: Cap. 1. De materijs ~~prima et ultima~~ ~~et Sulphure~~ spermaticis.
> (Chapter 1. On the spermatic materials.)
> 3v: Chap. 2. De materia prima. (Chapter 2. On the prime matter.)
> 5r: Chap. 3. De sulphure Phorum. (Chapter 3. On the sulfur of the philosophers.)
> 6v: Cap. 4. De agente primo (Chapter 4. On the first agent.)
> 11v: Chap. 5. Praxis (Chapter 5. Praxis.)

Despite these seemingly clear divisions, a preliminary reading of the text quickly leads one to sympathize with Westfall's claim that it reflects a disordered state of mind. It is not easy, at least initially, to make out the subjects described allusively in each of the first four chapters. Their disjointed snippets and quotations represent Newton's florilegium style at its densest and least approachable. But this is not the product of madness, however temporary. Rather, it is Newton's way of sifting through his sources and reassembling the disparate parts of a great puzzle distributed piecemeal among the diverse sons of art. As we will now see, the very materials that figured in Newton's laboratory notebooks, in his interpretations of Snyders (especially Keynes 58), and in his instructions to Fatio can be found in *Praxis*, though locating them may require a bit of digging.

Praxis: Chapter One, The Spermatic Materials

Chapter one of *Praxis* is perhaps the easiest of the four to decipher. In a word, it is devoted to the two serpents entwined on the caduceus of Mercury. Newton begins the chapter with a summary of Nicolas Flamel's comments concerning the images carved on his charnel house in the Cemetery of the Holy Innocents. Flamel equated the dragons guarding the golden apples of the Hesperides with the two serpents sent by Juno to strangle the

FIGURE 18.1. Plate from the *Livre de Nicolas Flamel* combining motifs from the putative *Book of Abraham the Jew* and the "charnel" house of Nicolas and his wife Perenelle. Reproduced from the 1672 edition of the *Bibliothèque des philosophes chymiques*.

infant Hercules, along with a winged and a wingless dragon painted on a sable field. All of these mythical creatures represent the two snakes wound about the caduceus, and these in turn are "the two male and female sperms of the metals" (*spermata duo metallorum masculinum et fæmininum*). It may be significant that Newton had by now acquired the French edition of Flamel found in the *Bibliothèque des philosophes chymiques*, for unlike the English translation that Newton owned, the Francophone text was accompanied by illustrations from the mysterious codex of Abraham supposedly written on rinds of "tender yong trees" (figure 18.1).[2] Of the seven illuminations representing Abraham's "hieroglyphics," five contain serpents or dragons, including the two snakes on the caduceus. Newton follows his summary of Flamel

[2]Nicolas Flamel, *Nicholas Flammel, his exposition of the hieroglyphicall figures which he caused to bee painted vpon an arch in St. Innocents Church-yard, in Paris* (London: Thomas Walkley, 1624), 6.

with additional comments taken from Philalethes, Maier, d'Espagnet, Grasseus, and several medieval sources, most of which also concern serpents or dragons. One seeming exception, though only an apparent one, occurs when Newton mentions the "King and Queen who putrefy together with the water-bearer at the beginning of the work."[3] A glance at several other manuscripts of the period reveals, however, that Newton equated the king and queen with the two serpents and the water bearer with the rod around which they were "writhen," a fact that he also implies in *Praxis* by adding in parentheses "caduceus."[4]

What does all of this mean, and why did Newton give the chapter the heading "On the Spermatic Materials"? We know from Keynes 58 that by his maturity Newton had determined the two serpents on the caduceus to represent the "double vitriol" or perhaps twin vitriols of copper and iron, the salts that he refers to there in a sort of shorthand as "Io⊖ + Co⊖." Just as these were the starting materials of the complicated series of operations described in Keynes 58, so they form the subject of the first chapter in *Praxis*. But why does Newton call them "spermatic materials"? Is there some reason other than the obvious fact that Flamel speaks of the two serpents as "two sperms" (*spermata duo*)? In fact, Newton did have a reason, and it is no surprise that the answer to our question can be found in the tradition of Basilius Valentinus, the muse of Snyders. The preface of the *Last Will and Testament* of Basilius announces that it will reveal "the first sperm" of metals and minerals. Later in the text, Basilius informs his reader that iron and copper vitriols contain the essence or *primum ens* of gold in an immature state.[5] It was not difficult, then, for Newton to see a primordial role for these "spermatic" materials, as they contained the seed of gold in a particularly rich, if undeveloped, condition. Indeed, in another manuscript Newton excerpted Basilius's claim that vitriol contains immature gold, along with the practical advice that the alchemist can work more cheaply and effectively with vitriol than with gold itself:

> But remember that these mineral spirits are found effectual in other metals also [besides Gold viz: in ♂ & ♂] & are found effectual in one mineral ⊕ from whence with more ease & less charges it may be had.[6]

But in order to operate on the latent gold in vitriol, Basilius then tells us that other materials are required, and these additional substances form the subject of the remaining three introductory chapters of *Praxis*. As we will soon see, one of these materials is the "spirit of mercury" that Newton had already decided at the beginning of his alchemical career derived from antimony.[7]

[3] Babson 420, 3r: "Rex et Regina qui una cum Aquario ~~Materiæ~~ (seu Caduceo)^initio operis putrefiunt."

[4] See Keynes 21, 16v: "The same thing is described in the figures of Abraham ye Iew where Mercury strikes on his helmet wth his rod & saturn wth wings displayed ~~comes~~ & ^an hour glass on his head comes running & flying against him as if he would cut off his leggs. ffirst ye serpents are twisted about ye rod by fermentation; ffor these three are ye King Queen & Water bearer or ye fire ye liquor of ye vegetable Saturnia & ye bond of ~~whe ☿ in Philaletha.~~ in Philaletha." See also Keynes 53, 2v, where the identification between the rod of the caduceus and the water bearer is again made.

[5] Basilius Valentinus, *Last Will and Testament* (London: T. Davis, 1658), pages A[5v] and 110.

[6] British Library, MS Add. 44888, 3v–5r.

[7] See chapter five of the present book, and Var. 259.11.6v.

Praxis: Chapter Two, The Prime Matter

The second chapter of *Praxis* is titled "On the Prime Matter," which could be taken in several ways. On the one hand, it could refer to the *primum ens* or specific matter into which a metal was thought to be capable of reduction by means of a powerful menstruum, or it could on the other hand refer to the "first matter" out of which metals and minerals in general were thought to be made. An examination of the various motifs collected by Newton here reveals that the second meaning is the correct one. Several powerful clues point to the conclusion that Newton was thinking of crude antimony, that is, the mineral stibnite, which Philalethes had explicitly equated with the first matter of the metals.[8] Early in the chapter, Newton recapitulates a passage from *Ripley Reviv'd* that gives an obvious description of the antimony ore:

> It's metalline but void of metallic ♄, fusible, fugitive, in no ways malleable, in colour sable wᵗʰ intermixed argent glittering branches composed of a pure ☿ & feculent ♄ the Green Lyon wᶜʰ easily destroys iron & devours also yᵉ companions of Cadmus.[9]

Of course thanks to Philalethes's occasional use of graduated iteration, the passage could in principle refer both to stibnite and to some material appearing at a later stage in the alchemical process. That Newton took it here simply to refer to crude antimony receives further support from his subsequent comments. Paraphrasing the anonymous *Instructio patris ad filium de arbore solari* found in the *Theatrum chemicum*, he says that it is necessary to extract a spirit "from the center of our Adamic, solar earth" with a white water.[10] Although this passage is quite opaque as it stands, Newton glosses it in a passage that is mostly canceled: "i.e., ~~regulus ♁ⁱˢ and of the metals from the mineral~~ from the mineral of ♂" (i.e., ~~Reg ♁ⁱˢ & metallorum~~ ~~ex miner‹illeg.›~~ ex minera ♂tis). The passage is still obscure, but one can see that Newton was thinking of the interaction between stibnite and iron that would lead to the reduction of metallic antimony, namely, regulus martis. This is confirmed by a parallel passage in another manuscript, where Newton says, "It is a virgin earth upon which the sun ∧[i.e., fused with Mars] has never launched its rays" (*Est terra virginea super quam sol* ∧[‹illeg.› i.e., marte fusus‹illeg.›] *radios suos nunquā lancinavit*)."[11] This gloss reveals without equivocation that the sun is the hidden sulfur within Mars or iron and the virgin earth is either stibnite or the regulus "hidden" within it. When stibnite and iron are combined at a high temperature in a crucible, the result is regulus martis, but in chapter two of *Praxis* Newton is focusing on the stibnite in its crude state rather than metallic antimony.

---

[8] This claim at least occurs in some versions of Philalethes's *Sir George Ripley's Epistle to King Edward Unfolded*. See Anonymous, *Chymical, Medicinal, and Chyrurgical Addresses: Made to Samuel Hartlib, Esquire* (London: Giles Calvert, 1655), 25.

[9] Babson 420, 3v.

[10] Anonymous, *Instructio Patris ad filium de Arbore Solari*, in *Theatrum chemicum* (Strasbourg: Heirs of Eberhard Zetzner, 1661), 6: 163–94, see 174.

[11] Keynes 21, 17v.

The white water used to extract the spirit from the center of the solar earth could therefore be interpreted as the aqua regia employed in making spirit or liquor of antimony.

Interestingly, Newton had still not finished with his consideration of stibnite in chapter two of *Praxis*. His next significant comments return again to the *Last Will and Testament* of Basilius. Just as he had done in his first notes on Basilius from the late 1660s, Newton recounts a process for subliming a product from previously prepared stibnite mixed with ground bole armeniac or tile meal.[12] The goal of this operation, as he had already deduced while a student at Trinity College, was to extract the "spirit of mercury" from crude antimony that had been previously prepared by a process that is not described in Basilius's text. Although *Praxis* is little more forthcoming on the preliminary preparation of the stibnite, chapter two gives a plenary description of the next steps in the process:

> ∧These Parables shew sufficiently how ye earth is to be prepared by turning it first into water & then into earth. But ∧after tis thus prepared it must be sublimed as Basil Valentine thus teaches. Saturn will put into your hand a deep glittering mineral wch in his mine is grown of ye first matter of all metalls. If this minera after its preparation wch he will shew unto thee is set in a strong sublimation mixed wth three parts of bole or tyle meal, then riseth to ye highest mount a noble sublimate like feather or alumen plumosum, wch in due time dissolveth into a strong & effectual water.[13]

The deep, glittering mineral that Saturn supplies is of course stibnite, which is often of a shiny black or silvery color. Such crude antimony was the "son" or sometimes the "daughter" of Saturn, so it was only appropriate that the titan should be the one to hand it over to the chymist. *Praxis* then goes on to say that the "strong & effectual water" into which the sublimate dissolves is the spirit of mercury, and that this spirit can extract the soul of common gold. Newton concludes chapter two by using the Basilian process to decode further alchemical *Decknamen*. Because the feathery deposit is the product of sublimation, which drives it to the top of the aludel or sublimatory, it "hangs over o^r heads," to use the words of the *Instructio de arbore solari*. Similarly, the sublimate is a "virgin earth" and also the "virginal, foliated earth" (*terra virginea foliata*), whose chaste status stems from the fact that the stibnite has never been subjected to the reducing action of iron, which would yield the regulus of antimony. Newton's idea here hearkens back to the earlier section of chapter two where he interpreted the *Instructio de arbore solari*'s instructions to take an earth on which the sun had never launched its rays to be a reference to crude antimony that had not yet felt the effect of intercourse with Mars (iron). Thus chapter two concludes with an unequivocal association between the prime matter of the chapter heading and stibnite.

---

[12] For Newton's recapitulation of the process in the 1660s, see Var. 259.11.8r and my discussion in chapter five.

[13] Babson 420, 4v–5r.

Given the many clues that chapter two of *Praxis* presents for the association of the prime matter and crude antimony, it comes as no surprise that the subject of the next chapter should be the "the sulfur of the philosophers," namely, the sulfurous principle in iron with which stibnite can be induced to release its reguline component (metallic antimony). Newton makes this identification quite clearly by deciphering several passages from Philalethes where the "American philosopher" speaks of the *chalybs* (steel) and its action on the *magnes* (stibnite). Thus an early passage in chapter three refers to,

> [t]hat Chalybs wch o^r Magnet chiefly attracts & swallows up in fusion to make ye starry Reg. of ♂ [Secr. Rev. p. 5, 7 16, 28. Comment. on Ripl. Pref. p. 7, 31. Marrow of Alk. p.17.].[14]

The explicit reference to the production of the star regulus means that the chalybs here can only be iron, or rather the putative sulfurous component within iron that is "eaten" by the magnes (stibnite), resulting in the reduction of metallic antimony. Newton had long since abandoned his early and idiosyncratic reading of the chalybs as lead that we found in Keynes 19, his "Summaries from the *New Chymical Light* which Pertain to Practice," as discussed in chapter nine of the present book. His interpretation of magnes and chalybs was now close to the obvious sense of the Philalethan text. Collecting further *Decknamen* from Philalethes, Newton says the chalybs is "our Cadmus, the God of war, Mars," and the sulfur hidden in the belly of the Ram, Aries. Since Aries was one of the two zodiacal houses in which the planet Mars was traditionally thought to exercise its maximum power, all of these *Decknamen* were intended by Philalethes to conjure up the metal iron, as Newton correctly understood.

But at the point where chapter three descends to practice, the obvious part quickly terminates. Beginning a new paragraph with the caveat, "Now this Sulphur must be also prepared," Newton launches into a somewhat obscure series of operations. Despite the bewildering language, we can in fact decipher the first series of processes with the help of Newton's glosses, which as usual are bracketed:

> For o^r crude sperm flows from a trinity of $^{\wedge immature}$ substances, in one essence of wch two <*illeg.*> (♂ & ♄) are extracted $^{\wedge out\ of\ ye\ earth\ of\ their\ nativity}$ by y^e third (☿) & then become a pure milky virgin-like Nature drawn from y^e menstruum of o^r sordid whore.[15]

Newton's comments reveal that he interpreted these words as an allusive instruction to combine the "immature substances" or ores of iron, lead, and antimony, but for the moment he does not reveal the method by which this combination is effected. He then goes on to identify the product, for which he uses the Philalethan *Deckname* "y^e menstruum of o^r sordid whore," with

---

[14] Babson 420, 5v.
[15] Babson 420, 5v.

Figure 18.2. Winged Saturn attacking Mercury with his scythe. Detail from the plate in the *Livre de Nicolas Flamel* found in the 1672 edition of the *Bibliothèque des philosophes chymiques*.

another alchemical cover name, the mythical *echeneis* fish or remora that could supposedly arrest the passage of ships at sea. In order to "extract" this fish from the ingredients mentioned, which is also the matter from which the philosophers' stone must be made, Newton says:

> To find this matter of o^{r} stone you must draw y^{e} moon [sp^{t} of ♁] from y^{e} firmament [<*illeg.*> in distilling] & bring it from heaven up y^{e} earth [of ♂] & turn it into water & then into earth [Instruct. de arb. Solar $^{\wedge c.\ 3.}$ p. 172, 173.

This tells us a great deal. We know what spirit of antimony meant to Newton from other sources; this was the same material that he described in Fatio's letter of August 1, 1693, and the same substance that appears countless times in Newton's laboratory notebooks, namely, the solution produced by reacting stibnite with aqua regia. Newton is simply telling us here that this spirit or liquor of antimony should be used to dissolve iron ore, then distilled and solidified.[16] The product will be a salt of iron, which Newton equates with the scythe of Saturn as depicted in the hieroglyphs of Abraham the Jew from Flamel: "This is the sharp spere of Mars & sith of Saturn." In the *Bibliothèque des philosophes chymiques* Saturn is pictured with an hourglass on his head and scythe in hand, ready to chop off Mercury's winged feet and thereby fix him to the ground, or render him nonvolatile (figure 18.2). Newton adds that this "metalliq fixt salt" of iron will be the "sharp spere of Mars" with which the god "gives ☿ work enough to do," again an allusion to

[16] Newton also describes a way of making an iron salt by subliming iron ore with his sophic sal ammoniac. See CU Add. 3973, 35r: "Iron ore ground fine 40^{gr} ✳ $^{\wedge well\ dried}$ 40^{gr} sublimed together left 36 1/2^{gr} below on wch rain water being poured, extracted 19^{gr} of salt of ♂ & there remained 18^{gr} of iron ore."

fixing mercury. Here Newton is combining a motif from Snyders with the scythe-bearing Saturn of Flamel. The spear of Mars appears in a passage from Snyders's *Metamorphosis* much beloved by Newton:

> Although the steely captain with his Spere gives Mercury work enough to do, yet can he not wholly over-power him if the old Saturn come not ^in to his help.[17]

In Newton's interpretation, Snyders is saying that the ferrous salt—the "Spere"—requires the aid of Saturn in order to achieve its aim of fixing Mercury. Thus Newton has not forgotten the role of Saturn or lead, which he already mentioned in his interpretation of the menstrual blood of the sordid whore. But since chapter three mainly concerns the philosophical sulfur, as its heading announced, he chooses to restrict his discussion for the remainder of the chapter to the sulfurous principle within iron. He therefore closes out chapter three by returning to a collection of *Decknamen* for the martial sulfur, equating it with Flamel's wingless dragon and the male serpent on the caduceus. There is no contradiction with chapter one, where the two serpents on the caduceus were either the vitriols of iron and copper or a single mixed vitriol of both, since in chapter three Newton is speaking of the hidden sulfur within iron, which according to his interpretation would also be found in the vitriol made from the metal.

Praxis: Chapter Four, The First Agent

So far we have chapters one through three of *Praxis* first describing the two vitriols making up the serpents on the caduceus, then crude antimony in various roles including its part in making "spirit of mercury," and finally the iron or its sulfurous component employed both in the reduction of antimony from its ore and in making a ferrous salt with liquor or spirit of antimony. We should recall that the serpents, antimony, and iron (within the male serpent) figured prominently in the first stages described by Keynes 58, where Newton provides directions for making a "dry water." It is no surprise, then, that the subject of chapter four proceeds to describe the central rod of Mercury's caduceus, since that followed the serpents and antimonial menstruum in Keynes 58 as well. There Newton presented a recipe that employed multiple imbibitions of "lead with its menstrue" and the vitriols, or a "double vitrol," of iron and copper. Similarly, the simplified instructions that Newton provided for Fatio's Francophone friend passed from a discussion of vitriol made from a copper compound by dissolving it first in liquor of antimony to directions for making sugar of lead and heating it with iron and "our earth" (probably stibnite). Chapter four of *Praxis* builds on similar operations after announcing that its subject is the rod in the following words:

---

[17]Cushing, 17v.

FIGURE 18.3. Detail from Newton's *Praxis* showing the derivation of the standard alchemical symbol for mercury from the winged caduceus of the god Hermes. As found in Huntington Library, MS Babson 420.

The rod of Mercury reconciles the two serpents ^& makes them stick toe it [Maier &
therefore is yᵉ medium of joyning their tinctures whence its called yᵉ bond
of Mercury [Secr. Rev. cap. 2][18]

Newton then provides an image of the original form of the caduceus as he
imagines it. The crossbar on the mercury symbol becomes two wings attached to the rod, connoting the volatility of the material hidden under
the name "caduceus," and the upper part of the glyph appears as the two
serpents (figure 18.3). Newton further underscores this volatility by adding
that the rod is a "fluxible Menstruum" and a "saline spirit." Taking a cue from
Michael Maier, he equates Mercury's winged, snake-bearing staff with the
grapevine that Dionysus used to kill the two-headed monster Amphisbaena
in Greek mythology.[19] At the same time, the rod of the caduceus is identical
with the golden bough of Virgil's *Aeneid* on which the two doves of Venus
landed in order to show Aeneas the way to Hades. Although Newton provides further synonyms for the rod, the golden bough with its doves leads
him to an allusive discussion in Philalethan language of how the substance
should be made. Unfortunately, this section lacks any parenthetical decoding on Newton's part, which make its meaning uncertain. If we consider the
canceled first draft of chapter four, however, a comprehensive attempt by
Newton to decode Flamel's hieroglyphics into practice emerges, and this will
allow us to see exactly how he viewed the "first agent."

Draft one of chapter four contains a page and a half where Newton provides bracketed interpretations of Abraham the Jew's hieroglyphics. Newton
thinks that these represent the making of the philosophers' stone, but in reverse order. Ever jealous of their secrets, the adepts were under no compunc-
tion to follow a plain order of exposition. At some point Newton canceled
these passages by drawing large "X's" across the pages, but this does not mean
that the deleted sections are irrelevant to an understanding of his exposition.
To the contrary, Newton's commentary on the fourth image in the *Bibliothèque des philosophes chymiques*, which shows the Massacre of the Innocents

[18] Babson 420, 6v.
[19] Babson 420, 11v. See Michael Maier, *Septimana philosophica* (Frankfurt: Lucas Jennis, 1620), 186. See
also Newton's synopsis of this passage from the *Septimana philosophica* in Keynes 32, 43v.

by King Herod, explains why he chose the title "On the First Agent" for this chapter. Speaking of the rod of Hermes, Newton says in a deleted passage, "^tis justly called by Flammel the first agent." He then provides an explanation of this material based on the image of King Herod and the Innocents:

> Now the making of this agent is thus described in the third ~~last~~ figuring of Abraham the Jew ~~in an inverted order. In ye 4th figure is~~ <*illeg.*> a King [o^r ♄] wth a great fauchion [♂] ~~who made to be killed~~ by fire wch is ye artificial death of metalls] in his presence [i.e., in ye crucible where he is] by some soldiers [i.e., firebrands or ignitions] a great multitude of little infants [particles of poudered Oars infants] whose mothers [particles of ye first matter] wept [by fusion] at ye feet of ye unpitiful soldiers: the blood of wch infants [i.e. ᵃ ♃ of metalls] was afterwards [in ye work of ye 3d & 2d ^& 1st ffigures] gathered up ^[sublimed] by other soldiers & put into a great vessel [ye r<*illeg.*> Caduceus ^& cold saturnal fire] wherein ye Sun and Moon [ye two serpents ^or Dragons] came to bath themselves.[20]

For us the most important thing about this passage is its beginning. There Newton clearly says that Abraham's hieroglyph describes the making of the first agent. Then, in parenthetical insertions, Newton equates the King, Herod, with "o^r ♄" and his great sword or "fauchion" with "♂." What could "our Saturn" refer to? The association with Saturn and Mars, or lead and iron, suggests that Newton is interpreting Herod and his sword as another expression for Saturn and his scythe, the latter being also the steely Captain's spear in chapter three that he had to employ with the aid of Saturn in order to fix mercury. We should recall that in chapter three of *Praxis* Newton explicitly gave the scythe or spear the meaning of a salt of iron extracted with the help of spirit or liquor of antimony, the material that we know well from his laboratory notebooks. Newton also said in chapter three that this ferrous salt needed to be drawn from iron ore rather than from the refined metal, and that an extract of lead ore should also be made, again by using antimony. This extract, I propose, is "our Saturn" and King Herod in Newton's interpretation of Flamel.

The possibility that "our Saturn" in the deleted draft of chapter four refers to a salt of lead extracted from its ore by liquor of antimony or possibly by Newton's antimonial "sophic sal ammoniac" receives support from a variety of sources. Perhaps most striking are the many instances in CU Add. 3973 and 3975 where Newton describes experiments for making a sweet salt of lead by either subliming the ore with sophic sal ammoniac and then extracting the salt by dissolution and crystallization, or by dissolving the ore directly in liquor of antimony. The product, he tells us, will be fusible, capable of being sublimed, and metallic or amber-like in appearance.[21] It would be rash to jump to a modern chemical identification of this material, although

---

[20] Babson 420, 7r–7v.

[21] For crystallization and recrystallization of Newton's proprietary "sugar of lead," see CU Add. 3973, 30r and 40r. The first of these experiments employs vitriol as well as sophic sal ammoniac when subliming the lead ore. For the fusibility and metallic or amber-like appearance of the salt, see 30r and 43v. In the second of these experiments, Newton compares a "spirit of lead ore drawn with Venus" (Sp^t of Le o. drawn w^th ♀) to

nitrates and chlorides of lead can indeed taste sweet. A further reason for identifying "our Saturn" with Newton's sweet salt of lead emerges from Fatio's August 1693 letter. There a "saccharum" or sweet compound was extracted from ceruse by sublimation with antimony regulus and vulgar sal ammoniac. As we have noted before, Newton seems to have been providing Fatio and his chymical friend with *succedanea*—simplified substitutes for the materials that he used in his laboratory. Thus Newton substituted ceruse for lead ore and a simple form of his sophic sal ammoniac for the version that he made from a white antimony calx in his experimental notebooks. In the recipe given to Fatio, Newton even suggests that ordinary sugar of lead (lead acetate) can be substituted for the version extracted from ceruse with the regulus and sal ammoniac, though he clearly prefers the latter form of saccharum.

That Newton meant his proprietary version of "sugar of lead" when he referred to "our Saturn" in chapter four seems fairly well established. But what was the role of Herod's "fauchion," the curved steel sword that Newton associated with Mars, or rather with a salt of iron? Returning again to Fatio's letter, we will recall that Newton advised the Francophone adept to add the sugar of lead to "two or three parts of our earth as well as one part of iron" and to let the material boil together in a "closed place." This is quite reminiscent of an experiment found in CU Add. 3973 and performed by Newton in February 1695/96 or possibly later. This section of Newton's experimental notes consists mainly of experiments with "oʳ ✳ freed from ♂," in other words, sophic sal ammoniac that had been liberated of its excess antimony. Among these experiments, however, one finds several that resonate with chapter four of *Praxis*, such as one that begins by melting stibnite with iron ore and observing the "fermentation" that occurs. This bubbling reaction between iron ore and antimony ore was a perennial interest of Newton's, which can be traced back to the winter of 1692/93.[22] In the process of describing this fermentation, Newton mentions that he also sublimed the mixture of iron ore and stibnite and then precipitated the sublimate in water. The ferrous precipitate was subsequently combined with more stibnite and heated with the addition of "oʳ saccharū ♄ⁿⁱ," in other words Newton's proprietary sugar of lead made with lead ore acted on by liquor of antimony. The product was a fluid, volatile salt, and one may suppose that the instructions for Fatio's friend aimed at a similar result.[23]

Assuming then that "our Saturn" in Newton's interpretation of Flamel meant his proprietary sugar of lead, how was this supposed to lead to the central rod of the caduceus? At this point, *Praxis* diverges from Keynes

---

his ordinary, though proprietary, sugar of lead. In none of these cases is he making the usual "vulgar" sugar of lead (lead acetate).

[22] CU Add. 3973, 27r–27v. The reaction could arise from the release of carbon dioxide if siderite, iron carbonate, was the ore involved. On the other hand, when Newton tried the experiment with stibnite that had been refined by being "sublimed precipitated edulcorated & melted in a glass vial luted," he got no swelling or bubbling. This suggests that the gas was coming from the crude antimony ore. Since Newton often complains of spar being mixed with his crude antimony, one wonders if the spar—perhaps calcite—might not have been the source of the bubbling.

[23] CU Add. 3973, 41r.

FIGURE 18.4. Flamel's "rose-tree" or bush growing up the side of a hollow oak tree at whose foot is "a most white water." Detail from the plate in the *Livre de Nicolas Flamel* found in the 1672 edition of the *Bibliothèque des philosophes chymiques*.

58. Keynes 58 advised that the two serpents and "lead with its menstrue" be digested together; this would lead to the formation of a black powder, which could then be sublimed to produce the two doves of Diana. A note in Keynes 58 shows that Newton identified this process with "yᵉ P̄hick calcination," namely, the first gate described in *Ripley Reviv'd*, which was supposed to result in a state of "compleat Rottenness." This stage does not appear in *Praxis* until chapter five of the text, where the two serpents and the rod are finally combined and the black powder produced.[24] Before that goal can be reached, the rod must still be made, and chapter four has more to tell us on how this is done. Newton explains the latter stages of the rod's production by invoking the third of Flamel's hieroglyphs, which describes a "rose-tree" or bush growing up the side of a hollow oak tree at whose foot is "a most white water" (figure 18.4).[25] The first draft of *Praxis* chapter four provides a detailed exposition of this image, terminating with the claim that it teaches the making of the rod:

> ~~In yᵉ 3d is~~ A fair rose tree [sharp ^pricking salt, ✳] flowred [i.e. sublimed into flowers] in yᵉ midst of a sweet Garden [of yᵉ Hesperides] climes<*illeg.*> up [by sublimation] against an hollow oak [yᵉ net of Vulcan, to elevate it] at yᵉ foot of wᶜʰ [oak] boiled [by heat] a fountain of most white water [oʳ sea] wᶜʰ ran headlong ~~into~~ down into yᵉ depths [of oʳ earth being dried upon it] notwithstanding that it first past yᵉ hands of infinite people [among yᵉ Chymists] who digged [by dissolving waters] in yᵉ [philosophic] earth but because they were blind none of them knew it but here & there one who considered yᵉ weight [it being invisible in yᵉ solvent]. And this is yᵉ making of yᵉ rod or first fire.[26]

[24] Babson 420, 11v.
[25] Flamel, *Nicholas Flammel, his exposition of the hieroglyphicall figures*, 13.
[26] Babson 420, 7v.

To Newton, the rose bush is a "sharp, pricking salt" for which he uses his habitual symbol for the sophic sal ammoniac, a "✳." This is followed by even more prosaic decoding; the flowers of the rose bush simply indicate that the salt must be elevated by heat into "flowers," the normal term for a sublimate (as in the expression "flowers of sulfur"). The fact that the rose bush grows up next to a hollow oak tree means that the sublimation must be made with the net, the alloy of copper and martial regulus that Newton had long equated with the oak in Ovid's story that Cadmus killed a dragon and pinned it to an oak tree with his spear, as related by Philalethes. Hence Flamel is telling the chymist to sublime the net or oak alloy with sophic sal ammoniac. The final feature in the hieroglyph is the white water that runs from the roots of the tree through the hands of bewildered chymists whose ignorance blinds them to its true use. In language reminiscent of *Praxis's* chapter two, this water, which is "our sea," must be heated and dried on "our earth."[27] The water is heavy because of the material dissolved in the solvent and hence invisible. If my interpretation up to this point is correct, Newton must be referring to the salt of lead, or else a lead-iron salt, dissolved in liquor of antimony. Hence the meaning of the entire passage can be compressed into a single sentence: the net must be sublimed with sophic sal ammoniac and the sublimate dissolved in hot salt of lead (or lead-iron) with liquor of antimony. The product, Newton says, will be the rod.

Conclusion to *Praxis*: The Return of Fatio

A comparison of the first four chapters of *Praxis* with Keynes 58, Fatio's August 1, 1693, letter, and Newton's laboratory notebooks has revealed with some degree of certainty that the subjects of these chapters are the two serpents (copper and iron vitriol made with liquor of antimony), stibnite and its products, the sulfur of iron, and the central rod of the caduceus. Newton describes these materials allusively, and their production is given in a nonlinear fashion, making their identification and extraction from *Praxis* something on the order of an archaeological dig. But the difficulties of interpretation pale by comparison to the problems presented in chapter five of the text, which is itself titled "Praxis" in the second draft. Here Newton presents two parallel sets of processes, both depending on the ingredients presented in the first four chapters but relying on different modes of operation that he refers to as "the dry way" and "the wet way." This division has little to do with the traditional chemical (and chymical) distinction between processes involving heat alone and those that require dissolution in a solvent or acid, for Newton's two paths each make use of processes involving fusion and solution.[28] For him, the two "ways" or paths were the

---

[27] Chapter two pronounces that "ye earth is to be prepared by turning it first into water & then into earth" and adds the following marginal gloss: "† Our Antimony sait Maier, is yᵉ King wᶜʰ cryes in the Sea Qui me liberabit ex aquis & in siccum [denuo] reducet, Ego hunc divitijs beabo. Maier." See Babson 420, 4v.

[28] Robert C. Kedzie, *Handbook of Qualitative Chemical Analysis* (Chicago: George K. Hazlitt, 1883), 4: "The fluid state may be secured either by solution or by fusion. Reagents, therefore, may be employed either

subject of intense speculation, and we cannot enter into the complexities of the subject here.[29]

What is clear, however, is that the dry way and wet way presented in chapter five of *Praxis* begin by combining materials that were themselves produced at a fairly late stage in the operations laid out by Keynes 58. The dry way starts by adding the rod to the two serpents, while the wet way uses the caduceus and the twin snakes. Hence although the two ways or paths diverge from each other, they each represent the later stages in two distinct but complementary series of operations intended to lead to the philosophers' stone. Since both the dry way and the wet way implicitly incorporate the earlier stages laid out in Keynes 58, they represent a gargantuan state of complexity that dwarfs even "Three Mysterious Fires." It would be a fool's errand to attempt a detailed explication of this material given the present state of our knowledge, especially since the many steps in the dry and wet ways employ the principle of graduated iteration. In other words, the same terms may mean different things at different stages of the overall process, as Philalethes said of the doves of Diana in *Ripley Reviv'd*. Instead of attempting the overwhelmingly tedious hermeneutical task of analyzing all of this, I have provided charts that capture the bulk of the stages in Newton's dry way and wet way (figures 2.4 and 2.5). These figures are somewhat simplified, particularly in that they terminate before the stages of multiplying the tincture and "whitening Latona" in the wet way. Even so, they comprise some fifty-one stages between them, which does not account for additional details that I may have elided inadvertently but which may have been highly significant to Newton.

One thing stands out beyond the remarkable proliferation of atomized operations in chapter five, and that is the striking role that Fatio de Duillier plays in them. Other scholars have noticed the reference to Fatio's letter of May 4, 1693, but the astonishing degree of importance that Newton grants it here has gone unmentioned. In a word, Newton assimilates Fatio's ground mineral that could purge mercury merely by being shaken with it in a glass to the black powder into which the sophic mercury is reduced during the regimen of calcination in *Ripley Reviv'd*. Since the Philalethan "rotten" black powder occupied a key position both in Keynes 58 and in the dry and wet ways of *Praxis*, the contribution of Fatio's French-speaking adept became a centerpiece of Newton's plan for acquiring the philosophers' stone. In order to see just how important Newton thought Fatio's powder was, it will be necessary to examine the beginning of the dry way as presented in chapter five:

---

in the wet way or in the dry way. In the wet way the reagent in solution is brought into contact with the substance to be analyzed, which is usually in the liquid form. In the dry way the substance to be analyzed and the reagent are brought together in the solid state and subjected to a heat sufficient to melt the reagent, or both the reagent and the assay."

[29] Newton's most significant sources for the terms "via sicca" and "via humida" seem to have been Johann Grasseus's *Arca arcani* and the *Mysterium occultae naturae anonymi discipuli Johannis Grassei*, as he reveals in the *Index chemicus* (Keynes 30/1), 25r and 89r.

This rod & yᵉ male & female serpents <illeg.> joyned in yᵉ proportion of 3, 1, 2 compose yᵉ three headed Cerberus wᶜʰ keeps yᵉ gates of Hell. For being fermented & <illeg.> digested together they resolve & grow dayly more fluid for 15 or 20 days & in 25 or 30 days begin to lack breath ^& thicken & put on a green colour & ~~then~~ in 40 days turn to a rotten black pouder. The green matter may be kept for ferment. Its spirit is yᵉ blood of yᵉ green Lion. The black pouder is our Pluto yᵉ God of wealth, oʳ Saturn who beholds himself in yᵉ looking glass of ♂, the calcination wᶜʰ they call yᵉ first gate, & yᵉ sympathetick fire of Snyders, composed of two contrary fires ♁ & ☉ by yᵉ mediation of his first fire. This pouder amalgams wᵗʰ ☿ & purges out its feces if shaken together in a glass [Epist. N. Fatij]. It mixes also wᵗʰ melted metalls & Regulus's & in a little quantity purifies yᵐ (as was ~~said~~ hinted) but in a greater, burns & calcines them & upon a certain sign, (vizᵗ in yᵉ beginning of yᵉ calcination before yᵉ resolved ♁ of yᵉ metal flys away & leaves yᵉ Reg. dead like an Electrum ^& relapsed into an hydrophoby if <illeg.> it be poured out into twice as much ☿ they amalgam & yᵉ feces of both are purged out wᶜʰ being well washed of & yᵉ matter sublimed wᵗʰ ✳ yᵉ Reg will be found resolved into ☿, ~~excep~~ that is its ♁ & ☿, for the salt of yᵉ metal will stay below, & may be eliviated.[30]

The rich profusion of alchemical motifs drawn from Philalethes and Snyders is obvious at the beginning of the dry way, but the succession of processes owes more to the "American philosopher" than to his German counterpart. Although there is considerable variation in the order of the regimens even in the Philalethan corpus, Newton had already mined his old favorite *Secrets Reveal'd* on this point by the early 1670s when he composed *Of Natures obvious laws & processes in vegetation*. *Secrets Reveal'd* says that if the sophic mercury and gold are sealed up in a flask and heated, "a most amiable greenness" will appear around the end of the fourth week, to be followed in another ten days by a coal-like blackness in which "all the members of thy Compound shall be turned into *Atomes*."[31] This vernacular use of "atoms," which refers to small bits like motes of dust, not the qualityless particles of Democritus and Leucippus, describes the minute grains into which the Philalethan sophic mercury divides gold. A few pages later, *Secrets Reveal'd* describes the green solution as ending "in a colour most black, and a powder discontinuous."[32] This product of philosophical calcination is the black powder that Newton would equate with the mysterious ground mineral of Fatio's talented friend.

The reader may well wonder at the fact that the blood of the green lion makes an appearance here as a derived product rather than an initial ingredient. In chapter five of *Praxis* it arises from the fermentation of the two serpents with the rod, while in Keynes 58 the blood is mixed with the two serpents in order to make the rod itself. Has Newton merely changed his

---

[30] Babson 420, 12r.
[31] Philalethes, *SR*, 81.
[32] Philalethes, *SR*, 87.

mind, or is something deeper happening here? The answer lies in Newton's continual reliance on graduated iteration as an interpretive tool. A glance at his laboratory notebooks shows that in them the blood of the green lion denoted either liquor of antimony or a solution of "green lion" itself, probably a copper-rich mineral, in that dissolvent. This emerges from the following passage in CU Add. 3975, stemming from a period between 1674 and 1681:

> The sp^t of ☿ once destilled grew warm also by mixing it wth water, & much more would ~~it after it is desti~~ it after a full separation from y^e flegm by y^e next preparation This sp^t onc<e> destilled draws y^e Salts of some metals (of ♂, ♀, Wismuth, Cobalt, ♃) but not of ♄ $^{\wedge \text{yet wth heat it works on}}$ ♄. The Lyons blood dissolved a præcipitate of saccarum saturni drawn out of Lead ore by vinegre, & perhaps will dissolve Lead ore.[33]

The blood of the green lion appears here as a menstruum for dissolving a precipitate of vulgar sugar of lead (lead acetate). The implication of this passage is that the lion's blood is either identical to the spirit (liquor) of antimony or possibly a solution of green lion itself in that menstruum. Elsewhere in CU Add. 3975 Newton treats the green lion as a naturally occurring mineral, which he calls a "seed metal" like the *Gur* identified by sixteenth-century German chymists, and describes experiments that he actually performed on it.[34] Hence it is clear that in the context of his own experimentation, he thought the lion and its blood were materials that could be obtained by simple mineral extraction and dissolution in strong acids. This usage occurs numerous times within the laboratory notebooks and cannot be dismissed as an early belief that Newton later rejected.[35] What we are witnessing in chapter five of *Praxis*, then, is an interpretation of the lion's blood based on graduated iteration. In the context of the dry way, the blood of the green lion is neither a mineral nor an extract of a mineral as it is in CU Add. 3975, but a product of fermentation and digestion induced by sealing up the two serpents with the rod.

After passing through the regimen represented by the blood of the green lion, the fermented serpents and rod will lead, according to *Praxis*, to the mineral powder of Fatio's friend. And this material, now equivalent to the rotten black powder of Philalethes, has acquired extraordinary new properties that Newton believes were only "hinted" at by Fatio's adept. A line-by-line analysis of the description that *Praxis* gives of the black powder would show exhaustively that most of its new powers stem from Newton's conviction that the material is identical to the black, "rotten atoms" of Philalethes.[36] Fortunately, we need not test the reader's limits of endurance, since one passage alone makes the complete assimilation of Fatio to Philalethes

---

[33] CU Add. 3975, 53v.

[34] CU Add. 3975, 52r, where Newton says "☿℞ acts not on Iron Ore, nor on Spar, nor on ♌ vir. or any seed metal."

[35] For example, folio 8v of CU Add. 3973 speaks of making a vitriol of green lion. This dated passage stems from January 1679/80. And of course on folio 5v of Keynes 58 itself, a passage that we have already examined, Newton reminds himself to "Extract ♀ from green Lyon wth ℀ diluted & make ye menstrue of this."

[36] See Philalethes, *RR*, 119, for the expression "rotten atoms."

sufficiently clear. Newton says that Fatio's powder can be combined with molten reguli in order to purify them, but that this may result in a destructive calcination and a burned product. If this turn of events should come to happen, "y^e resolved ♀ of y^e metal flys away & leaves y^e Reg. dead like an Electrum $^{\wedge\text{\& relapsed into an hydrophoby}}$." The language of a composite, electrum-like material that shuns moisture as a rabid dog does in the throes of hydrophobia stems from a famous passage in *Secrets Reveal'd*. Philalethes there describes the sublimation of quicksilver from martial regulus of antimony, intended to "acuate" or sharpen the quicksilver so that it becomes a sophic mercury, in startling terms. Since the regulus cannot form an amalgam without the help of an intermediary metal (silver), the metallic antimony is like a child who has been bitten by a mad dog, and who has developed hydrophobia as a result, making it unable to "drink up" the quicksilver. Only after being combined with the two doves of Diana—which meant silver to Philalethes but not to Newton—could the regulus form part of an amalgam. In *Secrets Reveal'd* this is accomplished by repeatedly distilling the mercury from the alloy of martial regulus and silver. The quicksilver, now having becoming a sophic mercury purged by the antimony and enriched with its martial sulfur, flies away and leaves behind "the dead *Doves* of *Diana*." The multiple repetition of the process, combined with intermediary washings, forestalls the possibility that the infant "again relapse into a *Hydrophoby*."[37] To summarize, in the Newtonian reading, Fatio's ground mineral first became the rotten black powder of Philalethes, and then, with Newton's insertion of the rabid, reguline child into the discussion, both the powder and the regulus receive the treatment that Philalethes in *Secrets Reveal'd* had used on quicksilver for making the sophic mercury. This literary transformation of Philalethes's method for making the sophic mercury into a means of purging reguli with Fatio's black powder was to Newton a natural way of reading the allusive text of *Secrets Reveal'd*. It was ever this way with the tricks of the adepts.

To conclude this chapter, *Praxis* should be seen neither as the product of a deranged intellect nor as the culminating denouement of Newton's alchemical career, but rather as the final product of his collaboration with Fatio de Duillier. The powder first described by Fatio in his letter of May 4, 1693, which his French friend reportedly used to purify quicksilver by squeezing the two materials through a chamois skin or shaking them in a bottle, became for Newton the black powder produced in the "philosophick calcination" of *Secrets Reveal'd* and *Ripley Reviv'd*. This was the material marked by "the intire Blackness and Cimmerian utter Darkness of compleat Rottenness" into which it was necessary that the initial ingredients of the philosophers' stone be dissolved.[38] In short, Newton thought that Fatio had given him the key to Philalethes's "first gate," the early regimen of putrefaction required in order to pass through the multiple stages denominated by

---

[37] Philalethes, *SR*, 16. The "electrum" mentioned by Newton appears to be an importation from Snyders. See Keynes 41, 13r, and Snyders, *Commentatio*, 12.

[38] Eirenaeus Philalethes, "The Vision of Sr George Ripley, Canon of Bridlington, Unfolded," in Philalethes, *RR*, 19.

the seven planets, each with its proper color, that would lead to the final prize, the transmutatory elixir. Yet this was not the only reason for Newton's excitement. Fatio's powder could also be assimilated to the black powder or "dry water" of Keynes 58, a complex product involving lead and other materials that was essential for the making of the caduceus. And the caduceus itself was the wonder-working staff of Hermes, which Newton again equated with the summum bonum of the adepts, the agency responsible for their stupendous powers. The goal of *Praxis* was precisely that of making the caduceus, whose manufacture required Fatio's black powder. The young Genevois had come bearing gifts, and Newton had made the most of them. Let us therefore leave Fatio to his marvelous black powder and pass to Newton's late alchemical collaborations after his transfer to London in 1696.

NINETEEN

The Warden of the Mint
and His Alchemical Associates

Newton and Captain Hylliard: An Alchemical Circle
in London at the End of the Seventeenth Century

In March 1695/96 Newton's attempts at gaining a new position, already underway in his correspondence with Fatio de Duillier from the beginning of the decade, finally bore fruit. On the nineteenth of that month, Newton received a letter from his longtime friend Charles Montagu, soon to become Baron Halifax. The letter offered Newton the position of warden of the Royal Mint, at the king's behest.[1] By the end of March, Newton had accepted the appointment, and on April 20 he had departed for London to take on the new post.[2] From a modern perspective the image of an alchemist in charge of the coin of the realm may seem utterly incongruous, but one must not forget that chrysopoeia was still an integral part of chymistry even in the final decade of the seventeenth century, taken seriously by the bulk of chymists including Newton's friends Boyle and Locke. Nonetheless, given Newton's penchant for secrecy when it came to the release of aurific information, it is a striking fact that he continued to be contacted by alchemists even after assuming his new, highly public role. It appears that Newton's attempts to acquire the caduceus of Hermes were sufficiently well known during his administration of the Mint that other practitioners of the transmutational art sought him out and attempted to share their knowledge with him. Two such figures, an otherwise unidentified "Captain Hylliard" and the slightly better known William Yworth, form the twin subjects of the present chapter.

The first of these petitioners is little more substantial than "Mr F." and "Mr Sl," but an examination of his activities and probable identity will have the added benefit of opening a narrow window on the alchemical scene in London near the beginning of the *siècle des lumières*. Captain Hylliard, whose first name is presently unknown, penned a brief alchemical *theorica* that is recorded in British Library Sloane 3711, a manuscript composed initially of writings in the hand of Starkey. At the very end of the manuscript, however,

---

[1] Montagu to Newton, March 19, 1695/96, in Newton, *Corr.*, 4: 195.
[2] Westfall, *Never at Rest*, 556. Westfall gives an excellent treatment of Newton's experience with the Mint in his chapter twelve.

one finds "A Copy of a paper which Captain Hylliard presented to M^r Is: Newton Warden of the Mint" recorded in an unknown hand.[3] Since Newton was warden of the Mint only for the span of years beginning in spring of 1696 and ending when he became master of the same institution in December 1699, it is likely that this Captain Hylliard transmitted his paper to the celebrated scientist during that period.

The contents of the captain's paper are fully in accord with the Helmontian chymistry then popular in England and abroad. The text begins with the claim:

> Water is y^t Genus generalissimum or all inclusive upon wch the Spirit of God moved in pduction of matter for creation of ye severall species.

In other words, water is a sort of "uniform catholic matter," as the mechanical philosophers would have said, which serves as the fundamental material out of which all other things are made by God. But how does nature act on this material in the created world as we know it? Hylliard argues that the prime instrument of nature is fermentation, and that evidence of this appears in the circulation of the blood, which is itself the product of a fermentation induced by alkaline and acid principles with the body. The alkaline principle in the blood is watery and mercurial, whereas the acid one is sulfurous. When the acid and alkaline principles meet each other, they produce a salt, which is the "domicile of nature." This salt can in turn be "anatomized," that is, analyzed, whereon it yields "a Water, a Spirit, a Tincture & an Earth." Moreover, the primordial status of water is demonstrated not only by analysis but also by synthesis, namely, by its stagnation and the subsequent deposition of sediment, which Hylliard views as a product of the water when acted on by the influence of the heavens. As Hylliard intimates in the following words:

> And this an every day demonstration is made appear in y^e: fermentation of water and other juices, from w^{ch}: by Art are produced Spirits, Tinctures, & Salts, and this nature likewise teacheth, when we see in the putrefaction of water agitated by y^e: power of Caelestiall Influence, or ♃ of nature a foetid sulphureous odour to ensue, and y^t: a Sediment then falls down being y^e: limus of production wherein lyeth hid as a Snake in y^e: grasse this great & hidden Mistery.[4]

Hylliard continues by saying that the way to extract and purify the great Arcanum that lies hidden in the sediment is by chymical analysis and resynthesis, which Hylliard identifies with the "clissus" of Paracelsus:

> And y^t secret & hidden Modus is ye Clissus Paracelsi w^{ch}: is nothing else but ye separation of y^e: principles their purification and reunion in a fusible and penetrating fixity.

What did Newton make of Hylliard's revelation that the slime deposited by stagnant water contains a great and hidden mystery? We cannot say with

---

[3] Sloane 3711, 84r.
[4] Sloane 3711, 84v.

certainty, but one thing is sure: he did not reject Hylliard's gift outright, for a recently discovered Newton manuscript transcribes the captain's paper verbatim. Royal Society MM/6/5, a late manuscript that lay buried unnoticed in the archives of the Society until 2004, contains a transcript in Newton's hand of Hylliard's paper.[5] Although Newton makes no comment about Hylliard's ideas, the fact that he sandwiched the captain's paper between authors whom we know he favored, namely, pseudo-Ramon Lull and the colorful cast of philosophers in the *Turba philosophorum*, means that there is every reason to think that he too was intrigued by the "Snake in y^c: grasse."[6] What more can we say about the Captain Hylliard whose work Newton felt obliged to transcribe? Although I have found nothing in the printed literature of chymistry, Hylliard's activities did not fail to leave a mark. In order to acquire more information on Hylliard, however, we must first introduce another figure who was involved in London alchemical circles, namely, the bookseller Richard Jones.

Richard Jones ran a bookstore in Little Britain, the bibliopolis of London, until his death late in 1722.[7] His stock was liquidated by auction in April and November of the following year, and it is perhaps as a result of this that two remarkable documents made their way into the Sloane Collection at the British Library. The first of these, Sloane 2574, is a catalog of Jones's manuscript inventory with careful annotations describing each work and indicating whether it was sold. Very likely because Jones was a bookseller as well as a practicing alchemist, he went out of his way to acquire and describe a great many chymical manuscripts. It is only thanks to Sloane 2574 that the scholarly world today knows of several manuscripts written by George Starkey that are now sadly lost.[8] The second document, Sloane 2573, is modestly titled "Rich. Jones / Collections ab^t Alchymy / &c." In reality, these "Collections" consist of Jones's laboratory records and experiments reflecting his experience at the bench over a period of at least ten years and terminating around 1705.[9] Most of the experiments are framed as blueprints for practice like Starkey's "conjectural processes." Thus a typical entry in his manuscript reads like the following: "If you add ☽ to ♂ in flux with or without ☿ then add ♀ often to the whole<.>"[10] The results of the experiment are not given, nor can we say definitely that Jones ever tried it. In some instances, however, Jones records his experiments in the past tense, and where he does, we can definitely see the footprints of Philalethes. A short series of

[5] John T. Young, "Isaac Newton's Alchemical Notes in the Royal Society," *Notes and Records of the Royal Society* 60 (2006): 25–34, see 31.

[6] RS MM/6/5, 9v–11r.

[7] Richard Jones's death was recorded by Humfrey Wanley, who said the bookseller had died by December 18, 1722. See Cyril Ernest Wright and Ruth C. Wright, *The Diary of Humfrey Wanley* (London: Bibliographical Society, 1966), 1: 178. See also Wright, *Fontes Harleiani* (London: British Museum, 1972), 207.

[8] See Newman, *GF*, 252–53.

[9] I base this dating on the twin facts that folio 51r of Sloane 2573, near the end of Jones's experimental entries, bears the date "Febr 25 ^in mane 1704/5," and this is preceded by a statement on 49r where Jones describes making the Philalethan sophic mercury followed by an amalgam with lead, which he then inspects nine or ten years later.

[10] Sloane 2573, 22r.

experiments probably dating from 1704 or 1705 attempts to replicate the Philalethan sophic mercury in the following words:

> I made an ♒ of ☽ dissolving the ☽ in Reg ♂ & mixing it with ☿
> I made an ♒ of ♀ the same way<.>[11]

The first of these experiments derives from Philalethes's directions to alloy regulus martis with two parts of refined silver (the doves of Diana) as a means of getting quicksilver to amalgamate with the regulus, and the second stems from the *Marrow of Alchemy*'s later discovery that an alloy of copper and the antimonial regulus (the net) could serve the same purpose. Earlier in the manuscript Jones describes the net by name and compares its properties to those of the star regulus.[12] Although he does not explicitly cite Philalethes as his source for the net or the sophic mercury, Jones refers elsewhere in his laboratory records to *Secrets Reveal'd* and *Ripley Reviv'd*, making it more than probable that his source for these desiderata was the New England adept.[13]

Toward the end of Sloane 2573, the impressive chrysopoetic experiments, ruminations, and projects left by Jones give way to medicinal recipes taken from different authors. Among these obscure figures are "Cossen Blackwell," "Dr Gower," "Capt Oudrad" (Ondrad?), and remarkably, "Capt Hilliard." The Captain Hilliard in question recommends that one make a medicine from "ffrogg Spane Water," presumably the expressed liquor from the jelly-like frogspawn found in ponds and swamps, which is to be "Swetened wth Sugar Candy" and ingested for "Severall Distempers."[14] Could this Captain Hilliard be the Captain Hylliard who informed Newton that the great secret of alchemy lay hidden in the slime left by stagnant water like a snake in the grass? Neither the dates in Jones's "Collections" nor the subject matter provide difficulties to this identification, but one would like to have more information before asserting that the two captains are one. In fact, precisely this information is forthcoming in Jones's other manuscript, Sloane 2574, his catalog.

One of the manuscripts carefully described by Jones's catalog appears to have been devoted mainly to "Processes on several Earths newbery & fullers Earth with Queries on the processes." From Jones's epitome, one can make out that a group of people associated in some way with one another were active in trying to extract salts and other active principles from earths and minerals, probably in the vicinity of Chelsea College, not far from Westminster Abbey. Among the names and titles that Jones recounts one finds "Mr Kemp," "Mr Bryant," "the Justice," "madm justice" (evidently the justice's wife), "mr Edwards the Chymist," "mr vivades<?>," "mr vaughan," "mr Barkeley," "Mr moreland," "Dr Lofflere<?>" "Dr. Savage," "Dr. Crell," "Dr Dickinson," "Mr Boyl,"

---

[11] Sloane 2573, 49r.
[12] Sloane 2573, folios 25r and 27r.
[13] Sloane 2573, folios 7r, 7v, and 22v. Jones also echoes the language of Philalethes in many places, as at 24r: "& this is the Hollow oak wch Cadmus fastens the serpent throug <sic> and through too for the tincture appears now in the ☿ on the outside these are Diana's doves wch asswage the lion and vanquish him."
[14] Sloane 2573, 60r.

and several times "Captain Hilliard." Some of these names are easily recognizable. "Mr Boyl" is surely Robert Boyle, on whose authority "Dr. Crell" relays the news that a disciple of Philalethes told the famous scientist that he "used Regulus." "Dr. Dickinson" must be Edmund Dickinson, the author of the well-known 1686 *Epistola ad Theodorum Mundanum*, which we have encountered already. And it is of course quite possible that "mr Vaughan" refers to Thomas Vaughan, the elegant writer whose *Fame and confession of the fraternity of R: C:* was given to Newton by Oliver Doyley. Vaughan composed treatises on the Sendivogian *sal nitrum* under the pseudonym of Eugenius Philalethes. According to Jones's notes, Vaughan "did extract a liquor from Gurr & sold a q^t to my Ld Yarmouth for 60^{lb}."[15]

It is very likely that the manuscript epitomized by Jones contained heterogeneous information, and that only some of the names mentioned actually associated with one another in life. Yet Jones's notes provide solid evidence that a few of these figures, including the mysterious Captain Hylliard, knew and worked with one another. The following passage is particularly revealing:

> pag. 7 The Justice an account of a justice that had an insipid liquor wch would dissolve ⊙ ^without noise or ebullition ... pag. 8 he showed ^Mr Bryant a salt in a gally pot that his Liquor was drawn from this salt pag. 9 an account of ~~Mr~~ Captain Hilliard's distilling from the Earth a volatile salt a white Liquor & a fetid oil wch he supposes to be the same of the Justices mr Bryant took a stiffe clayish earth from Chelsey as also the Capt says he put it in a retort did draw a whitish transparent liquor wch tasted pretty sharp of the volatile salt pag.

From this we learn that Captain Hylliard distilled a salt, liquor, and oil from an earth that he supposed to be the same material that a certain justice (presumably a justice of the peace) operated on. The justice had shown the material from which he distilled his insipid liquor to one Mr. Bryant. Additionally, Mr. Bryant distilled "a whitish transparent liquor" from a stiff clay found at Chelsea, "as also the Capt says." Evidently then, Captain Hylliard was familiar with the work of Bryant and the justice and probably with the men themselves. Further comments by Jones about a curative red liquor drawn from a mineral earth also support some sort of collaboration. He reports that "they took clay from Chelsey" as their starting material, and that Captain Hylliard "saies if the red oyl be drawn from sand it will burn." A subsequent comment speaks of "a clay from Chelsea College when building whence the liquor was drawn."

---

[15] The Lord Yarmouth mentioned here may be Robert Paston, first earl of Yarmouth (1631–1683). According to the *Oxford Dictionary of National Biography*, Paston's library contained numerous books on alchemy. See the online *Oxford Dictionary of National Biography*, accessed August 26, 2016. Moreover, Paston was the chymical patron of Thomas Henshaw, a friend and collaborator of Vaughan's. See Donald R. Dickson, "Thomas Henshaw and Sir Robert Paston's Pursuit of the Red Elixir: An Early Collaboration between Fellows of the Royal Society," *Notes and Records of the Royal Society* 51 (1997): 57–76. The fact that Thomas Vaughan died in 1666 does not impede his identification with the Vaughan of Sloane 2574, since Jones's catalog refers to manuscripts of varying dates.

Additional notes provide information that may be related to the paper that Hylliard gave Newton. Beyond their work on earths, the group of chymists active around Chelsea College were also operating on stagnant water collected from "moorish grounds." Some of them evidently thought that the oily matter floating on the ponds in moors actually consisted of the congealed beams of the sun and moon, and a soldier used to skim the "film" off of the water and treat it with gold and copper. Jones also mentions an experiment performed on "a green water in Salisbury court" near Saint Paul's Cathedral, and incidentally close to his shop in Little Britain. These starting materials are reminiscent of the putrefied water that Hylliard mentions in his paper for Newton. And the account that Jones gives of Hylliard's work mentions that the captain analyzed his materials into "a volatile salt a white Liquor & a fetid oil," which no doubt were thought of as close relatives of the three chymical principles salt, mercury, and sulfur. Hylliard's paper for Newton describes a similar "anatomy" or analysis of the sedimentary slime from putrefied water. Are these experiments further attempts by Hylliard to isolate the "Snake in y^e: grasse this great & hidden Mistery" that he promised the warden of the Mint?

Another focus of these London chymists lay in the minerals associated with lead ore, which of course was also a perennial interest of Newton's. Jones asserts that "several people" aimed to find "a liquor in lead mines & a clay in lead mines." One of these was the "justice" who had an insipid liquor that would quietly dissolve gold, thereby fulfilling one of the traditional desiderata of the sophic mercury, namely, the ability to melt the precious metal like ice in warm water. Jones reports that the justice "confesses his matter to come out of the mines," and a few lines later, he adds that "madm justice" claimed that the clay or earth "came out of Darby shire that it was Saturns Childe came out of a lead mine." It appears, then, that the group made up of the justice and his wife, "Mr. Bryant," and Captain Hylliard were all engaged in a similar set of operations aiming at the extraction of a sophic mercury or other menstruum from a variety of materials including stagnant water and several earths. The degree of association between these individuals cannot be determined from Jones's manuscript, although it appears that they at least knew one another.

From the account given by Jones we receive the impression that Captain Hylliard was interested both in chrysopoeia, the undoubted end point of the gold-dissolving liquor, and chymical medicine. The concern with iatrochemistry is attested not only by his concoction of frogspawn water and sugar candy but also by a final reference in Jones's catalog. Among further descriptions of the manuscript, one finds a casual mention of "Capt Hilliars Balm with the Figures of Furnaces." There is every reason to think that "Hilliar" is just another variant spelling of Hylliard, and that we are back in the company of the same captain. It appears, then, that he composed a manuscript on a medicinal balm complete with images of the required furnaces. Alas, neither this nor the "Processes on several Earths" has been identified among the manuscripts in the Sloane collection.

Can we say anything more of the Captain Hylliard who delivered his brief paper to Newton around the turn of the century? Unfortunately, the term "captain" is rather vague and could have either a military or nautical signification. The reference to "Chelsea College," however, undoubtedly refers to the institution that Charles II would turn into a home for military pensioners, namely, the Royal Hospital Chelsea, as it is called today. This fact, along with the reference to a soldier who experimented with the scum on stagnant water, suggests a military setting for Captain Hylliard rather than a maritime one. There is also a slim possibility that Captain Hylliard could be identical with Robert Hyllyard or Hildyard, who attended Trinity College in 1689, and who might thus have been an acquaintance of Newton's.[16] Hylliard, Hilliard, Hyllyard, and Hildyard are all early modern alternative spellings, and Robert Hyllyard received a commission as captain of the foot in 1700.[17] He was a prominent figure, however, and the second baronet of Winestead in Yorkshire. He did spend time in London, having been elected as MP of Hedon in 1701, which is interestingly the same year that Newton was elected MP of Cambridge for the second time.[18] But it seems unlikely that Jones or the anonymous copyist of Sloane 3711 would refer to a man of such high rank as a mere captain; Robert Hyllyard had acquired his baronetcy in 1688, long before Newton ascended to his position as warden of the Mint in 1696. Nor have I found any independent indication of alchemical interest on the part of Hyllyard, further reducing the likelihood of his being our captain. For now we must remain content with the knowledge that the chymical Captain Hylliard was a minor figure in London involved in the quest for chrysopoeia and in the preparation of medicines. Perhaps more on the circle of chymical practitioners in Chelsea around the end of the seventeenth century will emerge in the future.

From the evidence supplied by Jones, we can at least make some general comments about the activities of the group of London chymists who were active in the vicinity of Chelsea College. Although Hylliard's subjects ranged from frogspawn to various earths and clays, he may have been motivated by a single, underlying concern, namely, the extraction of the celestial influence that lay hidden within all of these materials. It is not hard to see in his focus on these materials the same desire to use chymical analysis for isolating the secret material that "lyeth hid as a Snake in y^e: grasse" that Hylliard describes in his paper to Newton. Very likely, when he coined this passage Hylliard was thinking of the Sendivogian aerial niter, in which the noble Pole had identified the life-giving and fertilizing *Ur*-principle of nature. The seventeenth century was rife with attempts to extract Sendivogius's aerial salt from the atmosphere and from various fluids and solids ranging from

[16] We learn the following about Hyllyard from the *Admissions to Trinity College*: "Hyllyard, Sir Robert, Bart. Son of Sir Christopher Hyllyard, Bart., of Wisted in Holderness, Yorkshire. School, Hull (Mr Pell). Age 18. Nobleman, June 13, 1688. Tutor, Mr Bynns. [Matriculated, as Robert Hildyard, 1689. Did not graduate.]" See W. W. Rouse Ball and J. A. Venn, *Admissions to Trinity College, Cambridge* (London: Macmillan, 1913), 2: 566.

[17] I have the following information from Helen Clark, Archives Supervisor of the East Riding Archives: Document DDHI/58/17/3 contains the "Commission by the Duke of Newcastle of Sir Robert Hildyard, baronet, as Captain of the 1st Company of the Regiment of Trained Bands of Foot in the East Riding 30 Aug 1700."

[18] Westfall, *NAR*, 623.

Maydew to excrement. Moreover, such attempts easily merged with chymical researches directed at the primordial material of metals, the Mathesian Gur. Just like the aerial niter, Gur was supposed to be a largely unformed progenitor of more specific matter, and if one could purify the still-living material of its dross, it might be possible to induce it to mature and develop into the philosophers' stone. This is probably the underlying motive behind Vaughan's experiment in which he "did extract a liquor from Gurr," as Jones reports. It is also the unstated reason for the excitement of "madm justice" about the clay found in Derbyshire lead mines for which she uses the alchemically charged expression "Saturns Childe."

Although it may be tempting to think that Newton, whose experimental research in alchemy focused heavily on processes involving antimony and a host of other metallic materials, had no interest in these attempts to extract the aerial niter from substances that were not metalliferous in the modern sense, this would not be entirely correct. His early theoretical treatise composed in the first half of the 1670s, *Of Natures obvious laws & processes in vegetation*, describes a method of extracting the "metalline fumes" from earth that has been deposited by stagnant water in a way that is highly reminiscent of Hylliard's technique.[19] Since Newton composed *Of Natures obvious laws* a quarter of a century before Hylliard delivered his chymical paper to the famous public figure of the 1690s, Newton's method of "unraveling" the structure of sediment in order to release its hidden, metallic spirit cannot have found its inspiration in the work of the obscure captain. Instead, both men were arriving at the same practical conclusion by drawing out the logical consequences of the Sendivogian theory of the aerial niter. The noble Pole had said that the life-giving niter was carried down from the upper atmosphere by rainwater, whereupon it joined to the "fatness of the earth." Water then provided a reasonable place to look for a matrix in which the aerial niter lay hidden, like Hylliard's snake in the grass. Although we have no record of Newton performing experiments of that sort, it is not unlikely that Hylliard's brief paper resonated, even in Newton's maturity, with a view that he had once vigorously held and perhaps still not altogether abandoned.

The group of London chymists whose activities Richard Jones recorded in his catalog seem to have been for the most part dedicated amateurs rather than professional refiners, distillers, apothecaries, or physicians. They reflect a particular subset of the field, and one that is very different from the world populated by central European figures with real mining and metallurgical interests whom we examined earlier, such as Sendivogius, Grasseus, Maier, and Thölde, not to mention the Lutheran pastors Mathesius and Solea. Despite Madame Justice's references to lead mines, it is unlikely that any of these London chymists had a strong connection with the mining world. Yet even at the end of the seventeenth century one can see the powerful effect that the earlier alchemist-metallurgists exercised by incorporating the miners' hylozoism and their very specific beliefs about the life cycle of the mineral world. The attempt to extract the starting material of the philosophers'

[19]Dibner 1031B, 3r.

stone from substances as varied as pond scum and lead ore made good sense to those who accepted the Sendivogian picture of a world constantly rejuvenated by the descent of the aerial niter and its incorporation with terrestrial matter. Nonetheless, Hylliard seems to have made little impact on Newton beyond the fact that the latter transcribed Hylliard's brief manifesto. Let us now pass from Hylliard to another resident of London who not only transmitted writings to Newton but even persuaded the now famous intellectual to collaborate in their composition.

Chymical Coworkers: Newton and William Yworth

William Yworth was a practicing distiller and chymical pharmacist who had moved from Rotterdam and established a shop in London called *Academia Spagirica Nova* by June 1691. Although his peculiar surname is expanded twice in his multiple surviving manuscripts to "Yarworth," it may have been an Anglicized Dutch name rather than an alchemical pseudonym, as suggested in an important article by Karin Figala and Ulrich Petzoldt.[20] His son Theophrastus, who carried on his pharmaceutical and distilling business, employed "Yworth," and this form of the name continued to be used by his descendants for several generations.[21] Yworth has an honored place in the history of alcoholic beverages, for he was an early distiller of gin in England.[22] Yet thanks to the assiduous detective work of Figala, we now know that Yworth not only published under his own name and in the capacity of an artisanal distiller but also wrote on the *arcana majora* of chymistry under the pseudonym "Cleidophorus Mystagogus," roughly translatable as "the Key-Bearer of the Mysteries."[23] Using this nom de guerre he published several books, including *Mercury's Caducean Rod* (1702, 1704) and *Trifertes Sagani* (1705). It was in his capacity as an aspirant to chrysopoeia that Yworth contacted Newton in the first decade of the new century. The two had become virtual neighbors by 1702, when Yworth moved to King Street, a few hundred yards from Newton's residence on Jermyn Street in London.[24] As I will now show, their physical separation was even less than that, for recently discovered documents provide us with the record of an interview that Newton held with Yworth.

[20] Figala and Petzoldt, "Alchemy in the Newtonian Circle," 182. The manuscript is Yale University, Mellon MS 80, 7r, 9v. Yworth's origins are obscure. He indicates that he was born in "Shipham" in several writings, but no such place has been identified in the Netherlands. There is also a village called Shipham in Somerset, opening up the possibility of an English birth for Yworth. Yet a consultation that I made of the Shipham parish records in a microfilm kept at the National Archives in Kew has not revealed any Yworths, Yarworths, or similar names for the relevant time period. For further details on Yworth's life, see Mandelbrote's entry in the online *Oxford Dictionary of National Biography*.

[21] See the genealogical information compiled in William Wallworth, *The Yworth Family* (s.l: Exile's Publications, 2016).

[22] Richard Barnett, *The Book of Gin* (New York: Grove Press, 2011), 36–38.

[23] Figala and Petzoldt, "Alchemy in the Newtonian Circle," 183–86.

[24] Karin Figala, "Zwei Londoner Alchemisten um 1700: Sir Isaac Newton und Cleidophorus Mystagogus," *Physis* 18 (1976): 245–73, see 258.

It has been claimed that Newton gave up his interest in alchemy on accepting his new position as warden of the Royal Mint in 1696. While it is true that no later dates of experiments are found in his laboratory notebooks, nearly half of CU Add. 3973—a remarkable twenty-nine closely written pages—consists of entries that the keepers of the manuscripts at Cambridge University Library have grouped after the final date entered there, "Feb. 1695/6."[25] However unlikely it may seem that all of this material stems from the period between February and late April of that year, when Newton moved to London, there is no way at present to determine the matter with certainty. What lies beyond doubt, however, is the fact that Newton continued his alchemical work in London via proxy. Just as Robert Boyle employed practitioners such as Frederic Slare and Ambrose Godfrey Hanckwitz to carry out chymical operations on his behalf, so Newton was paying Yworth for his chymical research. This is revealed unequivocally by an undated letter from Yworth, who identifies himself as "W. Y." The letter, which seems to have been written between 1702 and 1705, requests "the wanted Alowance" because Yworth is being deluged by taxes and a visit from his landlady.[26] Yworth also mentions that he has sent Newton a book,

> wch I hope upon thy Judicious Consideration will satisfy thee from Acetum to Elixer, and how long that may be justly said to Reign, sc. even to ye Production of Azoth<.>

Although the identity of the particular book that Yworth sent is not definitely known, Newton owned several of his titles, including two published in 1702.[27] The treatise of most concern to us, however, is one that Yworth never printed, namely, his *Processus mysterii magni*. As Figala and Petzoldt have shown, this work was the product of an ongoing collaboration between Newton and Yworth that must have taken place over a considerable span of time. The two scholars have identified three partial drafts of the *Processus* in Yworth's hand that he evidently gave to Newton, as well as a partial draft and a substantial fragment of the text that Newton copied (Keynes 66 and Keynes 91).[28] Most interestingly, they have shown that Newton was actively involved in Yworth's research, for there are multiple instances where the celebrated scientist added in the quantities of ingredients where Yworth left them out.[29]

---

[25] CU Add. 3973, 30v. I do not count empty pages or the final folio on which there is writing, since this concerns the making of an alloy for telescope mirrors. Nor do I count sheet fifteen, a small inserted leaf making up folio 48r–48v. The content reveals that this sheet corresponds to folios 65v and 66v of CU Add. 3975, both of which were composed in 1682. I owe this information to the latent semantic analysis tool developed by Wallace Hooper, which appears on the *CIN* site.

[26] "W. Y." to Newton, undated, Newton, *Corr.*, 7: 441. The editors, Hall and Tilling, arrive at the *terminus post quem* of 1702 from Yworth's reference to "ye Queen's Tax." Mandelbrote points out that Yworth was already living away from London in 1705. See Mandelbrote, online *Oxford Dictionary of National Biography*.

[27] Harrison, *Library*, 198–99. The two books in question are *Mercury's Caducean Rod* and *A Philosophical Epistle* (also printed as part of the former).

[28] The manuscripts of the *Processus mysterii magni* in Yworth's hand are Mellon 80, Keynes 65, and Hampshire Records Office NC 17. See Figala and Petzoldt, "Alchemy in the Newtonian Circle," 180 and 191.

[29] Figala and Petzoldt, "Alchemy in the Newtonian Circle," 188–89.

We can now add a great deal to the picture of this collaboration between Newton and Yworth, thanks in part to two manuscripts unknown to Figala and Petzoldt. The first of these, Royal Society MM/6/5, was discovered in the archives of that institution in 2004.[30] It contains a number of writings in Newton's hand, including the short alchemical manifesto given to him by Captain Hylliard when he was warden of the Mint. The second manuscript, Dibner 1041B, consists of a very late Latin florilegium by Newton into which he incorporates parts of Yworth's *Processus*. Both of these documents are highly significant in their own ways; I will focus first on the Royal Society manuscript for what it tells us about the personal interaction between Newton and Yworth.

As I argue in appendix four, there can be little doubt that the final two pages of Royal Society MM/6/5 (15r and 15v) represent Newton's report of an interview that he held with Yworth. The questions and answers in the Royal Society manuscript correspond closely to the content of the *Processus* in the draft found in Keynes 66, and the latter even incorporates changes that Newton made after having had various points clarified by his oral questioning of Yworth (see appendix four). Moreover, Yworth's high level of technical expertise as a distiller and maker of chymical medicinals shines through in both texts. In each of them he describes "superior" and "inferior" waters, for example, that appear at various stages in the operations. As the *Processus* makes clear, the two liquids were isolated from each other by the use of a special separatory funnel that Yworth employed, not a commonly described piece of apparatus in alchemical laboratories. But this was not enough for him; he also devised an additional "separatory glass" equipped with a cork in the top and a "ground stopper" beneath for further separations.[31] This is only the beginning of the specialized apparatus that Yworth describes. In another instance he says to take a "steen" (presumably a container like a German Beer stein), cut off the bottom, and build it into a small brick furnace. A metal rim (a "verge") is then provided for the steen by attaching a pan whose center has been cut out, and a retort whose bottom has been cut off is attached to that. All of this is luted together and used for a high-temperature distillation.[32]

Another feature that emerges from the discussion between Yworth and Newton is the concern with safety in the laboratory, a topic with some bearing on the issue of Newton's "derangement" of 1693, as we saw earlier. Both the interview on Royal Society MM/6/5 and the *Processus* ascribe toxic qualities to the "Green Lyon," which for Yworth is a liquid product produced in the distillation of an unnamed substance with which he begins the *Processus*. It is worth quoting the section of the *Processus* where Yworth describes the dangers this material posed:

> And continuing the destillation by degrees of fire untill a white fume came not changing the Receiver, there came a more yellow spirit & oyle & after

---

[30] John T. Young, "Isaac Newton's Alchemical Notes in the Royal Society," *Notes and Records of the Royal Society* 60 (2006): 25–34.

[31] Keynes 66, 4r and 7r.

[32] Keynes 66, 1v–2r.

that a blood red one but in such an exceeding wild fume as that 'twas like the clouds in the Air penetrating the Lute tho kept with all the diligence imaginable & the stink was such & so great that I can't describe it. The operation being over I took off the Receiver and holding my nose to it I was so suffocated & seized all over in my spirits that I thought I should have immediately dropped down dead.[33]

Work at the bench was not a trivial matter in the days before fume hoods and glassware with tight joints. Despite the special precautions that Yworth took with his apparatus, even wrapping the joints in animal bladders before smearing them with lute, the sealant failed.[34] Interestingly, Yworth takes up the green lion's stench again in the interview. This is in part because the green lion was a traditional desideratum of alchemy, and earlier authors such as Flamel and Philalethes, both of whom Yworth cites, had described the material as filthy and disgusting.[35] The dangerous stink was therefore a sign of impending success that could not be avoided:

In the end of every digestion, in destilling the <*illeg.*> spirits from the e<*illeg.*> egg after 3 or 4 or perhaps 5 days destillation, when the white fumes ~~begin to rise~~ or fumes of the green Lyon begin to rise, they will be apt to penetrate the lute & fill the room wth a stinking scent. When you see this signe you must put an end to the destillation.[36]

One cannot help but wonder about the identity of the mysterious material with which Yworth began his series of operations. The *Processus* begins with the oblique statement that "The matter you know well." What follows is a list of alchemical commonplaces that could describe any number of substances. It is a "known mineral," but that means little in the literature of alchemy, for even the term "mineral" could be used tropologically to mean anything that is "dug out" or extracted, not necessarily a product of subterranean origin. As for the claim that the initial matter is not only mineral but also animal and vegetable, this is an old riddle inherited from Arabic alchemy, and again could mean a wide variety of things. In the High Middle Ages, the scholastic alchemist Roger Bacon had gone so far as to equate the subject of this puzzle with human blood.[37]

The degree of Yworth's evasiveness about the identity of his initial ingredient is matched by the extreme precision of his directions for working with it. The "vile" and "contemptible" matter must be taken in the form of a powder in its cleanest, purest form, and then sifted finely. Being a commercial distiller, Yworth was used to working with considerable quantities of ingredients. Thus he says to take half a bushel of the powder and place it in a large iron pot on which an earthen "cap," meaning the capital or still

---

[33] Keynes 66, 5v.
[34] This detail appears in the version of the *Processus mysterii magni* found in Mellon 80, at 44v.
[35] Yworth cites Flamel and "other philosophers" on this point at Keynes 66, 5v.
[36] RS MM/6/5, 15v.
[37] William R. Newman, "The Philosophers' Egg: Theory and Practice in the Alchemy of Roger Bacon," *Micrologus* 3 (1995): 75–101.

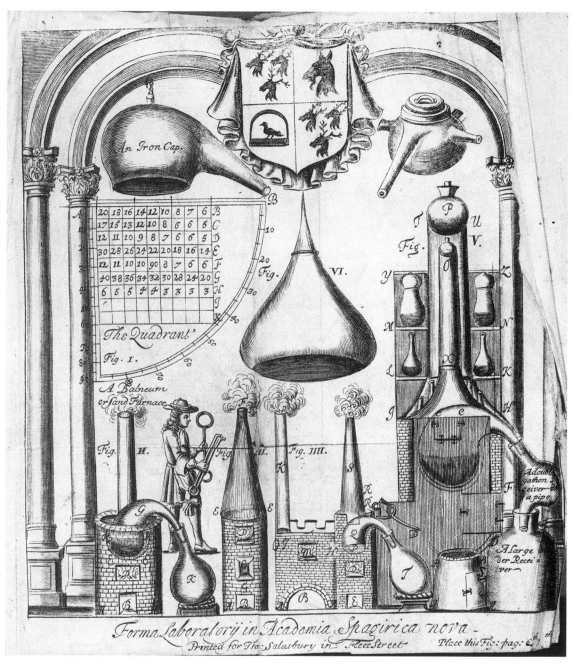

FIGURE 19.1. Plate from William Yworth's 1692 *Chymicus rationalis*. The image presents an idealized picture of Yworth's laboratory at his *Academia Spagyrica Nova* in London. The specialized equipment gives a sense of Yworth's expertise in distilling.

head, is placed, to which "a large ^glass Cooler with a pipe" is attached. By this Yworth means a glass condenser, perhaps similar to the intermediary apparatus pictured in an image from his *Chymicus rationalis* of 1692 (figure 19.1). The condenser is then attached to a twenty-gallon receiver in the shape of a globe, and all the foresaid apparatus is sealed together with founder's lute. Yworth's subsequent directions are complicated, but the over-all method involves "dissecting" or analyzing the unnamed initial ingredient into its components by processes of distillation, sublimation, and calcination, which produce such desiderata as the forenamed green lion along with various calces, waters, oils, vapors, and "wild gass." Rather than describing these operations in detail, let us now turn to the vexed problem of identifying Yworth's mysterious starting ingredient. As it turns out, within the menagerie of *Decknamen* and riddles that Yworth provides he has also interspersed some useful clues.

The first paragraph of the *Processus* informs us that the initial ingredient is not only vile and contemptible but also a "saturnal matter." Moreover, it is an immature substance, and it contains "the seed of ^all minerals & ^all metals in their first & unspecificated nature." Clearly Yworth equates his starting material with the *Ur*-matter from which all metals and minerals spring. Although this could mean a number of different things in practice, the term "saturnal" is significant, particularly in the light of several comments that Yworth makes in the draft of the *Processus* found in Yale University's manuscript Mellon 80. In the midst of describing a white, "lunar" sublimate, derived from his unnamed matter that "you know well," Yworth launches into a seeming criticism of Basilius Valentinus. This is somewhat anomalous, since elsewhere Yworth has only the highest praise for the supposed Benedictine. Among his comments we find one that unwittingly unveils the nature of Yworth's cryptic starting material. I reproduce it here:

> Neither did I find these reiterated Sublimations to add much Efficacy either to yᵉ Sublimate or Waters; nor indeed yᵉ Bole or Tile meal mentioned by Basil in pag. 150 of his Elucid. Where he speaks of yᵉ preparation of this Sublimate, except understood as a Parabolical Speech, to figure out yᵉ red Earth, in Colour alike to both: for either Bole or Tile-meal will Congeal in yᵉ Vessel like Stone, and will surely break your Vessel and tho' by great strength of fire a Sublimate may Come up, yet t'is very Corrosive, as wanting yᵉ Mortifying Power of yᵉ Red Earth, & yᵉ Benignant Fire it contains.[38]

As one can see, Yworth refers here to a method of subliming his initial ingredient with ground bole armeniac or tile meal, materials that were traditionally employed as a refractory filler to prevent a sublimandum from coalescing into a molten mass that would be difficult to volatilize. In the process of rejecting this method, the Dutch distiller points to Basilius's recommendation of it on page 150 of the "Elucidation." The pagination reveals that Yworth is referring to the 1672 printing of the Basilian *Last Will and Testament*. The page in question contains a passage that we already referred to earlier in the present book. In this part of the "Elucidation," Basilius said to "ask counsel of god *Saturn*," who will provide "a deep glittering *Minera* for an offering, which in his Myne is grown of the first matter of all Metals." Already by the late 1660s Newton had deciphered this correctly as a reference to stibnite,

[38] Mellon MS 80, 30v–31r.

which was often viewed as having its descent from Saturn. The passage in Basilius says that one can sublime feathery, alum-like "flowers" from crude antimony by mixing it with three parts of bole or tile meal and subjecting it to an intense heat. From all this it emerges with great clarity that Yworth's enigmatic matter that "you know well" must be stibnite, or crude antimony. Like Basilius, Yworth considered this Saturnine substance to be the origin of metals and minerals, and hence a logical place to look for a universal dissolvent that could act "without ebullition" just as "mercury penetrates metals in amalgamation."[39]

From a technical point of view, then, the *Processus* describes Yworth's attempt to "anatomize" stibnite into its putative components by means of heat. This project had been announced by the Basilian "Elucidation" found in the *Last Will and Testament*, and it attracted many followers. There was a logic to it, since stibnite is of course a sulfur compound, just as iron and copper vitriol are. The sulfurous properties of the mineral become obvious on its refining and in the manufacture of various antimonial compounds such as *Sulphur auratum* (antimony pentasulfide). And since destructive distillation of the vitriols can yield "oil of vitriol" or the modern sulfuric acid, why shouldn't the same approach with stibnite produce a menstruum as well? The answer is that the vitriols are hydrated sulfates, already containing oxygen bonded to their sulfur, and simple roasting releases gaseous sulfur dioxide that can combine with the water vapor released by heating them in several steps to form sulfuric acid. Pure stibnite, however, is a sulfide, a simple combination of elemental antimony and sulfur, with no oxygen or water. In order for the compound to decompose and form the desired sulfur dioxide, and eventually sulfuric acid, oxygen and water vapor must be introduced from another source. Yworth was well aware of the need for external atmosphere, and he therefore advises that vents be provided at various points in his apparatus.[40] Other contemporary reports of this process, such as the one found in the Academician Nicolas Lemery's 1707 *Traité de l'Antimoine*, also provide plenty of opportunity for the addition of fresh air and water vapor to the heated stibnite. Lemery describes putting four ounces of the ore into a large earthen retort heated red hot. The retort had a large aperture that the chymist would open every half hour during the heating in order to add more stibnite and to stir the hot powder with a spatula, thus exposing it to fresh air. After adding a total of twenty-four ounces and heating for five hours, Lemery tells us that he found only five and a half drachms of liquid in the receiver—a testimony to the difficulty of the procedure.[41]

Now that we have decoded Yworth's starting material and the outlines of his process for dissecting it, some further questions necessarily emerge. What influence—if any—did Newton exercise on the research of Yworth? And in a closely related vein, what role did Yworth's work play in Newton's own experimental project? The first question can be addressed by examining the

[39] RS MM/6/5, 15v.
[40] Such a vent is very clearly described in Keynes 66 at 2r.
[41] Nicolas Lemery, *Traité de l'antimoine* (Paris: Jean Boudot, 1707), 69–71.

books that Yworth published before adopting his alchemical pseudonym, Cleidophorus Mystagogus, in 1702. Already in his 1691 *New Art of Making Wines*, Yworth declared his serious interest in transmuting metals and announced his plan to publish a succession of works on the higher secrets of chymistry. Among the six forthcoming books that Yworth mentions in 1691 one finds *A Magical Magazine*, projected to consist of six parts. Parts two and four will bear the subtitles *Mercury's Caduce Rod* and *Trifertes Soladinis*, obvious variations on the titles *Mercury's Caducean Rod* and *Trifertes Sagani*, first published in 1702 and 1705, as separate books.[42] From all this it is clear that Yworth had a long-standing interest in chrysopoeia along with other *arcana majora*. Although we do not know when he first made Newton's acquaintance, it is even possible that these interests predated his immigration to England. Nevertheless, one will look in vain for the antimonial process so carefully described in the *Processus mysterii magni* among the books that Yworth published before 1702, despite their profusion of alchemical themes and *Decknamen*.[43] The absence of the antimonial procedure seems not to be a mere matter of genre since it does occur, albeit in somewhat veiled terms, in Yworth's 1705 *Compleat Distiller; or, the Whole Art of Distillation*. Chapter five of that text is devoted to the production of the "secret menstruums" of the ancients, among which we find a recipe that begins by taking the "known Animal, Vegetable, and Mineral Matter," in short the same mysterious ingredient that initiates the *Processus mysterii magni*.[44]

Given the late appearance of the stibnite distillation process in Yworth's corpus, it is far from impossible that it was Newton who directed the Dutch distiller to that procedure. Several other clues seem to point in that direction. First, as we saw earlier, Newton's *Praxis* incorporated the Basilian directions for distilling stibnite with bole or tile meal directly into the text as an integral part of Newton's quest for the philosophers' stone. We know from an independent source that Newton had experienced difficulties in making the operation succeed. An experiment recorded in CU Add. 3975 that probably stems from the time when he was writing *Praxis* says that when he tried to distill crude antimony mixed with fullers' earth, "ye retort melted & sunk down wthout making y^e ☿ rise."[45] Given the prominent place that *Praxis* devotes to this process, Newton may well have desired more successful experimental data on the subject than he was able to acquire during

[42] William Yworth, *A New Art of Making Wines, Brandy, and Other Spirits* (London: T. Salusbury, 1691), unpaginated "Advertisement" after page 153. Yworth mentions here that his plan is to publish these as parts of the *Magical Magazine* "if any considerable Subscriptions are made." Otherwise, he says, they will be published separately.

[43] I have consulted the following pre-1702 imprints by Yworth: *A New Treatise of Artificial Wines* (London: A. Sowle, 1690); *A New Art of Making Wines* (London: T. Salusbury, 1691); *Introitus apertus ad artem distillationis* (London: Joh. Taylor, 1692); *Chymicus rationalis* (London: Thomas Salusbury, 1692); *Cerevisiarii comes* (London: J. Taylor and S. Clement, 1692); and *The Britannian Magazine* (London: T. Salusbury, 1694 and 1700).

[44] William Yworth, *The Compleat Distiller; or, The Whole Art of Distillation* (London: J. Taylor, 1705), 235–38.

[45] CU Add. 3975, 136r. The experiment directly precedes Newton's description of Fatio's marvelous mercury-purifying powder. Newton refers to "Artificial depurated ☿" here, but this probably means stibnite that has been purified of its gangue by melting.

his last years in Cambridge. Further confirmation of this emerges from yet another Newtonian manuscript, which may contain the final traces of Newton's interaction with Yworth.

Although largely ignored by previous scholars, the Smithsonian Institution's Dibner manuscript 1041B preserves a text that may be Newton's last alchemical florilegium. Consisting of seven pages on six leaves, this interesting manuscript bears the Latin heading "Separatio Elementorum" (separation of the elements), but that is actually the title of the first chapter alone. A comparison of this "Separatio Elementorum" with the English text of Yworth's *Processus mysterii magni* shows that it is in fact an abridged Latin translation of the *Processus*'s first chapter. It is fascinating to see Yworth's work incorporated wholesale into Newton's florilegium, but what is even more interesting is what follows. There is a second chapter in Dibner 1041 B, which is titled "Reductio et Sublimatio" (reduction and sublimation). This chapter at first seems to bear no relation whatsoever to Yworth's *Processus*, as it is largely a compilation from the work of pseudo-Ramon Lull. The reader will remember from our chapter thirteen that Newton thought the Lullian corpus's heavy emphasis on "spirit of wine" or the quintessence to veil a discussion of the philosophers' stone. Under the influence of early modern chymists such as Edmund Dickinson and Adrian von Mynsicht, Newton decided that pseudo-Lull's discussion of ethyl alcohol distilled from wine was really about a mineral that had to undergo a barrage of processes in order to become a powerful menstruum. What then does this have to do with Yworth?

If one progresses further into the *Processus mysterii magni* it emerges that Yworth, like Newton, thought of such mineral extractions as leading to a "philosophical wine." In fact, chapter three of the *Processus* as found in Keynes 66 has the title, "Of the Philosophers spirit of wine or Aqua Vitae, the burning water, and the spiritus mundi & fire of union." A glance at the content shows that Yworth is speaking of his antimonial distillation products fabricated by the operations already described and other refinements.[46] Nonetheless, Yworth differs strikingly from Newton in his relative lack of interest in pseudo-Lull. The name does not appear in Keynes 66, nor is it prominent in any of the printed works by Yworth that I have consulted. We are now in a position to arrive at several conclusions. First, the fact that Yworth's project for anatomizing stibnite in the *Processus mysterii magni* does not appear in his earlier works leads to the inference that this was primarily Newton's project, and that Yworth was being paid to carry out the experimental work. Second, the heavy presence of pseudo-Lull in the florilegium making up Newton's manuscript Dibner 1041B (and its absence in Yworth's work) strongly suggests that Newton was reviving his earlier

---

[46] Keynes 66, 3v–4r. Despite the fact that Yworth misleadingly states that his spirit of wine "hath its descent from urine," his subsequent comments are unequivocal: "It is thus prepared. ℞ the superior waters or mercurial spirit & red oyle mentioned in the first chapter where I shew the separation of the Chaos, & after they have stood in a cold cellar for a month & the nethermost water is drawn off to a drop, put them into a strong double Glass Quart Retort & add to them first the vinegar well deflegmed mentioned in Chap. 2 & then the sublimat<e> mentioned in the same chapter." All of these references point back to the operations performed earlier in the text on the "matter you know well," namely, stibnite.

project for laying out the multiple *Opera* or stages necessary for the production of the philosophers' stone as we described them in chapter thirteen. It seems that by the early eighteenth century Newton had come to the conclusion that Yworth's process for dissecting stibnite was the correct path leading to the Lullian quintessence.

What then can we say about Newton's alchemical collaborations in general terms? It is clear that even though Yworth was receiving an "allowance," both he and Fatio (along with the latter's anonymous friend) were active research participants, not mere laboratory technicians. In Fatio's case this emerges from two sources: first, the August 1, 1693, letter to Newton and the Cantabrigian natural philosopher's undated reply show unequivocally that Newton was imparting key features of his process for making the caduceus of Mercury to Fatio's friend. The main materials involved, *viride æris*, stibnite, lead, and iron reemerge in the complicated series of operations presented by "Three Mysterious Fires," though that document makes the concession to Newton's normal experimental practice of using lead and iron in their unrefined form while at the same time introducing corrosive sublimate, a reagent that he seldom employed after the 1670s. Hence we can see the normal give-and-take of a collaborative effort in "Three Mysterious Fires." Yworth's *Processus mysterii magni*, on the other hand, seems to represent the impressive technical skill of the Dutch distiller without much operational input from Newton. Yet the project of dissecting antimony itself descends from the Basilian practice that Newton described in *Praxis*, and may well find its immediate origin in Newton's need to make Basil's process work. Thus the *Processus* also represents a collaborative project, though of a slightly different sort from the one in "Three Mysterious Fires."

In any event, both collaborations drive home a point that I have already made in this book. Newton's research into the higher arcana of chymistry was not the lone pursuit that Westfall and some other scholars have depicted. In addition to the collaborations with Fatio de Duillier and Yworth, the evidence of book and manuscript loans that we discussed in chapter five militate against any claim that Newton was a "solitary scholar" in his alchemical studies even in their early phase. The fact that Newton actually carried out at least one interview with Yworth on the subject of their joint research also opens the possibility that other similar exchanges lie buried in the voluminous Newton papers. Indeed, one such document may exist in a manuscript of uncertain date where Newton lists six questions, among which we find "How he contrives his Lamp" and "Whether y^e ~~matters~~ ^spt in ye first digestion stink & how soon & wth what odor."[47] Another testimony lies in a letter of March 2, 1682/3 sent from London by the otherwise unknown Francis Meheux. Meheux responds to Newton's previous questions about an earth or ore extracted from a depth of two feet along with an unspecified

[47] Mellon MS 78, 7v. Since 5v refers to the *Centrum naturæ concentratum* attributed to "Alipili," "now done in English 1696," at least part of the manuscript postdates that year. But the part of the manuscript with the six queries is found on a different sheet with a different watermark (or rather a countermark—"MC"). The fact that Newton refers to Philalethes as "Æyrenæus" on this sheet (at 6v) suggests that it may belong to an earlier period. The matter awaits resolution.

water and relates that "hee," an unnamed chymist, is attempting to carry out the great work with the water.[48] Adding this to the material that we have unearthed about the alchemical friend for whom Fatio was the intermediary, the project with Yworth, the "Londoner" who visited Newton in March 1695/6, and the interaction with Captain Hylliard, it emerges that the famous savant was interacting eagerly with a wide range of London chymists on the subject of the philosophers' stone. In the remainder of this book we will explore Newton's less secretive exchanges with the scientific community in the great commercial and intellectual center in order to determine the degree of porosity between his alchemical work and the discoveries that have led to his enduring fame.

[48] Francis Meheux to Newton, March 2, 1682/3, in Turnbull, *Correspondence of Isaac Newton*, 2: 386.

TWENTY

Public and Private

Introduction

Newton's chymical collaborations with Nicolas Fatio de Duillier and William Yworth were focused on discovering and exploiting the *arcana majora* of the art as expounded by such mysterious authors as Johann de Monte-Snyders, Basilius Valentinus, Michael Sendivogius, and Eirenaeus Philalethes. While involved in these secretive, high-stakes ventures, however, Newton was also engaging with chymistry in the public scientific sphere. He composed a variety of treatises and letters between 1670 and 1717 that brought his views on the nature and structure of matter to the attention of the scientific public. Although some of these documents did not find their way into print under their original form, almost all of the ideas expressed in them reemerged in works that were published or at least shared with prominent scientific figures during Newton's lifetime. We need to avoid falling into the old habit of drawing a hard line between "alchemy" and "chemistry" when placing Newton's private chrysopoetic ventures into juxtaposition with his public chymistry. As we have already seen, Newton's personal laboratory notebooks are filled with themes and goals taken from writers on chrysopoeia, and yet they also employ the published work of the Baconian "naturalist" Robert Boyle at great length, and even make use of David von der Becke's writings on the chymical affinities at work in sal ammoniac and salt of tartar to advance Newton's aurific project. The bifurcation to be drawn here is not one between a supposedly animistic, spiritual world of alchemy and a dry, factual realm of practical chemistry, as it has been portrayed by many historians in the past. Rather, it is a distinction between chymical projects that were thought to have potentially dangerous consequences, thereby making it imperative to keep them private, and other realms of chymistry that were deemed safe and even salubrious for public consumption.

Nothing could reveal the division between public and private more clearly than a well-known letter that Newton wrote to Henry Oldenburg, secretary of the Royal Society, on April 26, 1676. Although Newton held Boyle's chymical work in the highest regard, he was deeply concerned about a tract that the English "naturalist" had published in the *Philosophical*

Transactions on the first of January. Boyle's little treatise, "Of the Incalescence of Quicksilver with Gold, Generously Imparted by B. R.," described an experience with mercury that had been treated so that it would rapidly heat gold calx or leaf gold in the palm of one's hand. Modern research has revealed that this incalescent mercury, which Boyle says he first encountered around 1652, was the sophic mercury of George Starkey, produced during the period when Boyle and Starkey were engaged in a program of serious collaborative research.[1] At any rate, the numerous hints that Boyle throws out in "Incalescence" did not fail to hit their mark. Newton's ears must particularly have pricked up when he encountered Boyle's reference to "diverse *Philalethists*" who were making mercurial *arcana*.[2] Surely this was a hint that Boyle's strange quicksilver was related to the sophic mercury of Eirenaeus Philalethes, Newton's chymical avatar. Although Newton was quick to disavow any possible success to Boyle's incalescent mercury as an agent of chrysopoeia, for reasons that we will discuss in due course, he was nonetheless very worried that his older compatriot might have unwittingly revealed "an inlet to something more noble." The full text of his comments suggest that the anxious Newton was tearing his hair:

> But yet because ye way by wch ☿ may be so impregnated, has been thought fit to be concealed by others that have known it, & therefore may possibly be an inlet to something more noble, not to be communicated wthout immense dammage to ye world if there should be any verity in ye Hermetick writers, therefore I question not but that ye great wisdom of ye noble Authour will sway him to high silence till he shall be resolved of what consequence ye thing may be either by his own experience, or ye judgmt of some other that throughly understands what he speaks about, that is of a true Hermetic Philosopher, whose judgmt (if there be any such) would be more to be regarded in this point then that of all ye world beside to ye contrary, there being other things beside ye transmutation of metals (if those great pretenders bragg not) wch none but they understand.[3]

Newton's fears have nothing to do with a spiritual or "mystical" dimension of alchemy, but focus solely on the "immense dammage to ye world" that could result from a widespread dispersion of the secrets of the art. As we have learned from the foregoing chapters of this book, he was willing to entertain the most extravagant assertions of chymical writers, including the claims of Edwardus Generosus that the philosophers' stone in its lunar form could project deadly, freezing rays, while the solar stone could serve as a source of intense and dangerous heat. What other powers Newton imputed to these alchemical products we can only imagine, but his point remains clear: although Boyle's publications on subjects such as the redintegration of niter and the color changes wrought by acids and alkalis were laudable and

[1] Principe, *AA*, 159–65.
[2] Robert Boyle, "Of the Incalescence of Quicksilver with Gold, Generously Imparted by B. R.," *Philosophical Transactions* 10 (1675/76): 510–33, see 530.
[3] Newton to Oldenburg, April 26, 1676, Newton, *Corr.*, vol. 2, letter 157, pp. 1–2, see p. 2.

welcome, he should forego future discussion of *arcana majora* such as the sophic mercury and hold himself to "high silence."

Keeping this distinction between secret and open research in mind, we can now chart out some of the most important points of intersection between Newton's chymical work and the British scientific community in the period roughly spanning the last quarter of the seventeenth century and the first quarter of the eighteenth. As we shall see in the next three chapters, there is a steady leakage from the seemingly watertight privacy of Newton's chrysopoetic projects into his public chymistry. Important areas of overlap include Newton's study of the ether, other subtle media, combustion, the microstructure of matter, the causes of fermentation and putrefaction, and the all-important subject of elective affinity. To some degree, all of these topics intersect with the general domain of optics, an area whose relationship with chymistry we already introduced in chapter six. Hence we will have occasion to return to the study of light and its properties in the remaining chapters. Finally, as a further illustration of the highly charged relationship between public and private chymistry, we will revisit Newton's relationship with Boyle, particularly the involvement of the famous physicist at the end of the seventeenth century with the disposal of Boyle's literary remains.

Ether, Air, and the Aerial Niter

The place to start our discussion is Newton's fascinating little treatise *Of Natures obvious laws & processes in vegetation*. As the reader will recall from our chapter eight, this document provided Newton with an opportunity to examine the demarcation between mechanical and vegetative operations, both in the realm of nature and in that of art. Building on the theory of a cosmic circulation propounded by Sendivogius, Newton argued that the globe of the earth is a "great animall" or rather an unsouled "vegetable" that continually draws in ethereal breath for its "refreshment" and breathes out gross exhalations after condensing the ether within its depths. At the same time, Newton believed that the descending ether was interwoven with grosser matter in the bowels of the earth, and that this subtle substance served as a principle of activity. Hence Newton's ether served both to carry down ponderous bodies, thereby acting as a cause of gravity, and to activate and direct otherwise brute matter. Waxing eloquent on the subject of the ether, Newton called it "Natures universall agent, her secret fire," and "y^e ~~sole~~ onely ferment & principle of ^all vegetation."[4] The echoes of Sendivogius's *Novum lumen chemicum* are particularly clear when Newton goes on to say that the ether may provide the vehicle for an even more subtle spirit that is in turn "ye body of light," for the Polish alchemist had claimed that all matter contains a hidden "spark" (*scintilla*) that is a guiding *semen* or seminal principle.[5] Surely Newton had this in mind when he added that the active principle was an

[4] Dibner 1031B, 3v–4r.
[5] Dibner 1031B, 4r.

"inimaginably small portion of matter" hidden within gross substances and guiding their actions.[6]

Another prominent feature on display in *Of Natures obvious laws* is Newton's keen desire to distinguish among the different subtle media that permeate our atmosphere. As we just saw, the ether per se is not the finest of these; Newton postulates that there is a yet more subtle spirit within the ether that acts as the body of light. At the other end of the scale, Newton categorizes a variety of gases that will not condense to a fluid when cold as "air" (elsewhere he uses the term "true" air or "permanent" air). Thus atmospheric air is grouped with the gases released when salts or vitriols are mixed (one thinks of carbon dioxide released by adding an acid to salt of tartar), destructive distillation of vitriols, corrosion of metals in acids, and fermentation. In addition, *Of Natures obvious laws* explicitly refers to the Sendivogian aerial niter, calling it "the ferment of fire & blood &c." and saying that it has an affinity with ordinary saltpeter.[7] Given that this reference occurs in the midst of a discussion of the mineral fumes or exhalations that generate saltpeter and sea salt by association with water vapor, it is very likely that Newton is thinking of the volatile niter as one of these grosser materials rather than as "true" air. In other words, the aerial niter is an exhalation contained within the atmosphere (but distinct from air properly speaking) that is responsible both for combustion and for the heating and vivifying action of the blood within the body. Finally, as we saw above, Newton also speaks of other gross exhalations, presumably including fumes and vapors, that are also carried up by the air when the vapor is generated in the bowels of the earth-vegetable. According to *Of Natures obvious laws*, then, the ambient is a complex mixture of airs, various vapors, fumes, and exhalations, volatile niter, ether, and perhaps a still more subtle spirit that forms the body of light.

The same desire to distinguish among the various subtle media formed the pretext for another short document by Newton composed soon after *Of Natures obvious laws*, namely, *De aere et aethere* (On the Air and the Ether). Probably written between 1673 and 1675, this incomplete treatise of two chapters may well be a précis of a longer work that Newton was planning to write on the subject.[8] In *De aere et aethere* Newton tries to explain a large number of phenomena on the basis of the varying density of the air working in tandem with what we may call repulsive forces. The juvenile status of the work reveals itself, among other ways, from Newton's treatment of capillary action; he erroneously states that the rising of water in thin glass tubes does

---

[6] Dibner 1031B, 6r.

[7] Dibner 1031B, 2r.

[8] Hall and Hall, *UPIN*, 187–88. The Halls base their *terminus post quem* on the fact that Newton refers to Robert Boyle's *New Experiments to Make Fire and Flame Stable and Ponderable* (published in 1673). The *terminus ante quem* derives from the fact that *De aere et aethere* attributes various roles to air that had been transferred over to the ether in Newton's *Hypothesis of Light*, which he sent to Henry Oldenburg in late 1675. Westfall, on the other hand, dates *De aere et aethere* to 1679, but he seems to misunderstand or oversimplify the Halls' reason for dating it earlier. See Westfall, *NAR*, 374n116. I accept the Halls' dating for reasons stated in the present chapter.

not occur "in an exhausted glass vessel."[9] Thus capillary action in a piece of paper partially immersed in water is due to two causes. First, the air "seeks to avoid the pores or intervals" between the particles of the paper, resulting in a greater rarity of air in the pores than in the exterior air. Then, since the pressure of the exterior air thereby becomes more powerful than that of the rarefied air in the pores of paper, it pushes water into them, and the liquid gradually creeps up the paper. A similar explanation accounts for the rise of water in a narrow glass tube partially immersed in the fluid. Here, however, it is a repulsive power between the particles of air themselves that accounts for the increased rarity in the tube, which again allows for a rise of water. Although Newton is agnostic as to the nature and origin of this repulsive power found in particles, he uses it to explain the great difficulty of pressing two convex lenses into close mutual contact, the impossibility of reuniting powdered metals by mere pressure, and the inability of lead and tin, even when molten, to bond with the iron walls of the vessel in which they are fused. Even the fact that "flies and other small creatures" can walk on water without wetting their feet results from the repulsion between particles.[10]

None of these phenomena made any appearance in *Of Natures obvious laws*, nor do the ones that *De aere et aethere* attributes to the ether as opposed to the air. Influenced by Boyle's *New Experiments to Make Fire and Flame Stable and Ponderable* of 1673, Newton now argues that there must be a more subtle medium than air in order to account for the increased weight of metals when they are calcined in a sealed vessel. Although the surrounding air cannot penetrate the glass, Newton says that there is "a most subtle saline spirit" that enters through the pores of the vessel and combines with the metal to form a calx. Similarly, there must be a subtle medium that accounts for the gradual running down of a pendulum's motion in a container exhausted of air, and something of the same nature must be invoked to explain electrical and magnetic effects. We are again in very different territory from the alchemical cosmology presented in *Of Natures obvious laws*, despite the two texts' shared concern with the differences between air and ether. And yet there is another very significant point of convergence between them.

Like *Of Natures obvious laws*, Newton's *De aere et aethere* attempts to go beyond the mere bifurcation between ether and air, differentiating among different components of the air itself. It is extremely interesting to compare the two

---

[9] Hall and Hall, *UPIN*, 221 (for the original Latin, see p. 214). The erroneous idea that capillary action is due to air pressure may stem from Newton's reading of Boyle's 1660 *New experiments physico-mechanicall, touching the spring of the air*. On pages 267–72, Boyle describes experiments with capillary action of liquids in thin glass tubes. When he tried to carry out the experiment in a vessel evacuated by his air pump, Boyle found that the thickness of the glass prevented him from clearly making out the level of the red wine that he had employed in the tube, and he left open the possibility that the phenomenon of liquids rising in narrow tubes was actually due to unequal pressure between the air in the tube and the air outside of it. In 1669 Boyle stated unequivocally that such capillary action does occur in a vessel exhausted of air. See Boyle, *A Continuation of New Experiments Physico-Mechanical* (Oxford: Richard Davis, 1669), experiment 27, pp. 91–92. *Pace* Westfall (*NAR*, 374), it would beggar belief to think that Newton could have upheld his erroneous view as late as 1679, when he was in deep communication with Boyle about the nature of matter. In fact, in his famous letter of that year to Boyle, Newton explicitly attributes capillary action in thin tubes to the ether, not the air. See Newton, *Corr.*, 2: 289.

[10] Newton, *De aere et aethere*, in Hall and Hall, *UPIN*, 221–24.

texts on this point, for by doing so one can see how Newton's theory has progressed. In accordance with his newfound belief that particles of matter repel one another, *De aere et aethere* argues that the generation of air occurs whenever the corpuscles of a body are torn apart from one another. When this separation takes place, the particles will naturally experience a mutual repulsion and disperse into "air" (we would say a gas), as when gunpowder is deflagrated, or filings of lead, brass, or iron are dissolved in aqua fortis with an accompanying ebullition. But Newton also points out that "aerial substances" differ from one another, depending on the material from which they are generated:

> So the atmosphere is composed of many kinds of air, which nevertheless can be divided into three chief kinds: vapours, which arising from liquids seem to be the least permanent and the lightest; exhalations, which arise from thicker and more fixed substances, especially in the vegetable kingdom, are of a middle nature; and air properly so called whose permanence and gravity are indications that it is nothing else than a collection of metallic particles which subterranean corrosions daily disperse from each other. This is confirmed by the fact that this latter air serves (as the almost indestructible nature of metals demands) neither for the preservation of fire nor for the use of animals in breathing, as do serve some of the exhalations arising from the softer substances of vegetable matter or salts.[11]

Neither the emphasis on repulsion nor the clear, tripartite division into permanent air, exhalations, and vapors is found in *Of Natures obvious laws*. Of greatest interest, however, is Newton's apparently new theory in *De aere et aethere* that "permanent air" is merely a collection of metallic particles that have been separated from one another by corrosion of metals within the earth. This has a distinct resonance with *Humores minerales*, the text that accompanies *Of Natures obvious laws* in Dibner 1031B. The reader will recall that in *Humores minerales* Newton constructed a theory to explain the fact that metals and ores, despite their constant subterranean corrosion by acids within the earth, do not disappear from the earth's surface. His explanation is that the dissolved metals are volatilized after sinking down toward the center of our globe, where they are divided into the metallic principles sulfur and mercury. Then in a highly attenuated form they undergo fermentation with one another to generate new metals. At the same time, some of the metallic fumes escape to the surface, where they "wander over the earth and bestow life on animals and vegetables. And they make stones, salts, and so forth."[12] As for *Of Natures obvious laws*, a large part of that text is devoted to explaining precisely how these metallic fumes produce salts when exposed to water or water vapor, particularly saltpeter and sea salt. As Newton explains, these two salts both originate from the same metallic source "in a highly volatile & anomalus condition"; their differences are due to a dissimilarity in their mechanical texture alone.[13] Yet here too, as in *Humores minerales*, Newton

---

[11] Newton, *De aere et aethere*, in Hall and Hall, *UPIN*, 227.
[12] Dibner 1031B, 6r.
[13] Dibner 1031B, 1v.

emphasizes a role for metals in the maintenance of life, as shown by chymical medicine with its emphasis on mineral cures. The same thing that is true of iatrochemical drugs should also be true of metallic vapors; hence, "wee must of necessity have a great dependence on them," which Newton says is revealed by "healthfull & sickly yeares."[14]

Although the emphasis on a role for metals in maintaining life and health in *Humores minerales* and *Of Natures obvious laws* might at first seem to contradict *De aere et aethere*'s claim that permanent air serves neither for the "preservation of fire nor for the use of animals in breathing," this is not actually the case. A careful examination of the terminology in the two earlier texts shows that they speak only of "fumes," "vapors," and "exhalations" as bestowing life on animals and vegetables. Genuine air is something quite distinct from these, as shown in the following passage from *Of Natures obvious laws*:

> By minerall dissolutions & fermentations then is constantly a very great quantity of air generated wch perpetually ascends wth a gentle motion (as is very sensible in mines) <illeg.> being a vehicle to minerall fumes & watry ~~exhalations~~ vapors, boying up the clouds.[15]

The actual air buoys up mineral fumes and watery vapors and is therefore heavier than them. Newton already refers here to the fact that air itself is generated by dissolution of some metals by acids in *Of Natures obvious laws*. Nonetheless, there is no trace of the theory that true, "permanent air" is merely a congeries of unaltered metallic particles in either of the early texts preserved in Dibner 1031B. Indeed, *Of Natures obvious laws* explicitly says that materials other than metals can also generate air when corroded in an acid, and the text even claims that such corrosion by an acid in the human body is the reason why "poyson swells in a man." It appears that Newton's metallic theory of the air was a new concept when it appeared in *De aere et aethere*, and it may well have been suggested to him by his theory of repulsion between particles when they are separated from one another.

If we pass now to Newton's *Hypothesis of Light*, we will see a further evolution of these ideas. The *Hypothesis of Light* is the intricate treatise that Newton transmitted to Oldenburg on December 7, 1675, to satisfy "the heads of some great virtuoso's" that "run much upon Hypotheses," after the controversy produced by his 1672 *New Theory about Light and Colours*. Significantly, there are two versions of the treatise, one the original version sent to Oldenburg, and the other an emended copy that was actually read at the Royal Society. In the following, I will mostly refer to the original draft, since the second version seems to have been sanitized somewhat in order to avoid ruffling feathers.[16] The *Hypothesis* covers a staggeringly large range of

---

[14] Dibner 1031B, 1r.

[15] Dibner 1031B, 3v.

[16] In Turnbull's edition, the passages that Newton deleted in the second version of the *Hypothesis of Light* are placed within curly brackets, and the words that replace them are included within square brackets. In my quotations I include only the original passages sent to Oldenburg in the letter of December 7, 1675, unless noted otherwise. For a discussion of the various drafts, see Newton, *Corr.*, vol. 1, letter 145, pp. 386–87, n. 1.

phenomena, many of which we cannot treat here in detail. Instead, we will focus on those topics that relate to Newton's chymical interests, especially though not exclusively those that have already emerged from our foregoing discussion. Let us begin with several features that the *Hypothesis* shares with *Of Natures obvious laws* before passing to entirely new chymical subjects that appear in the 1675 treatise. Perhaps the most striking parallel with *Of Natures obvious laws* appears in the description of the earth's inner workings proffered by the *Hypothesis*. This rich passage must be presented here in its full form:

> So may the gravitating attraction of the Earth be caused by the continuall condensation of some other such like æthereall Spirit, not of the maine body of flegmatic æther, but of something very thinly & subtily diffused through it, perhaps of an unctuous or Gummy, tenacious & Springy nature, and bearing much the same relation to æther, wch the vitall æreall Spirit requisite for the conservation of flame & vitall motions (I mean not ye imaginary volatile saltpeter), does to Air. For if such an æthereall Spirit may be condensed in fermenting or burning bodies, or otherwise inspissated in ye pores of ye earth to a tender matter wch may be as it were ye succus nutritious of ye earth or primary substance out of wch things generable grow; the vast body of the Earth, wch may be every where to the very center in perpetuall working, may continually condense so much of this Spirit as to cause it from above to descend with great celerity for a supply. In wch descent it may beare downe with it the bodyes it pervades with force proportionall to the superficies of all their parts it acts upon; nature makeing a circulation by the slow ascent of as much matter out of the bowells of the Earth in an æreall forme wch for a time constitutes the Atmosphere, but being continually boyed up by the new Air, Exhalations, & Vapours riseing underneath, at length, (Some part of the vapours wch returne in rain excepted) vanishes againe into the æthereall Spaces, & there perhaps in time relents, & is attenuated into its first principle. For nature is a perpetuall circulatory worker, generating fluids out of solids, and solids out of fluids, fixed things out of volatile, & volatile out of fixed, subtile out of gross, & gross out of subtile, Some things to ascend & make the upper terrestriall juices, Rivers and the Atmosphere; & by consequence others to descend for a Requitall to the former. And as the Earth, so perhaps may the Sun imbibe this Spirit copiously to conserve his Shineing, & keep the Planets from recedeing further from him. And they that will, may also suppose, that this Spirit affords or carryes with it thither the solary fewell & materiall Principle of Light; And that the vast æthereall Spaces between us, & the stars are for a sufficient repository for this food of the Sunn & Planets. But this of the Constitution of æthereall Natures by the by. (Newton to Oldenburg, December 7, 1675, Newton, *Corr.*, vol. 1, letter 146, 3665–66)

The immediate stimulus for this extraordinary passage was Newton's experimentation with static electricity produced by rubbing a glass hemisphere under which was a space containing bits of paper. He interpreted the saltation of the paper fragments as the product of an ethereal wind emitted from the glass by his rubbing, returning to the glass when his hand was withdrawn and then recondensing within the glass. In the same fashion, an invisible, ambient ether—or rather a more subtle medium contained within the "flegmatic" ether—might continually undergo condensation within the earth; this would cause other ether to rush down and fill its place, and the down-rushing ether would push any interceding bodies toward the center of the earth, thus accounting for gravity. In the meantime, this subterranean inspissation of the ether would be counterbalanced by the emission of air, vapors, and exhalations that were also continually rising from "the bowells of the Earth." The resulting circulation of ether and less subtle atmospheric media cannot fail to bring to mind the Sendivogius-inspired world-vegetable described in *Of Natures obvious laws*. It is therefore surprising to find Newton making an explicit disclaimer in which he openly rejects the Sendivogian aerial niter in the following terms: "I mean not ye imaginary volatile saltpeter." What is the meaning and significance of this rejection?

The reasons for Newton's dismissal of the volatile niter in the *Hypothesis of Light* are neither straightforward nor simple, but as we shall see, they expose the inadequacy of facile claims that Newton continued to adhere to an unqualified belief in "the alchemical vegetable spirit" throughout his career.[17] In the *Hypothesis*, Newton clearly rejected the aerial niter, which was the main alchemical spirit of the seventeenth century concerned with life and vegetation. Our job is to explore the reasons for this dismissal and to determine what parts of chymical theory Newton actually kept. One obvious possibility for Newton's bluntly negative words may lie in the fact that the *Hypothesis* followed on the feet of his bitter exchange with Robert Hooke the nature of light, which began directly after Newton's presentation of the *New Theory* in 1672. The *Hypothesis* begins with a discussion of Hooke's theory of light and color, and even suggests that the ethereal theories expounded there by Newton may bear some relation to Hooke's wave-theory, as expressed in Hooke's 1665 *Micrographia*. Justly famous for its detailed images of microscopic observations ranging from fleas to the structure of cork, Hooke's *Micrographia* also presents a comprehensive theory of combustion that treats the air as a chymical menstruum that dissolves bodies during their burning.[18] Just as an ordinary chymical menstruum consists of an "acid salt" in an aqueous solution, so the menstruum of the air is made up of two parts—volatile niter in place of the acid salt, and an inactive aerial medium in place of the water. As Hooke puts it, the dissolution of "sulphureous" (combustible) bodies in the air is made by a subtle substance mixed into the air "that is like, if not the very same, with that which is fixt in Salt-peter."[19] Like many others in midcentury English scientific circles, Hooke has put the Sendivogian volatile niter to work.[20] Hence the fact that

[17] The expression belongs to Dobbs, *JFG*, 248.

[18] See Robert Frank, *Harvey and the Oxford Physiologists* (Berkeley: University of California Press, 1980), 137–38, for a brief discussion of Hooke's theory of combustion. Frank cites most of the relevant older scholarship on this subject in his n. 179.

[19] Robert Hooke, *Micrographia* (London: Io. Martyn, and Ia. Allestry, 1665), 103.

[20] Frank, *Harvey and the Oxford Physiologists*, 119, 137–39, 221–74.

Newton explicitly asked Oldenburg to remove the passage dismissing the aerial niter from the version of the *Hypothesis* read at the Royal Society; "least it should give offence to somebody" automatically suggests that the potentially aggrieved party might be Hooke.[21]

Given the rancor between Newton and Hooke, it may be tempting to suppose that the disparaging passage on the volatile niter in the *Hypothesis* was originally inserted as a means of needling the author of *Micrographia*, and then removed when Newton realized that his comment was likely to generate more ill will than he bargained for. We know it as a fact that Newton was aware of Hooke's theory of combustion, for the former's early notes on *Micrographia* survive, and they contain extracts to the effect that the air contains a volatile saltpeter responsible for burning as well as respiration.[22] Another possible explanation for Newton's rejection of the volatile niter has been raised by A. R. Hall, who supposes that Newton preferred the original theory in its Sendivogian guise and found its reworking by midcentury mechanists such as Hooke to savor too much of "post-Cartesian particulate physics." A preference for the "noble Pole" over Hooke finds support in the fact that Newton possessed multiple copies of Sendivogius and continued citing him in his alchemical writings until their cessation at some point after his move to London, while he neither owned a copy of *Micrographia* nor referred to it in his mature notes.[23] Although Hall's explanation has its attractions, as does the supposition that Newton originally inserted the dismissive passage in the *Hypothesis* to belittle Hooke, both approaches are ultimately unsatisfactory. Neither of them takes into account other evidence showing that Newton did in fact abandon the volatile niter as the component of the atmosphere consumed in burning and breathing, though he may have retained other aspects of Sendivogius's theory.

The first unequivocal evidence that Newton was developing his own theory of combustion appears in the unfinished "Conclusio" that he wrote for the 1687 edition of the *Principia* but then suppressed.[24] The "Conclusio" is particularly rich in its treatment of chymical topics, and in several important ways prefigures the celebrated *Query 31* of Newton's 1717 *Opticks*. For the moment, we will restrict ourselves to Newton's discussion of combustion, but we will have cause to return to the other topics in the "Conclusio" later. What we will see here is that Newton has for the most part substituted sulfur or a putative component of sulfur for the aerial niter in the process of burning. He begins his treatment with the claim that flame is nothing but a glowing vapor accompanied by heat. The incandescence of flame results from a "fermentation" that leads to the emission of extremely tiny particles from the vapor, which "are transformed into light." This transformation of bodies into light would be a famous theme of *Query 30* in the 1717 *Opticks*, but it already plays an important role in the "Conclusio." By "fermentation"

---

[21] Newton to Oldenburg, January 25, 1675/76, Newton, *Corr.*, vol. 1, letter 153, p. 414.

[22] Newton, "Out of Mr Hooks Micrographia," in Hall and Hall, *UPIN*, 407.

[23] A. Rupert Hall, "Isaac Newton and the Aerial Nitre," *Notes and Records of the Royal Society of London* 52 (1998): 51–61, see especially 56–57.

[24] For discussion of the "Conclusio," see Hall and Hall, *UPIN*, 198–202 and 320–21.

Newton probably means to suggest both the spreading of flame from an ignited source and the rapid motion of tiny corpuscles. As the motion increases, ever smaller particles are released until the body in question begins to glow or burst into flame, as in the case of axles overheated by friction. Bodies do not "feed" flame unless they emit a "sulphureous vapor," and the igniting of this vapor results in the production of flame "by the propagation of fermentation." Examples Newton provided include the flammable vapor of spirit of wine and the exhalation rising from a recently extinguished candle. But what is the precise role of the "sulphureous vapor?" Is it merely a combustible material found within bodies and consumed when they burn, or does sulfur have a more fundamental role here? The following passage, in which sulfur appears multiple times, provides some preliminary answers:

> In coals and ignited materials, however, heat seems to be excited and conserved by the action of a sulphureous spirit. For fire can hardly burn and be supported without fatty and sulphureous matter; with the addition of Sulphur it generally becomes intense. For the fume of Sulphur abounds in an acid spirit, which makes the eyes smart, and when condensed under the bell runs down as a corrosive liquid of the same kind as spirit and oil of vitriol. These only differ by the phlegm in the spirit, and when mixed with other bodies whether dry or fluid excite heat in them, and not infrequently vehement heat. Therefore spirit of Sulphur meeting with the particles of coals and fumes heats them till they glow; for the encounter of hot bodies is the more vehement. And inasmuch as air abounds in sulphureous spirits, it also makes ignited matter grow hot and because of the subtlety of the spirit is required for the maintenance of fire. Whence I suspect that the heat of the Sun may be conserved by its own sulphureous atmosphere.[25]

There is no mention of the aerial niter in this discussion of combustion. Instead, Newton argues that the incandescence of red-hot coals is maintained by a "sulphureous spirit." This spirit is apparently identical to the material produced by burning ordinary sulfur under a glass bell, the seventeenth century's spirit of sulfur *per campanam*, which today we would call sulfuric acid, and which many early modern chymists recognized to be identical with oil of vitriol or the more dilute spirit of vitriol. But why would sulfuric acid have a role in combustion? The key to understanding Newton's reasoning lies in the fact that dissolution in strong acids is often accompanied by rapid warming. As he puts it, oil and spirit of vitriol "excite heat" in dissolving bodies, "and not infrequently vehement heat." In a similar fashion, the putative sulfurous

[25]Newton, "Conclusio," in Hall and Hall, *UPIN*, 343. Although I have followed the Halls' translation here for the most part, I have corrected one significant error. They supply the Latin word "nitri" after "spiritu" in their transcription of the Latin text on p. 329. But a consultation of the original manuscript, CU Add. 4005, shows that "nitri" was lacking in Newton's own draft, though it may have been added by his amanuensis in a later copy. At any rate, "niter" makes no sense in context. Newton is actually saying something rather obvious, that spirit and oil of vitriol are merely different dilutions of the same material (we would say sulfuric acid in an aqueous solution). Adding "niter" here as the Halls do leads to the absurd claim that spirit of niter (nitric acid) and oil of vitriol "only differ by the phlegm in the spirit."

spirit in the air engages in a vigorous chymical reaction with the particles of the coals and smoke, leading to heat and flame. Moreover, it is the subtlety of the sulfurous spirit that accounts for the fact that air is required to maintain flame; the tenuous material is rapidly being used up as it combines with the particles of the coals, and so it must be constantly replenished. Newton goes so far here as to suggest, in a variant on "the solary fewell & materiall Principle of Light" that he had already introduced in the *Hypothesis*, that the sun itself must have a sulfurous atmosphere in order for its heat and light to continue unabated. Although the explicit references here to the role of a sulfurous spirit in the atmosphere help to illuminate the bases of Newton's developing theory of combustion, serious problems remain. In particular, what is the relationship of the sulfur resident in combustible bodies to the sulfurous spirit in the air that is required for burning to take place? The earlier passage spoke of a "sulphureous vapor" emitted by bodies such as alcohol and hot wax; is this what the acidic spirit in air is supposed to act on? The unfinished "Conclusio" provides little more information on this point. For a clearer picture of Newton's theory of combustion, we must therefore turn to another document.

In March 1691/2, Newton would expand on these ideas in a little treatise that he partially wrote in his own hand for his acolyte the Scottish physician Archibald Pitcairne. The product, *De natura acidorum* (On the Nature of Acids), was later printed in John Harris's *Lexicon technicum* (1710), but without some of the important information appearing in the version that Newton transmitted to Pitcairne.[26] If Newton's "Conclusio" to the *Principia* left any remaining doubts about his abandonment of the aerial niter, a glance at *De natura acidorum* will quickly dispel them. With the following words, Newton unambiguously substitutes sulfur for the aerial niter as the component in the atmosphere responsible for combustion and respiration:

> Sulphur seems to be what is deposited on the lungs from the air, and what is supplied from the air to maintain fire seems to be the same. Here to mind that in embryos the blood of the lungs is devoid of . . . sulphur.[27]

While this passage lays to rest any possibility of a lingering affection on Newton's part for the aerial niter, it does not clear up the questions that we previously posed. Although burning and breathing rely on a sulfurous component in the atmosphere, we are left in the dark as to how this aerial sulfur interacts with terrestrial matter. Later in the text, however, the mystery is laid to rest. Newton expands his acidic theory of combustion in the following passage:

> The spirit of sulphur agitates and corrodes all liquids such as water, spirit of wine, spirit of nitre etc and all spongy or fine solids such as clayey earths, nitre, iron, copper etc. (that is substances whose particles of ultimate order of composition are so disjoined or else small that that spirit can quite at

[26] John Harris, *Lexicon technicum* (London: Daniel Brown, Timothy Goodwin, J. Walthoe, John Nicholson, Benjamin Tooke, Daniel Midwinter, M. Atkins, and T. Ward, 1710), vol. 2, "Introduction."

[27] Newton, *Corr.*, 3: 210.

once make its way between them) and effervesces with them because of its acid being attracted by them. By the power of this acid it feeds a flame. For sulphur applied to a lighted coal in which it finds a related acid fume rouses <*exagitat*> the smoke up with such force that light is emitted.[28]

As in the "Conclusio," "spirit of sulphur" refers to the acidic liquid, again our sulfuric acid or spirit of sulfur, produced by burning sulfur under a bell jar and allowing the product to sit until all of it is converted to H_2SO_4. Rather quaintly, Newton says that this spirit "corrodes" water, spirit of niter (nitric acid), and spirit of wine (ethanol), but his meaning is clear. All three liquids heat up immediately when mixed with concentrated sulfuric acid, indicating to Newton that their corpuscles have been set into rapid motion by their attraction for the particles of the spirit of sulfur. In order to show that flame as well as heat is somehow generated by the action of this acid material, Newton then passes from spirit of sulfur made *per campanam* to the spirit of sulfur that is (according to late seventeenth-century chymistry) contained within ordinary sulfur. As he puts it, "By the power of this acid it feeds a flame." Thus when common sulfur is placed on a lit coal, the resulting increase in flame and heat results from the activity of the hidden acid spirit within it, just as the spirit made and collected under a glass bell heats up the various aforementioned liquids. The spirit of sulfur "feeds a flame" because of its attraction for the volatile corpuscles in the coal that have an affinity for it and rush toward it with great speed. In other words, the combustion is due to a chymical affinity between the acid spirit in the sulfur and inflammable particles in the coal. The same principle of chymical affinity is at work when the sulfurous, acidic particles in atmospheric air combine with heated material bodies to produce combustion more generally.

A Chymical Theory of Light

If we continue our examination of Newton's developing ideas about sulfur, it becomes clear that the material plays an increasingly important role not only in his thoughts about combustion but also in his mature optical theory. The first English edition of the *Opticks*, appearing in 1704, connected sulfur, inflammability, and refraction in a highly significant fashion. Newton observed that inflammable bodies exhibited considerably more refraction than noncombustible materials of a similar density. In order to reveal this fact, he even constructs a table relating the refractive powers and densities of twenty-two different materials, thus deriving the refractive power of each substance "in respect of its density." By the compositional reasoning descending from Paracelsus and other chymists, substances that burned contained the principle sulfur, which was released during the process of combustion. On the

---

[28] Newton, *Corr.*, 3: 212. I have followed Turnbull's translation for the most part, but as he misunderstood the expression "invenit acidum fumum congenerem" and several technical terms, the translation has been modified.

basis of this chymical theory, then, Newton argued for a correlation between the amount of sulfur in a given body and its ability to refract light. As he put it in proposition ten of the 1704 *Opticks'* second book,

> the refraction of Camphire, Oyl-Olive, Lintseed Oyl, Spirit of Turpentine and Amber, which are fat sulphureous unctuous Bodies, and a Diamond, which probably is an unctuous substance coagulated, have their refractive powers in proportion to one another as their densities without any considerable variation. But the refractive powers of these unctuous substances are two or three times greater in respect of their densities than the refractive powers of the former substances in respect of theirs.[29]

The purely phenomenological claim that these "unctuous" bodies are much more refractive for their density than glass, crystal, pseudo-topaz, and other "stony" concretes in turn supports a much more fundamental feature of Newton's mature optical theory. From the linkage between greater refractive power and the chymical principle sulfur, he infers that refraction in general is caused by the sulfurous component of materials:

> All Bodies seem to have their refractive powers proportional to their densities, (or very nearly;) excepting so far as they partake more or less of sulphurous oyly particles, and thereby have their refractive power made greater or less. Whence it seems rational to attribute the refractive power of all Bodies chiefly, if not wholly, to the sulphurous parts with which they abound. For it's probable that all Bodies abound more or less with Sulphurs. And as Light congregated by a Burning-glass acts most upon sulphurous Bodies, to turn them into fire and flame; so, since all action is mutual, Sulphurs ought to act most upon Light.[30]

Again following existing chymical theory, Newton assumes that it is not just inflammable bodies that contain sulfur. Instead, in accordance with the view of sulfur as a chymical principle, "it's probable that all Bodies abound more or less with Sulphurs," though obviously combustible materials will contain more of the principle than incombustible ones. Thus he is able to argue that the sulfur in bodies is what acts on light to produce refraction in general. Moreover, since Newton already demonstrated in the previous proposition that "*bodies reflect and refract Light by one and the same power,*" it follows that his sulfurous theory of refraction also accounts for reflection.[31]

Several things are worthy of note in this emphasis on sulfur as a source of refractive power. First, this new information about sulfur and refraction is found among the propositions of the *Opticks* rather than among the more speculative queries. Newton is stating it as a matter of high probability or fact, not as a hypothesis or unproven theory. Furthermore, since the sulfurous theory of refraction is found in all the later editions of the *Opticks* as well, including the Latin *Optice* of 1706, we can consider it an integral part

---

[29] Newton, *Opticks* (London: Sam. Smith and Benj. Walford, 1704), book 2, part 3, proposition 10, p. 75.
[30] Newton, *Opticks* (1704), book 2, part 3, proposition 10, p. 76.
[31] Newton, *Opticks* (1704), book 2, part 3, proposition 9, p. 70.

of Newton's mature optical theory. If we do turn to Newton's expanding list of queries, however, other features of his sulfurous theory of refraction emerge. *Query 7*, for example, asks the following:

> Is not the strength and vigor of the action between Light and sulphureous Bodies observed above, one reason why sulphureous Bodies take fire more readily, and burn more vehemently, then other Bodies do?[32]

The powerful activity observed between light and sulfurous bodies refers primarily to the correlation between "unctuosity" and refractive power outlined by Newton's table in proposition ten. His point is that there is chymical affinity between light and the sulfur principle that causes a mutual attraction between the two. Although he does not use the term "affinity," his meaning is clear: the tiny material corpuscles making up light are attracted by the sulfur in bodies. This affinity results in refraction and reflection, and when it is extremely powerful, as in the case of light concentrated by a burning glass, it leads to the ignition of bodies that are particularly rich in sulfur. In the case of such concentrated light, one can see the full fruition of the ideas expressed seventeen years earlier in the unfinished "Conclusio" to the *Principia*. Newton had already suggested there that sulfur was necessary for combustion, that combustion resulted in the release of light particles from bodies, and that this release was due to the transformation of gross matter into the extremely subtle medium of light. Several years later, in *De natura acidorum*, he had gone so far as to present combustion explicitly in terms of chymical affinity. What is new in the 1704 *Opticks* is Newton's explicit insistence on the mutual character of the activity between light and the sulfur within bodies. Although earlier treatises, such as his 1675 *Hypothesis of Light*, presented a model whereby the ether in the vicinity of a body's surface and within its pores bent or reflected light, and at the same time the light heated the ether by setting it into vibration, the sulfurous theory of refraction is absent there.[33] In short, what we see in the 1704 *Opticks* is a fully fledged chymical theory of light apparently intended to supplant or perhaps complement ethereal models such as that of the *Hypothesis*.[34]

It is important to recognize that Newton's sulfurous theory of refraction was not, although it treated particles of light as though they were chymical corpuscles subject to attractive forces, a theory of color. Even though a given body might refract or reflect light more than another body of the same or different density, the refractive power exercised by a particular medium was continuous over the entire spectrum, not specific to particular colorific rays. This fact, which confused some theorists in the nineteenth century, meant that another explanation had to be employed in order to explain the differing

---

[32] Newton, *Opticks* (1704), book 3, query 7, p. 133.

[33] Newton, *Corr.*, 1: 371.

[34] Nonetheless, Newton famously brought the ether back in the 1717 edition of the *Opticks* in queries 17–24, and uses unequal gradients of ethereal density there to explain refraction. His reasons may stem from experiments performed with two thermometers, one in a vessel exhausted of air, carried out by Jean-Théophile Desaguliers in 1716. See Henry Guerlac, "Newton's Optical Aether," *Notes and Records of the Royal Society* 22 (1967): 45–57.

colors of bodies.[35] That explanation lay in Newton's work on the colors produced by thin films and plates of glass, in other words in the phenomenon that has come to be known as Newton's rings. Already in the 1660s, inspired by Hooke's *Micrographia*, Newton had begun researching the phenomenon of colored rings that appear when two glasses are pressed closely together.[36] What Newton saw was a black center within a series of periodic colored concentric circles; some of the colored bands presented an entire visible spectrum until one moved out toward the periphery of the circle where the colors began to merge and eventually give way to whiteness. Newton realized that the thickness of the layer of air between the two glasses was thinnest at the common center of the rings, and that the blackness there was due to the light being transmitted rather than reflected. Substituting a convex lens with a constant curvature for one of the glass plates, Newton was able to link the different spectral colors that appeared to the gradually increasing thickness of the film of air between the two pieces of glass. Applying Euclid's formula for the sagitta of an arc to the relation between the radius of the convex lens (extended vertically to form a conceptual sphere) and its horizontal diameter, he was even able to measure the thickness of the film at various points, which allowed him to relate specific colors to specific thicknesses.[37] This breakthrough would provide Newton with his theory of the colors of bodies.

By correlating the different bands of repeated colors with the varying thickness of the film between the lens and the flat glass, Newton came up with the idea that differently sized corpuscles acted on light to produce different colors. Assuming that the microparticles out of which opaque bodies are made are themselves transparent, Newton could then argue that the size and density of the corpuscles were all that mattered in the production of different colors. The fact that the colored bands repeat themselves in Newton's rings at differing thicknesses of the interceding film was dealt with by dividing the separate groups of colors into different orders corresponding to their increasing particle size. The green of vegetables belonged to the third order, Newton thought, because the third green ring was particularly vivid, whereas the blue of the sky was probably produced by the dimmer blue of the first order.[38] Once one had determined the order of the color produced by a material, it was possible in principle to deduce the size of the particles making it up from the sagittal relationship used to find the thickness of the film corresponding to that order and color.

Newton's explanation was a physical theory that in itself left no fundamental role for chymistry in deriving the color of bodies. Although chymical operations such as the addition of acids to syrup of violets to turn it red could induce a new color, the resulting color was simply the product of attenuation or thickening of the invisible corpuscles involved.[39] This point has

[35] See Alan Shapiro's excellent treatment of the late eighteenth- and nineteenth-century attempt by chemists to capitalize on his sulfurous theory of refraction in Shapiro, *FPP*, 247–50.

[36] CU Add. 3975, 5v–7r.

[37] Shapiro, *FPP*, 52–55.

[38] Newton, *Opticks* (1704), book 2, part 3, proposition 7, pp. 59–60.

[39] Newton, *Opticks* (1704), book 2, part 3, proposition 7, p. 60.

been made with great clarity by the historian Alan Shapiro, whose words are well worth quoting here:

> We should pause here to grasp fully what Newton has wrought. He has reduced the property of the color of a body solely to the size and density of its corpuscles and made it completely independent of its chemical composition.[40]

Shapiro's point is all the more important in light of previous claims that have been made for the role of alchemy in formulating Newton's ideas about the colors of bodies, particularly by Dobbs. In her *Foundations of Newton's Alchemy*, Dobbs makes the argument that Newton's view of the colors corresponding to corpuscles of different sizes was determined by his knowledge of the alchemical regimens described by Philalethes and other chrysopoetic authors. As the reader will recall from the foregoing chapters of this book, the first stage in the alchemical series leading to the philosophers' stone after Philalethes sealed the sophic mercury and gold up in a flask and subjected them to heat was putrefaction, which produced "the intire Blackness and Cimmerian utter Darkness of compleat Rottenness."[41] According to Dobbs, the blackness of the *putrefactio* regimen was what led Newton to claim that the color of the smallest particles composing matter was black. As she puts it:

> That assumption was drawn directly from the alchemical doctrine that the black matter of putrefaction was in a relatively unformed condition, or in mechanical terms, that it was composed of matter in particles smaller than those produced later in the alchemical process as the matter "matured" or was shaped into various complex substances. There was really no justification for equating black with the smallest particles, except that unquestioned assumption from alchemy.[42]

Unfortunately, Dobbs's claim runs counter to the facts, as Shapiro has pointed out.[43] Newton's idea that the smallest particles were responsible for blackness derived from his observation that the central spot in the colored rings produced by thin films was black. Since this color appeared only at the center where the convex lens lay closest to the plate of glass beneath it, the particles of air allowing for the transmission of light had to be smaller than those producing colors in the concentric rings. Although Newton may have seen the correlation between blackness and small particle size as corresponding conveniently with the alchemical claim that during the stage of putrefaction "thy Compound shall be turned into *Atomes*," his theory of colored bodies did not derive from chymistry.[44]

---

[40] Shapiro, *FPP*, 121.

[41] Eirenaeus Philalethes, "The Vision of Sr George Ripley, Canon of Bridlington, Unfolded," in Philalethes, *RR*, 19.

[42] Dobbs, *FNA*, 225.

[43] Shapiro, *FPP*, 116n48.

[44] Philalethes, *SR*, 81.

In addition to clarifying the debt that Newton's mature optics owed to his chymical research, the present chapter has revealed the surprising fact that despite his ongoing chrysopoetic project, Newton had abandoned the Sendivogian theory of the aerial niter by the mid-1670s and moved in the direction of sulfur as an explanans for phenomena as widely divergent as burning and breathing. Newton employed the ostensibly composite nature of sulfur, the fact that it contained a hidden, but active, acid spirit conjoined to a nonflammable component, to account for its reaction with other materials. The new prominence that Newton gave to sulfur was also evident in his developing theory of light, where the content of this flammable and "unctuous" substance in other materials mapped directly onto their relative refrangibility. The acid, sulfurous spirit had an affinity for numerous material constituents and even for the incredibly small particles making up ordinary light. What were the causes underlying this abrupt shift in Newton's thought? Was he motivated purely by his own work at the bench, or did the new prominence of sulfur reflect trends within the larger chymical community of which Newton was aware? In the next chapter we will examine this issue in the context of Newton's relationship to the republic of chymistry both in England and in Europe at large.

TWENTY-ONE

The Ghost of Sendivogius

NITER, SULFUR, FERMENTATION, AND AFFINITY

Sulfur and the Emerging Phlogiston Theory

Even though Newton's color theory had little to do with the colors of the alchemical regimens, his attribution of refractive power to the sulfur content of illuminated materials fully justifies the view that he held a chymical theory of light. Nor did this fact escape his successors. In the years directly before the Chemical Revolution of the late eighteenth century, European chymists tried to push Newton's chymistry of light further by attaching his linkage of refractivity and sulfur to the phlogiston theory championed by Georg Ernst Stahl.[1] This raises an interesting question: was Newton himself influenced by the phlogiston theory? Although I have found no evidence that Newton read the work of Stahl, he was acquainted with a number of continental chymists who were writing at the period when the Paracelsian sulfur as an inflammable component of metals was gradually metamorphosing into phlogiston. Newton's *Index chemicus* cites the *Physica subterranea* of Stahl's hero Johann Joachim Becher, for example, and the multiple editions of the *Opticks* make use of the work being published at the time by the various chymists of the Parisian *Académie royale des sciences*.[2] Several features of Newton's theory closely parallel developments in pre-Lavoisian chymistry that were taking place in the second half of the seventeenth century, particularly the displacement of the aerial niter theory with sulfur or a putative component of sulfur and the emphasis on the acid produced by burning sulfur *per campanam*. Before dispensing with Newton's ideas about sulfur and its role in combustion and light, it will therefore be useful to say something about the fate of the aerial niter theory in the chymical literature of the late seventeenth century and its sulfurous replacement.

As Robert Frank has demonstrated ably in his work on British scientists who were trying to determine the role of air in respiration, the Sendivogian theory of the aerial niter achieved remarkable popularity as a means

---

[1] For the outlines of this story, see Shapiro, *FPP*, 242–53.

[2] See Newton's citations of Becher on folio 11r of the *Index chemicus* (Keynes 30/1) in the *CIN* edition. For Newton's debts to the chymists of the *Académie royale des sciences*, see Lawrence M. Principe, "Wilhelm Homberg et la chimie de la lumière," *Methodos* 8 (2008), online edition, paragraph no. 27, at https://methodos.revues.org/1223?lang=en, accessed December 10, 2016.

of explaining the chymical properties of the atmosphere. In addition to its prominent use in Hooke's *Micrographia*, the aerial niter was a major research interest of Thomas Henshaw, a founding member of the Royal Society who wrote about saltpeter in the first volume of the *Philosophical Transactions*.[3] It also made a star appearance in the work of the cavalier-scientist Kenelm Digby, and perhaps reached its apogee in the celebrated *Tractatus quinque* of John Mayow, published in 1674.[4] But a brief examination of two authors whom Newton definitely read—Robert Boyle and Nicolas Lemery—shows that the aerial niter was already falling out of favor in some chymical circles by the 1660s and 1670s.

Despite basing his important work on analysis and resynthesis or redintegration on experimentation with saltpeter, Boyle was singularly taciturn on the subject of the aerial niter. The few comments in his voluminous corpus that do mention it are extremely reserved. One sees this caution already in *New Experiments Touching Cold* (1665), where Boyle says that he is unsure of the claim that the "aerial salt, which some moderns call volatile Nitre," actually consists of "true and perfect Salt-petre."[5] This does not mean, of course, that Boyle denied the composite character of the atmosphere, but rather that he refused to identify the combustible, respirable part of it with niter. His *Suspicions about Some Hidden Qualities of the Air*, published nine years later, affirms that there is "some vital substance, if I may so call it, diffus'd through the Air." Importantly, Boyle is also willing in this text to entertain a concept like Hooke's that the air is a menstruum that becomes "glutted" by solutes, but he still hesitates to identify this material with the aerial niter.[6] Instead, he merely says that the air may contain "some secret powerful substance, that makes it a *Menstruum*." The reasons for Boyle's reserved attitude probably stem from his own extensive research on saltpeter. His posthumous *General History of the Air* (1692) presents a stinging rebuke of the claims made for the aerial niter in the following terms:

> I know that divers learned Men, some Physicians, some Chymists, and some also Philosophers, speak much of a *Volatile Nitre*, that abounds in the Air, as if that were the only Salt wherewith it is impregnated. But though I agree with them, in thinking that the Air is in many Places impregnated with Corpuscles of a Nitrous Nature; yet I confess I have not been hitherto convinc'd of all that is wont to be delivered about the Plenty and Quality of the Nitre in the Air: For I have not found, that those that build so much upon this volatile Nitre, have made out by any

[3] Thomas Henshaw, "Some Observations and Experiments upon May-Dew," *Philosophical Transactions* 1 (1665–66), 33–136. For Henshaw's interest in the aerial niter, see Alan B. H. Taylor, "An Episode with May-Dew," *History of Science* 32 (1994): 163–84; see also Donald R. Dickson, "Thomas Henshaw and Sir Robert Paston's Pursuit of the Red Elixir: An Early Collaboration between Fellows of the Royal Society," *Notes and Records of the Royal Society* 51 (1997): 57–76.

[4] For Digby, Mayow, and the aerial niter, see Frank, *Harvey and the Oxford Physiologists*, 126–27, 142, and 258–74.

[5] Boyle, *New Experiments and Observations Touching Cold*, in *Works*, 4: 380; 1665, p. 460.

[6] Boyle, *Tracts containing I. suspicions about some hidden qualities of the air: with an appendix touching celestial magnets and some other particulars: II. animadversions upon Mr. Hobbes's Problemata de vacuo: III. a discourse of the cause of attraction by suction*, in *Works*, 8: 123, 129–30; 1674, pp. 8, 27, and 31.

competent Experiment, that there is such a volatile Nitre abounding in the Air. For having often dealt with Salt-peter in the Fire, I do not find it easy to be raised by a gentle Heat; and when by a stronger Fire, we distil it in close Vessels, 'tis plain that what the Chymists call Spirit of Nitre, has quite differing Properties from crude Nitre, and from those that are ascribed to the volatile Nitre of the Air; these Spirits being so far from being refreshing to the Nature of Animals, that they are exceeding corrosive.[7]

As Boyle points out, the standard method of producing spirit of niter, or nitric acid, was destructive distillation of saltpeter. The result was a choking, poisonous gas, the modern nitrogen dioxide, hardly the salubrious, life-giving principle of respiration and combustibility envisioned by the proponents of the aerial niter. It is entirely possible that Boyle held these views early in his career, for the *General History of the Air* is a pastiche of notes compiled over a period of decades, as Michael Hunter and Edward Davis have shown.[8] Hence if Newton had an opportunity to discuss the aerial niter with Boyle before his rejection of it in the *Hypothesis*, he would not have received a rosy endorsement of the theory. At any rate, Newton's reading notes on *New Experiments Touching Cold* survive in CU Add. 3975, so we know that he was exposed to Boyle's critical evaluation of the aerial niter at an early period.[9]

Similarly, Newton had read the *Course of Chymistry* by the French apothecary and academician Nicolas Lemery in its 1686 printing, though our only evidence of his reading stems from the early 1690s.[10] Lemery, like Boyle, invoked experimental evidence to cast doubt on the idea that niter is a principle of inflammability, which was a key feature of the aerial niter theory. Although other chymists had claimed that niter is inherently combustible, Lemery argues, in reality "Saltpeter is not at all Inflammable by nature." As the French chymist correctly points out, saltpeter will not ignite by itself in a red-hot crucible, though it will deflagrate when placed directly on the coals heating the crucible. According to Lemery, the reason the niter ignites on a glowing coal is because of the

---

[7] Boyle, *The General History of the Air Designed and Begun by the Honble. Robert Boyle*, in *Works*, 12: 32; 1692, p. 41.

[8] Boyle, *Works*, 12: xi–xxiv.

[9] CU Add. 3975, particularly folios 18v–19r where Boyle's discussion of niter is reprised. This early section of the notebook was probably composed by or before 1670.

[10] The testimony that Newton read Nicolas Lemery's *Course of Chymistry* descends from two sources. First, he owned a copy of the 1698 third English edition, which exhibits signs of Newton's characteristic dog-earing. See Harrison, *Library*, 177, no. 938. This edition was published too late for Newton to have read it before writing his "Conclusio" or *De natura acidorum*, but there is independent evidence that he had also read the 1686 second edition. Newton's *Index chemicus* contains a number of references to Lemery's *Course*, and the pagination given by Newton can only refer to the 1686 edition. This is particularly evident from folio 36r of the *Index chemicus*, where Newton provides an entry on the "Essentia vegetabilium," followed by a string of page numbers. If one consults the first, second, and third English editions of Lemery's *Course*, it becomes quite clear that these page numbers, especially the later ones, can only belong to the second edition of 1686. The latest citation in this draft of the *Index chemicus*, found on 16r, belongs to Boyle's *Strange Reports*, published in 1691. See the *Index chemicus* (Keynes 30/1) in *CIN*, accessed December 10, 2016.

[s]ulphureous Fuliginosities of the coals, which are violently raised and rarified by the Volatile nature of the Niter, as I shall prove in the Operation upon fixt Niter.[11]

This important passage tells us that Lemery attributes the flaming of the saltpeter to sulfurous fumes that the molten niter somehow raises from the burning coal. This sounds suspiciously like Newton's account of combustion in the "Conclusio" and *De natura acidorum*, where he said that sulfur placed on hot coals "rouses up" or "drives out" the sulfurous smoke from them. Nor does the similarity stop there. Lemery also theorizes about the origin of volcanoes in the *Course of Chymistry*, suggesting that these arise from the same causes as the great heat generated by leaving a moistened paste of sulfur and iron filings together for four or five hours. According to Lemery, this stems from the friction of the tiny corpuscles of "the *acid* part of the Sulphur" rubbing against those of the iron.[12] Lemery's theory of subterranean heat caused by sulfur would resurface almost verbatim in *Query 23* of Newton's 1706 *Optice* and again in *Query 31* of the 1717 *Opticks*, as recently shown by Lawrence Principe.[13] What is of most interest to us, however, is Lemery's assertion that it is a putative acid component within sulfur that agitates the iron and leads to its incalescence. Something like this claim possibly underlay Newton's own theory that the spirit of sulfur *per campanam* was the agency leading to combustion in general, even if his immediate source was not Lemery.

Long before Lemery and Newton, chymists had identified spirit of sulfur *per campanam* and oil of vitriol as the same material. Some, like the Wittenberg medical professor Daniel Sennert in his posthumous *Paralipomena* of 1642, had even gone so far as to argue that common sulfur consisted of an acid spirit identical to the spirit produced *per campanam* and an oily or resinous component.[14] This increasingly common view would resurface as a mainstay of the phlogiston theory in the early eighteenth century, particularly in the famous work on sulfur by Stahl. In his *Treatise on Sulphur*, Stahl argued that during the combustion of sulfur, both phlogiston and an acid were released; the acid was identical to that which is obtained from the destructive distillation of vitriol.[15] Stahl's basic position, which carefully distinguished the flammable component or phlogiston

---

[11] Nicolas Lemery, *A Course of Chymistry* (London: Walter Kettilby, 1686), 290. The same information is found in Lemery, *An Appendix to a Course of Chymistry* (London: Walter Kettilby, 1680), 76.

[12] Lemery, *A Course of Chymistry* (1686), 139–40.

[13] Principe, "Wilhelm Homberg et la chimie de la lumière," online edition, paragraph no. 27, at https://methodos.revues.org/1223?lang=en, accessed December 10, 2016.

[14] I have used the 1643 imprint: Daniel Sennert, *Paralipomena* (Lyon: Huguetan, 1643), 198–99. Sennert argues first that spirit of sulfur and of vitriol are the same thing: "spiritus sulphuris & vitrioli essentia nullo modo differant, sed ex eadem re generentur & parentur," and then says that sulfur consists of two parts—one resinous or bituminous "ob quam facile concipit flammam," the other saline, "e qua iste spiritus acidus destillando provenit."

[15] Jon Eklund, "Chemical Analysis and the Phlogiston Theory, 1738–1772: Prelude to Revolution" (PhD diss., Yale University, 1971), 155. See Eklund's illuminating discussion of the prehistory of the phlogiston theory on pp. 1–39.

of sulfur from the acid supposedly contained within it, is already found in the earlier writers. Lemery, for example, despite his claim that sulfur's acidic component contributes to the heating up of iron in volcanoes, views the acid in sulfur as a positive hindrance to its flaming combustion, saying that "the Oily part" would soon produce "a great white flame" except for the fact that "the Acid part" is more fixed and "so forces it to cast but only a small blue flame."[16]

If we compare the ideas of the proto-phlogiston chymists to Newton's "Conclusio" and *De natura acidorum*, then, both elements of similarity and serious divergence emerge. Like his chymical peers, Newton accepted the identity of spirit of sulfur *per campanam* and oil or spirit of vitriol, and like them he believed this acid substance to be a component of sulfur. But Newton differed from them in explicitly locating the inflammable power of sulfur in its acid. His association between combustibility and the acidic, sulfurous spirit is evident even in the multiple editions of the *Opticks*, beginning with *Query 23* of the 1706 *Optice*, though with important new modifications based on experimental evidence from the current chymical literature. These experiments corroborate existing features of Newton's theory of combustion rather than altering it fundamentally. The first new evidence relates to the passage from heat to flame, a problematic feature of his theory in the "Conclusio" and *De natura acidorum*. Newton had already argued that flame is nothing but incandescent smoke or vapor, and that the state of incandescence was simply due to an extremely rapid motion of corpuscles that caused tiny light particles to be emitted from the hot matter. But the examples that he had of fire arising from motion alone were all at the gross mechanical level, as when rubbing axles burst into flame. In order to make the claim that something similar was happening at the microlevel in chymical operations, Newton needed examples of spontaneous combustion taken from the realm of chymistry. Moreover, to support his theory that flame and fire resulted from the interaction of an acidic, sulfurous spirit and a combustible material, Newton required experimental evidence that acid reactions could produce actual flame, not merely heat. Such examples are patently absent from the "Conclusio," where vapors from spirit of wine and hot wax are ignited "by the propagation of the fermentation" supplied by a match or other source of open flame. Nor do *De natura acidorum*'s examples of effervescence generated by the interaction of acids with spirit of wine, metals, and other materials lead him nearer to his goal, since the acid reactions known to Newton in 1692 led to heat but not to flame.[17]

The problem posed by the absence of spontaneous combustion from acids was solved in 1694, with a publication in the *Philosophical Transactions* by Boyle's former laboratory assistant Frederic Slare. Slare had been working for some years on the attempt to make "two Liquors kindle" although individually "they are actually cold," a problem that he had inherited from the

---

[16] Lemery, *Course of Chymistry*, 12.
[17] Newton, "Conclusio," in Hall and Hall, *UPIN*, 342, and Turnbull, *Correspondence*, 3: 212.

Danish chymist and physician Olaus Borrichius.[18] In 1683, Slare was only able to produce heat and smoke without flame from turpentine mixed with aquafortis and spirit of wine added to spirit of niter. But in 1694 he published the results of another series of experiments based on a more powerful "Compound Spirit of Nitre" produced by distilling equal parts of saltpeter and oil of vitriol together. Slare considered oil of vitriol to be a "liquid sort of Fire" and ordinary spirit of niter to have "many Effects of Fire"; the compound spirit was therefore thought to contain a "much greater quantity of igneous Matter" than either acid did separately. In order to produce actual flame by adding his compound spirit to turpentine, Slare also found that he had to put in a little "Balsam of Sulphur." When he mixed the fortified turpentine with his compound spirit of niter, the result was actual flame. An even more spectacular result occurred when Slare mixed oil of caraway seeds with his compound spirit in a glass vessel emptied of air by a vacuum pump. "In the twinkling of an eye," Slare says, "the Receiver was blown up" and the remaining oily material ignited. The "stupendious" result "surprized and frightned" Slare and the other observers, who had expected the vacuum to have an inhibiting effect on the course of the reaction.[19]

The fact that Newton recapitulated Slare's account of spontaneous combustion almost verbatim in *Query 23* of the 1706 *Optice* and its successor *Query 31* of the 1717 *Opticks* reveals the importance of the results for his own theory of combustion. Newton even reproduces the precise quantities of materials given by Slare and passes on the caveat that the turpentine must be thickened with balsam of sulfur, all without the slightest mention of Slare or his publication.[20] Newton now had the evidence he needed to assert that acid reactions could lead to flame as well as heat, and since Slare's compound spirit was thought to contain oil of vitriol along with spirit of niter, the sulfurous spirit of the "Conclusio" and *De natura acidorum* was implicated as well.

The role of the acidic, sulfurous spirit was further corroborated by a new publication in which Nicolas Lemery fleshed out his earlier observation that moistened sulfur and iron filings produce what we would call an intense exothermic reaction. In 1700 Lemery published his "Explication physique et chimique des feux souterrains, des tremblements de terre, des ouragans, des éclairs et du tonnerre" in the *Mémoires* of the *Académie royale des sciences*.[21] This essay expanded on Lemery's earlier observation that the iron-sulfur reaction could account for subterranean heat by saying that it also brought about the emission of a vapor or "sulfurous wind" (*vent sulfureux*). This exhalation in turn led to events ranging from subterranean heat and volcanoes

---

[18] Frederic Slare, "An Account of Some Experiments Made at Several Meetings of the Royal Society by the Ingenious Fred. Slare M.D.," *Philosophical Transactions* 13 (1683): 289–302, see 292–94.

[19] Frederic Slare, "An Account of Some Experiments Made by the Mixture of Two Liquors Relating to the Production of Fire and Flame, Together with an Explosion; Actually Cold. By Frederick Slare, M.D.," *Philosophical Transactions* 18 (1694): 201–18.

[20] Newton, *Optice* (1706), 324–25, *Opticks* (1718), 353–54.

[21] Nicolas Lemery, "Explication physique et chimique des feux souterrains, des tremblements de terre, des ouragans, des éclairs et du tonnerre," *Mémoires de mathématique et de physique de l'Académie royale des sciences* (1700), 101–10. For Newton's debt to Lemery, see Principe, "Wilhelm Homberg et la chimie de la lumière," paragraph 27, at https://methodos.revues.org/1223?lang=en, accessed December 10, 2016.

to thunder, lightning, and even hurricanes. Basing himself on an analogy with gunpowder, Lemery argues that thunder and lightning occur when the sulfurous wind encounters a "subtle niter" found in the air. He conceives of the resulting flash and bang as a literal explosion. As in the case of Slare's work, Newton reprised Lemery without attribution, to which he added some observations showing how he managed to adapt the French chymist's comments to his own theory of combustion and respiration:

> Also some sulphureous Steams, at all times when the Earth is dry, ascending into the Air, ferment there with nitrous Acids, and sometimes taking fire cause Lightening and Thunder, and fiery Meteors. For the Air abounds with acid Vapours fit to promote Fermentations, as appears by the rusting of Iron and Copper in it, the kindling of Fire by blowing, and the beating of the Heart by means of Respiration.[22]

Significantly, the "subtle niter" that produces a gunpowder-like explosion with sulfurous winds in Lemery's version becomes "nitrous Acids" in Newton's reworking, thus removing any possible echoes of the aerial niter. Instead of modeling thunder and lightning directly on gunpowder as Lemery had done, Newton presents what he considers a more fundamental explanation based on the intense heat generated by mixing sulfuric and nitric acids, the "sulphureous Steams" and "nitrous acids" of his account. This does not mean that Newton has abandoned the gunpowder approach to meteorological phenomena, however. Elsewhere in *Query 23* of the *Optice* he explains the deflagration of gunpowder itself as a fermentation induced when "the acid spirit of the Sulphur," which he explicitly equates with "that which distils under a Bell into Oil of Sulphur," enters the fixed body of the saltpeter and rarifies it "into the spirit of the Nitre."[23] Furthermore, it is the action of acids such as these that produces the fermentation leading to rusting of metals, kindling of fires, and respiration in animals.

In addition to the new chymical material supplied by Slare and Lemery, Newton was also able to capitalize on recent discoveries made by Wilhelm Homberg, like Lemery a member of the *Académie royale des sciences*. *Query 23* of the 1706 *Optice* and its successor *Query 31* tacitly appropriated Homberg's celebrated analysis of sulfur by means of turpentine or fennel oil, which had appeared in the *Mémoires* of the *Académie* for 1703.[24] Like Homberg, Newton says that sulfur is composed of an "inflammable thick Oil or fat Bitumen" and "an acid Salt," along with fixed earth and a little metal.[25] Newton devotes the section of *Query 23* (and *Query 31*) in which this borrowing occurs to the affinities between some materials that are so great as

[22] Newton, *Opticks* (1718), 355. See *Optice* (1706), 326.
[23] Newton, *Optice*, 295–96; *Opticks* (1718), 317.
[24] Wilhelm Homberg, "Essai de l'analyse du soufre commun," *Mémoires de mathématique et de physique de l'Académie royale des sciences* (1703), 31–40. Newton, *Optice* (1706), 330–31; *Opticks* (1718), 359–60. See Principe, "Wilhelm Homberg et la chimie de la lumière," paragraphs 25–27, at https://methodos.revues.org /1223?lang=en, accessed December 10, 2016.
[25] Newton, *Opticks* (1718), 359. See Principe, "Wilhelm Homberg et la chimie de la lumière," paragraphs 25–27, at https://methodos.revues.org/1223?lang=en, accessed December 10, 2016.

to allow them to bond and sublime together. Hence his immediate concern is the sublimation of sulfur in a closed vessel where the material rises intact without burning and separating into its putative components. Nonetheless, it seems odd that Newton detects no dissonance between Homberg's explicit claim that the component of sulfur responsible for its inflammation is its "thick Oil or fat Bitumen" rather than its "acid Salt" as in his own explanation of combustion. Possibly, Newton saw no contradiction with his theory, since for him the acid component in sulfur produced combustion by the swift motion of its invisibly small corpuscles due to their vigorous affinity for other sulfurous materials such as that supposedly found in charcoal. Newton had long believed that common sulfur itself contained both an acid part and another constituent to which the acid was bonded; Homberg's paper provided him with a more fine-grained analysis confirming this idea. Yet the fact remains that Homberg's essay says nothing of a role for the acid spirit of sulfur in combustion. As in the case of Lemery's meteorological comments about niter, Newton seems to have appropriated Homberg's results and adapted them to his preexisting theory. In short, Newton's sulfurous theory of combustion was actually an acid theory of combustion, owing at least as much to the Hookean and Boylean notion of the air as a menstruum as it borrowed from the growing emphasis on sulfur promoted by the proto-phlogiston chymists. In order to probe more deeply into the reasons why Newton differs from the proto-phlogiston school in locating the inflammable component of sulfur within its acid salt, we must now consider his theory of the microstructure of matter.

Sulfur and the Microstructure of Matter: Newton's Shell Theory

Newton is famous for his belief that there is actually very little solid matter in the universe. As the eighteenth-century phlogiston theorist Joseph Priestley would say in recapitulating Newton's view, "all the solid matter in the solar system might be contained within a nut-shell."[26] Although Newton expressed his views on void space and matter in multiple contexts, perhaps the most celebrated exposition of it occurs in Book 2 of the 1717 *Opticks* (and in the 1706 *Optice*). There Newton presents a corpuscular schema of matter consisting of multiple stages of composition in which each corpuscle of a larger stage is made up of smaller corpuscles belonging to the next stage down plus a volume of void equal to the volume of the smaller corpuscles immediately making up the larger one.[27] Such a framework allows for great porosity in bodies, and Newton uses the fact that water can be made to penetrate even the seemingly dense metal gold, a metal whose invisible pores also

[26] Joseph Priestley, *Disquisitions Relating to Matter and Spirit* (London: J. Johnson, 1777), 17.
[27] Newton, *Opticks* (1718), book 2, part 3, proposition 8, pp. 242–44. For graphic illustrations of Newton's schema, see Shapiro, *Fits, Passions, and Paroxysms*, 132, and Arnold Thackray, *Atoms and Powers* (Cambridge, MA: Harvard University Press, 1970), 64. The proportion of void to matter follows the formula $2^n - 1 : 1$, where n is the stage of composition.

allow the passage of magnetic "effluvia." The same ideas are found already in *De natura acidorum* in the following words:

> Gold has particles which are mutually in contact: their sums are to be called sums of the first composition and their sums of sums, of the second composition, and so on. Mercury can pass, and so can Aqua Regia, through the pores that lie between particles of last order, but not others. If a menstruum could pass through those others or if parts of gold of the first and second composition could be separated, it would be liquid gold. If gold could ferment, it could be transformed into any other substance.[28]

The chrysopoetic echoes in this famous passage are obvious. The idea of a universal transmutability of matter induced by a menstruum that makes gold ferment cannot help but suggest the opening words of *Secrets Reveal'd*, where Philalethes asserts that "our Gold-making POWDER (which we call our *Stone*)" consists of nothing but "Gold digested unto the highest degree." In *Secrets Reveal'd* the gold is first dissolved by a menstruum in the form of the sophic mercury, whereupon it undergoes subsequent putrefaction and fermentation.[29] Newton contrasts the extreme subtlety required of such a radical menstruum with the comparatively gross action of aqua regia and ordinary quicksilver, which pass between the larger particles of the metal and dissociate them from one another without dividing them into their constituent corpuscles.

As if the alchemical overtones of fermenting gold were not conspicuous enough, the version of *De natura acidorum* preserved by Pitcairne presents the following extraordinary statement:

> Note that what is said by chymists, that everything is made from sulphur and mercury, is true, because by sulphur they mean acid, and by mercury they mean earth.[30]

This striking claim not only affirms the medieval alchemical theory that metals and minerals consist of the principles sulfur and mercury but also equates these materials respectively with "acid" and "earth." Moreover, Newton has expanded the alchemical theory to account for the composition of "all things," not merely metals and minerals. Paracelsus had extended the domain of the chymical principles in a similar fashion over a century before, but it is highly significant that Newton has here tacitly dropped the third Paracelsian principle salt. For Newton, salts are themselves compounds, not fundamental types of matter like sulfur and mercury. One may well wonder, however, why sulfur and mercury, or rather acid and earth, would occupy any such primordial status in Newton's system, given the hierarchical system of composition that we previously described, which seems to imply that the only fundamental differences between particles lie in their size and the

---

[28] Newton, *Corr.*, 3: 211.
[29] Philalethes, *SR*, 1.
[30] Newton, *Corr.*, 3: 210. Turnbull translates Newton's "chimici" as "chemists," whereas I have employed the contemporary term "chymists."

amount of void they contain. The response to this puzzle resides in the fact that the acid and earth, or sulfur and mercury of *De natura acidorum*, exist at a higher level of composition than that of the primordial particles of matter, though they are still smaller than the corpuscles of the metals and other materials that they make up.

Newton's theory of material composition in *De natura acidorum*, the *Optice* and successive editions of the *Opticks*, and other works has sometimes been called his "shell theory" of matter (not to be confused with Priestley's famous quip that all the matter in the Newtonian solar system could be fit into a nut shell).[31] As he explains it in *De natura acidorum*, water, acid, and earth consist of particles belonging to three different size ranges: the water corpuscles are the smallest, followed by the larger acid, and then earth. The acid particles have an attractive power that allows them to pull both water and earth corpuscles to themselves, and this, along with the intermediate size of the bits of acid, allows them to act as an intermediary between the other two substances. A major goal of this discussion is to explain the apparent conundrum presented by the fact that materials with a greater specific weight than either water or the mineral acids, such as dissolved metals, distribute themselves evenly in an acid solution and do not sink to the bottom. Newton's idea is that the corpuscles of a metal (or other corrodible material) dropped in an acid will immediately be surrounded by the acid particles. The corpuscles of acid will form a sort of shell around a kernel made up of the larger, earthy particles of the metal. The resulting composite corpuscles composed of acid and earth will float in the solution because of the force of attraction that the acid shell of each particle has for the particles of water. But if a material having greater affinity for the acid than the acid has for the earth is then added, the principle of elective affinity kicks in; the acid particles making up the shell of the composite corpuscle will abandon their earthy kernel and run to the newly added material.[32] This is what happens when salt of tartar is used to precipitate an acid solution containing a metal. Freed from the acid that held them in solution because of its affinity for water, the metallic corpuscles now mass together and fall to the bottom in the form of a powder or muddy sediment.

Newton's shell theory accounts not only for the floating and precipitation of metals in acid solutions but also for certain other phenomena of a more fundamental nature. If a compound consists of acid particles that have been dominated and suppressed by the earthy kernel at the center of the corpuscle, then a fatty, sticky, insipid material will be formed, which is insoluble in water. Examples include common sulfur (on the assumption that

---

[31] Important elements of Newton's shell theory already appear in a well-known letter of February 28, 1678/79, that he wrote to Robert Boyle. I will consider this letter later in the present chapter. See Newton, *Corr.*, 2: 288–97.

[32] Newton does not, to my knowledge, use the actual term "elective affinity," which achieved popularity in the second half of the eighteenth century. Nonetheless, he employs the principle of relative interactivity between different substances, which is the essential characteristic of affinity theory. This was widely recognized by eighteenth-century writers on affinity tables who read *Query 23* of the Latin *Optice* or *Query 31* of the 1717 *Opticks*. Hence I employ the term "elective affinity" advisedly.

it is a compound) along with *mercurius dulcis* or mercurous chloride made by subliming quicksilver with corrosive sublimate, horn silver made by adding spirit of salt (hydrochloric acid) to common silver, and a famous compound of copper made by subliming the metal with corrosive sublimate. In each case, the acid corpuscles have become buried in the central core of the compound corpuscle, reducing their perceptible activity accordingly.

Nonetheless, if such "fatty bodies" as we have just encountered meet with another material that has a greater affinity for their acid than their own earthy kernel possesses, then the acid will follow the principle of elective affinity and forsake its own earth for the new material. If this happens rapidly and with great force, powerful heat and inflammation occur, an explanation that we have encountered already in Newton's theory of combustion. But if the attraction is gentler, a fermentation will result; this can even lead to a generalized putrefaction in which the decomposing material entirely loses its previous character. Here Newton reveals his perennial fascination with fermentation and putrefaction, presenting them as keys for effecting transmutation in general:

> This putrefaction arises from this, that the acid particles which have for some time kept up the fermentation do at length insinuate themselves into the minutest interstices, even those which lie between the parts of the first composition, and so, uniting closely with those particles, give rise to a new mixture which may not be done away with or changed back into its earlier form.[33]

Here one can see the integration of Newton's shell theory and his hierarchical construction of matter from particles of descending size. If the acid particles ferment with a substance for a long time and manage to work their way between the particles of the first composition, they can induce an irreversible transformation. For Newton this was a hallmark of "vegetation" as opposed to mere mechanism. As he had said as long ago as *Of Natures obvious laws*, "mechanicall coalitions ^or seperations of particles" will "returne into their former natures if reconjoned," as in Boyle's famous redintegration of saltpeter.[34] Such simple, building-block redintegration cannot take place when the building blocks themselves have been disseuered.

But what of the acid that works its way into "the minutest interstices" of bodies? Did Newton not explicitly say in another passage of *De natura acidorum* that aqua regia and other mineral acids penetrate only between "particles of last order, but not others?" When Newton speaks of "acid particles" that can enter into the very smallest of pores, he cannot mean the common mineral acids. Rather he is thinking of something like the menstruum alluded to earlier in the text, which might be able to liquefy gold at room temperature and make it ferment. Remembering Newton's identification of "acid" with "sulfur," it is not difficult here to see the active, sulfurous principle of the Philalethan tradition, which provided the "fermental virtue" to

[33] Newton, *Corr.*, 3: 210.
[34] Dibner 1031B, 5v.

Starkey's sophic mercury that allowed it to act on gold.[35] When such a sulfur acts at the most basic level of material composition, working its way into the "minutest" pores of a body, the result is a type of change that Newton had already identified in *Of Natures obvious laws* as a nonmechanical operation.

Newton's shell theory, with its alternating layers of earth (mercury) and acid (sulfur), brings to mind earlier ruminations on the structure of matter whose origins lie in medieval alchemy. The school of Isma'īlī alchemists associated with the semifabulous Persian Jābir ibn Ḥayyān had already laid the foundations, albeit quite vaguely, for thinking of matter in corpuscular terms where a given portion of matter was said to have an external, "manifest" part and an internal, "occult" or hidden part. The occult and the manifest were often viewed as bearing opposed qualities, so that a material that was internally cold and dry, for example, would be externally hot and wet. These ideas became an integral part of medieval and early modern Western alchemy and were importantly reworked by the Flemish chymist Joan Baptista Van Helmont in the early seventeenth century. Van Helmont explicitly treated water and other materials as consisting of structured atoms with sulfur and mercury layers that could be inverted on occasion to produce radically different properties, such as the conversion of a liquid to a vapor (or as Van Helmont says, "gas").[36] The echoes of this theory are already heard in *De natura acidorum*, but they resonate more sharply in *Query 23* of the 1706 *Optice* and its 1717 descendent *Query 31*:

> As Gravity makes the Sea flow round the denser and weightier Parts of the Globe of the Earth, so the Attraction may make the watry Acid flow round the denser and compacter Particles of Earth for composing the Particles of Salt. For otherwise the Acid would not do the office of a Medium between the Earth and common Water, for making Salts dissolvable in the Water; nor would Salt of Tartar readily draw off the Acid from dissolved Metals, nor Metals the Acid from Mercury. Now as in the great Globe of the Earth and Sea, the densest Bodies by their Gravity sink down in Water, and always endeavour to go towards the Center of the Globe ; so in Particles of Salt , the densest Matter may always endeavour to approach the Center of the Particle : So that a Particle of Salt may be compared to a Chaos; being dense, hard, dry, and earthy in the Center; and rare, soft, moist, and watry in the Circumference.[37]

The appealing notion of a corpuscle of salt as a miniature simulacrum of the terrestrial globe and sea allows Newton to explain the solubility of the salt. The layer of acid particles around each earthy kernel in a saline corpuscle, with the acid's shared attraction for water and earth, causes the individual salt particles to separate from a gross mass and distribute themselves evenly in the surrounding water. If the acid particles are then pulled away from the

---

[35] See Starkey to Boyle, May 1651, in Newman and Principe, *LNC*, 23 and 25.

[36] William R. Newman, "The Occult and the Manifest among the Alchemists," in *Tradition, Transmission, Transformation*, ed. F. Jamil Ragep, Sally P. Ragep, and Steven Livesy (Leiden: Brill, 1996), 173–200.

[37] Newton, *Opticks* (1718), 361–62. See *Optice* (1706), 332–33.

saline corpuscles by an alkali or even a metal for which they have greater affinity than they do for their earthy kernel (as in the case of dissolved quicksilver), the dissolved material precipitates. When Newton goes on to employ an antithesis-rich vocabulary in speaking of the "dense, hard, dry, and earthy" center of each salt corpuscle and its "rare, soft, moist, and watry" circumference, one cannot fail to think of the internal occult and external manifest of his alchemical sources.

The critical, intermediary role played by acids in Newton's theory underscores the importance of sulfur for him if we recall his claim in *De natura acidorum* that acid and earth are identical to the sulfur and mercury principles of the chymists. As he also states there, "acid" is what "attracts and is attracted strongly," which characteristics must therefore be transferrable to the sulfur principle.[38] If we return to Newton's thoughts about combustion, the same ideas recur. It is the affinity of the sulfurous, acidic fumes in the atmosphere with the sulfur latent in combustible materials that allows for the occurrence of heat and inflammation, which are both the products of intestine motion among minute particles. The generalizing, universal character of Newton's theory is striking. The activity of sulfur-acid accounts for burning, fermentation, dissolution, precipitation, and a host of other properties including apparently sapidity, stickiness, and of course the ability of bodies to refract and reflect light, as we saw in Book 2 of the *Opticks*. It is no surprise that Newton wrote a work on the nature of acids (which might as easily have been called a treatise on sulfur), and perhaps even less of a surprise, once we understand his layered, planetary theory of matter, that he upheld a modified menstruum theory of combustion rather than simply replicating the proto-phlogiston theory of contemporaneous European chymists.

Fermentation, Putrefaction, and the Electrical Spirit

Newton follows his famous comparison of a particle of salt to the terraqueous globe in the 1717 *Opticks* with comments on putrefaction, an activity that occurs when the watery acid of the corpuscle soaks into "the Pores of the central Earth by a gentle Heat." The seemingly biological language of rotting and decay here reflects that of early modern alchemy in general and brings to mind the process of fermentation that typically enters into Newton's discussions of putrefaction. In order to pursue the connections between these fundamental operations and Newton's chymistry more generally, we need to consider the evolution of the thoughts presented in *De natura acidorum* and the 1706 *Optice* a bit more closely than we have done so far. As Cesare Pastorino has recently shown, draft copies deriving from both texts reveal that between the appearance of the *Optice* and the second edition of the *Principia* published in 1713, Newton undertook a reworking of his chymical ideas. Many of these extended back at least as far as *Of Natures obvious laws*, but Newton integrated them with new experimentation on the nature

[38] Newton, *Corr.*, 3: 207: "acidum enim dicimus quod multum attrahit et attrahitur."

of electricity carried out by the Royal Society's curator of experiments Francis Hauksbee.[39] These suppressed drafts include versions of a *Query 24* and *25* intended to follow the final query in the 1706 *Optice*, as well as a short treatise titled *De vita et morte vegetabili* (On Vegetable Life and Death), both found in Cambridge University Library Additional Manuscript 3970. *De vita et morte vegetabili* in particular builds on and extends the ruminations found in *De natura acidorum*, and even contains many passages virtually identical to those found in the earlier treatise.[40]

De vita et morte vegetabili begins at once with a discussion of an "electrical force" that is excited by friction, just as the static electricity produced in Hauksbee's experiments by the rubbing of a glass container emptied of air. Newton had long before performed experiments with static electricity, which provided evidence in the *Hypothesis of Light* for an ethereal wind circulating between the earth and the heavens. But one of Hauksbee's achievements had been to show that electricity can not only attract small bodies but also repel them, and this observation allowed Newton to assimilate the experimental curator's work to his own concept of a fundamental, short-range repulsive force active in matter.[41] Newton was excited by the implications of Hauksbee's work, and his new enthusiasm for electricity appears in the discussion of a "very subtle spirit pervading gross bodies" in the famous "General Scholium" accompanying the 1713 second edition of the *Principia*.[42] The electrical spirit there becomes a generalized agent responsible for short-range attraction and cohesion, repulsion, the emission of light, and the transmission of nervous impulses in animals. Although his enthusiasm for electricity as a polyvalent cause of phenomena eventually waned after 1713, for a time it became the basic agency that for Newton lay behind chymical phenomena of association and dissociation.[43] Acid menstrua, for example, now operated "by means of an electrical force" (*vi electrica*), which caused their corpuscles to attack the particles of the tongue and stir up the sensation of sourness.[44] Apart from the new role for electricity as a causal factotum, however, many of the ideas in *De vita et morte vegetabili* clearly stem from Newton's chymistry and can be used to flesh out his thoughts on fundamental chymical operations.

Like *De natura acidorum*, *De vita et morte vegetabili* discusses the dissolution of "bodies" (material substances in general) by means of the mineral

[39] Cesare Pastorino, "Alchemy and the Electric Spirit in Isaac Newton's *General Scholium*," forthcoming.

[40] CU Add. 3970, folios 238r, 239r, and 240v contain numerous passages that are verbatim identical to the notes accompanying the text of *De natura acidorum* as recorded by Pitcairne on March 2 and 3, 1691/2. See Pastorino, "Alchemy and the Electric Spirit," p. 11 of typescript.

[41] Roderick W. Home, "Francis Hauksbee's Theory of Electricity," *Archive for History of Exact Sciences* 4 (1967): 203–17.

[42] For the history of this phrase and the appearance of the qualifiers "electric and elastic" in older English translations, see I. Bernard Cohen, ed., *Isaac Newton: The Principia* (Berkeley: University of California Press, 1999), 283–92 and 943–44.

[43] For Newton's more restricted role for the electrical spirit after 1713, see Roderick W. Home, "Newton on Electricity and the Aether," in *Contemporary Newtonian Research*, ed. Zev Bechler (Dordrecht: D. Reidel, 1982), 191–214.

[44] CU Add. 3970, 237r.

acids and other, more subtle menstrua. Newton says here again that bodies dissolved by the mineral acids can easily be returned to their pristine state because the largest corpuscles, those of the "final composition," are undamaged by the menstruum, but merely separated from one another. He then presents the following passage, which is particularly rich in the insight that it offers concerning the relation of fermentation to putrefaction:

> If a dissolution is of the sort that by the action of a menstruum and the reaction of the body, certain more subtle spirits are excited, which can enter the pores of the parts of final composition, then these spirits gradually enter and dissolve those particles, and they separate them into particles of the penultimate composition, just as the acid menstruum dissolved the whole body and separated <it> into particles of the final composition in the foregoing example. And the body has now lost its old form, for the particles of the penultimate composition do not return into particles of the ultimate composition except by generation. And the same is true of the dissolution of particles of the penultimate composition into particles of the antepenultimate composition, etc. But we are accustomed to call these dissolutions the corruption and putrefaction of the body.
>
> Sometimes putrefaction is induced by a ferment, and the ferment is a vegetal body abounding in spirits which can enter the pores of the particles of final composition and dissolve those particles, and by dissolving, gradually excite new spirits of the same genus, by which the putrefaction is completed. The nutriment of animals is fermented by juices in the stomach and above all by bile.[45]

Unlike the related passages in *De natura acidorum*, *De vita et morte vegetabili* carefully spells out the relationship between fermentation and putrefaction. Fermentation is brought about by a "ferment," or as we would now say, a leavening agent, though Newton of course conceives of this in much more general terms than moderns do. In the second paragraph quoted above, the ferment is said to enter between the large particles of the final composition, namely, those involved in "vulgar" chymical operations, just as a common mineral acid would do. Instead of merely dissociating these large corpuscles from one another, however, the ferment dissevers them and simultaneously excites "new spirits of the same genus" as itself. These are the "more subtle spirits" alluded to in the first paragraph that cause the "corruption

[45] CU Add. 3970, 237r: "Si dissolutio ejusmodi est ut per actionem Menstrui et reactionem corporis spiritus aliqui subtiliores excitentur qui poros particularum <illeg.> compositionis ultimæ pervadunt possint ingredi possint, tunc spiritus illi ^paulatim permeant & dissolvunt has partes perinde u et in particulas compositionis penultimæ separant, perinde ut Menstruum acidum dissolvebat corpus totum et in partes compositionis ultimæ, in casu priore separabat. Et corpus formam veterem jam amisit. Nam particulæ compositionis penultimæ in partes compositionis ultimæ non nisi per generationem redeunt. Et par est ratio dissolutionis particularum compositionis penultimæ in particulas compositionis antepenultimæ &c. Hasce vero dissolutiones corruptionem corporis et putrefactionem dicere solemus. ~~fermentum est substantia quae spiritibus sub~~ Putrefactio per fermentum quandoq; inducitur, et fermentum est corpus vegetabile spiritibus abundans qui per poros partium <illeg.> compositionis ultimæ permeare possent et dissolvere & partes illas dissolvere & dissolvendo novos ejusdem generis spiritus ^paulatim excitare quibus ~~dissolutio~~ putrefactio compleatur. Nutrimentum animalium per succos in stomacho et maxime per bilem fermentatur."

and putrefaction of the body." Fermentation, then, is a process that precedes putrefaction proper; the latter, when carried to completion, results in a total loss of form making it impossible for the body to be returned to its previous state by simple redintegration.

An important element of the above passage lies in Newton's insistence on the fact that during fermentation, a subtle spirit can be "excited" by the breakdown of grosser matter into smaller corpuscles, and that this spirit in turn excites yet more active spirits like itself. The Latin infinitive *fermentare* derives from *fervere*, to boil, so it is no surprise that fermentation should result in the release of a spirit or vapor.[46] To Newton, however, it is this expelled, subtle material that leads to additional fermentation until all the susceptible material has been fundamentally altered. This is a key feature of fermentation in his mind, as it helps to explain how a small quantity of an active material can effect great changes on another substance by something akin to a chain reaction. We already encountered something like this in Newton's short text *Humores minerales* from the first half of the 1670s, in which he described a subterranean, circulatory process in which descending minerals dissolved by ordinary acids encountered the rising fumes of other minerals and underwent a process of putrefaction. The passage is so similar to the one just quoted from *De vita et morte vegetabile* that it is worth reproducing here:

> Indeed, these spirits meet with metallic solutions and will mix with them. And when they are in a state of motion and vegetation, they will putrefy and destroy the metallic form and convert it into spirits similar to themselves.[47]

The fermentation and putrefaction that permeate *Humores minerales* and its sister text *Of Natures obvious laws* had long been used by alchemists to account not only for natural processes but also for the spectacular transmutations that they claimed to effect with their elixir or philosophers' stone.[48] Newton himself had pursued this topic in his experimental notebooks: two pages of CU Add. 3973 dating from 1692 and early 1693 discuss the action of barm, the agent of fermentation in beer making; this discussion falls squarely in the midst of Newton's notes on making volatile Venus, Neptune, and other chrysopoetic desiderata.[49] And of course Philalethes had argued in chapter one of *Secrets Reveal'd* that the philosophers' stone is merely "Gold digested unto the highest degree," so a metallic transmutation could be seen in terms of an assimilation of base matter into the same material as the aurific agent itself. In Newton's draft queries to the *Opticks* also found in

[46] See the online *OED*, s.v. "ferment." Accessed December 23, 2016.
[47] Dibner 1031B, 6v.
[48] See Antonio Clericuzio, "Mechanism and Chemical Medicine in Seventeenth-Century England: Boyle's Investigation of Ferments and Fermentation," in *Early Modern Medicine and Natural Philosophy*, ed. Peter Distelzweig, Benjamin Goldberg, and Evan Ragland (Dordrecht: Springer, 2016), 271–94. Clericuzio overgeneralizes, however, when he says on page 277 that "Most alchemists maintained that the philosophers' stone transmuted metals by means of fermentation."
[49] CU Add. 3973, 25r–25v.

CU Add. 3970, he adds further comments that expand on the ability of a ferment to assimilate other material into its own substance. Thus he says that "leaven turns past to leaven," meaning that yeast converts unleavened flour and water into dough that has the power of converting yet more "paste" into a leavening agent. In the same fashion, "a magnet turns iron to a magnet," and "fire turns its nourishment to fire." Indeed, it is such fermentation that allows the living corpuscles of an animate body to transform food into their own substance.[50] As Newton would add in the same draft notes, when describing the growth of an embryo in the womb, "By fermentation the nourishment is subtilized & replenished with spirit & put into motion" so that the embryo can assimilate it. Given their long-standing prominence in the literature of alchemy, it is no surprise that such thoughts should exercise a powerful grip on Newton's mind, though it is perhaps striking that his fundamental ideas had changed so little between the composition of Dibner 1031B and the first or second decade of the eighteenth century.

Evidence of the role that chrysopoetic transmutation continued to play in Newton's thoughts on processes of decay and transformation can be seen in some additional comments from his draft of *Query 25* in CU Add. 3970 that probably stem from his reading of the Philalethes corpus. After again discussing the role of fermentation in breaking bodies down into particles smaller than those of their final stage of composition, he adds:

> [And by this means bodies ^must lose their old form & texture ~~before they~~ & be ^destroyed & broken ~~to pieces~~ into the last parts before they can be formed.] ffor as an old house must be pulled down & its stones separated before ~~they can be put together in another manner~~ a new house can be be <*sic*> built out of ~~them~~ its materials: So ^natural bodies must be ^dissolved broken & separated into their least parts by fermentation & putrefaction & lose their ^old form ^& texture before ~~they can be formed anew~~ a new ^natural body can be formed out of them.[51]

In this paragraph Newton uses the disassembly of an old house into its stones as an analogy for the dissolution of bodies by fermentation and putrefaction. Once the house has been broken down into its "least parts," namely, the stones making it up, the building can be rebuilt into an entirely different structure, but not before such analysis has taken place. A very similar analogy appears in a source that we know Newton to have read, namely, the *Enarratio methodica trium Gebri medicinarum* attributed to Eirenaeus Philalethes (though not actually by Starkey). To the question, "What remains after putrefaction" in the making of the philosophers' stone, the author replies:

> Just as an integral whole such as a house consists of integral, united parts, and when one part is destroyed or disjoined from the place that it formerly had in the whole, the quantity and form of the entire house is destroyed,

[50]CU Add. 3970, 241v.

[51]CU Add. 3970, 235v. A slightly fuller Latin version of this paragraph is found in *De vita et morte vegetabili* at 238r (written vertically in the margin).

and yet the stones, planks, and foundations from which the house was made remain, so in our work.[52]

The author of the *Enarratio methodica* uses the destruction of a building to illustrate the fact that putrefaction breaks a body into its small parts but does not destroy it altogether. Since the overall gist of his argument is that the same parts go on to form a different metal by means of transmutation, his point is not far removed from Newton's.

To conclude this section, the remarkably fundamental role that Newton allocated to fermentation is particularly evident in *Query 23* of the 1706 *Optice* and in English form, in *Query 31* of the 1717 *Opticks*. Although much has been written on the "active principles" that Newton believed to be the origins of observable phenomena such as the falling of bodies, insufficient notice has been given by scholars to the equal billing that Newton grants to the causes of fermentation.[53] At various points in *Query 31*, he refers to fermentation as though it is an independent phenomenon on an equal footing with electricity, magnetism, and gravity, in other words, something akin to what we would today call a fundamental force. This is particularly evident near the end of *Query 31*, where Newton famously distinguished between forces and the occult qualities of the scholastics:

> And the Aristotelians gave the Name of occult Qualities not to manifest Qualities, but to such Qualities only as they supposed to lie hid in Bodies, and to be the unknown Causes of manifest Effects: Such as would be the Causes of Gravity, and of magnetick and electrick Attractions , and of Fermentations, if we should suppose that these Forces or Actions arose from Qualities unknown to us, and uncapable of being discovered and made manifest.[54]

Newton admits here that the causes of the falling of bodies, the attraction exercised by magnets and statically charged bodies, and the heating and effervescence sometimes presented during fermentation are inaccessible to the senses, and at least for now unknown. He goes on to deny that this sensory inaccessibility makes his "active principles" identical to scholastic occult qualities, however, because unlike the Aristotelians he can derive principles from the phenomena and generalize from them. But for us the important thing is the primordial status that Newton grants to fermentation here,

---

[52] [Pseudo-]Eirenaeus Philalethes, *Enarratio methodica trium Gebri medicinarum* (London: William Cooper, 1678), 31–32: "Sicut totum integrale, puta domus, consistit ex suis partibus integralibus unitis, & destructa seu disjuncta una parte a loco suo quem prius habebat in toto, destruitur quantitas & forma totius domus, & tamen remanent Lapides, ligna & fundamenta, ex quibus constabat domus: Ita etiam fit in proposito nostro." Newton cites the Philalethan *Enarratio methodica trium Gebri medicinarum* in the *Index chemicus* at Keynes 30/1, 4v and 58r, for example.

[53] An exception is J. E. McGuire, "Force, Active Principles, and Newton's Invisible Realm," *Ambix* 15 (1968): 154–208, though in McGuire's complicated treatment it is easy to lose sight of the fundamental distinction that Newton means to draw between observable phenomena such as the weight and tendency of bodies to fall (their gravity) and the invisible cause of their weight and falling (the active principle responsible for gravity). For more on fermentation, see Dobbs, *JFG*.

[54] Newton, *Opticks* (1718), 377.

alongside gravity, electricity, and magnetism. For him, fermentation appears to be a sort of fundamental force. Newton's view of fermentation as a physical primitive is implied in various other parts of *Query 31*, but the full implications of its foundational status are spelled out in a remarkable passage where Newton explicitly contrasts it to gravity. The passage comes at the end of his rebuttal of the Cartesian claim that motion is conserved in the universe, a position that Newton discredits by considering a host of examples, including the humble cases of pitch, oil, and water, which all soon stop revolving after being stirred. Since Descartes built his vortical theory of planetary motion on the conservation principle, Newton's rebuttal occupies a particularly prominent place in *Query 31*. A corresponding significance lies in his immediately subsequent discussion of the active principles that he views as most important in supplanting Cartesian mechanism:

> Seeing therefore the variety of Motion which we find in the World is always decreasing, there is a necessity of conserving and recruiting it by active Principles, such as are the cause of Gravity, by which Planets and Comets keep their Motions in their Orbs, and Bodies acquire great Motion in falling; and the cause of Fermentation, by which the Heart and Blood of Animals are kept in perpetual Motion and Heat; the inward Parts of the Earth, are constantly warm'd, and in some places grow very hot; Bodies burn and shine, Mountains take Fire, the Caverns of the Earth are blown up, and the Sun continues violently hot and lucid, and warms all things by his Light. For we meet with very little Motion in the World, besides what is owing to these active Principles.[55]

The contrast that Newton draws here between the two types of phenomena classed under gravity and fermentation is striking. Gravity is the realm of planetary orbits and falling bodies, obviously belonging to the domain of what we would today call physics. Fermentation, on the other hand, is chymical. Fermentation is present in the circulation and warmth of the blood, the subterranean heating of the earth, the action of volcanoes and earthquakes, and even in the heat and light of the sun. Newton goes so far as to say that fermentation is the process by which "bodies burn and shine" in general. We have encountered most of these examples already either in the *Opticks* or in the works that preceded it, such as the suppressed "Conclusio" to the *Principia* and *De natura acidorum*. It is clear that fermentation belongs to the domain of chymistry just as gravity belongs to physics, but there is more to Newton's words than this straightforward demarcation into physical and chymical phenomena.

A careful look at Newton's language shows that his emphasis lies in the continued operation of the phenomena classed under gravity and fermentation. This accords with his anti-Cartesian goal of demonstrating that conservation of motion as a fundamental principle is a will-o'-the-wisp. Although the planets remain in their orbits over the *longue durée* and the sun continues to heat, both of these phenomena require the presence of "active principles,"

[55] Newton, *Opticks* (1718), 375.

the hidden causes underlying gravity and fermentation. Hence Newton's treatment of fermentation is not concerned here with singular, transitory phenomena as, for example, the striking of a match. Instead he emphasizes that the heart and blood of animals are kept in "perpetual" motion and fervor, the subterranean globe is "constantly" warmed, and the sun "continues" in its heat and shining. In all these cases, the fermentation is a self-sustaining process, much like what we saw in *De vita et morte vegetabili*. Just as Newton said there that subtle menstrua can act on matter and "excite new spirits of the same genus" as themselves, which in turn operate on further matter in a continued process of action and reaction, so the sun assimilates its "sulphureous atmosphere," to use the words of the "Conclusio," fermenting its fuel and converting it into its own substance while giving off heat and light.

Does the self-sustaining nature of the fermentation described in these examples near the end of *Query 31* mean that this is a necessary component of fermentation as such? Other comments in *Query 31* make it clear that for Newton the term "fermentation" did not necessarily imply continued activity. If provided with sufficient material on which to act, at least some ferments can continue their action indefinitely, but there are also fermentations of short duration. Thus when describing Lemery's theory of earthquakes, tempests, and other violent meteorological phenomena, Newton attributes them to a fermentation between various acids in the atmosphere. This leads him to one of the rare instances where he generalizes his comments about the process:

> Now the above mention'd Motions are so great and violent as to shew that in Fermentations, the Particles of Bodies which almost rest, are put into new Motions by a very potent Principle, which acts upon them only when they approach one another, and causes them to meet and clash with great violence, and grow hot with the Motion, and dash one another into pieces, and vanish into Air, and Vapour, and Flame.[56]

The root idea here is that fermentation is a process in which the corpuscles of a body are set into a state of motion by "a very Potent Principle," namely, the active principle responsible for the fermenting. The forcefulness of the corpuscles' motion and their clashing on impact result in their dissolution accompanied with the release of "air," along with vapor and inflammation. Although one might be tempted at times to read Newton's use of the term "fermentation" in a purely phenomenological sense, referring merely to the violent heating and bubbling such as one often encounters in the action of acids, this passage makes it clear that for him the term has a more general meaning. Like his contemporary Thomas Willis, whose 1681 *Medical-Philosophical Discourse of Fermentation* famously defined the process as "an intestine motion of Particles," Newton thought of fermentation as the internal movement of invisibly small corpuscles within a given material.[57] The

---

[56] Newton, *Opticks* (1718), 355.

[57] Thomas Willis, *A Medical-Philosophical Discourse of Fermentation* (London: T. Dring, C. Harper, J. Leigh, and S. Martin, 1681), 9.

hidden active principle causes the particles to be attracted to one another, and if the motion produced thereby is sufficiently powerful, the phenomena of bubbling, heating, and so forth will appear. But the fermentation per se does not consist in these signs; rather, it is the motion itself induced by the active principle. It is this sense that Newton has in mind when he contrasts fermentation to gravity near the end of *Query 31* and sets it up as an independent "fundamental force." The mysterious active principle behind the corpuscular motion of fermentation is a hidden cause of chymical affinity in general; hence, our discussion must now turn to that subject.

Chymical Attraction and Elective Affinity

Any consideration of Newton and the subject of chymical affinity must consider, however briefly, the tortured historiography of the subject. It was once popular among historians, and even among eighteenth-century figures themselves, to see Newton's *Query 31* as the origination point of the great vogue for affinity tables evident particularly in the second half of the century. These graphic displays of solutions and precipitations typically embodied the principle of elective affinity, according to which one material has a greater tendency to combine with another than with a third party; consequently, if the third material is already combined with the first, it will be displaced by the introduction of the second. *Query 23* of the *Optice* and *Query 31* of the 1717 *Opticks* had famously spelled out the same principle with scores of examples such as the following:

> And so when a Solution of Iron in *Aqua fortis* dissolves the *Lapis Calaminaris* and lets go the Iron, or a Solution of Copper dissolves Iron immersed in it and lets go the Copper, or a Solution of Silver dissolves Copper and lets go the Silver, or a Solution of Mercury in *Aqua fortis* being poured upon Iron, Copper, Tin or Lead, dissolves the Metal and lets go the Mercury, does not this argue that the acid Particles of the *Aqua fortis* are attracted more strongly by the *Lapis Calaminaris* than by Iron, and more strongly by Iron than by Copper, and more strongly by Copper than by Silver, and more strongly by Iron, Copper, Tin and Lead, than by Mercury?[58]

As the century progressed, attempts to find new elective affinities and to determine the factors behind their variations swelled into a dominant, pan-European research project. Perhaps the culmination of the eighteenth-century concern with affinity can be seen in the *Dissertation on Elective Attractions* of Torbern Bergman, a Swedish scientist whose important work was first published in 1775 and translated into English a decade later. In its most complete form, Bergman's table of affinities consisted of fifty-nine columns, with the dissolving agent at the top and the respective *solvenda* in descending order below. Bergman would begin his *Dissertation* by explicitly echoing Newton's famous distinction at the beginning of *Query 31* between

---

[58] Newton, *Opticks* (1718), 355–56; *Optice* (1706), 327.

gravitation and the short-range attraction exercised by "the small Particles of Bodies."[59] Well before Bergman's declaration of allegiance to the author of the *Opticks*, the famous Glasgow and Edinburgh professor of medicine William Cullen had explicitly linked affinity tables to "the Great Newton," and this acknowledgment of Newton's role became a commonplace among contemporary scientists.[60]

Neither Cullen nor Bergman drew up the first affinity table, however. The originator of these graphic schemata was Etienne-François Geoffroy, a member of the *Académie royale des sciences* who published his influential *Table des differens rapports observés en Chymie entre différentes substances* in 1718. Over the last half century, the issue of Newton's possible influence on Geoffroy has become a contested issue.[61] Geoffroy was a Fellow of the Royal Society and a regular correspondent with Hans Sloane; he knew English, and had a strong interest in British science.[62] He may well have been acquainted with the 1717 edition of the *Opticks* before publishing his *Table*, and it is of course possible that he was stimulated by it to put the phenomenon of elective affinity into graphic form.[63] On the other hand, Geoffroy studiously avoided the use of Newton's term "attraction," instead employing the more neutral French term *rapport* for the elective affinities among substances. Nor

[59] Torbern Bergman, *A Dissertation on Elective Attractions* (London: J. Murray, 1785), 2: "It has been shewn by Newton, that the great bodies of the universe exert this power directly as their masses, and inversely as the squares of their distances. But the tendency to union which is observed in all neighbouring bodies on the surface of the earth, and which may be called contiguous attraction, since it only affects small particles, and scarce reaches beyond contact, whereas remote attraction extends to the great masses of matter in the immensity of space; seems to be regulated by very different laws."

[60] Manuscript passage quoted from Cullen quoted in Georgette Nicola Lewis Taylor, "Variations on a Theme: Patterns of Congruence and Divergence among 18th Century Chemical Affinity Theories" (PhD diss., University College London, 2006), 279; the passage is dated by Taylor to the 1760s. For more on Cullen's acknowledgment of Newton's role in the science of affinity, see also pp. 16, 47, and 70–71. As Taylor also points out on p. 47, in response to the now-unpopular view of Newton's preponderating role, "It is unfair, however, to condemn modern historiography for the widespread assertion that Newton 'invented' affinity. Many 18th century chemists asserted something remarkably similar."

[61] The controversy seems to have originated in an article by Bernard Cohen that refers to Geoffroy as one of Newton's "chemical disciples." See I. Bernard Cohen, "Isaac Newton, Hans Sloane and the Académie Royale des Sciences," in *Melanges Alexandre Koyré*, ed René Taton and I. B. Cohen (Paris: Hermann, 1964), 61–116, especially p. 80. A similar though more subtle approach to the Newton-Geoffroy relationship is found in Arnold Thackray, *Atoms and Powers* (Cambridge, MA: Harvard University Press, 1970). The positions of Cohen and Thackray were vigorously rejected in W. A. Smeaton, "E. F. Geoffroy Was Not a Newtonian Chemist," *Ambix* 18 (1971): 212–14. The issue is discussed more recently in Mi Gyung Kim, *Affinity, That Elusive Dream* (Cambridge, MA: MIT Press, 2003), 142, and in the other sources referred to in the notes following. Interestingly, Geoffroy reappears once again in blanket terms as a disciple of Newton in J. B. Shank, *The Newton Wars and the Beginning of the French Enlightenment* (Chicago: University of Chicago Press, 2008), 114–20.

[62] Bernard Joly, "Etienne-François Geoffroy, entre la Royal Society et l'Académie royale des sciences: Ni Newton, ni Descartes," online in *Methodos* at https://methodos.revues.org/2855#bodyftn9, accessed December 25, 2016. See paragraphs 30–31 and 38–41, where Joly discusses Geoffroy's interest in English science and the possible influence of Newton on him.

[63] See Sloane 3322, fol. 101r, a letter from Geoffroy to an unnamed recipient, presumably Sloane. The letter, dated March 15, 1718, refers to what may be the 1717 edition of the *Opticks* in the following terms: "J'ai remis a M^r. L'abbé Bignon les Transact. Philosoph. L'optique de M. Newton, et la dessein de la machine pour elever l'eau." We do not know, of course, how long Geoffroy possessed the *Opticks* before sending it to Bignon, or whether he read *Query 31*, but the letter does raise the possibility of a debt to Newton. And of course there was also *Query 23* of the *Optice*, which Geoffroy could have consulted at any time between 1706 and 1718.

did Geoffroy need the empirical information provided by *Query 31* for his *Table*, which systematizes information already widely available among chymists.[64] Whether Newton himself provided an impetus at the beginning of the century-long fascination with affinity is not yet subject to a definitive answer.[65] What can be said with certainty is that he became the patron saint of chymical attraction as the century progressed.

If we attempt to trace a generalized concept of chymical affinity back to its earliest appearance in Newton's thought, *Of Natures obvious laws & processes* emerges as a likely point of origin. Already in this text of around 1670 Newton presents fundamental interparticular association as a nonmechanical operation governed by a hidden principle:

> Yet those grosser substances are very apt to ~~bee~~ put on various external appeanes <*i.e.* appearances> according to the present state of the <*illeg.*> invisible inhabitant as to appear like bones flesh <*illeg.*> wood fruit &c Namely they consisting of ~~heterogeneous~~ ^{differing} particles watry earthy saline airy oyly spirituous &c those parts may bee variously moved one among another according to the acting of the <*illeg.*> ^{latent} vegetable substances & be ~~put~~ variously associated & concatenated together by their influence<.>

The "latent" or hidden "vegetable substances" refer to the unimaginably small portions of matter endowed with active powers that Newton inherited from chrysopoetic writers such as Sendivogius and Philalethes. But Newton's emphasis here is on association and concatenation, not on the parallel displacement characterizing *Query 31*'s presentation of elective affinity. The 1675 *Hypothesis of Light* develops the idea of a hidden principle of activity further, referring to a "secret principle of unsociablenes" between different substances. The context for this is a discussion of the voluntary bunching and subsequent relaxation of muscles, which Newton wants to explain in terms of an ether, or rather multiple ethers. His comments must be seen as an attempt to avoid an absurd explanation in terms of ethereal pressure where the thin and delicate nerves are thought to transmit highly pressurized subtle media to the muscle. In order to escape this exigency his explanation assumes that the "common æther" of the atmosphere enters into the muscle directly at the locus of its swelling rather than passing through the nerves. An "Animal Spirit" or internal ether within the muscle is counterpoised by the external ether when the muscle is in a state of relaxation. The

---

[64] Some of this information has been gathered in Ursula Klein, "E F Geoffroy's Table of Different Rapports Observed between Different Chemical Substances—A Reinterpretation," *Ambix* 42 (1995): 79–100. Unfortunately, however, Klein links her useful spadework to inflated claims about Geoffroy's significance in the history of chemistry over the *longue durée*, based on a misunderstanding of previous chymical theory. See William R. Newman, "Elective Affinity before Geoffroy: Daniel Sennert's Atomistic Explanation of Vinous and Acetous Fermentation," in *Matter and Form in Early Modern Science and Philosophy*, ed. Gideon Manning (Leiden: Brill, 2012), 99–124.

[65] Although I do not affirm that Geoffroy's *Table des rapports* was influenced by Newton, I see no way to deny categorically that after reading the 1706 *Optice* or the 1717 *Opticks*, the French academician might have seen an opportunity to capitalize on the interest in elective affinity generated by *Query 31* (or/and *23*), and that this could have encouraged him to compile a table.

nerves running from the brain to the muscle also contain the animal spirit mixed in with the "Animal juices," and the animal spirit acts as a "mediator" between the external ether and the animal juices in much the same way that Newton would later claim that acid particles act as a mediator between the earthy core and external water in a corpuscle of salt. Just as the acid allows salt particles to mix with water, so the animal spirit allows the vegetable juices in the muscle to mix more freely with the external ether. If the soul directs even a tiny bit more of the subtle, animal spirit through the nerves, the "sociability" between it and the external ether will therefore encourage the external ether to enter into the muscle, mix with the animal juices, and cause the muscle to swell.

Newton's convoluted theory also requires him to explain how the internal ether or animal spirit, despite being incredibly subtle, can be retained within the porous human body. This is where the flip side of his principle of mediation comes in, namely, in the form of an "unsociablenes" between different materials. The animal spirit has such a lack of sociability, for which we may employ the term affinity even though Newton does not, with the "Coats of the braine, Nerves & muscles." Despite the fact that these organic materials are quite permeable, they do not permit the escape of the particles of animal spirit because of the unsociableness between them, which keeps the corpuscles of spirit from entering the pores in bodily tissues. Newton now brings in a variety of examples drawn from the world of chymistry to support his case:

> you may consider, how liquors & Spirits are disposed to pervade or not pervade things on other accounts then their Subtility; water & Oyle pervades Wood & Stone wch Quicksilver does not; & Quicksilver, Metalls, wch water & Oyle doe not. Water and Acids Spirits pervade Salts, wch Oyle <&> Spirit of Wine do not, & oyle and Spirit of Wine pervade Sulphur wch water <&> acid Spirits do not. So some fluids (as Oyle and water) though their pores are in freedome enough to mix with one another, yet by some secret principle of unsociablenes, they keep asunder, & some that are Sociable may become unsociable by adding a third thing to one of them, as water to Spirit of Wine by dissolving Salt of Tartar in it.[66]

At the end of his list of unsociable and sociable materials, Newton adds a phrase about spirit of wine and salt of tartar to illustrate the mutability of these relations. If salt of tartar (our potassium carbonate) is added to spirit of wine (ethyl alcohol) containing some water, the salt of tartar will attract water out of the alcohol and leave the alcohol in a state of higher concentration. Newton could easily have described this phenomenon in terms of elective affinity; the "secret principle" of sociability between the water and the salt of tartar would thus be greater than that between the spirit of wine and the water. Hence the water would "elect" to abandon the spirit of wine and combine with the salt of

---

[66] Newton, *Corr.*, 1: 368. I have added angle brackets to indicate corrections made against Turnbull's transcription, based on a consultation of CU Add. 3970, 540v, found on the Cambridge University facsimile of the manuscript at https://cudl.lib.cam.ac.uk/view/MS-ADD-03970/1100 consulted 26 December 2016.

tartar. But significantly, he does not use the language of combination and displacement here, nor does he do so when he continues his discussion by adding in examples where unsociable substances are made sociable. Instead, his focus in the above example lies in the action of a "third thing" (here salt of tartar) to create unsociability by a sort of negative "mediation."

The tight proximity between Newton's concept of mediation and his later ideas about chymical attraction and affinity is on even greater display in the famous letter that he wrote to Robert Boyle on February 28, 1678/9. Here Newton builds on the same ideas that he used in the *Hypothesis of Light*, but now to explain chymical interactions more generally:

> When any metal is put into common water, ye water cannot enter into its pores to act on it & dissolve it. Not yt water consists of too gross parts for this purpose, but because it is unsociable to metal. For there is a certain secret principle in nature by wch liquors are sociable to some things & unsociable to others. Thus water will not mix with oyle but readily wth spirit of wine or wth salts. It sinks also into wood wch Quicksilver will not, but Quicksilvers sinks into metals, wch, as I said, water will not. So Aqua fortis dissolves ☽ not ☉; Aqua regis ☉ & not ☽, &c. But a liquor wch is of it self unsociable to a body may by ye mixture of a convenient mediator be made sociable. So molten Lead wch alone will not mix wth copper or wth Regulus of Mars, by ye addition of Tin is made to mix wth either.[67]

This list is practically identical to the examples of unsociability and sociability in the *Hypothesis*, but Newton follows it with comments that reveal more clearly the connection with elective affinity. Just as molten lead can be made to mix more easily with martial regulus by adding in some tin, so a metal can be made to mix with water by the mediation of "saline spirits," meaning the mineral acids. In such a case, the acid spirits will first "by their sociableness enter into its pores & gather round its outside particles." The corpuscles of acid spirit then "hitch themselves" into the pores of the metal, separate its corpuscles from one another, and encompass them "as a coat or shell does a kernel." It is not difficult to see an early version of Newton's shell theory of matter in this description, though without the generalized supposition of "acid" and "earth" that characterize *De natura acidorum*. Newton's subsequent comments to Boyle show how his corpuscular shell theory, in conjunction with the principle of sociability, could work to explain the displacements characteristic of elective affinity:

> If into a solution of metal thus made, be poured a liquor abounding with particles, to wch ye former saline particles are more sociable then to ye particles of ye metal, (suppose with particles of salt of Tartar:) then so soon as they strike on one another in ye liquor, ye saline particles will adhere to those more firmly then to ye metalline ones, & by degrees be wrought of from those to enclose these.[68]

[67] Newton, *Corr.*, 2: 291–92.
[68] Newton, *Corr.*, 2: 292.

In this example there is a greater degree of sociability, or affinity if one prefers, between the "saline particles" of the mineral acid and the salt of tartar than there is between the menstruum and the metal. As a result, the particles of the acid shell are "wrought" off of their metallic kernel and go on to form another shell around a corresponding corpuscle of salt of tartar. This classic description of elective affinity is followed by Newton's explanation of the precipitation of the liberated metal. The tartareous corpuscles now floating in the solution will crowd the freed particles of metal together, and the latter will "cohere & grow into clusters," whereupon their weight will result in their precipitating to the bottom of the vessel.

We have seen then how the generalized notion of corpuscular attraction and coherence presented in *Of Natures obvious laws* evolved over the 1670s into Newton's chymical "mediation" and then became a full-scale principle of elective affinity. The concept of elective affinity is presented at greater length in the unfinished "Conclusio" to the first edition of the *Principia*. After again describing the precipitation of metals in acid solutions by salt of tartar, Newton passes to further examples to illustrate the same principle:

> Thus also the acid spirit in mercury sublimate, acting on metals, leaves the mercury. That spirit in butter of antimony coalesces with water poured on it, and allows the antimony abandoned by it to be precipitated. And the acid spirit, joined with common water in aqua fortis and spirit of vitriol, by acting on metals dissolved in those menstruums, leaves the water and allows the water to ascend by itself with a merely gentle heat, whereas before it could not be separated from the spirit by distillation. And spirit of vitriol, meeting with the fixed particles of salt of nitre, looses the spirit of nitre which was formerly joined to those fixed particles, so that the latter spirit can be more easily distilled than before.[69]

These examples clearly reflect the experimentation in Newton's laboratory notebooks. One of Newton's very earliest chymical experiments resulted in the "acid spirit in mercury sublimate" leaving the compound in order to react with another metal. This involved his attempt to extract the mercuries of the metals by "baking" successive metals with mercury sublimate and sal ammoniac. As he says, "yᵉ salts will act upon yᵉ metals," with the result that "you shall have their ☿ ruining <*i.e.* running> at yᵉ bottom."[70] Although he would only later recognize that the quicksilver released by the process was identical to the original mercury in the sublimate, such attempts to arrive at the mercury of the metals may well have been Newton's first exposure to reactions that could be classed under the rubric of elective affinity. As for the next operation described above, the formation of insoluble, white *mercurius vitae*, the oxychloride of antimony produced when butter of antimony is dropped into water, was a standard method Newton employed to "liberate" various sublimates from antimony, as we discussed in the context of his laboratory notebooks. The third process, consisting of the separation of

---

[69] Newton, "Conclusio," in Hall and Hall, *UPIN*, 335–36.
[70] CU Add. 3975, 41v.

water from an "acid spirit" by first combining a metal with the acid and then distilling off the water of crystallization, is also present in his experimental records.[71] The final example, the production of spirit of niter by distilling saltpeter with oil of vitriol, was a standard artisanal practice that Newton no doubt carried out as well.

Although the list of chymical examples from the "Conclusio" illustrates Newton's growing interest in displacement reactions, it also reveals another important fact: each of the above examples describes a single, individual case of elective attraction. An acid spirit combines with metals to liberate mercury that was combined with the acid, water associates with another acid spirit to precipitate antimony, yet another acid spirit combines with a metal to release water that can then be distilled off, or oil of vitriol combines with the fixed part of saltpeter to release the spirit of niter. These are not series of displacements such as one finds in *Query 31* and in the affinity tables, but rather independent examples. Where then do we first encounter actual replacement series of multiple substances such as those that first appear in the 1706 *Optice* and reemerge in the 1717 *Opticks*? Interestingly, the answer may lie in Newton's chrysopoetic reading notes and florilegia.

Newton's *Opera* florilegium, which we dated in chapter thirteen to a period between 1686 and the early 1690s, contains an analysis of Sendivogius's *Novum lumen chemicum* in terms of solutions and precipitations of metals. In tractate nine of his book, Sendivogius had used the geocentric system as an illustration of the fact that the planetary powers descend to the earth but do not ascend from it. The same thing is true, he says, of the metals that correspond to the individual planets. Hence a superior planet like Mars sends its virtues down to its inferior counterpart Venus, and so iron can be transmuted to copper, but not vice versa. By the same logic, Jupiter (tin) becomes Mercury (quicksilver), and Saturn (lead) becomes Luna (silver).[72] In the course of expounding this passage, Newton explicitly links it to the precipitation of metals dissolved in an acid:

> Venus mates with Mars, Luna with Saturn, and Mercury with Jove, because an acid spirit <deserts> Venus so that it may enter Mars, and deserts Luna so that it may penetrate Saturn, and deserts Mercury so that it may work on Jove.[73]

There is no reason to think that Newton read this as a single series of reactions instead of as paired examples, but the important thing is that he clearly did identify Sendivogius's series of "transmutations" as solutions and precipitations. It is likely that he read similar passages in other chrysopoetic writers in the same way. Some alchemical authors, for example Philalethes, explicitly presented such "transmutations" as a single series. In his *De*

[71] BML B MS c41 c, 1r: "Sal ♀ᶜ impregnatū, non potest destillari sed a ⊕ˡᵒ fixatur ut ni præter aquam insipidam destillaverit."

[72] Michael Sendivogius, *Novum lumen chemicum* (Geneva: Joannes de Tournes, 1639), 45–47.

[73] Keynes 41, 15r: "Coit Venus cum ♂, Luna cum ♄ & ☿ Mercurius cum ♃ quia spiritus acidus Venerem <deserit> ut Martem ingrediatur & Lunam deserit ut saturnum penetrat & Mercurium deserit ut operetur in Iovem."

metallorum metamorphosi (On the Metamorphosis of the Metals), a work that Newton read and commented on soon after its appearance in 1668, the "American philosopher" has the following relevant comments:

> I could here recount diverse mutations, such as that of Mars into Venus by means of the acid dripping of vitriol, of Venus into Saturn, of Saturn into Jove, of Jove into Luna, which operations (foreign to the apex of our art) very many vulgar chymists know how to produce.[74]

Philalethes explicitly begins this passage with a reference to the supposed transmutation of iron into copper by means of copper vitriol, which he obviously considers specious. As a devoted follower of Van Helmont, the author beneath the mask of Philalethes, Starkey, would have known that the Flemish chymist had revealed the supposed transmutation of iron into copper by vitriol to be a mere displacement in his *Ortus medicinae*.[75] Since Philalethes then immediately launches into successive, equally commonplace mutations of the "transmuted" copper into lead, lead into tin, and tin into silver, a natural way to read the passage is as a series of solutions and displacements. Newton's early abstract of the passage suggests that he too read it in that fashion: "It is vulgarly known," he says, that one can "mutate ♂ into ♀ by means of the dripping of vitriol, ♀ into ♄, ♄ into ♃, ♃ into ☽ &c."[76] Eighteenth-century affinity charts agree with Philalethes about the series so far as the solution of iron, copper, and silver go, though lead and tin are more problematic.[77]

Newton's discovery of elective affinity in chrysopoetic writers once again points to the arbitrary character of modern divisions between early modern chemistry and alchemy. The reader might be tempted to point out, however, that even if Newton's reading of Sendivogius and Philalethes was correct, the material transmitted obscurely by these authors was already available in more open form. Christophle Glaser's 1663 *Traité de chymie*, for example, gives a very clear description of the replacement series for "fixed niter" (primarily potassium carbonate), zinc or calamine, iron, copper, and silver in aqua fortis.[78] A similar though less detailed series including iron, copper, and silver, may be found in Boyle's *Experiments, notes, &c. about the mechanical origine or production of divers particular qualities* published in 1676.[79]

[74] Eirenaeus Philalethes, *Tres tractatus de metallorum transmutatione* (Amsterdam: Johannes Janssonius à Waisberge and the widow of Elizeus Weyerstraedt, 1668), 9: "Possim hic mutationes diversas metallorum recensere, ut nempe Martis in Venerem per acidum Vitrioli stalagma, ♀ in Saturnum, ♄ in Jovem, ♃ in Lunā, quas quidem operationes plurimi (ab Artis apice alieni) norunt præstare Chemici vulgares."

[75] Joan Baptista Van Helmont, *Ortus medicinae* (Amsterdam: Ludovicus Elsevir, 1648), 692.

[76] Newton's abstract of this passage appears in an early part of the Jerusalem manuscript Var. 259, at 8.2r: "Vulgaritur notū est mutare ♂ in ♀ per acidum Vitrioli Stalagma, ♀ in ♄, ♄ in ♃, ♃ in ☽ &c." The early date of Var. 259.8 is revealed from the large number of Latin diacritics found there, along with the unbarred Saturn symbol.

[77] See for example the 1730 table of Jean Grosse, reproduced in Kim, *Affinity, That Elusive Dream*, 223. The affinity series placed beneath the symbol for nitric acid corresponds to the series of transmutations in Philalethes. Lead and tin are absent for the series grouped under sulfuric acid, however.

[78] Christophle Glaser, *Traité de la chymie* (Paris: Glaser, 1663), 76–77.

[79] Boyle, "Of the Mechanical Causes of Chymical Precipitation," in *Experiments, notes, &c. about the mechanical origine or production of divers particular qualities*, in *Works*, 8: 492; 1676, 32–34. This passage is discussed in Taylor, *Variations on a Theme*, 43–44.

But there is no evidence that Newton read Glaser, and his exposure to Sendivogius and Philalethes preceded the publication of Boyle's book by over half a decade. It is therefore likely that Newton's own chrysopoetic research and experimentation, coupled with his wide chymical reading, led him to compile the lists of solutions and displacements that characterize *Query 31*. The chymical affinities described by such staunch Newtonians as Cullen and Bergman and attributed to their illustrious avatar were actually, in good part, a gift of Hermes.

Conclusion

In this chapter and the previous one we have observed how Newton's substitution of sulfur for the Sendivogian aerial niter dovetailed with developments in public chymistry across England and the Continent, and we have seen how this trend reflected the widely accepted belief that sulfur itself was composed of an acid component coupled with an unctuous principle. At the same time, however, Newton's theory of combustibility depended primarily on the acid principle rather than the oily one, which linked him to an older British tradition and distinguished him from the proto-phlogistonists. Newton's emphasis on the acid thought to be bound up in sulfur is integrally related to his comments to Archibald Pitcairne in which he identified sulfur with acid tout court, and mercury with earth. He expressed these views in the context of his theory that corpuscles of salt displayed a complex shell structure, and that their supposed putrefaction resulted from the penetration of the sulfurous, acid particles into their central earthy (or mercurial) core. As the notes in his *De vita et morte vegetabili* and draft versions of the *Opticks* queries also make clear, Newton saw a close link between putrefaction and fermentation. The latter process resulted from the motion of minute corpuscles within a material, which could either lead to its complete decay or to other developments such as its combination with other substances. Fermentation and putrefaction had also long been key processes for chrysopoetic writers such as Newton's old favorite, Philalethes. But for Newton, one could argue, fermentation acquired an even more crucial importance: it became the process governing chymical activity in general. For this reason Newton made the radical step of granting fermentation, or rather the cause behind it, the status of a fundamental force alongside the principles that govern gravity, magnetism, and electricity.

This of course led to another question. Why do some substances react with one another more vigorously than others? What decided the course of a particular reaction was the relative affinity that the corpuscles of a given material had either for one another or for a third party. This was what Newton expressed verbally in *Query 31* of the *Opticks*, and what the compilers of affinity tables soon were depicting by means of graphic symbols. If one follows the development of Newton's ideas from his early *Hypothesis of Light* through his 1678/9 letter to Boyle, the unfinished "Conclusio" to the 1687 *Principia*, the *De natura acidorum* of 1691/2, and the successive editions of the *Opticks*, the relationship between his growing emphasis on relative affinity and the increasing importance of the acid-sulfur principle becomes evident. The vague "principle of unsociablenes" described in the *Hypothesis* gradually gives way to a principle of relative affinity as in the letter to Boyle, where Newton describes the release of a metal from its "coat" of acid particles when salt of tartar is added to the solution and the acid corpuscles are "wrought off" because of their greater affinity for the alkali. In a similar fashion, Newton's views on combustion come to be dominated by the idea that if the acid, sulfurous spirit in the atmosphere has a greater affinity for a given material than the particles of that substance have for one another, flame and fire will ensue. While the power that these ideas exercised on Newton kept him from abandoning the older tradition of combustion as the action of an acid menstruum, they led him, nonetheless, to the forefront of the eighteenth century's growing appreciation of elective affinity. Whether Geoffroy's *Table des differens rapports* owes any debt to Newton or not, the fact remains that both men were at the leading edge of a new movement in chymistry that would come to dominate the field in the half century before the Chemical Revolution inaugurated by Lavoisier.

TWENTY-TWO

A Final Interlude

NEWTON AND BOYLE

Throughout the previous chapter we saw multiple examples where Newton's private chrysopoetic research overlapped with his public chymistry and contributed in important ways to its development. A particularly cogent case of this intersection may be seen in Newton's relationship with Robert Boyle. The English "naturalist," as Boyle referred to himself, was at the same time one of the most famous scientists in Britain, known for his experimental expertise and for his prominent role in the Royal Society, and also a semicloseted seeker of the philosophers' stone.[1] Newton would interact with Boyle on both levels. Let us return now to Newton's anxious and slightly exasperated comments to Oldenburg in response to the publication of Boyle's "Inacalescence" article of January 1, 1675/6. Boyle had discovered a way of impregnating quicksilver so that gold would heat up spontaneously when mixed with it and melt, "like ice in warm water," as the adepts were wont to say about the dissolution of the noble metal in their sophic mercury. Newton was concerned that Boyle, whom he described to Nicolas Fatio de Duillier in 1689 as "too open & too desirous of fame," had provided an indirect path for the vulgar that might lead to a dangerous dispersion of alchemical secrets.[2] Newton's personal relationship with Boyle was clearly a tense one, despite the high regard that Newton held for the older man's scientific work and the demonstrated influence that Boyle had on his own. As he also said to Fatio, Boyle had at "divers times offered to communicate & correspond wth me in these matters but I ever declined it," again because of Boyle's excessive openness. Yet there was another element to Newton's chariness beyond his habitual horror of popularizing the *arcana majora*. Newton thought that Boyle was on the wrong path to alchemical success, and the material that we covered in the previous chapters allows us to see why.

In his letter of early 1676 to Oldenburg, Newton explained the nature of his disagreement with Boyle's methods in some detail. The gist of the

[1] This dichotomy is brought out quite clearly in Principe, *AA*.
[2] Newton to Fatio de Duillier, October 10, 1689, in Newton, *Corr.*, 3: 45.

problem as Newton portrayed it, lay in a conflict with his own hierarchical theory of matter:

> Not that I think any great excellence in such a ☿ either for medical or Chymical operations: for it seems to me yt ye metalline particles wth wch yt ☿ is impregnated may be grosser yn ye particles of ye ☿ & be disposed to mix more readily wth ye ☉ upon some other account then their subtilty, & then in so mixing, their grossnes may enable them to give ye parts of ye gold ye greater shock, & so put ym into a brisker motion then smaller particles could do: much after ye manner that ye saline particles wherewith corrosive liquors are impregnated heate many things wch they are put to dissolve, whilst ye finer parts of common water scarce heat any thing dissolved therein be ye dissolution never so quick; & if they do heat any thing; (as quick lime) one may suspect that heat is produced by some saline particles lying hid in ye body wch ye water sets on work upon ye body wch they could not act on whilst in a dry form. I would compare therefore this impregnated ☿ to some corrosive liquor (as Aqua fortis) the ☿ial part of ye one to ye watry or flegmatic part of ye other, & ye metallick particles wth wch ye one is impregnated to ye saline particles wth wch ye other is impregnated, both wch I suppose may be of a middle nature between ye liquor wch they impregnate & ye bodies they dissolve & so enter those bodies more freely & by their grossness shake ye dissolved particles more strongly then a subtiler agent would do. If this analogy of these two kinds of liquors may be allowed, one may guess at ye little use of ye one by ye indisposition of ye other either to medicine or vegetation.[3]

Newton bases his analysis of Boyle's incalescent mercury on a straightforward analogy between the impregnated quicksilver and a corrosive menstruum such as aqua fortis. Presumably it was Boyle's insistence on the considerable heat emitted by the mercury that led Newton to his dismissive conclusion. In his essay, Boyle suggested that the incalescent mercury could grow so hot as to burn one's hand or even crack a glass vessel.[4] In Newton's view, a proper sophic mercury should work gently, without the excessive heat often produced by the dissolution of metals in acids. The incalescent mercury, in his analysis, is a mixture of ordinary quicksilver corpuscles and the larger, unspecified "metalline particles" with which the quicksilver is impregnated, just as a common, acid menstruum is merely a mixture of water and "saline particles." The impregnating, metalline particles are larger than those of the quicksilver and act as a mediator between it and the gold to be dissolved, just as the acidic corpuscles are bigger than those of water and serve as a mediator between it and a metal that is undergoing corrosion. If we recall Newton's shell theory of matter, it is clear that the acid particles are too large to penetrate beyond the outermost pores of a metal; they can only separate metallic corpuscles and surround them "as a coat or shell does a kernell."

---

[3] Newton, *Corr.*, 2: 1–2.

[4] Boyle, "Of the Incalescence of Quicksilver with Gold, Generously Imparted by B. R.," *Philosophical Transactions* 10 (1675/76): 510–33, see 524–25.

Hence Boyle's incalescent mercury is doomed to failure as an agent of transmutation precisely because it operates in the same way as a mineral acid and does not penetrate into the deeper pores that would allow it to make gold ferment, a process that Newton would later outline in *De natura acidorum*. In order to accomplish that feat, the impregnating particles would have to be small enough to penetrate gently into the deeper pores of the gold and to induce its particles gradually to putrefy rather than separating them by imparting a rude shock as the mineral acids do.

It is now known that Boyle's incalescent mercury was actually a version of Starkey's sophic mercury, which the young New Englander had taught Boyle to make in the early 1650s. Boyle himself alludes indirectly to this fact when he says, in "Of the Incalescence of Quicksilver with Gold," that he first acquired the impregnated mercury "about the year 1652."[5] It may at first seem surprising that Newton would reject a product stemming from the same hand that wrote the Philalethes corpus, a body of writings that captivated him over his generation-long involvement with chrysopoeia. The mystery is only deepened by the work of previous Newton scholars, who have uniformly argued that the *Clavis*, the portion of Starkey's 1651 letter to Boyle containing a recipe for the sophic mercury and copied out at some point by Newton (in MS Keynes 18), played a major role in Newton's alchemical endeavor.[6] Did Newton simply fail to recognize the Philalethan sophic mercury in Boyle's description? Or was Newton perhaps being hypocritical or disingenuous in his dismissal of it to Oldenburg? Surprisingly, we can answer both questions, the first by means of recourse to Boyle's treatise itself, and the second by a return to Newton's reading notes and experimentation.

A close inspection of Boyle's 1676 "Of the Incalescence" reveals that the English "naturalist" dropped some very broad hints about the preparation of his wonderful mercury.[7] Boyle begins by making the principal claim to which Newton would object, that his mercury is "more subtle and penetrant" than garden variety quicksilver.[8] How was this remarkable subtlety and penetration to be attained? Although he does not provide an explicit recipe, Boyle as much as says that his preparation involves antimony on two occasions. The first assertion occurs in the midst of a discussion of *mercurii corporum* (mercuries of the bodies), meaning the so-called running mercuries that chymists had long been attempting to extract from metals. We already saw Newton attempting to carry out this very operation in CU Add. 3975 at the beginning of his alchemical career.[9] In his usual tortuous prose, Boyle wonders whether such extracted mercuries would also grow hot, just as "*Antimonial Mercury*" does:

[5] Boyle, "Of the Incalescence of Quicksilver with Gold," 521. See Principe, *AA*, 155–79.

[6] Dobbs, *FNA*, 133–34, 175–86, 229–30; Dobbs, *JFG*, 15–17; Westfall, *NAR*, 370–71. Even Karin Figala, who correctly rejected Newton's authorship of the *Clavis*, still saw Newton's "ductus" in the document. See Figala, "Newton as Alchemist," *History of Science* 15 (1977): 102–37, especially 108.

[7] This point is established by Principe, *AA*, 155–79.

[8] Boyle, "Of the Incalescence of Quicksilver with Gold," 517.

[9] CU Add. 3975, 41v.

I would much scruple to determine thence, whether those <mercuries> that are *Mercurii corporum*, and were made, as Chymists presume, by extraction only from Metals and Minerals, will each of them grow hot with Gold, as, if I much mistake nos <*sic*>, I found *Antimonial Mercury* to do.[10]

Admittedly, Boyle does not go so far as to identify his own incalescent mercury with the antimonial one described here, but the fact that the antimonial mercury is clearly his standard of referent for incalescence makes it an obvious way to read the passage. Moreover, Boyle reinforces this impression a few pages later and even provides additional information about the preparation of antimonial mercury. In the context of suggesting that the incalescent property may be imparted by more than one preparation of quicksilver, Boyle adds, almost as an aside, that the antimonial variety employs "solid metals as *Mars*." Any self-respecting chymist of the seventeenth century would have recognized this as an allusion to martial regulus, or metallic antimony produced from stibnite by means of iron:

> Such a Mercury may be (I say not, easily or speedily, but successfully) prepar'd, not only by employing Antimony and solid Metals as *Mars*, but without any such Metal at all, or so much as *Antimony* it self.[11]

Hence Boyle's treatise on incalescence makes it sufficiently clear that his process employs an amalgam involving the martial regulus of antimony and quicksilver, even if the regulus is not strictly required. Newton himself was of course aware of the amalgamation of quicksilver and regulus that forms the technical basis of the Philalethes corpus. The *Clavis* openly describes the means of uniting quicksilver and antimony regulus by first alloying the antimony with silver and even identifies the two portions of silver with the doves of Diana. Moreover, Boyle hinted that Philalethes was involved in his process when he alluded to "divers *Philalethists*" who might not believe that there are other mercuries besides the antimonial one that heat up with gold.[12] Hence we can answer the first question that we posed: Newton would almost certainly have recognized that Boyle's antimonial mercury was based on an interpretation of Philalethes, even if that interpretation did not square with Newton's own idiosyncratic understanding of the American adept.

This leaves us with our second question still unanswered, however. Was Newton's dismissal of the obviously Philalethan mercury described by Boyle a disingenuous move on his part? As we have determined, Newton probably understood the nature and origin of Boyle's incalescent mercury. But Newton, who neither knew that Starkey was the real author of the Philalethes writings nor that the *Clavis* was taken from Starkey's 1651 letter to Boyle, never accepted the validity of the process described there. The reader who has worked through the previous chapters of the present book will understand that the issue stemmed from Newton's earliest interpretations

---

[10] Boyle, "Of the Incalescence of Quicksilver with Gold," 525.
[11] Boyle, "Of the Incalescence of Quicksilver with Gold," 530.
[12] Boyle, "Of the Incalescence of Quicksilver with Gold," 529.

of Philalethes, such as those in Keynes 19, where the young Cantabrigian decodes *Secrets Reveal'd*, using Sendivogius, d'Espagnet, and Philalethes to explicate one another. Newton decides there that the secret to the philosophers' stone lies in a process wherein lead and crude antimony are melted together and digested to release a "mercury" (presumably a regulus). Already in this very juvenile interpretation, Newton deciphers the doves of Diana in an idiosyncratic way, where they become the "sulfur floating on the mercurial water." The doves also have feathers, of course, which the young chymist decodes as a white powder that must be extracted from the doves and used in further processes. Although Newton's interpretation of Diana's doves would change over time as he moved from the simple lead-antimony process of Keynes 19 to the fantastic panoply of operations in such late texts as "Three Mysterious Fires" and *Praxis*, he never equated the doves with the silver of Starkey's process for making the sophic mercury. In short, Newton thought that the recipe for the sophic mercury presented in the *Clavis* and implied in Boyle's article, "Of the Incalescence of Quicksilver with Gold," was false.

The realization that Newton rejected Starkey's actual process for making the sophic mercury in favor of his own idiosyncratic interpretations both answers our second question and opens several further avenues of investigation. Newton was not prevaricating when he denied any "great excellence in such a ☿ either for medical or Chymical operations." Just as he explained to Oldenburg, the incalescent quicksilver was a pseudosophic mercury that could penetrate no farther than ordinary corrosives did into the corpuscular microstructure of gold. This was not the way to make the noble metal ferment and putrefy; for that one needed to understand the true nature of Diana's doves and their role in making Mercury's caducean rod, the end goal of Newton's long alchemical quest. To think that the genuine Philalethan sophic mercury could be made by such a simple process as the one described in the *Clavis* was for him a laughable delusion of neophytes, cheats, and fools.

But if that was truly Newton's attitude, then why was he exercised about the possibility that Boyle's little treatise might provide "an inlet to something more noble" that could in turn unleash "immense damage" on the world? This seeming inconsistency can be explained if we consider the use that Newton himself was making of antimonial reguli and quicksilver from the earliest years of his experimentation. Newton was already making reguli of the various metals known to him by using them to reduce stibnite to metallic antimony, just as in the standard method with iron, by 1674 if not earlier. He was also experimenting with the amalgamation of these reguli and quicksilver, as in the following early record: "If Reg ♄ melted bee dropped upon ☿ it will amalgam but noe other Reg."[13] Such simple experiments would soon give way to Newton's attempts to ferment mercury with the hollow oak or net and other antimonial alloys, and these in turn formed the basis of the complicated solutions, precipitations, and sublimations that we surveyed when examining Newton's laboratory notebooks. Similar operations still make up an essential part of the fully mature *Praxis*, where

[13] CU Add. 3975, 43r.

Newton builds on atomized fragments of the Philalethes treatises to develop his grand, master process for arriving at the philosophers' stone, as we have seen throughout the present book. Hence even if Boyle's incalescent mercury as a final product incorporated Starkey's "erroneous" interpretation of Philalethes, it might still put the vulgar on a path that could lead them to the true understanding that Newton believed he had attained. This was something that had to be avoided.

The same anxieties about divulging the secrets of alchemy emerge again in even stronger form a decade and a half after Newton's exchange with Oldenburg, this time on the occasion of Boyle's death on New Year's Eve, 1691. On January 26, less than a month after Boyle's death, Newton would begin making inquiries to their mutual friend John Locke about the fate of a mysterious "red earth" that had belonged to Boyle as well as the process for making "ye red earth & ☿."[14] Boyle had appointed Locke, along with Edmund Dickinson and the physician Daniel Coxe, to sort through his posthumous papers and determine their fate.[15] Apparently Locke was surprised that Newton knew about this sensitive subject, for the latter felt the need to explain the source of his knowledge in another letter to Locke sent three weeks later. Here Newton says that he had heard "you had writ for some of Mr Boyles red earth & by that I knew you had ye receipt."[16] The next extant letter, sent from Cambridge on the seventh of July, begins a delicate dance in which Newton tries to work out the obligations entailed by promises of secrecy that he and Locke had previously made to Boyle. One can discern Newton's eagerness to learn more about Boyle's process, but at the same time, his letter reveals a distinct reserve. The letter begins abruptly, as the top of the sheet has been torn off:

> . . . as I can. You have sent much more earth then I expected. For I desired only a specimen, having no inclination to prosecute ye process. For in good earnest I have no opinion of it. But since you have a mind to prosecute it I should be glad to assist you all I can, having a liberty of communication allowed me by Mr B. in one case wch reaches to you if it be done under ye same conditions in wch I stand obliged to Mr B. For I presume you are already under ye same obligations to him. But I feare I have lost ye first & third part out of my pocket. I thank you for what you communicated to me out of your own notes about it.[17]

From this we learn that Locke sent Newton a sample of Boyle's red earth, and that Newton has no intention of carrying out "ye process," though he is willing to help Locke do so. Newton's professed reticence is undercut to some degree by the postscript, however, which adds that as soon as "ye hot weather is over I intend to try ye beginning, tho ye success seems improbable." Newton also mentions that he has lost the "first & third part " of the recipe "out of my

---

[14] Newton, *Corr.*, 3: 193.
[15] Newton, *Corr.*, 3: 216.
[16] Newton, *Corr.*, 3: 195. See also Principe, *AA*, 11–12 and 176–78.
[17] Newton, *Corr.*, 3: 215.

pocket." Remarkably, the lost first and third sections have recently resurfaced in the library of the Science History Institute, along with some additional material that Newton acquired from Boyle. At the top of the manuscript, Newton has written "Roth Mallors Work."[18] This was the obscure chymist Erasmus Rothmaler, who was working with or for Boyle in 1685.[19] Boyle also owned a manuscript of Rothmaler's *Consilium philosophicum*, which the English scientist planned to bequeath to Newton, but the 170 pages of this text have all the content of an elaborate calling card intended to titillate the recipient with the names of famous authorities.[20] The first part or "period" of Rothmaler's process is also contained in Locke's response, which is extant in a letter that he sent to Newton on July 26. The process began by "cleansing" quicksilver with sulfur and then shaking it with "Mineral Soap," either mineral or refined antimony.[21] The "soap" is first embodied with the quicksilver and will then "by further agitation be spued out by it." As indicated in notes that Locke or his amanuensis copied from Boyle's remains, this operation will make mercury "somewhat incalescent," a fact that will acquire considerable significance in Newton's subsequent response.[22]

Newton replied to Locke's message on August 2 with an interesting and convoluted letter. His words make the reasons for his disparaging attitude toward Boyle's process quite clear. The copy of the recipe that Boyle had given Newton contained the information that the mercury produced in the first set of operations was incalescent, and Newton therefore identified it with the material that he had already rejected in 1676:

> This ℞ I take to be ye thing for ye sake of wch Mr B procured ye repeal of ye Act of Parl. against Multipliers, & therefore he had it then in his hands. In ye margin of ye ℞ was noted yt the ☿ of ye first work would grow hot wth ☉ & thence I gather that this ℞ was ye foundation of what he published many years ago about such as would grow hot wth ☉ & therefore was then known to him, that is sixteen or 20 years at least. And yet in all this time I cannot find that he has either tried it himself or got it tried wth success by any body els. For when I spake doubtingly about it, he confest

[18] Lawrence M. Principe, "Lost Newton Manuscript at CHF," *Chemical Heritage* 22 (2004): 6–7. Principe identifies the manuscript as Newton's and places it in the context of the exchange with Locke. The manuscript is found in the Neville Collection at the Science History Institute and bears the shelfmark QD14.N498.

[19] Principe, "Lost Newton Manuscript at CHF," 6. For Rothmaler's activities in Europe, see Hjalmar Fors, *The Limits of Matter* (Chicago: University of Chicago Press, 2015), 58–59. Two letters from Rothmaler to Boyle survive in the archives of the Royal Society. They have been edited and translated in Michael Hunter, Antonio Clericuzio, and Lawrence M. Principe, eds., *The Correspondence of Robert Boyle* (London: Pickering and Chatto, 2001), 6: 147–50.

[20] RS, Boyle Papers 23. An interleaved slip imparts the following message: "Mr Rothmaler's Booke yt I had from himselfe. by way of gift as I / understood it. Wch I bequeath to Mr Newton ^the mathematitian of Cambridge." See also Principe, *AA*, 112.

[21] For the "soap" *Deckname*, see Principe, *AA*, 176.

[22] The first part of the recipe is very similar to one found in Locke's Bodleian Locke MS C44, which he had copied from Boyle's manuscripts. On p. 27 one finds the following words: "To purifye Mercury & make it somewhat incalescent. Rx [antimonium] opt: lbi pulverisa subtilissime et statim uncias duas super injice [mercurii] per furfur depurati. In retorta terra agita fortiter per horam integram. Destilla per gradus in igne reverberii et tandem urge igne fortissimo." The "furfur" in the recipe might either be a misreading for "sulfur" or a *Deckname*.

that he had not seen it tried but added yt a certain Gentleman was now about it & it succeeded very well so far as he had gone & yt all ye signes appeared so yt I needed not doubt of it. This satisfied me yt ☿ by this ℞ may be brought to change its colours & properties: but not that ☉ may be multiplied thereby.[23]

Just as Newton had spurned the heat-producing quicksilver of Boyle's 1676 essay as a violent, superficial menstruum, so he now rejected the incalescent mercury of Rothmaler's process. Although he was quite wrong in identifying the two, since in 1676 Boyle was thinking of Starkey's sophic mercury, the presence of incalescence was enough for Newton to mark Rothmaler's product as a mere sophistication rather than a true means of making gold putrefy and vegetate. The interesting claim that it was Rothmaler's mercury that led Boyle to throw his weight behind the 1689 repeal of Henry IV's anti-alchemical act against multipliers is meant to underscore the fact that Boyle had possessed the mercury without any further success for a number of years. In subsequent comments Newton goes on to say that he had himself made inquiries about a "company" attempting Boyle's recipe in London, but that they too "could not make the thing succeed."

Newton therefore based his doubts about Rothmaler's process on the twin facts that it produced heat like a mineral acid and that no one, including Boyle, had succeeded in using it for chrysopoeia. But here another very interesting thing surfaces. At the end of the letter, after casting doubt on Boyle's recipe, Newton makes a deeply incongruous claim. Despite his having spent over twenty years in the quest for the philosophers' stone, and with the frenzied summer of collaborative research that led to "Three Mysterious Fires" and *Praxis* still in the future, he seems abruptly to dismiss the entire enterprise:

> In diswading you from too hasty a trial of this ℞ I have forborn to say any thing agt multiplication in general because you seem perswaded of it: tho there is one argumt against it wch I could never find an answer to & wch if you will let me have your opinion about it, I will send you in my next.[24]

Alas, we do not possess any further comments from Newton to Locke on the subject of multiplication. But the dates in his laboratory notebooks show that he avidly continued his chrysopoetic experiments for at least another four years, before he departed Cambridge for London. Was he actually trying to delude Locke, with whom he had shared other sensitive information on subjects ranging from religion to patronage, into thinking that he had no real interest in the subject? Before replying, one should recall that Locke belonged to the same circle of Newtonian friends as Fatio de Duillier, who had already been in deep discussion with Newton about alchemy since 1689. Indeed, the young Genevois would relay an invitation from Locke in 1693 suggesting that he and Newton join the philosopher to live at the home of Lady

[23] Newton, *Corr.*, 3: 217.
[24] Newton, *Corr.*, 3: 219.

Masham at Oates.[25] At the beginning of the 1690s, these men had entered into a relationship with Newton as close as anyone ever managed to attain. Moreover, only four months before his August letter to Locke, Newton had affirmed the alchemical sulfur-mercury theory of the metals to Archibald Pitcairne and argued that if "gold could ferment, it could be transformed into any other substance." It does not seem that Newton was shy about making his alchemical interests known to his circle of close acquaintances. Before writing off Newton's comments to Locke as sheer duplicity, we should therefore consider another possibility, and one that has not heretofore been raised.

The term "multiplication," which appears in various forms three times in Newton's letter, could be used as a blanket term for all alchemical transmutation, but it did not necessarily have such a wide meaning in every case. Thomas Norton's fifteenth-century *Ordinall of Alchimy*, for example, a text that Newton knew and used in the popular 1652 *Theatrum chemicum britannicum* of Elias Ashmole, makes a sharp distinction between multiplication and legitimate alchemy.[26] No doubt attempting to avoid the shadow cast by the 1404 act against multipliers, Norton asserts that the *Multiplyers* are purveyors of "false illusions" who beguile the unsuspecting with "fals othes." Norton explains his rejection of multiplication in the following terms:

> When such men promise to Multiplie,
> They compasse to doe some Villony,
> Some trew mans goods to beare awaye;
> Of such fellowes what shulde I saye?
> All such false men where ever thei goe,
> They shulde be punished, thei be not so.
> Upon *Nature* thei falsely lye
> For Mettalls doe not Multiplie;
> Of this Sentence all men be sure,
> Evermore Arte must serve *Nature*.
> Nothing multiplieth as Auctors sayes,
> But by one of theis two wayes,
> One by rotting, called Putrefaction,
> That other as Beasts, by Propagation;
> Propagation in Mettalls maie not be,
> But in our Stone much like thing ye may see.
> Putrefaction must destroy and deface,
> But it be don in its proper place.[27]

Norton accuses the multipliers of attempting to make metals multiply in the biblical sense of procreation. As he says later in the text, they do indeed grow beneath the surface of the earth, but only under the influence of "the vertue Minerall," which is not found aboveground. It is folly to think that humans

---

[25] Fatio to Newton, April 11, 1693, in Newton, *Corr.*, 3: 391. See also Fatio to Newton, February 24, 1689/90, in Newton, *Corr.*, 3: 390.

[26] See, for example, the many references to Norton in the *Index chemicus*, Keynes MS 30/1.

[27] Thomas Norton, *Ordinall of Alchimy*, in Elias Ashmole, *Theatrum chemicum britannicum* (London: Nathaniel Brooke, 1652), 6, 17–18.

can induce metals to undergo propagation in their laboratories, but this does not mean that alchemy is futile. Instead, as Norton suggests in the passage above, "Putrefaction must destroy and deface" the existing metal before it can be transmuted into another. This does not occur, he says a page later, when "corrasive waters have made dissolucion," that is, when the mineral acids corrode metals. The dissolution and putrefaction of metals must be of a deeper sort than vulgar menstrua can achieve, and this type of fundamental transformation is what Norton promises to teach in his *Ordinall*.

Norton's views about the distinction between superficial corrosion and genuine putrefaction were not unusual in the literature of alchemy, nor was his blanket dismissal of multiplication.[28] Indeed, a similar admonishment may be found in works by Newton's contemporaries, such as the 1698 text by one Hortulanus Junior, *The Golden Age; or, The Reign of Saturn Review'd*. A self-styled follower of Philalethes, the "last and best Interpreter of all the Ancient Philosophers," this pseudonymous author recounts the admonitions against "Deluders, and Cheating *Multipliers*" stemming from such august authorities as Geoffrey Chaucer and the fifteenth-century alchemist George Ripley.[29] Moreover, Hortulanus Junior refers to a warning from the utopian Interregnum agrarian and alchemical writer Gabriel Plattes or Plat aimed at "sophistical Multipliers and Imposters."[30] Turning to Hortulanus's source, the 1655 "Caveat for Alchemists" by Plattes, one will encounter a rejection of multiplication expressed in terms that would have warmed the cockles of Newton's heart:

> To sum up all, Let men beware of all books and receipts, that teach the multiplication of gold or silver, with common quicksilver by way of animation or *minera*, for they cannot be joyned inseparably by any *medium*, or means whatsoever.[31]

From all this it is clear that even in the second half of the seventeenth century, "multiplication" could bear the pejorative sense of a sophistical alchemical operation, particularly one that employed vulgar, commonplace materials such as quicksilver and corrosives. It is therefore not unlikely that Newton was using the term in this derogatory sense when he expressed his doubts to Locke about multiplication in general. This was the realm of

---

[28] See, for example, *Bloomfields Blossoms*, in Ashmole, *Theatrum chemicum britannicum*, 315, which links superficial corrosion to multiplication in the following terms: "Dissolve not with Corrosive nor use Separacion With vehemence of Fire, as Multipliers doe use." See also the text identified by Ashmole merely as *Anonymi*, at *Theatrum chemicum britannicum*, 414, where the author prays that "noe Multiplyer meete with my Booke, Nor noe sinister Clerkes."

[29] Hortulanus Junior, *The Golden Age; or, The Reign of Saturn Review'd* (London: Rich. Harrison, 1698), 10, 191, 200.

[30] Hortulanus Junior, *Golden Age*, 82.

[31] Gabriel Plattes, "A Caveat for Alchemists," in the anonymous *Chymical, medicinal, and chyrurgical addresses: Made to Samuel Hartlib, Esquire* (London: Giles Calvert, 1655), 81–84 (mispaginated). For Plattes, see Charles Webster, *The Great Instauration* (London: Duckworth, 1975), 47–51. See also Webster, "The Authorship and Significance of *Macaria*," *Past and Present* 56 (1972): 34–48. Also cf. Webster, "Macaria: Samuel Hartlib and the Great Reformation," *Acta Comeniana* 26 (1970): 147–64, and Webster, *Utopian Planning and the Puritan Revolution: Gabriel Plattes, Samuel Hartlib, and Macaria*, Wellcome Unit for the History of Medicine Research Publications 2 (Oxford, 1979).

sophistical alchemy, not far removed from the counterfeiting and coin clipping that would soon be exercising Newton as warden and master of the Royal Mint. If we briefly look at Newton's use of the term again, such an interpretation seems entirely likely. A few lines after referring to Henry IV's act against multipliers, he says to Locke that Boyle's assurances convinced him of the fact that Rothmaler's mercury "may be brought to change its colours & properties: but not that ☉ may be multiplied thereby." But what respectable alchemist of Newton's day would have said that the sophic mercury itself was a means of increasing the quantity of one's gold? The object of the Philalethan sophic mercury was to putrefy gold, lead it through the different planetary regimens, and arrive ultimately at the philosophers' stone. Only after the philosophers' stone had been acquired could there be a transformation of base metal into gold, and as for the term "multiplication," to Philalethes this referred primarily to an increase in the stone's transmutative power achieved by combining it with mercury and repeating the series of regimens again.[32] There is good reason to think, then, that Newton's repeated references to multiplication in the August letter refer to a specific, delusional type of chrysopoeia, not to the entire enterprise of transmutation. In short, he is referring to the "corrosive" character of incalescent antimonial mercuries, which causes them to grow hot on encountering gold. Thus if we possessed the "argumt against it wch I could never find an answer to," which Newton offered to Locke in order to steer him away from multiplication, it might well have consisted of the same corpuscular reasoning that he employed against Boyle's essay on incalescence in his message to Oldenburg of 1676.

Although my argument to this point might seem to absolve Newton of the imputation of outright deceit toward Locke, there is no denying the tortured quality of his reasoning, particularly in the matter of promises made to Boyle. This emerges particularly in the second paragraph of his August letter, where Newton goes into fascinating detail about the strictures that Boyle imposed on him before revealing his secrets. The impression that Newton carefully tries to convey is one of a delicate game where the possessor of the secret was the cat and Newton a reluctant mouse. Although the passage is long and intricate, no other writing by Newton captures so well the guarded character of chymical exchanges in the seventeenth century:

> But besides if I woud try this ℞, I am satisfied that I could not. For Mr B has reserved a part of it from my knowledge. I know more of it then he has told me, & by that & an expression or two wch dropt from him I know that what he has told me is imperfect & useless wthout knowing more then I do. And therefore I intend only to try whether I know enough to make a ☿ wch will grow hot wth ☉, if perhaps I shall try that. For Mr B. to offer his secret upon conditions & after I had consented, not to perform his part looks odly; & that ye rather because I was averse from medling wth his ℞ till he perswaded me to do it, & by not performing his part he

[32] See the chapter titled "The Multiplication of the Stone" in Philalethes, *SR*, 114–15. Newton also uses the term "multiplication" in a similar sense throughout his chymical corpus.

has voided ye obligation to ye conditions on mine, so yt I may reccon my self at my own discretion to say or do what I will about this matter tho perhaps I shall be tender by using my liberty. But that I may understand ye reason of his reservedness, pray will you be so free as to let me know the conditions wch he obliged you to in communicating this ℞ & whether he communicated to you any thing more then is written down in ye 3 parts of ye ℞. I do not desire to know what he has communicated but rather that you would keep ye particulars from me (at least in ye 2d & 3d part of ye ℞) because I have no mind to be concerned wth this ℞ any further then just to know ye entrance. I suspect his reservedness might proceed from mine. For when I communicated a certain experimt to him he presently by way of requital subjoined two others, but cumbered them wth such circumstances as startled me & made me afraid of any more. For he expected yt I should presently go to work upon them & desired I would publish them after his death. I have not yet tried either of them nor intend to try them but since you have the inspection of his papers, if you designe to publish any of his remains, You will do me a great favour to let these two be published among ye rest. But then I desire that it may not be known that they come through my hands. One of them seems to be a considerable Expt. & may prove of good use in medicine for analysing bodies, the other is only a knack.[33]

Newton begins the passage by again excusing himself from replicating Rothmaler's experiments, this time because he does not have all of the information necessary to carry them out. Then he backtracks and says, "I intend only to try whether I know enough to make a ☿ wch will grow hot wth ☉," in other words the first part of the recipe. Newton then states that he agreed to all of Boyle's conditions and promises, but crucial information was still not forthcoming. He laments Boyle's secretiveness and even suggests that the genial "naturalist" acted in bad faith: "after I had consented, not to perform his part looks odly." And Boyle's bad behavior was exacerbated by the fact that Newton had not even wanted to meddle with his recipe in the first place! Hence Boyle "voided ye obligation to ye conditions on mine"; in other words, Boyle's failure to comply with the very terms that he set forth frees Newton to do as he wishes with the material that the older man divulged to him. After all of these complaints against Boyle's acquisitive secrecy, it is then astonishing that Newton explicitly requests Locke to withhold the second and third parts of Rothmaler's recipe from him, the very recipe that he just accused Boyle of partially reserving. Newton has no mind "to be concerned wth this ℞ any further then just to know ye entrance," namely, the production of the incalescent mercury. The affable Locke must have been reminded of Odysseus tied to the mast at his own request so that he could not succumb to the song of the Sirens.

What are we to make of this punctilious combination of self-pity, irritation, and cunning? Newton's cold absence of personal affection for the recently deceased Boyle may not fill the modern reader with sympathy for the towering intellectual. But we need not leap to the tempting conclusion that Newton was frantically pursuing Rothmaler's mercury while at the same time attempting to cover his tracks. In fact, we have Newton's laboratory records for the period from December 1692, only five months after his exchange with Locke, and these concern the fermentation of barm to make beer, not the incalescent mercury.[34] The experimental notes then pass without interruption to a record of Newton's ongoing project for making volatile Venus, the net, and the other standard products of his laboratory that he had been trying to perfect since the 1670s. The records give us no reason whatsoever to suppose that he broke off this long-standing research project in order to attempt Rothmaler's process. To the contrary, all his reasoning led him to conclude, as he said to Locke, that the recipe was unlikely to succeed in producing a genuine sophic mercury. What then was the origin of Newton's obvious exasperation with Boyle in the August letter?

The answer shines forth from the remainder of the paragraph quoted above: Newton was annoyed because he felt that Boyle was wasting his time. This emerges after Newton's surprising admission that his own caution was the origin of Boyle's chariness: "I suspect his reservedness might proceed from mine." Newton adds that after giving Boyle some experimental information, the older scientist replied in kind by presenting him with two additional recipes. Newton was surprised by the stern conditions that Boyle levied on this exchange, however: he "cumbered them wth such circumstances as startled me & made me afraid of any more." Thus Newton admits some culpability for Boyle's subsequent reticence in the matter of the Rothmaler recipe. Yet he feels justified all the same. The conditions that Boyle tried to impose on the two earlier recipes given to Newton were not merely the usual extraction of promises to secrecy, but rather the twin requirements that Newton "should presently go to work upon them" and that he should publish them after Boyle's decease. The first of these conditions would have put Newton practically in the role of Boyle's laboratory technician, a position that the author of the Philalethes treatises himself, George Starkey, had declined some forty years before.[35] Newton had no more desire than Starkey did to abandon his own research and become Boyle's operator; to the contrary, investing the time and considerable effort involved in carrying out Boyle's research would have interrupted Newton's own very intensive chymical work in the laboratory. The same thing was true of Boyle's request that Newton publish the two recipes that he had bequeathed upon him; hence, we see Newton trying to pass on this obligation to Locke.

What were the two recipes that Boyle delivered with the injunction that Newton replicate and then publish them? A comparison of the Science History Institute manuscript and a fragmentary draft letter by Newton published in the supplemental material to his *Correspondence* reveals that

---

[34] CU Add. 3973, 25r–28r.
[35] Newman, *GF*, 62–78.

the two recipes bequeathed by Boyle have survived.[36] The first of them suggests that sal ammoniac be mixed with quicklime, and that a salt should be extracted from the product. This is then added to spirit of niter or aqua fortis to yield a menstruum that will dissolve gold. One can readily believe that chloride from the sal ammoniac would yield a type of aqua regia. I suspect that this recipe is the one that Newton referred to as "only a knack." At any rate, he did not bother to copy it for the letter that he was preparing to send out for publication. The other recipe, presumably the one that Newton calls "a considerable Expt. & may prove of good use in medicine for analysing bodies," directs one to digest butter of antimony that has liquefied in humid air at the temperature of hot blood.[37] Boyle says that the liquid butter of antimony will first putrefy, turning black, and then clarify again. At this point it will be "a menstruum for resolving bodies like the Alkahest but not so potent."[38]

In the end, the 1692 exchange with Locke reveals a Newton who was happy to receive secrets from Boyle so long as they did not impede the progress of his own research. It is true, of course, that it was he who initiated the hunt for Boyle's red earth on January 26, less than a month after the death of the famous chymist. The elaborate secrecy in which Boyle veiled the process must have made Newton's ears prick up, especially because it was out of character for the famously scrupulous Boyle to behave "oddly" by failing to deliver on his promises. Hence he contacted Locke in order to learn more details about what Boyle knew and to "understand ye reason of his reservedness." There was a possibility that the incalescent mercury of Rothmaler might be an "an inlet to something more noble," as Newton had put it in 1676, even though he had serious doubts. His refusal to accept Boyle's authority in the matter of the sophic mercury fits closely with the image of Newton that we have encountered over the course of the present book. From his earliest days as a student of chymistry, Newton believed himself to belong to an elite group. He was on the road to becoming an adept, a path that required the scintillating intellect and intuitive apprehension of a *filius doctrinae*, a son of art. The gift of God, the stone of the philosophers, was close at hand, almost within his grasp. No one, not even the author of *The Sceptical Chymist* and *The Origin of Forms and Qualities*, could move Newton from the path of the adepts that he, on the basis of his decades-long interpretation of the masters, had learned to follow. Convinced of the correctness of his own alchemy, Newton did not need the advice of Boyle; if anything, the contrary was the case. Boyle, after all, had failed to understand that the self-heating mercury of Rothmaler was probably a superficial knack

[36] See Newton, *Corr.*, 7: 393, for the undated letter, which mentions no recipient. Although the letter refers to two recipes given to Newton by Boyle, and requests that they be published, Newton only copied one of them. This recipe corresponds, though not verbatim, to the second of the two recipes found in Science History Institute MS QD14.N498, 1v.

[37] In the undated draft letter, Newton appends the following words, which are very similar to the ones in his August 2, 1692, letter to Locke, to the recipe: "The menstruum prepared by the first of these two experimts was proposed by Mr Boyle as a thing wch might be of good use in medicine for analysing & subtiliating bodies." See Newton, *Corr.*, 7: 393. I thank Michael Hunter for directing me to this letter.

[38] Science History Institute MS QD14.N498, 1v.

like the gold-dissolving menstruum made with quicklime and sal ammoniac. And yet there was the possibility, however slight, that Boyle's mercury was the real thing. God works in mysterious ways, and the philosophers' stone was his special gift, the *Donum Dei*. Had not Geber himself said that God "extends it to and withdraws it from whomever He wills?"[39]

[39] William R. Newman, *The Summa perfectionis of Pseudo-Geber* (Leiden: Brill, 1991), 640.

Epilogue

The picture of Newton that we receive from his exchange with Locke after the death of Boyle is not entirely flattering, but it does confirm a pattern the famous physicist displayed from his earliest attempts to decode the riddles of his alchemical masters. Newton believed himself to be a member, or at least an apprentice, in an exclusive brotherhood, the school of Sendivogius, Philalethes, Snyders, and the other inaccessible intellects whom he thought to have attained the higher reaches of the hermetic art. The self-confidence that allowed Newton to overturn two millennia of optical theory and to dismantle Cartesian physics with the *Principia* was the same awareness of his special status as a novitiate in the fraternity of the adepts. It was this certainty that allowed him to dismiss Boyle's hermetic legacy as "only a knack" and to write off Rothmaler's mercury as a product of inferior quality. It is worth reviewing the origins of Newton's belief in his special status, as this will speak to his mainstream science as well. If we consider the earliest origins of Newton's interest in the natural world and its manipulation, the works of Bate and Wilkins immediately come to mind. Both *Mysteries of Nature and Art* and *Mathematicall Magick* belong to the capacious genre of books of secrets and natural magic, in the spirit of Giambattista della Porta's sixteenth-century *Magia naturalis*.

Now one might dismiss Newton's interest in these topics as a boyish dalliance, and it is indeed true that he showed little interest in what passed for magic in the seventeenth century as he matured. Yet both Bate and Wilkins consider chymical topics, albeit in passing, and Newton's notes on *Mathematicall Magick* testify to the interest that the Wadham College warden's comments on alchemical marvels stimulated. The next step in Newton's peregrination appears to have been supplied by the anonymous *Treatise of Chymistry* that served as the primary basis of his early chymical dictionaries. Newton may have been working through this text as early as the first half of the 1660s, since it displays no explicit knowledge of Boyle's work in the field. Only with Don. b. 15, the dictionary that he compiled after acquiring a copy of Boyle's 1666 *Origin of Forms and Qualities*, does the aristocratic English "naturalist" enter directly into Newton's chymical studies. From this work Newton learned of the impressive menstrua described by Basilius Valentinus, which apparently led him to make the detailed, early studies of the supposed Benedictine monk that we examined in chapter five. And by 1669 Newton had jumped headfirst into the bottomless ocean of alchemical enigmata, devouring the works of Philalethes, Sendivogius, and the profusion of authors massed together like sardines in the *Theatrum chemicum*.

If one views this early evidence of chymical interest as forming a continuum, then it follows that Newton's first sustained aspiration to the experimental understanding of nature fell within the broad domain of chymistry. Although his interests had evolved far beyond Bate and Wilkins by 1669, there is still something of Wilkins's *thaumatopoiētikē* or thaumaturgy in Newton's ongoing approach to alchemy. Newton wanted not only to understand nature but also to use his knowledge to perform wonders. The goal of acquiring the mirific *arcana majora* of the sages still gripped Newton in his late endeavors to recreate the caduceus of Mercury, an end supremely evident in *Praxis* and still at work in the early eighteenth century when he was collaborating with William Yworth. Yet by no means should we see Newton's private, chrysopoetic project as a mere quest for the production of effects. To the contrary, as the queries to the *Opticks* make clear, there was a seamless boundary between Newton's public thoughts about the deep structure of matter and his attempts to put the relative affinities governing corpuscles to work in the privacy of his laboratory. Indeed, as the present book has demonstrated, a constant bleed-through from Newton's chrysopoetic reading and research to his publicly revealed chymistry occurred at numerous levels.

Nor can we argue that Newton's attempts to decipher the enigmas of the adepts were anything but rational. Never does he appeal to the evidence of his own dreams, for example, despite the fact that some, like Starkey and his hero Van Helmont, occasionally relied on an oneiric epistemology to account for the fact that God could reveal alchemical knowledge in the form of nocturnal visions.[1] Yet in other respects, Newton's attempts to extract the secrets of the sages from written texts were not markedly different from those of his alchemical predecessors. Like them, he was keenly aware of literary tricks such as *dispersa intentio*, parathesis, syncope, and the graduated iteration that trumped them all. Other chrysopoetic chymists before Newton had kept laboratory notebooks and commonplace books, and among them Starkey was not the only one who learned to use the classic figures of speech as part of his university education. The difference between Newton's approach to decipherment and that of his predecessors was not one of quality but of degree. No one but Newton compiled the *Index Chemicus* with its ninety-eight folios devoted to the understanding of what were for the most part traditional *Decknamen*. This master concordance was yet another tool in his unyielding attempt to unravel the gnomic *dicta* of the chymists, serving alongside his extracts, abridgements, commonplace entries, double-column analyses, and *florilegia* in the interest of finding the royal road to the elixir.

Nowhere else in Newton's scientific work can one see the same degree of combined textual scholarship and experiment that we encounter in his

---

[1] Despite their claim to sporadic dream revelations, however, even Starkey and Van Helmont were hardnosed chymists who believed that success in the laboratory was "bought with sweat." For their references to dreams, see Newman and Principe, *ATF*, 56–58, 97, 197–205.

alchemy. His biblical hermeneutics and deep studies of chronology married his skill in astronomy to his understanding of humanist scholarship, but without anything approaching the generation-long labors of the laboratory elicited by chymistry. The unique and often idiosyncratic nature of Newton's alchemical quest calls for one final question. Can we see something of his reclusive, aloof character, his reliance on a brilliant, innate acumen to the exclusion of any personal warmth, and above all his belief that he stood above the herd of common intellects, fully justified by his scientific discoveries, as stemming from his early and ongoing idealization of the adepts? Did he model himself on the unerring and inaccessible sons of art? No certain answer can be given, but Newton may well have felt himself an alchemist in the most essential part of his nature even if his long pursuit of the philosophers' stone did not yield the results that he had hoped for when, as a callow youth, he set his aim at "somthing beyond ye Reach of humane Art & Industry."

APPENDIX ONE

The Origin of Newton's Chymical Dictionaries

Sloane 2206, titled "A Treatise of Chymistry" on the flyleaf, is a clearly written manuscript with an unusual format: it is 170 mm high by 235 mm wide, and written only on the recto sides of the folios. The manuscript consists of twenty-one folios, including a sheet of "The Usual Chymical Characters" and "The Usual Medicinal Characters" at the back. The text is in brown ink, and the furnaces are drawn in red ink with brown hatching and lettering. To judge by the scribal abbreviations and the forms of the letters, the manuscript seems to have been copied around the end of the seventeenth century. Unfortunately, it was rebound on stubs by the British Library at some point, and the original binding was lost along with whatever information there may have been concerning provenance and ownership.

It is obvious that Newton's chymical dictionaries existing in the manuscripts Schaffner Box 3, Folder 9 and Bodleian Don. b. 15 derive from an ancestor of Sloane 2206. The Schaffner manuscript is an imperfect copy of the text with significant errors and lacunas, whereas Don. b. 15 supplements the "Treatise of Chymistry" with material taken from Robert Boyle. One might suppose prima facie that the "Treatise of Chymistry" could be an original text by Newton, which he then copied onto a single sheet in Schaffner Box 3, Folder 9. But there are powerful reasons for discounting this possibility, which I present here.

First, as mentioned in chapter five of the present book, Newton has made a mess of things at the very beginning of the Schaffner manuscript by trying to combine all of the braces used in the "Treatise of Chymistry" to dichotomize apparatus and operations onto a single sheet. Thus he writes "In Chymistry (Pyrotechny Spargyry &c) are considerable," followed by the deleted bifurcation beginning abruptly "Operation it selfe" and "To the Operation." These fragments are all that is left of Sloane 2206's beginning, which runs thus: "In Chymistry otherwise called Pyrotechny and Spagiry, are considerable the—" followed by "Subservients to the Operation, where consider the—" and "The Operation it selfe, which is—." It appears that Newton meant to carry the beginning of the first dichotomy over to the verso side of the sheet, but then thought better of it and deleted the entire bifurcation instead.

There is also other evidence that clearly reveals the "Treatise of Chymistry" to have been anterior to the Schaffner manuscript. Numerous lacunas in the Schaffner manuscript are not found in Sloane 2206. The following five lines from table 5 in Sloane 2206 (8r) are omitted, for example in the Schaffner MS:

> ♂ into a red Pouder or Crocus by the fume of Aqua fortis. So the Sharp Nitrous Spirit in the Air brings ♂ into Rust. Hither refer the bringing of the Metalls [especially the Noble, vizt. ⊙ and ☽] to be pulverizable by hanging them over the Fume ^{of} ☿ or ♄. Note. The acid Liquor must be made to send some Vapour by easy Heat.

A similar situation is found on Sloane 2206, 14r. Comparing this to the Schaffner MS, one sees that Newton has left out four lines of 2206's introduction to "Dissolution." This is probably because he previously left out 2206's dichotomy between "Calcination" and "Dissolution." Instead, Newton jumps right into "Sublimation" without any introductory lines or any dichotomizing braces. Nor does Newton provide the preceding bracketed division between "Solution" and "Coagulation." There are several other similar instances of a missing line or two in Schaffner Box 3, Folder 9, but the most telling example of omission is found at the end. Here Newton has entirely omitted folios 17r–21r of the "Treatise of Chymistry." About halfway through copying the treatise's chapter on distillation, Newton breaks off abruptly with no explanation.

This pattern of lacunas again shows that Sloane 2206 represents an earlier state of the text than that found in the Schaffner manuscript, but does not prove in itself that Newton was copying another author's work rather than his own. Yet there is evidence for this as well. For example, Sloane 2206 describes the method of extracting salts from "terra damnata" on folio 13r. "Terra damnata" was a standard technical term for any residue left behind after a calcination, yet on Schaffner 1v, Newton has misread the expression twice as "Terra Dameta." One might write this off as an oversight except that the same mistake occurs in Bodleian Don. b. 15 where Newton again writes, on 2v, "Sal Terræ Dametæ is $^{a\ fixt\ salt}$ got by pouring hot water on the l̶a̶t̶$_{Terra}$ to imbibe its salt, y^n filtrating & evaporating it." There is no evading the fact that the chymical noviciate mistook a trivial term of art intended to refer to leached calcination products in general for a specific type of fixed salt, whether the error originated with him or derived from copying a defective manuscript. An error of this sort could only stem from the misapprehension of another author's meaning.

The evidence is strong, then, for the "Treatise of Chymistry" stemming from another author rather than from the pen of Newton. When might the text have been written, and who might its author have been? The second part of the question must remain open for the moment, but we can at least supply a *terminus post quem* for the "Treatise" by considering another passage that

Newton omitted in his hasty copying of the text. After describing a water bath on 5r, Sloane 2206 supplies the following comments:

> It is sometimes called Balneum Mariæ, from Mary or Miriam the sister of Moses, under whose name goes a Treatise de Lapide Philosoph: Or Maris, because it boyls with Waves like the Sea. Of this are severall Contrivances, Glaubers is the best in his 3. ^Part of his Philosophicall Furnaces.

It is likely that the author here refers to the English translation of Johann Rudolph Glauber's *Furni novi philosophici* that appeared in 1651 as *A Description of New Philosophical Furnaces* (London: Tho. Williams, 1651). Hence we have not only a *terminus post quem* but also one of the sources of the "Treatise." Still more dating information can be gleaned from the author's reference to *ens veneris* on 14r, where he states that the medicament "is ✳ sublim'd from dulcify'd colcothar." This is the method for making *ens veneris* that George Starkey invented in the early 1650s while working with Robert Boyle. The method of production was subsequently published by Boyle in *Some considerations touching the usefulnesse of experimental naturall philosophy* (London: Ric. Davis, 1663), 163–66. But because the author of the "Treatise" may have already known of *ens veneris* from Starkey's own activities as a prominent medical practitioner in London, 1663 cannot serve as a firm chronological marker. The best we can say is that the "Treatise" is unlikely to have been written before the second half of the 1650s, when Starkey had acquired some fame as a Helmontian chymist.

APPENDIX TWO

Newton's "Key to Snyders"

Newton's short "Key to Snyders," part of Sotheby lot 103, is currently in private possession. The following transcription, on which my translation is based, was made from a photocopy. Since I have not had access to the original manuscript, it is impossible to say anything about its watermarks or other physical characteristics. The manuscript is described in a catalog prepared by Sotheby's for a 2004 auction (*Sir Isaac Newton: Highly Important Manuscripts*, New York, December 3, 2004, p. 20). My transcription follows the foliation supplied with the photocopy, since I am unable to correct it by means of personal inspection. I have retained Newton's erratic hyphenation at line breaks throughout the transcription.

The manuscript contains one barred Saturn symbol and no unbarred ones. Aside from this, and the rather hesitant state of Newton's interpretation, there are no other clues to provide a date of composition. If we build on the clue provided by the barred Saturn symbol, the *Key to Snyders* is unlikely to have been written before 1674, since Newton typically used the barred version of the symbol after that date (see chapter five).

<1r> 14 A Key to Snyders

Humida solutio fit per astrale semen quod est siccus liquor ceri
fluus. Hic liquor est primus ignis quo metallum adigi debet in
fluxum. (p 10). Secundus est sal præparatum ex ♀ᶜ et corniculata
Diana sine semine auri. Tertius spiritus ☿ⁱʲ, vel potius Venus
Philosophica: ~~instar mercurij currens~~ ^{quia dicitur ☿o fere similis}. fforte tamen aqua
sicca ^{imprægnata ♂te et ♀e quia dicitur metallicus}.

Mineralis metallicus♃ ignis materia prima est, quae repe
ritur in minera Saturni tanquam in domo sua universali.
Ex hac domo discedere debet præ angustia ignei volantis Dra
conis qui domicilium frigidi saturni taliter incendit ut in
eo mori & spiritum suum exhali cogatur. Si possis reci
piente capere hunc spiritum, habes universale menstruum,
astralem ignem, qui effigiem habet aquæ siccæ simul ac humidæ quæ nihil humectat nisi metalla. In pondere
ante omnia gravis levis♃ est. Est verus separator ^impuritatum metalli-

corum sulphurum. Similis est duplici mercuriali aquæ.
diciturⓎ sp^{tus} acidus et duplex corrosivum. Per hunc solum
anima Regis in oleum reducibilis est. p 15

Reductio sive destructio in materiam primam præcipuū
punctum ad Universale generalissimum. Ego autem, hoc præs
tare possum spatio trium dierum, idⓎ per materiam primam.
p. 33.

Mercurius Saturni de divite quadam nondum fusa minera
magna dexteritate est educendus, ~~Qui hoc non apprehendit~~ ope
Aquilæ ac Draconis, quod totum idem est habens solummodo
diversa nomina. Qui hoc non intelligat, nihil unquam pro-
fectus obtinebit. p. 48

Sub Vitriolo Sulphur ardens intelligitur. Hoc vocamus
aliquando igneum draconem. Qui frigidum meum Draconem
[tertium ignem] intelligit, & ipsum in præfatis figuris offenderit
nihilo ulterius eget quam duntaxat rubeum quendam igneum
volantem Draconem hunc suo fratri adjungere & —— AbsⓎ
hoc nequit Universale generalissimum confici p 56.

Humida solutio fit per ☿ saturni quem animam mundi vocavi
p 66.

<1v> Primum metallum est sperma ^et radix omniū metallorum. Reperitur in mi
nera Saturni. Apparet ut minera Mercurij. Vocatur sapien
tum plumbum, de quo lac virginis distillare solemus habetⓎ
Veneream proprietatem. Ultimum metallum est quod ad naturam
auri pervenit. p. 69.

ffrigidus metallicus ignis de minerali quodam nondum fuso
& immalleabili Saturno extrahi oportet, qui mercurius Saturni
nuncupatur. P 70.

AbsⓎ frigido illo mercurio Saturni, & infernali illo Ma
gico Elemento igneo^ igne secundo ac tertio, Chymia nihil utile est expedire.
Ille frigidus metallicus ~~ignis~~ mercurius ex non fuso quodam imma
leabili minerali Saturno extrahitur. p 70, 71.

Universale generalissimum ex duplici mercurio ut supra
narravi producitur, & cum solari sulphure animatur & fer
mentatur & cum perdurante auri sale figitur, & ulterius in
infinitum per duo alia sulphura augmentatur in quantitate
et qualitate. Qualitas augetur per ♀, quantitas per jam dictum ☿^{ium} qui naturam tam proprietatis Venereæ quam
^mineralis frigidi saturni: unde duplici ☿^{io} assimilatur, duplatusⓎ ☿^{ius}
dicitur, cui omnes qualitates appropriantur, nam habet quali
tatem veneris ratione calidi ♀^{is}, frigiditatem autem ex par
te saturni. p. 72

Ex Luna [☽] et similiter ex frigido illo Arietino ♄^{no}
mercurius potest fieri: ut et ex minera Veneris [⊕] Solaris
quidam ☿^{ius} qui solari ♀^{e} præditus sit, unde ipsum ☿^{ium} ☉^{is}

nominavi, siquidem ad generationem solis usurpandus sit. Ex his [saltem duobus posterioribus] fit Universale generalissimum p. 72

Mercurius vivus ex Antimonio et Bismutho potest fieri qui unicè ad medicinam conducunt. Hi sunt nihilominus salva sua excellenti medicina, pro ☿^{io} Ph̄orum neutiquam censendi, cum ille ☿^{ius} sit menstruum universale & bis de ☿^{io} natus primò de Lunari postea de Solari, unde duplatus ☿^{ius} dicitur. Lunaris ☿^{ius} [^{forte} tertius ille metallicus frigidus ignis ∧$^{sed\ dubito}$] leviori opera ex Satur- <*1rB*> nino corpore educitur ac destillatur, & habet naturam Lunæ. Solaris ☿^{ius} extrahitur de minera ♀^{eris} per tartarum & salem Ammoniacum. Qui novit ☿^{ium} frigidi Saturni et ☿^{ium} cali dæ Veneris in Oleum redigere, is habet menstruum universale firmamq; clavem adigendi omnia 🜍^{a} in potabilitatem. p 67 68.

Saturnus est Lunaris, sed Sapientum Saturnus insigniendus cha ractere Solis ∧$^{nam\ materia\ sapientum\ solaris\ est.}$ p 56.

Neptune & Venus make to fly.

Th

<*1vB*> Neptune & Venus make to fly
The snake wch els beneath must ly.

Thou who by y^{e} evaporated Neptune & Venereal property art become an Eagle.

Materiam tuam singulari et occulto artificio in aquam convertes & postquam evaporavit, occulto medio in terram mutabis, quæ est terra virginea Sapientum. Ex hac terra sapientes suum ☿^{ium} & suum duplatum ☿^{ium} parunt & aquam suam vitæ siccam hauriunt quæ corpora omnia radicaliter solvit. Instruct Patris ad fil. c 4.

Hic est sanguis leonis viridis

<*Translation of "A Key to Snyders"*>

The wet solution comes about through an astral seed which is a dry liquor flowing like wax. This liquor is the first fire, by which the metal is forced into flux. (p. 10). The second fire is a salt prepared from ♀ and horned Diana without the seed of gold. The third fire is the spirit of ☿ or rather the philosophical Venus ~~running like mercury~~ ∧$^{because\ it\ is\ referred\ to\ as\ practically\ like}$ ☿. But perhaps the dry water $^{impregnated\ with}$ ♂ ^{and} ♀ $^{because\ it\ is\ called\ metallic}$.

The metallic and mineral fire is the first matter, which is found in the mineral of Saturn as in its universal house. It must withdraw from this house due to the torment of the fiery, flying dragon who ignites the home of cold Saturn so that he is forced to die in it and his spirit is forced to exhale. If you can capture this spirit in a receiver, you have the universal menstruum, the astral fire, which has the likeness of dry water and at the same time wet, which wets nothing but metals. It is light and heavy in weight beyond all other things. It is the true separator of the metallic ∧^{impurities} of sulfurs. It is similar to the double mercurial water. And it is called acid spirit and double corrosive. The soul of the King is reducible into oil by this alone.

The reduction or destruction into the prime matter is a particular point for the most general Universal. But I can produce this in the space of three days, and that through the prime matter. p. 33.

The mercury of Saturn must be drawn with great dexterity from a certain rich mineral, not yet fused, ~~Whoever does not understand this~~ with the help of the Eagle and the Dragon, which is all the same, only having different names. Whoever does not understand this will never obtain anything of perfection. p. 48.

Burning sulfur is understood under Vitriol. We sometimes call this a fiery dragon. Whoever understands my cold dragon [the third fire], and hits upon it in the foregoing figures, needs nothing more than just to join this particular red, fiery, flying Dragon to its brother & ——. Without this the most general Universal cannot be made. p. 56.

The wet solution is performed by the ☿ of Saturn, which I have called the soul of the world. p. 66.

<*1v*> The first metal is the sperm ^and root of all metals. It is found in the mineral of Saturn. It appears like the mineral of Mercury. It is called the lead of the wise, from which we are accustomed to distill the milk of the virgin, and it has a Venereal property. The final metal is that which arrives at the nature of gold. p. 69.

The cold, metallic fire must be extracted from a certain mineral Saturn, not yet fused, and non-malleable, which is called mercury of Saturn. p. 70.

Without that cold, mineral Saturn & that infernal, Magical, fiery Element ^the second and third fire, Chymistry can prepare nothing. That cold, metallic ~~fire~~ mercury is extracted from a certain unfused, non-malleable, mineral Saturn. p. 70, 71.

The most general Universal is made from a double mercury, as I said above, & it is animated & fermented with a solar sulfur, & it is fixed with an enduring salt of gold, & it is further augmented into infinity through two other sulfurs, both in quantity and quality. The quality is augmented by ♀, the quantity by the foresaid ☿, which has the nature both of the Venereal property and of ^mineral, cold Saturn: whence it is likened to a double ☿, and is called doubled ☿, to which all qualities are appropriated, for it has the quality of Venus by reason of hot ♀, but cold from the side of Saturn. p. 72

From Luna [☾] and likewise from that cold, Arietine ♄ a mercury can be made: just as a certain ☿ which is gifted with a solar ♀ can be made from the mineral of Solar Venus [⊕], whence I have called it the ☿ of the ☉, since it must be taken for the generation of Sol. From these [at least from the two latter], the most general Universal is made. p. 72.

A living mercury can be made from Antimony and Bismuth, which are uniquely suitable for a medicine. These are, despite their excellence as medicine, by no means to be recommended for the philosophers' stone, since that ☿ is a universal menstruum, & born twice from ☿, first from the Lunar and then from the Solar, whence it is called doubled ☿. The Lunar ☿ [^perhaps that third cold, metallic fire ^but I doubt it] can be extracted and distilled by a rather easy operation from a Saturnine *<1rB>* body, & it has the nature of Luna. The Solar ☿ is extracted from the mineral of ♀ with tartar and sal Ammoniac. Whoever knows how to reduce the ☿ of cold Saturn and the ☿ of hot Venus into an Oil, will have the universal menstruum and a secure key for forcing all ♃ˢ to be potable. p. 67, 68.

Saturn is Lunar, but the Saturn of the Wise must be signified by the character of the Sun ^for the matter of the wise is solar. p. 56.

Neptune & Venus make to fly.
Th
<1vB>   Neptune & Venus make to fly
              The snake wᶜʰ els beneath must ly.
Thou who by yᵉ evaporated Neptune & Venereal property art become an Eagle.
You will convert your matter into water by a singular and hidden artifice, & after it has evaporated, you will change it into earth by a hidden means, which is the virgin earth of the Wise. From this earth the wisemen prepare their ☿ & their doubled ☿, & they draw up their dry water of life which radically dissolves all bodies. *Instructio Patris ad filium*. c. 4.

This is the blood of the green lion.

"Three Mysterious Fires"

Newton's autograph manuscript "Three Mysterious Fires" consists of two folios. It is found in the Smith Historical Manuscripts Collection of the Rare Book and Manuscript Department of the Columbia University Library. The manuscript is described thus in the Columbia University online "Digital Collections" page: "The Three Mysterious Fires: Commentary on Monte-Snyder's Tractatus de Medicina Universali."[1]

I have not examined the manuscript in situ but have made the following transcription from a photocopy found in the papers of Richard Westfall at Indiana University (QC3.N512.folder 15).

The manuscript bears the clear traces of Newton's collaboration with Nicolas Fatio de Duillier, the unnamed Francophone alchemist who was a friend of Fatio's, and possibly another party. It cannot be earlier than May 18, 1693, for reasons discussed in chapter seventeen of the present book, and it is unlikely to be later than November 14 of the same year, a point at which Fatio's chymical interests appear to have shifted to another topic (see chapter seventeen). The preliminary English section is written in an assertive and cocksure tone that is alien to the cautious approach found in Newton's other alchemical manuscripts of the period. This is unlikely to be his own composition. The authorship of the Latin text is more ambiguous, since it employs Newton's idiosyncratic symbols for the ores of various metals, and it displays other Newtonian characteristics such as his emphasis on repeated sublimations with differing proportions of ingredients. Nonetheless, formal considerations make it highly unlikely that the Latin section as we have it represents an original composition by Newton alone. For example, the Latin text contains a six-word passage of alternate readings in curly brackets, suggesting that Newton was copying something in a hand that he could not make out fully. This impression is corroborated by the five clear lacunas in the Latin section, where Newton again could not make sense of the script that he was copying. The fourth of these lacunas even retains the superscript inflectional suffix "*mam*," meant to go at the end of a Latin word that Newton obviously could not read. Finally, the manuscript breaks off abruptly in mid-sentence at the end, without a necessary verb.

The best way to account for the fact that Newton was copying a manuscript in someone else's hand that contained some material originating from himself lies in the conclusion that "Three Mysterious Fires" was a collaborative effort where drafts were passed back and forth between the different contributors. In fact, we know that Newton was engaged in this sort of collaborative practice from another source. The letter from Fatio to him dated August 1, 1693, and found in the William Andrews Clark Memorial Library (MS F253L 1693) contains seventeen lines of Latin text that Fatio had copied from a previous letter sent to him by Newton. His anonymous alchemical friend had questions concerning the recipe found therein, so Fatio was querying Newton about the passage. Given the fastidious clarity of Fatio's handwriting, it seems odd that Newton would have experienced the obvious transcribing difficulties that emerge unequivocally in "Three Mysterious Fires." One may therefore speculate that the text copied by Newton stemmed from another hand, possibly that of Fatio's Francophone alchemist or even from another party involved in the collaboration. I reproduce the text below in its entirety, followed by my translation of the Latin section.

<Text of "Three Mysterious Fires," 1r>

The first thing wch must be understood are the three mysterious fires. The first ought to render metal fusible & this without any enigma is ye regulus of antimony. The other ought to sympathise wth ye metallick fire, & altho Snyders doth declare that it is double yet he will considere it as one; tho they have a contrary nature in their qualities. But it is enough for him that they perform the same effect in his designe. He calls it a sympathic burning Hermaphroditick fire. He says that sulphur & Niter are two violent fires but y^t if one knows how to reconcile them nothing but God can hinder us from obtaining <illeg.> health & riches & that it is the only thing wch he had reserved <illeg.> ~~kept secret~~ to himself & to those whom God has elected to it. He does not dissemble, for the truth is that ♃ & niter are the two contrary fires wch being united are able to penetrate any metal whatsoever, to incend its soul & to extract it, being joyned wth the cold metallick fire wch he calls the soul of Saturn & wch doth amalgam it wth all metals, & ~~let~~ suffers it self to be calcined in the fire wth ye help of ye double igneous element. Now that cold fire is regulus of ♂ [i.e., the same wth ye first fire] He saith one must begin where nature <illeg.> has ended & by that igneous magical element composed of two infernal & contrary matters calcine the otherwise <illeg.> ^{inexpugnible} doores of y^e fortress of sol. By that & in all the extent of his book he denotes y^t you must use gold, & joyne to it the soul of saturn, Which ought to be taken from the ~~saturnal~~ mineral & <illeg.> unmelted $^{\wedge Saturn}$ because it doth not burn as common ☿ but has a terrestrial & dry quality by wch it is able to defend the sulphur of Sol least it be burned & fly away wth its mercury. Gold being amalgamed wth y^e mercury of Saturn becomes porous & then the infernal fire can sooner & better calcine the strong body & reduce it into ashes. ffrom whence is drawn by the clear dew of heaven the sulphur & from yᵉ remaining body ~~the~~ is drawn by a lixivium after a due reverberation the most pretious medicinal salt wᶜʰ yᵉ sages have said to be yᵉ P͞hers stone. He advertises that yᵉ separation of yᵉ ♃ from yᵉ salt can be made in a little while wᵗʰ an open fire: But that you must take care least the fire of yᵉ metals be burnt, & that you must for that purpose have a guardian or keeper wᶜʰ may hinder it. That he has named that guardian. It <illeg.> is Tartar wᶜʰ he declares to be much favourable to metals & to have a great affinity wᵗʰ them.

    To reduce then regularly all metals & minerals unto the first matter since it is the ground of all radicall mineral & metallick destruction. That reduction is made wⁿ you incorporate the mineral starrs <or> ᵃˢᵗʳᵃ to yᵉ philosophical heaven. This <illeg.> [heaven] is regulus of ♂ joyned wᵗʰ <illeg.> gold & all the other metals. After wᶜʰ the sympathick fire can easily teare its members. That sympathetick fire is a part of the magical elements ffor it is composed of an aereal salt of an oleaginous substance & of a vegetable earth. By the composition of those three you may by a dry way open the internal parts of all metallick bodies in order to draw the soul & afterwards yᵉ salt. In hard metals you must have more of that infernal thunderbolt <or>ᶠᵒᵘᵈʳᵉ then in others. In a little time you may destroy a great quantity of sol. To do that take eight parts of yoʳ aereal salt wᶜʰ is niter, of your dry oleaginous ~~matter~~ ˢᵘᵇˢᵗᵃⁿᶜᵉ wᶜʰ is sulphur, four parts, of yoʳ vegetable earth wᶜʰ is tartar, two parts. Reduce yᵉ whole into an impalpable powder & mix it wᵗʰ care. After wᶜʰ melt one part of pure gold & when it is throughly hot throw upon it three parts of yoʳ first magical fire (wᶜʰ is your Regulus of ♂.) Leave it in the fire till a pellicle or thin skin appears then throw it into a Cone. After wᶜʰ make it to melt again in a very violent heat. Throw in some of yoʳ composition ~~&~~ ᵒʳ infernal thunderbolt till all your Gold & Regulus be consumed into a precious scoria <or> ˢᶜᵒʳⁱᵘᵐ. You <illeg.> ^ᵐᵘˢᵗ ᵗʰᵉⁿ grind them warm & if there was a part of yoʳ Regulus not consumed, you must add some fresh regulus & <illeg.> begin again to fulminate. Put them into ^ᵛᵉʳʸ clear water till all be dissolved Philtrate yᵉ whole. There will pass a very clear water. Put it ~~by its~~ apart & that is the drink of wᶜʰ Mars cannot drink & into wᶜʰ throwing some vinegre of white wine he saw that out of water fire did come, & yᵗ yᵉ water was immediately changed & became a thick essence of a deep red. Then he said, O Venus, my lovely Venus thy beauty belongs to none other but <illeg.> me. There will remain some feces in the philtre wᶜʰ you must well wash & even cause to boyle that there may remain none of the salts; & throw again some Vinegre till nothing more will precipitate, & the feces that remain after you have

<1v>

well dried & grownd them, you must must <sic> reverberate with ~~the~~ hallf as much flowers of ♃: after wᶜʰ the salt may be easily extracted even wᵗʰ the spirit of vinegar. It is better to do it wᵗʰ yᵉ mercurial spirit. The <illeg.> sulphur of metals is wholy combustible when separated from its salt.

A <*illeg.*> Min ♀is <*illeg.*> ^& ♄i ana (a qua ☿⌒ ^in proport 3 ad 1 elevatus fierit) 2 partibus per deliquum madentibus sublimetur ☿ 3 ꝑts elevetur sublimatum ^adhuc semel vel bis ~~vel ter~~ ab eadem min madente. Sublimetur ^lento tridui igne (quo materia tantum fluat<)> sublimatum hoc (pts 3) a Reg of ♄ ♁, ♂, ♀, ☉ in proport 2, 1, 1, 4/3 vel 4/5 vel 4/7~~)~~ mistis) 1 ꝑt & habebitur in fundo cap. mort. in medio <*illeg.*> cinab. in summitate subl. album. Misceantur ōia et sublimentur iterum ac tertiò ~~ut quarto~~ ut prius. Tunc ~~rejectis fæcibus seu~~ cap mortuo et cinabari rejectis, sublimetur sublimatum ^bis vel ter ~~bis~~ per se. Dein digeratur ^per 3 vel 4 dies in B. M. cujus calore (quem manu ferre vix possis) materia statim liquescat. Materia ~~die 3o vel 4to~~ cito putrescet et die tertio vel 4to iterum clarescet decidentibus fæcibus & <*illeg.*> ruber ut sanguis apparebit nec magis diaphanus. Continuando digestionem materia albescet, sed sumenda est ubi proximè post putrefactionem rubescit. Destilletur ~~&~~ <*illeg.*> & rejectis fæcibus siquæ sint ~~destilletur iterum~~ <*illeg.*> rectificetur destillando donec fæces nullæ relinquantur, id quod ~~nonnumquam prima~~ nonnumquam 2da vel tertia vel ^ferte certe quarta vice eveniet. Hic est sps ☿ij.

Ex corporum reductorum fæcibus ^cum dimidia sulphuris parte calcinatis extrahatur sal cum hoc ☿ij spiritu ^vice spiritus aceti & abstrahatur sptus ut sal maneat in fundo. Sali affundatur 🜄 ^rectificatissimus quo 🜍 omne et quicquid impurum est dissolvatur. Dein exiccetur <*illeg.*> ~~et~~ sal et ~~ad usum serventur~~ cum ejus parte una digerantur ^duæ vel tres vel quinqꝫ partes <*illeg.*> spiritus ☿ij per dies 🜍i ~~ex corporibus extracto o~~ 10 vel plures donec optime uniantur.

Sulphuri ex corporibus ^per fulminationē extracto ^et optime loto et leniter arefacto affundatur tantum spiritus <*illeg.*> ☿ij in quo sal nondum dissolutus fuit vel potius triplo <*illeg.*> vel quintuplo plus, et digerantur per dies 10 vel plures, dein distilletur & rejectis fæcibus destilletur iterum donec nullæ amplius fæces restent, id quod venire solet 2da 3a vel 4ta vice. Nam sps <*illeg.*> elevabit et secum rapiet 🜍 totum demptis illis fæcibus.

Conjungantur duæ partes hujus liquoris in quo 🜍 est cum una parte liquoris alterius in quo Sal est, et abstrahatur spiritus donec materia in fundo spissa sit instar mellis sed paulo liquidior. Ponantur in ovo Phico ut ovi pars quarta plena sit, & digerantur usqꝫ ad complementum. Nam die 30mo putrescent, deinde albescent et rubescent.

<2r>

1. ℞ ☿⌒ 3, ♂° 1, ♀° 1. ^Addi etiam potest pars 1 ^min ♄ni. Digerantur per dies 3 vel 4 & Abibit Caput Mortuum in loco frigido in mucilaginem. Sublimentur. Ascendet totum ☿⌒ti pondus. Misceatur ~~Cap~~ <*illeg.*> mucilago cum ☿ ~~^in calido~~ & ffiet fermentatio, & ☿ incorporabitur cum <*illeg.*> Muc. Addatur ^cito ☿ ^novus & si materia indurescat teratur iterum & misceatur cum novo ☿ donec nihil amplius incorporabitur. Sublimetur Sublimatum ^3 ptes<?> cum ☉e per ♁ purificato et in pulverem redacto 1 parte ponatur in Alembico cum capite globoso & fistula aperta in summitate globi <*illeg.*> & digeratur in furno <*illeg.*> undiqꝫ clauso in cujus operculo sunt foramina tria vel quatuor ad regendum calorem ^et transmittendam fistulam Sit calor tantus ut materia fluat & ascendet pars materiæ ad latera vitri in forma annuli colorati & tandem separabitur annulus a reliqua materia (jam concava existente) et ascendet in fistulam usqꝫ ~~et~~ eamqꝫ claudet.

Calor jam non ultra augendus est ne annulus ulterius ascendat aut vitrum frangatur. Tum *<illeg.>* ubi annulus fistulam *<illeg.>* occluserit *<illeg.>* decidet inde quasi pluvia perpetua (ex guttulis ^minimis constans) in materam inferiorem. Vbi motus cessaverit {ponatur materia —— <or> distilletur et ponatur spiritus} in globo vitreo & *<illeg.>* materia digerendo in calore *<lacuna>* putrescet & *<illeg.>* colores omnes successive induet ac tandem pellucida manebit. Si sub initio pars septima vel nona ~~materia~~ auri *<illeg.>* ♃is vel alterius ~~materia~~ metalli masculi addatur materiæ in Alembico, incipiet metallum ascendere ad latera vasis in forma annuli, & a ~~materia~~ liquore decurrente perpetuo deorsum feretur donec metallum totum dissolvatur.

2. Sublimatum quod a mineris duobus ascenderit post debitam *<illeg.>* præparationem ponatur in Retorta cum Recipiente magno intercedentibus duobus Aludellis. *<image of apparatus>* Destiletur et implebitur Retora *<sic>* cum fumis albis. Vbi cessaverit *<illeg.>* fumus ille & Retorta clara est sine albedine *<illeg.>* auferatur ab igne, & Recipiens (in quo nihil apparebit) *<illeg.>* dextre auferatur & citissime claudatur ut nihil exeat fumorum & *<illeg.>* in frigido loco ponatur *<illeg.>* *<illeg.>* post dies tres vel quatuor vapor incipiet condensari et per totam Recipientis concavitatem in aquam pinquem decurrere. ~~Et~~ Ex libris ^autem quin℥ sublimati præpati *<sic>* habebis libras quatuor cum semisse illius aquæ.

Digeratur aqua in globis vitreis parvis calore temperato donec putrescat & post nigredinem albescat & circulus albus in circuitu vasis ^supra materiam appareat. Separetur quod clarum est a fecibus. Iterum digeratur & separetur a fæcibus donec post quatuor vel quin℥ repetiones *<sic>* nullæ amplius ^sint fæces. Arescat ~~materia~~ fæx & lento calore torreatur donec *<lacuna>* <.> Infundatur spiritus abstractus & digeratur donec spiritus rubescat. Abstrahatur spiritus donec maneat oleum rubrum. Reaffundatur spiritus fæcibus donec extraxerit quicquid rubedinis extrahi potest et servetur tota rubedo seu oleum rubrum abstracto spiritu. Reverberentur fæces. Reaffundatur spiritus & extr^ahetur sal fixus. Et nota quod sal oleum et sps sunt ad invicem ut *<lacuna>*. Imbibatur sal cum spiritu suo paulatim addendo singulis vicibus *<lacuna>*mam partem salis et interponendo digestionem dierum *<lacuna>* et spiritus coagulabitur in sale & cum eo unietur componendo materiam fusibilem. Et nota quod una pars salis *<illeg.>* retinebit 9 partes spiritus. Deinde oleum in 9 vel 10 æquales partes divisum addatur gradatim interpositis digestionibus dierum trium, & habebitur materia in frigido coagulata sed calore levissimo fluens in qua aurum instar glaciei in aqua tepida liquescit.

Præparatur autem sublimatum ponendo in globo ligneo cujus orificium superne clauditur cum choclea & ^coquendo hoc vas in aqua pluviali *<illeg.>* bulliente ~~coquendo~~ ^hoc *<illeg.>* in alio vase cujus pars inferior terrea est, superior vitrea. Coquatur *<illeg.>* autem pr horas plus minus octodecim, & sublimatum emollescet, et per lignu *<text breaks off abruptly>*

<English Translation of Latin Section>

<1v> From two parts of the ore of Venus ^and an equal amount <of the ore> of Saturn previously wetted per deliquium (from which mercury sublimate has been

elevated in the proportion 3 to 1), let three parts of mercury be sublimed; let the sublimate be elevated from the same mineral, wetted again by deliquescence, once or twice. Let this sublimate (three parts) be sublimed from one part of regulus of Saturn, antimony, Mars, Venus, Sol ^on a mild fire for three days (in which the matter just flows <)> mixed together in the proportions 2, 1, 1, 4/3 or 4/5 or 4/7); in the bottom will be had a caput mortuum, in the middle a cinnabar, and in the top a white sublimate. Let all be mixed together and sublimed again and a third time as before. Then, once the caput mortuum and cinnabar have been removed, let the sublimate be sublimed twice or thrice per se. Then let it be digested three or four days in a balneum mariae in whose heat (which you can hardly bear to touch) the matter liquefies immediately. The matter will quickly putrefy and on the third or fourth day will again clarify with the dregs falling down and it will appear as red as blood and no more transparent. With a continuation of the digestion the matter will whiten, but it should be taken where it has reddened soon after putrefaction. Let it be distilled with the dregs removed if there are any and let it be rectified by distillation until no dregs are left behind, which sometimes on the second or third, or certainly by the fourth time will happen. This is the spirit of mercury.

From the calcined dregs of the bodies reduced ^with a half part of sulfur a salt is extracted by means of this spirit of mercury ^in place of spirit of vinegar and a spirit is extracted so that the salt remains in the bottom. Highly rectified spirit of wine is poured on the salt, by which all the sulfur and anything impure is dissolved. Then the salt should be dried out and with one part of it, two or three or five parts of the spirit of mercury should be digested for ten days or more until they are well united.

On the sulfur extracted from the bodies by fulmination and washed well and slowly dried, as much spirit of mercury in which salt has not been dissolved, or rather three or five times as much, is poured and they are digested for ten days or more, then it should be distilled, and with the dregs removed distilled again until no further dregs remain, which usually happens at the second, third, or fourth time. For the spirit will rise up and will carry all the sulfur up with it, with the dregs left behind. Two parts of this liquor in which the sulfur is present should be conjoined with one part in which the salt is present, and the spirit is abstracted until the matter in the bottom is thick like honey but a little more liquid. They should be put in the philosophical egg so that a fourth part of the egg is full, and they should be digested until completion. For by the thirtieth day they will putrefy, then whiten and then redden.

<*2r*> 1. Take three parts of corrosive sublimate, one part of ore of Mars, one part of ore of Venus ^1 part of ore of Saturn can also be added. Let them be digested for three or four days and the caput mortuum will pass into a mucilage in a cold place. Let them be sublimed. The whole weight of the corrosive sublimate will ascend. Let the mucilage be mixed with mercury and a fermentation will occur, and the mercury will be incorporated with the mucilage; new mercury should be quickly added ^and if the matter hardens it should be ground again and mixed with new mercury until nothing further will be incorporated. Let it be sublimed. Let 3 parts<?> of the sublimate be

put ^{with one part of gold purified through antimony and reduced to powder} into an alembic with a spheri-cal head and an open tube in the top of the sphere and let it be digested in a totally sealed-up furnace in whose lid there should be three or four holes for regulating the heat ∧and for allowing the tube to pass through. The heat should be so great that the matter flow and part of the matter will ascend to the sides of the glass in the form of a colored ring, and finally the ring will be separated from the remaining matter ∧(which is already in a hollow form) and it will ascend up into the tube and seal it shut. The heat should be augmented no further lest the ring ascend upwards or the glass be broken. Where the ring has then clogged the tube, a sort of perpetual rain will fall down consisting ∧of very small drops onto the matter below. When the motion has stopped, {the matter should be put —— <or> the spirit should be distilled and put} into a glass sphere, and the matter upon digesting in the heat <*lacuna*> will putrefy and will successively put on all colors and it will finally remain transparent. If near the beginning a seventh or ninth part of Sol or Jupiter or of another masculine metal be added to the matter in the alembic, the metal will begin to ascend to the sides of the alembic in the form of a ring and it will be borne downwards by the continually descending liquor until all the metal is dissolved.

2. The sublimate that will have ascended from the two ores after a proper preparation should be put in a retort with a big receiver and with two inter-vening aludels. <*image of apparatus*> Let it be distilled and the retort will be filled with white fumes. When the fume has ceased and the retort is clear without whiteness it should be removed from the fire and the receiver (in which nothing will appear) should be skillfully removed and very quickly sealed up so that no fume escapes, and it should be put in a cold place; after three or four days the vapor will start to be condensed and run down the in-side of the globe in the form of a fat water; from five pounds of the prepared sublimate you will have four and a half pounds of this water.

Let the water be digested in small glass spheres at a temperate heat until it putrefies, and after blackness it whitens and a white circle appears on the circuit of the glass above the matter. What is clear should be separated from the dregs. Let it be digested again and separated from the dregs until after four or five repetitions there be no more dregs. Let the dregs then be dried out in a moderate heat and torrified until <*lacuna*> The abstracted spirit should be poured on and digested until the spirit reddens. The spirit should be ab-stracted until a red oil remains. Let the spirit be poured back on the dregs until it has extracted whatever redness can be extracted and the whole redness will be preserved or a red oil with the abstracted spirit. The dregs should be reverberated. The spirit should be poured back on and the fixed salt extracted. Note that the salt, oil, and spirit are alternately as <*lacuna*>. Let the salt be gradually imbibed with its spirit multiple times by adding a <*lacuna*>^{mam} part of the salt and interposing a digestion of <*lacuna*> days, and the spirit will be coagulated in the salt and will be united with it in composing a fusible mate-rial. Note that one part of salt will retain nine parts of spirit. Then the oil di-vided into nine or ten equal parts should be gradually added with interposed digestions of three days, and the matter will be had coagulated in the cold but flowing in a very modest heat in which gold liquefies like ice in warm water.

The sublimate is prepared by putting it in a wooden globe whose upper orifice is sealed with a screw and by cooking this vessel in boiling rainwater in another vessel whose lower part is earthen and upper part is glass. Let it be cooked for about eighteen hours, and the sublimate will soften and through the wood <text breaks off abruptly>

APPENDIX FOUR

Newton's Interview with William Yworth

Acomparison of Royal Society manuscript MM/6/5 and Newton's partial transcript of William Yworth's *Processus mysterii magni* in Keynes 66 provides convincing evidence that MM/6/5 contains the product of an oral interview held between Newton and Yworth. In the following I present a few of Newton's questions and the answers he received to the parallel points in the *Processus* that served as the sources of his investigation. I have chosen passages that highlight Yworth's characteristic chymical terminology. Although the individual *Decknamen* employed are of course not unique to him, their aggregate is nonetheless indicative of his idiosyncratic interpretation of the terms. The fact that Newton's questions and Yworth's answers are interspersed among four lines of comments drawn from an unidentified book dealing with classical mythology is probably an artifact of their being an apograph rather than the original transcript.

The first passage that we will examine is the third of the queries that Newton presents on 15r. I reproduce it here:

> Q. 3? after ye eagles are over, is the corrosive red heterogeneous spirit separated from ye green Lyon or from ye black body. And at that time is any thing distilled from ye black body after the Gr. Lyon is poured off.

The "eagles" that Newton refers to are an expression for sublimations borrowed from Eirenaeus Philalethes. In *Secrets Reveal'd*, for example, the "American philosopher" says, "every sublimation of the ☿ of Philosophers let be one *Eagle*."[1] This expression was commonly used by a multitude of English chymists by the late seventeenth century, but the corrosive, red spirit that may be separated from the green lion or from the black body signifies something much more particular. As for the green lion, it is a liquid here as the following sentence reveals, since it is "poured off." The source of these peculiarities emerges quickly if we turn to Newton's partial transcript of the *Processus* in Keynes 66. Yworth describes an elaborate distillation of "the matter you know well," which leads to the following results:

[1] Eirenaeus Philalethes, *Secrets Reveal'd* (London: W. C., 1669), 15.

Take off the Helme & you will find a heavy black & unctuous body in great quantity, & the vessel bring moved to one side there will flow from it an heavy oyley substance ∧of a dark redish colour wᶜʰ may be poured off by gentle inclination the body remaining dry as a field doth after a mighty torrent. Pour it off gently & put it into another vessel for a further operation. ffor it is ~~the Green Lyon~~ that Fiery Dragon that overcomes all things.[2]

Here we find a "heavy black & unctuous body," which is the black body referred to in Newton's question. Similarly, a "heavy oyley substance ∧of a dark redish colour" is poured off from the black body, which corresponds to Newton's "red heterogeneous spirit" that is separated either from the green lion or from the black body—Newton was not sure which. As for the possible objection that Yworth has changed his meaning by deleting "Green Lyon" and replacing it with "Fiery Dragon," this is vitiated by the fact that a few lines later the *Processus* adds that the green lion is a foul and heterogeneous component of the fiery dragon that must be separated from it. It is also a "sharp aquaeity," that is, an acidic liquid, which corresponds to Newton's reference to corrosiveness. Thus we have Newton's third question and the *Processus* both speaking of a "black body," a red liquid, and the green lion in virtually the same language. The similarity continues if we examine other portions of the two texts as well.

A number of times, for example, Newton's questions refer to a "lunar sublimate" that is produced along with the distillate when "the matter you know well" is treated. When Newton comes to the point of writing down the answers that he received to his questions, he has quite a lot to say about this lunar sublimate and even refers to the textual passage in which he has found its description:

> The Lunar subl of Ch. 1 is cold & earthy. By resublimation from the red earth it becomes hot & active & fiery so as to fume & fret & burn the flowers ∧of the matter unless cooled by a due proportion of yᵉ inferior waters. In yᵉ putrefaction it melts into an oyle, & increases yᵉ white oyle.[3]

According to this response, the lunar sublimate is found in chapter one of an unnamed text. Not surprisingly, we can show that this text is the *Processus* in the version that Newton transcribed. In chapter one, which bears the heading "Of the Preparation of the crude matter & separation of Elements," Yworth describes the flowers that sublime in the following terms:

> Only (by the way) I let thee know that they are a dry fume which helps to coagulate the moist, as being a saline yet sulphureous and combustabile earth wᶜʰ may not improperly be called Antimonial, yet are of a Lunary nature and are the foundation of or Lune central.[4]

From Newton's response given above we learn that the lunar sublimate should be resublimed from a red earth, but this operation must be performed

[2] Keynes 66, 5r.
[3] Royal Society MM/6/5, 15r.
[4] Keynes 66, 2r.

with the aid of "yᵉ inferior waters" to avoid creating a product that will fume and burn the flowers. If the process is successfully carried out, the red earth can then be putrefied, whereon it will melt into a white oil. In locating the parallel passage in Yworth's *Processus*, it will behoove us to consider the order in which the composition of that text and the interview with Newton took place. From Newton's reference to chapter one of the *Processus*, we know that his reading of the text is what precipitated his queries. Thus he had already read a draft of the *Processus* before interviewing Yworth. As we will now see, it was in all probability the very transcript that has come down to us as Keynes 66, for we see Newton modifying the text there in accordance with the answer that he received from Yworth. Keynes 66 is a tangled mass of deletions and interlinear insertions as a result of Newton's newfound knowledge:

> In the meane time ~~to the <illeg.>~~ take ~~the vitriolic salt &~~ ^the red earth finely sifted^ & the white lunar ~~that is about 10 ounces of red earth 10 ounces of sublimate~~ sublimate <illeg.> & the inferior waters prepared & separated from its volatile salts, each of ten destillations ^that is about 10 ounces of red earth 10 ounces of sublimate & ten or twelve pounds of water. Dissolve^ ~~Separate the waters. Dissolve the vitriol in~~ the sublimate in the water & then put in also the red earth, & ~~either with or without its vinegre~~ some of the ~~the waters & then put in also the sublimate & dissolve of & evaporate~~ vinegar of the first chapter. Draw of the flegm gently to a dryness & then sublime wᵗʰ a strong fire & a ~~to a dryness. When the salt is dry mix it with some of the red earth~~ white sublimate will ascend, & what remains will be black light combust feces. The red earth dulcifies ~~& put the mixture into a Quart Retort, & destill as in Chap. II first~~ the sublimate & takes away its corrosiveness. The Vinegre purifies it & makes it white & the <illeg.> spirit of ~~drawing off & setting aside the flegm as in Chap. II &~~ ^<illeg.>^ ~~putting on a clean~~ the red waters cool it. ffor if there be not red water enough the sublimate will fume after it is cold. ~~Receiver, & increasing the heat. And~~ when ~~the operation is over you will find a white sublimate in the neck of the Retort & a vinegar in the Receiver. Dissolve this sublimate in the Vinegar & keep it for use.~~[5]

A patient reading of this passage reveals all the elements in the answer that Newton had gleaned from his interview with Yworth. The lunar flowers must be dissolved in the inferior waters and then the red earth is to be added. The waters cool the sublimate, and the red earth, once conjoined with them, serves to dulcify it and prevent fuming. This new information supplants the deleted passages, which say nothing of the cooling role of the inferior waters. Newton had found his answer, and he was busy incorporating it into Keynes 66. At this point it seems unnecessary to continue comparing parallel passages: 15r and 15v of Royal Society MM/6/5 clearly contain the transcript of Newton's interview with Yworth, which he then incorporated into his copy of the *Processus mysterii magni*.

[5] Keynes 66, 7r.

Index

Footnotes are indexed only when they contain information not found in the main text. If an author or work is mentioned both on a page of text and in a footnote on that page, only the main text is indexed.

Philalethes, Eirenaeus (*continued*)
277, 282, 285–91, 293–94, 303, 306, 318, 346–47, 358–66, 371, 379–80, 382, 399–400, 402, 405, 409–13, 417–19, 426, 432, 434–35, 450, 460, 462, 467–69, 474, 478–80, 484–87, 491–92, 494, 497–98, 517; *Brevis manuductio ad caelestem rubinum,* 137; *De metallorum metamorphosi,* 86, 478–79; *Enarratio methodica trium Gebri medicinarum,* 468–69; evolution in Newton's spelling of his name, 247; *Experiments for the Preparation of the Sophick Mercury,* 39, 220, 285, 288; *Introitus apertus ad occlusum regis palatium,* 21, 50, 83, 85, 197; as a key to Sendivogius, 196; *The Marrow of Alchemy,* 23, 33, 84, 197–203, 208–9, 226, 232, 254, 259, 282n61, 286–90, 289, 318, 358, 360, 364, 402, 418; Newton's early, Sendivogian reading of, 190; *Opera omnia,* 34–35; *Opus tripartitum,* 210; "Philalethists," 435; *Ripley Reviv'd,* 23n4, 27, 29, 32–35, 50, 111, 210, 251, 290, 293n88, 379, 400, 402, 408, 410, 413, 418, 450n41; *Secrets Reveal'd,* 21, 27–29, 32, 35, 98nn27, 29, 148, 150–51, 154, 160, 164, 168, 186, 190–91, 194–202, 209, 232, 247, 259, 288, 293, 402, 411, 413, 418, 450n44, 460, 467, 486, 517; "Sr George Ripley his Epistle to K. Edward unfolded," 110, 400n8; supposed biographical details of, 84. *See also* Starkey, George
Philalethes, Eugenius, 38. *See* Vaughan, Thomas
Philolaus, 55
Philoponus, John, 117
philosophers' stone (*passim*), 10; defined, 1, 13, 21; and enthusiasm, 20n2; as elixir, 62; lunar or white versus solar or red, 154; as medicine for humans and metals, 154; Newton's failure to attain it, 148; for producing visions and communicating with spirits, 107; relation to "concentrated extract of gold," 225; as "the stone," 31
Philotis. See *Turba philosophorum*
phlogiston theory, 18, 452, 455–56, 459, 464, 480
Phoebus, 235–36, 244, 388
phosphorus, 112
Pierpont Morgan Library, Newton MS, 89–91
Pirithous, 52
Pitcairne, Archibald, 445, 460, 465n40, 480, 489
pitch, 273
planets, seven, 32, 56, 61, 281; and alchemical regimens, 151
Plattes, Gabriel, 491
Plessner, Martin, 203n56, 215n11
Pluto, 52, 411
Poitou, 384
pond slime, analysis of, 416–20
Pope, Alexander, 1–2
Porta, Giambattista della, 89, 497
potassium bisulfate, 105
potassium carbonate. *See* salt of tartar
potassium silicate. *See* oil of glass
potassium sulfate, 105
Pounds, C. A., 394n65

Priesner, Claus, 80n39
Priestley, Joseph, 459
prima mixta, 121
prima naturalia, 121
prime matter, 68, 118, 228, 397, 400–401
primum ens, 79, 204–6, 212, 215, 217, 264, 293, 401
Principe, Lawrence, 6nn19–20, 9n24, 25nn16, 17, 30n25, 39nn44–46, 45n2, 81n43, 83nn44, 46, 84n47; 98n26; 107n49, 112nn71, 73, 121n18, 132n42, 163n24, 202n51, 223n1, 264n9, 271n51, 287n70, 313n26, 317n34, 318n38, 365n47, 369n5, 371n18, 435n1, 452n2, 455n13, 457n21, 458n25, 484n7, 487n16, 488nn18–21, 498n1
principles, alchemical, 7, 29, 64–66, 68, 70–71, 74, 78, 101–5, 116, 141, 143, 228, 439, 460, 490; and Gur, 76; separation of, 116; three, 101–6, 116, 416, 420; as vital spirits, 144–45. *See also* sulfur and mercury
Prinke, Rafał T., 66n5, 67n6
prisca sapientia, 55, 57
prisms, 114–15, 120, 123–32
pristine state. *See* reduction to the pristine state
projection, in alchemical transmutation, 224–25
prophecy, 45–51; 62
Proserpina, 52, 192
prytaneum, 57, 60
pseudonyms: for Newton, 11, 36–37, 85; for Sendivogius, 36, 83; for Starkey, 83; for Thölde, 83
putrefaction, 14, 53, 62, 104, 140, 142–43, 153, 164, 174, 179, 214, 253, 379, 462; of fat water, 392; of Gur, 217; of mercury, 292, 371; its relationship to fermentation for Newton, 466–67; of salts, 326. *See also* fermentation
pyrotechny (as chymistry), 95
Pythagoras, 54–56, 59, 61, 295

Queen, 291, 399
quicklime, 495–96
quintessence, 16, 58, 60–61, 262–80, 291–95, 375; attracts birds and humans, 274–75

rain: combines with metallic vapors to produce salt, 159
Rampling, Jennifer, xii, 6n19, 34n33, 242n30, 263n6; 268n18
Rattansi, Piyo M., 54–57, 59, 109n58
Reck, Cathrine, 26, 316, 338–39, 345, 359, 378, 381
red earth, 59; Boyle's, 487, 495, Yworth's, 428, 518–19
redintegration, 14, 18, 125, 132–34; as Newton's means of testing for mechanical versus vegetative generation, 176–78; of niter, 158, 462; of stibnite, 132–34; of turpentine, 132–34; use of the term by Newton, 132–34
reduction to the pristine state, 122–23, 125, 134, 466
reduction: uses in alchemy, mineralogy, and modern chemistry, xvi